ISBN 978-1-5284-7504-4
PIBN 10104920

NOTICES

OF THE

PROCEEDINGS

AT THE

MEETINGS OF THE MEMBERS

OF THE

Royal Institution of Great Britain

WITH

ABSTRACTS OF THE DISCOURSES

DELIVERED AT

THE EVENING MEETINGS

✦

VOLUME XVII
1902—1904

LONDON
PRINTED BY WILLIAM CLOWES AND SONS LIMITED

1906

CONTENTS.

1902.

1903.

CONTENTS.

1904.

PLATES.

————•◊•————

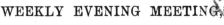

Royal Institution of Great Britain.

WEEKLY EVENING MEETING,

Friday, January 17, 1902.

His Grace the Duke of Northumberland, K.G. D.C.L. F.R.S.,
President, in the Chair.

The Right Hon. Lord Rayleigh, M.A. D.C.L. LL.D. Sc.D.
F.R.S. *M.R.I.*, Professor of Natural Philosophy, R.I.

Interference of Sound.

For the purposes of laboratory or lecture experiments it is convenient to use a pitch so high that the sounds are nearly or altogether inaudible. The wave lengths (1 to 3 cm.) are then tolerably small, and it becomes possible to imitate many interesting optical phenomena. The ear as the percipient is replaced by the high pressure sensitive flame, introduced for this purpose by Tyndall, with the advantage that the effects are visible to a large audience.

As a source of sound a "bird-call" is usually convenient. A stream of air from a circular hole in a thin plate impinges centrically upon a similar hole in a parallel plate held at a little distance. Bird-calls are very easily made. The first plate, of 1 or 2 cm. in diameter, is cemented, or soldered, to the end of a short supply-tube. The second plate may conveniently be made triangular, the turned down corners being soldered to the first plate. For calls of medium pitch the holes may be made in tin plate. They may be as small as $\frac{1}{2}$ mm. in diameter, and the distance between them as little as 1 mm. In any case the edges of the holes should be sharp and clean. There is no difficulty in obtaining wave-lengths (complete) as low as 1 cm., and with care wave-lengths of ·6 cm. may be reached, corresponding to about 50,000 vibrations per second. In experimenting upon minimum wave-lengths, the distance between the call and the flame should not exceed 50 cm., and the flame should be adjusted to the verge of flaring.* As most bird-calls are very dependent upon the precise pressure of the wind, a manometer in immediate connection is practically a necessity. The pressure, originally somewhat in excess, may be controlled by a screw pinch-cock operating on a rubber connecting tube.

In the experiments with conical horns or trumpets, it is important that no sound should issue except through these channels. The horns end in short lengths of brass tubing which fit tightly to a short length of tubing (A) soldered air-tight on the face of the front

---

* 'Theory of Sound,' 2nd ed., § 371

plate of the bird-call. So far there is no difficulty ; but if the space
between the plates be boxed in air-tight, the action of the call is interfered with. To meet this objection a tin-plate box is soldered
air-tight to A, and is stuffed with cotton-wool kept in position by a
loosely fitting lid at C. In this way very little sound can escape
except through the tube A, and yet the call speaks much as usual.
The manometer is connected at the side tube D. The wind is best
supplied from a gas-holder.

With the steadily maintained sound of the bird-call there is no
difficulty in measuring accurately the wave-lengths by the method of
nodes and loops. A glass plate behind the flame, and mounted so
as to be capable of sliding backwards and forwards, serves as reflecting wall. At the plate, and at any distance from it measured by an

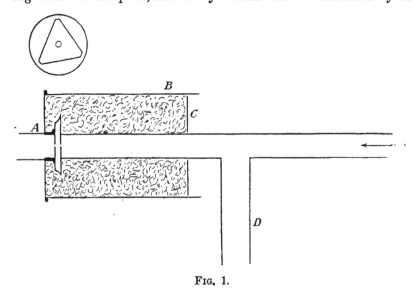

FIG. 1.

even-number of quarter wave-lengths there are nodes, where the
flame does not respond. At intermediate distances, equal to *odd*
multiples of the quarter wave-length, the effect upon the flame is a
maximum. For the present purpose it is best to use nodes, so adjusting the sensitiveness of the flame that it only just recovers its height
at the minimum. The movement of the screen required to pass over
ten intervals from minimum to minimum may be measured, and gives
at once the length of five complete progressive waves. For the bird-
call used in the experiments of this lecture the wave length is 2 cm.
very nearly.

When the sound whose wave length is required is not maintained,
the application of the method is, of course, more difficult. Nevertheless, results of considerable accuracy may be arrived at. A steel
bar, about 22 cm. long, was so mounted as to be struck longitudinally

every two or three seconds by a small hammer. Although in every position the flame shows some uneasiness at the stroke of the hammer, the distinction of loops and nodes is perfectly evident, and the measurement of wave length can be effected with an accuracy of about 1 per cent. In the actual experiment the wave length was nearly 3 cm.

The formation of stationary waves with nodes and loops by perpendicular reflection illustrates interference to a certain extent, but for the full development of the phenomenon the interfering sounds should be travelling in the same, or nearly the same, direction. The next example illustrates the theory of Huyghens' zones. Between the bird-call and the flame is placed a glass screen perforated with a circular hole. The size of the hole, the distances, and the wave length are so related to one another that the aperture just includes the first and second zones. The operation of the sounds passing these zones is antagonistic, and the flame shows no response until a part of the aperture is blocked off. The part blocked off may be either the central circle or the annular region defined as the second zone. In either case the flame flares, affording complete proof of interference of the parts of the sound transmitted by the aperture.

From a practical point of view the passage of sound through apertures in walls is not of importance, but similar considerations apply to its issue from the mouths of horns, at least when the diameter of the mouth exceeds the half wave-length. The various parts of the sound are approximately in the same phase when they leave the aperture, but the effect upon an observer depends upon the phases of the sounds, not as they leave, but as they arrive. If one part has further to go than another, a phase discrepancy sets in. To a point in the axis of the horn, supposed to be directed horizontally, the distances to be travelled are the same, so that here the full effect is produced, but in oblique directions it is otherwise. When the obliquity is such that the nearest and furthest parts of the mouth differ in distance by rather more than one complete wave-length, the sound may disappear altogether through antagonism of equal and opposite effects. In practice the attainment of a complete silence would be interfered with by reflections, and in many cases by a composite character of sound, viz. by the simultaneous occurrence of more than one wave-length.

In the fog-signals established on our coasts the sound of powerful sirens issues from conical horns of circular cross-section. The influence of obliquity is usually very marked. When the sound is observed from a sufficient distance at sea, a deviation of even 20° from the axial line entails a considerable loss, to be further increased as the deviation rises to 40° or 60°. The difficulty thence arising is met, in the practice of the Trinity House, by the use of two distinct sirens and horns, the axes of the latter being inclined to one another at 120°. In this way an arc of 180° or more can be efficiently guarded, but a more equable distribution of the sound from a single horn remains a desideratum.

Guided by the considerations already explained, I ventured to recommend to the Trinity House the construction of horns of novel design, in which an attempt should be made to spread the sound out horizontally over the sea, and to prevent so much of it from being lost in an upward direction. The solution of the problem is found in a departure from the usual circular section, and the substitution of an elliptical or elongated section, of which the short diameter, placed horizontally, does not exceed the half wave-length; while the long diameter, placed vertically, may amount to two wave-lengths or more. Obliquity in the *horizontal plane* does not now entail much difference of phase, but when the horizontal plane is departed from such differences enter rapidly.

Horns upon this principle were constructed under the supervision of Mr. Matthews, and were tried in the course of the recent experiments off St. Catherine's. The results were considered promising, but want of time and the numerous obstacles which beset large-scale operations prevented an exhaustive examination.

On a laboratory scale there is no difficulty in illustrating the action of the elliptical horns. They may be made of thin sheet brass. In one case the total length is 20 cm., while the dimensions of the mouth are 5 cm. for the long diameter and $1\frac{1}{4}$ cm. for the shorter diameter. The horn is fitted at its narrow end to A (Fig. 1), and can rotate about the common horizontal axis. When this axis is pointed directly at the flame, flaring ensues; and the rotation of the horn has no visible effect. If now, while the long diameter of the section remains vertical, the axis be slewed round in the horizontal plane until the obliquity reaches 50° or 60°, there is no important falling off in the response of the flame. But if at obliquities exceeding 20° or 30° the horn is rotated through a right angle, so as to bring the long diameter horizontal, the flame recovers as if the horn had ceased sounding. The fact that there is really no falling off may be verified with the aid of a reflector, by which the sound proceeding at first in the direction of the axis may be sent towards the flame.

When the obliquity is 60° or 70° it is of great interest to observe how moderate a departure from the vertical adjustment of the longer diameter causes a cessation of effect. The influence of maladjustment is shown even more strikingly in the case of a larger horn. According to theory and observation a serious falling off commences when the tilt is such that the difference of distances from the flame of the two extremities of the long diameter reaches the half wave length—in this case 1 cm. It is thus abundantly proved that the sound issuing from the properly adjusted elliptical cone is confined to a comparatively narrow belt round the horizontal plane and that in this plane it covers efficiently an arc of 150° or 160°.

Another experiment, very easily executed with the apparatus

already described, illustrates what are known in Optics as Lloyd's bands. These bands are formed by the interference of the direct vibration with its very oblique reflection. If the bird-call is pointed toward the flame, flaring ensues. It is only necessary to hold a long board horizontally under the direct line to obtain a reflection. The effect depends upon the precise height at which the board is held. In some positions the direct and reflected vibrations co-operate at the flame and the flaring is more pronounced than when the board is away. In other positions the waves are antagonistic and the flame recovers as if no sound were reaching it at all. This experiment was made many years ago by Tyndall who instituted it in order to explain the very puzzling phenomenon of the "silent area." In listening to fog-signals from the sea it is not unfrequently found that the signal is lost at a distance of a mile or two and recovered at a greater distance in the same direction. During the recent experiments the Committee of the Elder Brethren of the Trinity House had several opportunities of making this observation. That the surface of the sea must act in the manner supposed by Tyndall cannot be doubted, but there are two difficulties in the way of accepting the simple explanation as complete. According to it the interference should always be the same, which is certainly not the case. Usually there is no silent area. Again, although according to the analogy of Lloyd's bands there might be a dark or silent place at a particular height above the water, say on the bridge of the *Irene,* the effect should be limited to the neighbourhood of the particular height. At a height above the water twice as great, or near the water level itself, the sound should be heard again. In the latter case there were some difficulties, arising from disturbing noises, in making a satisfactory trial; but as a matter of fact, neither by an observer up the mast nor by one near the water level, was a sound lost on the bridge ever recovered.

The interference bands of Fresnel's experiment may be imitated by a bifurcation of the sound issuing from A (Fig. 1). For this purpose a sort of T-tube is fitted, the free ends being provided with small elliptical cones, similar to that already described, whose axes are parallel and distant from another by about 40 cm. The whole is constructed with regard to symmetry, so that sounds of equal intensity and of the same phase issue from the two cones whose long diameters are vertical. If the distances of the burner from the mouths of the cones be precisely equal, the sounds arrive in the same phase and the flame flares vigorously. If, as by the hand held between, one of the sounds is cut off, the flaring is reduced, showing that with this adjustment the two sounds are more powerful than one. By an almost imperceptible slewing round of the apparatus on its base-board the adjustment above spoken of is upset and the flame is induced to recover its tall equilibrium condition. The sounds now reach the flame in opposition of phase and practically neutralise one another. That this is so is proved in a moment. If the hand be introduced between either orifice and the flame, flaring ensues, the sound not intercepted being free to produce its proper effect.

The analogy with Fresnel's bands would be most complete if we kept the sources of sound at rest and caused the burner to move transversely so as to occupy in succession places of maximum and minimum effect. It is more convenient with our apparatus and comes to the same thing, if we keep the burner fixed and move the sources transversely, sliding the base-board without rotation. In this way we may verify the formula, connecting the width of a band with the wave-length and the other geometrical data of the experiment.

The phase discrepancy necessary for interference may be introduced, without disturbing the equality of distances, by inserting in the path of one of the sounds a layer of gas having different acoustical properties from air. In the lecture carbonic acid was employed. This gas is about half as heavy again as air, so that the velocity of sound is less in the proportion of $1 : 1\cdot25$. If l be the thickness of the layer, the *retardation* is $\cdot25\,l$; and if this be equal to the half wave-length, the interposition of the layer causes a transition from complete agreement to complete opposition of phase. Two cells of tin plate were employed, fitted with tubes above and below, and closed with films of collodion. The films most convenient for this purpose are those formed upon water by the evaporation of a few drops of a solution of celluloid in pear-oil. These cells were placed one in the path of each sound, and the distances of the cones adjusted to maximum flaring. The insertion of carbonic acid into *one* cell quieted the flame, which flared again when the second cell was charged so as to restore symmetry. Similar effects were produced as the gas was allowed to run out at the lower tubes, so as to be replaced by air entering above.*

Many vibrating bodies give rise to sounds which are powerful in some directions but fail in others—a phenomenon that may be regarded as due to interference. The case of tuning forks (unmounted) is well known. In the lecture a small and thick wine-glass was vibrated, after the manner of a bell, with the aid of a violin bow. When any one of the four vibrating segments was presented to the flame, flaring ensued; but the response failed when the glass was so held at the same distance that its *axis* pointed to the flame. In this position the effects of adjacent segments neutralise one another and the aggregate is zero. Another example, which, strangely enough, does not appear to have been noticed, is afforded by the familiar open organ pipe. The vibrations issuing from the two ends are in the same phase as they start, so that if the two ends are equally distant from the percipient, the effects conspire. If, however, the pipe be pointed towards the percipient, there is a great falling off, inasmuch as the length of the pipe approximates to the half-wave length of

---

* In a still atmosphere the hot gases arising from lighted candles may be substituted for the layers of CO_2.

the sound. The experiment may be made in the lecture-room with the sensitive flame and one of the highest pipes of an organ, but it succeeds better and is more striking when carried out in the open air with a pipe of lower pitch, simply listened to with the unaided ear of the observer. Within doors reflections complicate all experiments of this kind.   [R.]

At the conclusion of the discourse His Grace the DUKE OF NORTHUMBERLAND, the President, unveiled a bust by the late Mr. Onslow Ford of Sir Frederick Bramwell, Bart., D.C.L. LL.D. F.R.S. M. Inst. C.E., formerly Honorary Secretary of the Royal Institution, and formally presented it to the members, as a token from the Managers and their friends of their esteem for him personally, and their appreciation of the way in which he had served the Institution.

Sir JAMES CRICHTON-BROWNE having accepted the bust on behalf of the members,

Sir FREDERICK BRAMWELL said he found great difficulty in making any adequate reply. At his age—he was well into his 84th year—many things were denied him, among others the faculty of blushing; otherwise he would not have been able to sit and listen to the laudatory remarks of the Duke of Northumberland and Sir James Crichton-Browne. The fifteen years he had served as secretary had been years of hard work, but of great enjoyment, for he had been brought into intimate contact with many men whom it was a pleasure and an honour to know. He was glad the bust had been given, for he believed the work of the world was done by average mediocrity pegging on, and it was well to encourage men in that sort of thing. There was one matter which he could not pass over, but of which he could scarcely trust himself to speak. He had anticipated the process of making this bust with horror, but it became an actual enjoyment to sit to the genius and genial man who had made it, and watch the clay growing into a likeness of himself, and the death of Mr. Onslow Ford was to him a source of infinite sorrow.

WEEKLY EVENING MEETING,

Friday, January 24, 1902.

Sir James Crichton-Browne, M.D. LL.D. F.R.S., Treasurer
and Vice-President, in the Chair.

H. G. Wells, Esq., B.Sc.

The Discovery of the Future.

It will lead into my subject most conveniently to contrast and
separate two divergent types of mind, types which are to be distinguished chiefly by their attitude towards time, and more partien-
larly by the relative importance they attach and the relative amount
of thought they give to the future of things.

The first of these two types of mind—and it is, I think, the
predominant type, the type of the majority of living people—is that
which seems scarcely to think of the future at all, which regards it
as a sort of black non-existence upon which the advancing present
will presently write events. The second type, which is, I think, a
more modern and much less abundant type of mind, thinks constantly,
and by preference, of things to come, and of present things mainly
in relation to the results that must arise from them. The former
type of mind, when one gets it in its purity, is retrospective in habit,
and it interprets the things of the present, and gives value to this and
denies it to that, entirely with relation to the past. The latter type
of mind is constructive in habit; it interprets the things of the
present and gives value to this or that, entirely in relation to things
designed or foreseen.

While from that former point of view our life is simply to reap
the consequences of the past, from this our life is to prepare the
future. The former type one might speak of as the legal or sub-
missive type of mind, because the business, the practice, and the
training of a lawyer dispose him towards it; he of all men must most
constantly refer to the law made, the right established, the precedent
set, and most consistently ignore or condemn the thing that is only
seeking to establish itself. The latter type of mind I might for contrast call the legislative, creative, organising, or masterful type,
because it is perpetually attacking and altering the established order
of things, perpetually falling away from respect for what the past
has given us. It sees the world as one great workshop, and the
present is no more than material for the future, for the thing that is
yet destined to be. It is in the active mood of thought, while the
former is in the passive; it is the mind of youth—it is the mind most

manifest among the Western nations; while the former is the mind of age—the mind of the Oriental.

Things have been, says the legal mind, and so we are here. And the creative mind says, We are here, because things have yet to be.

Now I do not wish to suggest that the great mass of people belong to either of these two types. Indeed, I speak of them as two distinct and distinguishable types mainly for convenience, and in order to accentuate their distinction. There are probably very few people who brood constantly upon the past without any thought of the future at all, and there are probably scarcely any who live and think consistently in relation to the future. The great mass of people occupies an intermediate position between these extremes: they pass daily and hourly from the passive mood to the active— they see this thing in relation to its associations and that thing in relation to its consequences, and they do not even suspect that they are using two distinct methods in their minds.

But for all that they are distinct methods—the method of reference to the past, and the method of reference to the future; and their mingling in many of our minds no more abolishes their difference than the existence of piebald horses proves that white is black.

I believe that it is not sufficiently recognised just how different in their consequences these two methods are, and just where their difference and where the failure to appreciate their difference takes one. This present time is a period of quite extraordinary uncertainty and indecision upon endless questions—moral questions, æsthetic questions, religious and political questions—upon which we should all of us be happier to feel assured and settled; and a very large amount of this floating uncertainty about these important matters is due to the fact that with most of us these two insufficiently distinguished ways of looking at things are not only present together, but in actual conflict in our minds, in unsuspected conflict; we pass from one to the other heedlessly, without any clear recognition of the fundamental difference in conclusions that exists between the two; and we do this with disastrous results to our confidence and to our consistency in dealing with all sorts of issues.

But before pointing out how divergent these two types or habits of mind really are, it is necessary to meet a possible objection to what has been said. I may put that objection in this form—Is not this distinction between a type of mind that thinks of the past and of a type of mind that thinks of the future a sort of hair-splitting— almost like distinguishing between people who have left hands and people who have right? Everybody believes that the present is entirely determined by the past, you say; but then everybody believes also that the present determines the future. Are we simply separating and contrasting two sides of everybody's opinion? To which I would reply that we are not discussing what we know and believe about the relations of past, present, and future, or of the relation of cause and effect to each other in time at all. We all

know the present depends for its causes on the past, and that the
future depends for its causes upon the present.   But this discussion
concerns the way in which we approach things upon this common
ground of knowledge and belief.   We may all know there is an east
and a west; but if some of us always approach and look at things
from the west, if some of us always approach and look at things from
the east, and if others, again, wander about with a pretty disregard of
direction, looking at things as chance determines, some of us will get
to a westward conclusion of the journey, and some of us will get to
an eastward conclusion, and some of us will get to no definite conclusion at all about all sorts of important matters.

And yet those who are travelling east, and those who are travelling west, and those who are wandering haphazard, may be all upon
the same ground of belief and statement, and amidst the same assembly of proven facts.

Precisely the same thing will happen if you always approach
things from the point of view of their causes, or if you approach
them always with a view to their probable effects.   And in several
very important groups of human affairs it is possible to show quite
clearly just how widely apart the two methods, pursued each in its
purity, take those who follow them.

I suppose that three hundred years ago all people who thought
at all about moral questions—about questions of Right and Wrong—
deduced their rules of conduct absolutely and unreservedly from the
past, from some dogmatic injunction, some finally settled decree.
The great mass of people do so to-day.   It is written, they say,
"Thou shalt not steal," for example—that is the sole, complete, and
sufficient reason why you should not steal; and even to-day there is
a strong aversion to admit that there is any relation between the
actual consequences of acts and the imperatives of right and wrong.
Our lives are to reap the fruits of determinate things; and it is
still a fundamental presumption of the established morality that one
must do Right though the heavens fall.   But there are people coming
into this world who would refuse to call it Right if it brought the
heavens about our heads, however authoritative its sources and sanc-
tions; and this new disposition is, I believe, a growing one.   I suppose in all ages people, in a timid, hesitating, guilty way, have
tempered the austerity of a dogmatic moral code, by small infrac-
tions, to secure obviously kindly ends; but it was, I am told, the
Jesuits who first deliberately sought to qualify the moral interpreta-
tions of acts by a consideration of their results.   To-day there are
few people who have not more or less clearly discovered the future
as a more or less important factor in moral considerations.   To-day
there is a certain small proportion of people who frankly regard
morality as a means to an end, as an overriding of immediate and
personal considerations out of regard to something to be attained in
the future, and who break away altogether from the idea of a code
dogmatically established for ever.   Most of us are not so definite as

that, but most of us are deeply tinged with the spirit of compromise between the past and the future; we profess an unbounded allegiance to the prescriptions of the past, and we practise a general observance of its injunctions, but we qualify to a vague, variable extent with considerations of expediency. We hold, for example, that we must respect our promises. But suppose we find unexpectedly that for one of us to keep a promise must lead to the great suffering of some other human being—must lead to practical evil: would a man do right if he broke such a promise? The practical decision most modern people would make would be to break the promise. But suppose it was not such *very* great suffering we were going to inflict, but only some suffering? And suppose it was a rather solemn promise? With most of us it would then come to be a matter of weighing the promise, the thing of the past, quite apart from its effect upon our credit, against the unexpected bad consequence, the thing of the future. And the smaller the evil consequences the more most of us would vacillate. But neither of the two types of mind we are contrasting would vacillate at all. The legal type of mind would obey the past unhesitatingly; the creative would unhesitatingly sacrifice it to the future. The legal mind would say that whoever breaks the law at any point breaks it altogether, while the creative mind would say, Let the dead past bury its dead.

I have taken this simple case of a promise as my illustration for many reasons, but it is in the field of sexual morality that the two methods are most in conflict.

And I would like to suggest that until you have definitely determined to adhere to one or other of these two types of mental action in these matters, you are not even within hope of a sustained consistency in the thought that underlies your acts, that in every issue of principle that comes upon you, you will be entirely at the mercy of the intellectual mood that happens to be at that particular moment ascendant in your mind.

In the sphere of public affairs also these two ways of looking at things work out into equally divergent and incompatible consequences. The legal mind insists upon treaties, constitutions, legitimacies, and charters; the legislative incessantly assails these. Whenever some period of stress sets in, some great conflict between institutions and the forces in things, there comes a sorting between these two types of mind. The legal mind becomes glorified and transfigured in the form of hopeless loyalty; the creative mind inspires revolutions and reconstructions. And particularly is this difference of attitude accentuated in the disputes that arise out of wars. In most modern wars there is no doubt quite traceable on one side or the other a distinct creative idea—a distinct regard for some future consequence. But the main dispute even in most modern wars, and the sole dispute in most mediæval wars, will be found to be a reference not to the future but to the past; to turn upon a question of fact and right.

The wars of Plantagenet and Lancastrian England with France, for example, were based entirely upon a dummy claim, supported by obscure legal arguments, upon the crown of France. And the arguments that centre about the present war in South Africa ignore any ideal of a great united South African state almost entirely, and quibble this way and that about who began the fighting, and what was or was not written in some obscure revision of a treaty a score of years ago. Yet beneath the legal issues the broad creative idea has been very apparent in the public mind during this war. It will be found more or less definitely formulated beneath almost all the great wars of the past century; and a comparison of the wars of the nineteenth century with the wars of the Middle Ages will show, I think, that in this field also there has been a discovery of the future, an increasing disposition to shift the reference and values from things accomplished to things to come.

Yet, though foresight creeps into our politics and a reference to consequences into our morality, it is still the past that dominates our lives. But why? Why are we so bound to it? It is into the future we go; to-morrow is the eventful thing for us. There lies all that remains to be felt by us and our children, and all those that are dear to us. Yet we marshal and order men into classes entirely with regard to the past, we draw shame and honour out of the past, against the rights of property, the vested interests, the agreements and establishments of the past, the future has no rights. Literature is for the most part history, or history at one remove; and what is culture but a mould of interpretation into which new things are thrust, a collection of standards, a sort of bed of Procrustes, to which all new expressions must be lopped or stretched. Our conveniences, like our thoughts, are all retrospective. We travel on roads so narrow that they suffocate our traffic; we live in uncomfortable, inconvenient, life-wasting houses out of a love of familiar shapes and familiar customs and a dread of strangeness; all our public affairs are cramped by local boundaries impossibly restricted and small. Our clothing, our habits of speech, our spelling, our weights and measures, our coinage, our religious and political theories, all witness to the binding power of the past upon our minds.

Yet we do not serve the past as the Chinese have done. There are degrees. We do not worship our ancestors, nor prescribe a rigid local costume; we venture to enlarge our stock of knowledge, and we qualify the classics with occasional adventures into original thought. Compared with the Chinese, we are distinctly aware of the future; but compared with what we might be, the past is all our world.

The reason why the retrospective habit, the legal habit, is so dominant, and always has been so predominant, is of course a perfectly obvious one. We follow the fundamental human principle and take what we can get. All persons believe the past is certain, defined, and knowable, and only a few people believe that it is possible to know anything about the future. Man has acquired the habit of going to the past because it was the line of least resistance for his mind. While a certain variable portion of the past is serviceable matter for knowledge in the case of everyone, the future is, to a mind without an imagination trained in scientific habits of thought, non-existent. All our minds are made of memories. In our memories each of us has something that, without any special training whatever, will go back into the past and grip firmly and convincingly all sorts of workable facts—sometimes more convincingly than firmly. But the imagination, unless it is strengthened by a very sound training in the laws of causation, wanders like a lost child in the blackness of things to come, and returns—empty.

Many people believe therefore that there can be no sort of certainty about the future. You can know no more about the future, I was recently assured by a friend, than you can know which way a kitten will jump next. And to all who hold that view, who regard the future as a perpetual source of convulsive surprises, as an inpenetrable, incurable, perpetual blackness, it is right and reasonable to derive such values as it is necessary to attach to things from the events that have certainly happened with regard to them. It is our ignorance of the future and our persuasion that that ignorance is absolutely incurable, that alone gives the past its enormous predominance in our thoughts. But through the ages the long unbroken succession of fortune-tellers—and they flourish still—witnesses to the perpetually smouldering feeling that after all there *may* be a better sort of knowledge, a more serviceable sort of knowledge than that we now possess.

On the whole, there is something sympathetic for the dupe of the fortune-teller in the spirit of modern science; it is one of the persuasions that come into one's mind, as one assimilates the broad conceptions of science, that the adequacy of causation is universal; that in absolute fact, if not in that little bubble of relative fact which constitutes the individual life, in absolute fact, the future is just as fixed and determinate, just as settled and inevitable, just as possible a matter of knowledge, as the past. Our personal memory gives us an impression of the superior reality and trustworthiness of things in the past, as of things that have finally committed themselves and said their say; but the more clearly we master the leading conceptions of science, the better we understand that this impression is one of the results of the peculiar conditions of our lives, and not an absolute truth. The man of science comes to believe at last that the events of the year A.D. 4000 are as fixed, settled, and unchangeable as the events of the year 1600. Only about the latter he has some material for belief, and about the former—practically none.

And the question arises, how far this absolute ignorance of the future is a fixed and necessary condition of human life, and how far some application of intellectual methods may not attenuate, even if it does not absolutely set aside, the veil between ourselves and things to come. And I am venturing to suggest to you, that along certain lines, and with certain qualifications and limitations, a working knowledge of things in the future is a possible and practicable thing.

And in order to support this suggestion, I would call your attention to certain facts about our knowledge of the past, and more particularly I would insist upon this, that about the past our range of absolute certainty is very limited indeed. About the past, I would suggest, we are inclined to over-estimate our certainty, just as I think we are inclined to under-estimate the certainties of the future. And such a knowledge of the past as we have is not all of the same sort, nor derived from the same sources.

Let us consider just what an educated man of to-day knows of the past. First of all, he has the reallest of all knowledge, the knowledge of his own personal experiences, his memory. Uneducated people believe their memories absolutely, and most educated people believe theirs with a few reservations. Some of us take up a critical attitude even towards our own memories ; we know that they not only sometimes drop things out, but that sometimes a sort of dreaming, or a strong suggestion, will put things in. But, for all that, memory remains vivid and real, as no other knowledge can be, and to have seen, and heard, and felt, is to be nearest to absolute conviction. Yet our memory of direct impressions is only the smallest part of what we know. Outside that bright area comes knowledge of a different order, the knowledge brought to us by other people. Outside our immediate personal memory, there comes this wider area of facts or quasi-facts, told us by more or less trustworthy people, told us by word of mouth or by the written word of living and of dead writers. This is the past of report, rumour, tradition, and history—the second sort of knowledge of the past. The nearer knowledge of this sort is abundant, and clear, and detailed ; remoter, it becomes vaguer ; still more remotely in time and space, it dies down to brief, imperfect inscriptions and enigmatical traditions, and at last dies away, so far as the records and traditions of humanity go, into a doubt and darkness as black, just as black, as futurity. And now let me remind you that this second zone of knowledge, outside the bright area of what we have felt, and witnessed, and handled for ourselves—this zone of hearsay, and history, and tradition, completed the whole knowledge of the past that was accessible to Shakespeare, for example. To these limits man's knowledge of the past was absolutely confined save for some inklings and guesses, save for some small, almost negligible beginnings, until the nineteenth century began.

Beside the correct knowledge in this scheme of hearsay and history, a man had a certain amount of legend and error that rounded off the picture in a very satisfying and misleading way—according to Bishop Ussher, just exactly four thousand and four years B.C. And that was man's Universal History—that was his all, until the scientific epoch began. And beyond those limits—? Well, I suppose the educated man of the sixteenth century was as certain of the non-existence of anything before the creation of the world as he was, and as most of us are still, of the practical non-existence of the future, or at any rate, he was as satisfied of the impossibility of knowledge in one direction as in another.

But modern science—that is to say, the relentless systematic criticism of phenomena—has in the past hundred years absolutely destroyed the conception of a finitely distant Beginning of Things, has abolished such limits to the past as a dated creation set, and added an enormous vista to that limited sixteenth century outlook.

And what I would insist upon is, that this further knowledge is a new kind of knowledge obtained in a new kind of way. We know to-day quite as confidently, and in many respects more intimately than we know Sargon, or Zenobia, or Caractacus, the form and the habits of creatures that no living being has ever met, that no human eye has ever regarded, and the character of scenery that no man has ever seen or can ever possibly see ; we picture to ourselves the Labyrinthodon raising its clumsy head above the waters of the carboniferous swamps in which he lived, and we figure the Pterodactyls, those great bird-lizards, flapping their way athwart the forests of Mesozoic age with exactly the same certainty as that with which we picture the rhinoceros or the vulture. I doubt no more about the facts in this further picture than I do about those in the nearer. I believe in the Megatherium, which I have never seen, as confidently as I believe in the hippopotamus that has engulfed buns from my hand. A vast amount of detail in that further picture is now fixed and finite for all time. And a countless number of investigators are persistently and confidently enlarging, amplifying, correcting and pushing further and further back the boundaries of this Greater Past, —this pre-human past that the scientific criticism of existing phenomena has discovered and restored and brought for the first time into the world of human thought. We have become possessed of a new and once unsuspected history of the world, of which all the history that was known, for example, to Dr. Johnson, is only the brief concluding chapter. And even that concluding chapter has been greatly enlarged and corrected by the exploring archæologist working strictly upon the lines of the new method ; that is to say, the comparison and criticism of suggestive facts.

I want particularly to insist upon this—that all this outer past— this non-historical past—is the product of a new and keener habit of inquiry, and no sort of revelation. It is simply due to a new and more critical way of looking at things. Our knowledge of the geological past, clear and definite as it has become, is of a different and lower order than the knowledge of our memory, and yet of a quite practicable and trustworthy order—a knowledge good enough to go upon. And if one were to speak of the private memory as the Personal Past, and the next wider area of knowledge as the Traditional or Historical Past, then one might call all that great and inspiring background of remoter geological time the Inductive Past.

And this great discovery of the Inductive Past was got by the discussion and rediscussion and effective criticism of a number of existing facts, odd-shaped lumps of stone, streaks and bandings in quarries and cliffs, anatomical and developmental details, that had always been about in the world, that had been lying at the feet of mankind so long as mankind had existed, but that no one had ever dreamt before could supply any information at all, much more reveal such astounding and enlightening vistas. Looked at in a new way, they became sources of dazzling and penetrating light—the remoter past lit up and became a picture. Considered as effects, compared and criticised, they yielded a clairvoyant vision of the history of interminable years.

And now, if it has been possible for men—by picking out a number of suggestive and significant-looking things in the present, by comparing them, criticising them, and discussing them, with a perpetual insistence upon *why?* without any guiding tradition, and, indeed, in the teeth of established beliefs—to construct this amazing searchlight of inference into the remoter past,—is it really, after all, such an extravagant and hopeless thing to suggest that, by seeking for operating causes instead of for fossils, and by criticising them as persistently and thoroughly as the geological record has been criticised, it may be possible to throw a searchlight of inference forward instead of backward, and to attain to a knowledge of coming things as clear, as universally convincing, and infinitely more important to mankind than the clear vision of the past that geology has opened to us during the nineteenth century?

Let us grant that anything to correspond with the memory, anything having the same relation to the future that memory has to the past, is out of the question. We cannot imagine, of course, that we can ever know any personal future to correspond with our personal past, nor any traditional future to correspond with our traditional past. But the possibility of an inductive future to correspond with that great inductive past of geology and archæology is an altogether different thing.

I must confess that I believe quite firmly that an inductive knowledge of a great number of things in the future is becoming a human possibility. I believe that the time is drawing near when it will be possible to suggest a systematic exploration of the future. And you must not judge the practicability of this enterprise by the failures of the past. So far nothing has been attempted, so far no first-class mind has ever focussed itself upon these issues. But suppose the laws of social and political development, for example, were given as many brains, were given as much attention, criticism and discussion as we have given to the laws of chemical combination during the last fifty years,—what might we not expect?

To the popular mind of to-day, there is something very difficult in such a suggestion, soberly made, but here, in this Institution, which has watched for a whole century over the splendid adolescence of science, and where the spirit of science is surely understood, you will know that, as a matter of fact, prophecy has always been inseparably associated with the idea of scientific research. The popular idea of scientific investigation is a vehement, aimless collection of little facts, collected as the bower-bird collects shells and pebbles, and the systematic, unreasonable stowing away of these little facts in methodical little rows; and out of this process, in some manner unknown to the popular mind, certain conjuring tricks—the celebrated wonders of science—in a sort of accidental way, emerge. The popular conception of all discovery is accident.

But you will know that the essential thing in the scientific process is not the collection of facts, but the analysis of facts: facts are the raw material, and not the substance, of science; it is analysis that has given us all ordered knowledge; and you know that the aim, and the test, and the justification of the scientific process is not a marketable conjuring trick, but prophecy. Until a scientific theory yields confident forecasts, you know it is unsound and tentative; it is mere theorising, as evanescent as art talk, or the phantoms politicians talk about. The splendid body of gravitational astronomy, for example, establishes itself upon the certain forecast of stellar movements, and you would absolutely refuse to believe its amazing assertions if it were not for these same unerring forecasts. The whole body of medical science aims, and claims the ability, to diagnose. Meteorology constantly and persistently aims at prophecy, and it will never stand in a place of honour until it can certainly foretell. The chemist forecasts elements before he meets them—it is very properly his boast; and the splendid manner in which the mind of Clerk Maxwell reached in front of all experiment, and foretold those things that Marconi has materialised, is familiar to us all. All applied mathematics resolves into computation to foretell things which otherwise can only be determined by trial. Even in so unscientific a science as political economy there have been forecasts.

And, if I am right in saying that science aims at prophecy, and if the specialist in each science is, in fact, doing his best *now* to prophecy within the limits of his field, what is there to stand in the way of our building up this growing body of forecast into an ordered picture of the future that will be just as certain, just as strictly science, and perhaps just as detailed as the picture that has been built up within the last hundred years to make the geological past? Well, so far and until we bring the prophecy down to the affairs of man and his children, it is just as possible to carry induction forward as back; it is just as simple and sure to work out the changing orbit of the earth in the future until the tidal drag hauls one unchanging face at last towards the sun, as it is to work back to its blazing and molten past. Until man comes in, the inductive future is as real and convincing as the inductive past. But inorganic forces are the smaller part and the minor interest in this concern. Directly man becomes a factor the nature of the problem changes, and our whole present interest centres on the question whether man is indeed, individually and collectively, incalculable, a new element which entirely alters the nature of our inquiry and stamps it at once as vain and hopeless, or whether his presence complicates indeed, but does not alter, the essential nature of the induction. How far may we hope to get trustworthy inductions about the future of man?

Well, I think, on the whole, we are inclined to underrate our chance of certainties in the future just as I think we are inclined to be too credulous about the historical past. The vividness of our personal memories, which are the very essence of reality to us, throws a glamour of conviction over tradition and past inductions. But the personal future must in the very nature of things be hidden from us so long as time endures, and this black ignorance at our very feet, this black shadow that corresponds to the brightness of our memories behind us, throws a glamour of uncertainty and unreality over all the future. We are continually surprising ourselves by our own wills or want of will; the individualities about us are continually producing the unexpected, and it is very natural to reason that as we can never be precisely sure before the time comes what we are going to do and feel, and if we can never count with absolute certainty upon the acts and happenings even of our most intimate friends, how much the more impossible is it to anticipate the behaviour in any direction of states and communities.

In reply to which I would advance the suggestion that an increase in the number of human beings considered may positively simplify the case instead of complicating it, that as the individuals increase in number they begin to average out. Let me illustrate this point by a comparison. Angular pit-sand has grains of the most varied shapes. Examined microscopically, you will find all sorts of angles and outlines and variations. Before you look, you can say of no particular grain what its outline will be. And if you shoot a load of such sand from a cart, you cannot foretell with any certainty where any particular grain will be in the heap that you make. But you can tell—you can tell pretty definitely—the form of the heap as a whole. And further, if you pass that sand through a series of shoots, and finally drop it some distance to the ground, you will be able to foretell that grains of a certain sort of form and size will for the most part be found in one part of the heap and grains of another sort of form and size will be found in another part of the heap. In such a case, you see, the thing as a whole may be simpler than its component parts, and this, I submit, is also the case in many human affairs. So that because the individual future eludes us completely, that is no reason why we should not aspire to, and discover and use, safe and serviceable generalisations upon countless important issues in the human destiny.

But there is a very grave and important-looking difference between a load of sand and a multitude of human beings, and this I must face and examine. Men's thoughts and wills and emotions are contagious.

An exceptional sort of sand grain—a sand grain that is exceptionally big and heavy, for example—exerts no influence worth considering upon any other of the sand grains in the load. They will fall and roll and heap themselves just the same, whether that exceptional grain is with them or not. But an exceptional man comes into the world—a Cæsar, or a Napoleon, or a Peter the Hermit—and he appears to persuade and convince and compel and take entire possession of the sand heap—I mean the community—and to twist and alter its destinies to an almost unlimited extent. And if this is indeed the case, it reduces our project of an inductive knowledge of the future to very small limits. To hope to foretell the birth and coming of men of exceptional force and genius is to hope incredibly, and if indeed such exceptional men do as much as they seem to do in warping the path of humanity, our utmost prophetic limit in human affairs is a conditional sort of prophecy. *If* people do so and so, we can say, then such and such results will follow, and we must admit that that is our boundary.

But everybody does not believe in the importance of the leading man. There are those who will say that the whole world is different by reason of Napoleon. But there are also those who will say the whole world of to-day would be very much as it is now if Napoleon had never been born. There are those who believe entirely in the individual man, and those who believe entirely in the forces behind the individual man; and, for my own part, I must confess myself a rather extreme case of the latter kind. I must confess I believe that if, by some juggling with space and time, Julius Cæsar, Napoleon, Edward IV., William the Conqueror, Lord Rosebery, and Robert Burns had all been changed at birth, it would not have produced any serious dislocation in the course of destiny. I believe that these great men of ours are no more than images and symbols and instruments taken, as it were, haphazard by the incessant and consistent forces behind them ; they are the pen-nibs Fate has used for her writing, the diamonds upon the drill that pierces through the rock. And the more one inclines to this trust in forces, the more one will believe in the possibility of a reasoned inductive view of the future, that will serve us in politics, in morals, in social contrivances, and in a thousand spacious ways. And even those who take the most extreme and personal and melodramatic view of the ways of human destiny, who see life as a tissue of fairy godmother births and accidental meetings and promises and jealousies, will, I suppose, admit there comes a limit to these things—that at last personality dies away and the greater forces come to their own. The great man, however great he be, cannot set back the whole scheme of things ; what he does in right and reason will remain, and what he does against the greater creative forces will perish. We cannot foresee him ; let us grant this. His personal difference, the splendour of his effect, his dramatic arrangement of events, will be his own ; in other words, we cannot estimate for accidents and accelerations and delays

—but if only we throw our web of generalisation wide enough, if only we spin our rope of induction strong enough, the final result of the great man, his ultimate surviving consequences, will come within our net.

Such, then, is the sort of knowledge of the future that I believe is attainable, and worth attaining. I believe that the deliberate direction of historical study, and of economic and social study towards the future, and an increasing reference, a deliberate and courageous reference to the future in moral and religious discussion, would be enormously stimulating and enormously profitable to our intellectual life. I have done my best to suggest to you that such an enterprise is now a serious and practicable undertaking. But at the risk of repetition, I would call your attention to the essential difference that must always hold between our attainable knowledge of the future and our existing knowledge of the past. The portion of the past that is brightest and most real to each of us is the individual past, the personal memory. The portion of the future that must remain darkest and least accessible is the individual future. Scientific prophecy will not be fortune-telling, whatever else it may be. Those excellent people who cast horoscopes, those illegal fashionable palm-reading ladies who abound so much to-day, in whom nobody is so foolish as to believe, and to whom everybody is foolish enough to go, need fear no competition from the scientific prophets. The knowledge of the future we may hope to gain will be general, and not individual; it will be no sort of knowledge that will either hamper us in the exercise of our individual free will, or relieve us of our personal responsibility.

And now, how far is it possible at the present time to speculate on the particular outline the future will assume, when it is investigated in this way?

It is interesting, before we answer that question, to take into account the speculations of a certain sect and culture of people who already, before the middle of last century, had set their faces towards the future as the justifying explanation of the present. These were the Positivists, whose position is still most eloquently maintained and displayed by Mr. Frederic Harrison, in spite of the great expansion of the human outlook that has occurred since Comte. If you read Mr. Harrison, and if you are also, as I presume your presence here indicates, saturated with that new wine of more spacious knowledge that has been given the world during the last fifty years, you will have been greatly impressed by the peculiar limitations of the Positivist conception of the future. So far as I can gather, Comte was for all practical purposes totally ignorant of that remoter past outside the past that is known to us by history; or, if he was not totally ignorant of its existence, he was, and conscientiously remained, ignorant of its relevancy to the history of Humanity. In the narrow and limited past he recognised, men had always been men like the men of to-day; in the future he could not imagine that they would be anything more than men like the men of to-day. He perceived, as we all perceive, that the old social order was breaking up, and after a richly suggestive and incomplete analysis of the forces that were breaking it up, he set himself to plan a new static social order to replace it. If you will read Comte, or what is much easier and pleasanter, if you will read Mr. Frederic Harrison, you will find this conception constantly apparent—this conception that there was once a stable condition of society, with Humanity, so to speak, sitting down in an orderly and respectable manner; that Humanity has been stirred up and is on the move ; and that finally it will sit down again on a higher plane, and for good and all, cultured and happy, in the reorganised Positivist state. And since he could see nothing beyond man in the future, there, in that millennial fashion, Comte had to end. Since he could imagine nothing higher than man, he had to assert that Humanity, and particularly the future of Humanity, was the highest of all conceivable things.

All that was perfectly comprehensible in a thinker of the first half of the nineteenth century. But we of the early twentieth, and particularly that growing majority of us who have been born since the 'Origin of Species' was written, have no excuse for any such limited vision. Our imaginations have been trained upon a past in which the past that Comte knew is scarcely more than the concluding moment; we perceive that man, and all the world of men, is no more than the present phase of a development so great and splendid that, beside this vision, epics jingle like nursery rhymes, and all the exploits of Humanity shrivel to the proportion of castles in the sand. We look back through countless millions of years and see the great Will to Live struggling out of the intertidal slime, struggling from shape to shape, and from power to power, crawling, and then walking confidently, upon the land; struggling, generation after generation, to master the air, creeping down into the darkness of the deep ; we see it turn upon itself in rage and hunger, and reshape itself anew ; we watch it draw nearer and more akin to us, expanding, elaborating itself, pursuing its relentless, inconceivable purpose, until at last it reaches us, and its being beats through our brains and arteries, throbs and thunders in our battleships, roars through our cities, sings in our music, and flowers in our art. And when from that retrospect we turn again towards the future, surely any thought of finality, any millenial settlement of cultured persons, has vanished from our minds.

This fact, that man is not final, is the great, unmanageable, disturbing fact that rises upon us in the scientific discovery of the future; and to my mind, at any rate, the question, What is to come *after* man ? is the most persistently fascinating and the most insoluble question in the whole world.

Of course we have no answer. Such imaginations as we have refuse to rise to the task.

But, for the nearer future, while man is still man, there are a few general statements that seem to grow more certain. It seems to be pretty generally believed to-day, it has become a commonplace with cabinet ministers now—though it was a mere irresponsible suggestion two years ago—that our dense populations are in the opening phase of a process of diffusion and aëration. It seems pretty inevitable, also, that the mass of white and yellow population in the world will be forced some way up the scale of education and personal efficiency in the next two or three decades. It is not difficult to collect reasons for supposing, and such reasons have been collected, that in the near future—in a couple of hundred years, as one rash optimist has written —or in a thousand or so, humanity will be definitely and consciously organising itself as a great world-state : a great world-state that will purge from itself much that is mean, much that is bestial, and much that makes for individual dulness and dreariness, greyness, and wretchedness in the world of to-day. And although we know that there is nothing final in that world-state; although we see it only as something to be reached and passed; although we are sure there will be no such sitting down to restore and perfect a culture as the Positivists foretell, yet few people can persuade themselves to see anything beyond that except in the vaguest and most general terms. That world-state of more efficient, more vivid, beautiful, and eventful people is, so to speak, on the brow of a hill, and we cannot see over—though some of us can imagine great uplands beyond, and something—something that glitters elusively, taking first one form, and then another, through the haze. We can see no detail; we can see nothing definable; and it is simply, I know, the sanguine necessity of our minds that makes us believe that those uplands of the future are still more gracious and splendid than we can either hope or imagine. But of things that can be demonstrated we have none. ·

Yet I suppose most of us entertain certain necessary persuasions, without which a moral life in this world is neither a reasonable nor a possible thing. All this paper is built finally upon certain negative beliefs that are incapable of scientific establishment. Our lives and powers are limited; our scope in space and time is limited; and it is not unreasonable that for fundamental beliefs we must go outside the sphere of reason, and set our feet upon Faith. Implicit in all such speculations as this is a very definite and quite arbitrary belief, and that belief is that neither humanity nor, in truth, any individual human being, is living its life in vain. And it is entirely by an act of faith that we must rule out of our forecasts certain possibilities— certain things that one may consider improbable, and against the chances; but that no one, upon scientific grounds, can call impossible. One must admit that it is impossible to show why certain things should not utterly destroy and end the entire human race and story ; why night should not presently come down and make all our dreams and efforts vain. It is conceivable, for example, that some mechanism of electrolysis, or perhaps I should say the progress that has been made towards an explanation of the phenomena.

The earlier theories, from Grotthuss * in 1806, all assume that the decomposition is caused by the attraction of the electrodes or by the passage of the current, and that a definite electromotive force, different for each electrolyte, is required in order that decomposition shall take place. According to these theories, if the electromotive force is below that definite minimum no decomposition can occur and no current can pass.

And indeed at one time it was supposed that this was so. But Faraday, in a series of ingeniously devised and carefully executed experiments, showed that with electromotive force below the minimum necessary for the production of bubbles of gas on the electrodes, a perceptible current could pass for many days. He supposed that this small current was due to non-electrolytic conduction by the electrolyte. But the study of the phenomena of the polarisation of the electrodes led ultimately to the complete explanation by Helmholtz † in 1873 of this apparently metallic conduction by the electrolyte, and to a proof that any electromotive force, however small, sends a current through an electrolyte and gives rise to separation of the ions proportional to the amount of electricity transmitted.

The phenomena of the polarisation of the electrodes may be described shortly as follows. In the electrolysis of water (or rather of dilute sulphuric acid) it had been observed so long ago as 1802 that platinum or silver plates which had been used as electrodes acquired peculiar properties, so that for a short time the plate that had been the anode acted like the silver, and the plate that had been the cathode like the zinc of a voltaic cell, producing a short-lived and rapidly diminishing current. This observation was first made by Gautherot ‡ a teacher of music in Paris, who notes the effect of the current on the tongue and states that he had succeeded in decomposing water by means of his apparatus. Shortly after, J. W. Ritter, apparently without knowing anything of Gautherot's work, made a great many observations on the same subject. I cannot refrain from reading to you a passage from a letter from Christoph Bernoulli to van Mons. I take it from the translation published in *Nicholson's Journal*, October 1805. "As Mr. Ritter at present resides in a village near Jena, I have not been able to see his experiments with his grand battery of two thousand pieces, or with his battery of fifty pieces, each thirty-six inches square, the action of which continues very perceptible for a fortnight. Neither have I seen his experiments with the new battery of his invention, consisting of a single metal, and which he calls *the charging pile.*

---

* Grotthuss, Annales de Chimie, lviii. p. 54 (1806).

† Helmholtz. Pogg. 150, p. 483 (1873); Faraday Lecture, Chem. Soc. Trans. 39, p. 287 (1881); Wied. 34, p. 737 (1888).

‡ Gautherot, Annales de Chimie, xxxix. p. 203 (1801).

"I have frequently, however, seen him galvanise louis-d'or lent him by persons present.   To effect this, he places the louis between two pieces of pasteboard thoroughly wetted, and keeps it six or eight minutes in the chain of circulation connected with the pile.   Thus the louis becomes charged, without being immediately in contact with the conducting wires.   If this louis be applied afterwards to the crural nerves of a frog recently prepared, the usual contractions will be excited.   I had put a louis thus galvanised into my pocket, and Mr. Ritter said to me a few minutes after, that I might find out this louis from among the rest, by trying them in succession upon the frog. Accordingly I made the trial, and in reality distinguished among several others a single one, in which the exciting quality was very evident.   This charge is retained in proportion to the time that the piece has remained in the circuit of the pile.   It is with metallic discs charged in this manner, and placed upon one another with pieces of wet pasteboard alternately interposed, that Mr. Ritter constructs his charging pile, which ought in remembrance of its inventor to be called the *Ritterian pile*.   Mr. Ritter made me observe, that the piece of gold galvanised by communication exerts at once the action of two metals, or of one constituent of the pile ; and that the half which was next the negative pole while in the circle became positive, and the half toward the positive pole became negative."

Brugnatelli * suggested that the polarisation of the plate which during the electrolysis had given off hydrogen was due to a compound of hydrogen with the metal of the electrode.   But it was not until Schönbein discussed the question in 1839 † that a systematic attempt was made to settle it by experiment.   Schönbein's results were in favour of the view that the polarisation is due to the formation, on the surfaces of the electrodes, of thin sheets of the products of the electrolysis.

Now the old theories assume that if we begin with very small electromotive force and gradually increase it, we have at first a state of tension, the electromotive force so to speak pulling at the ions, that this tension increases as the electromotive force increases till it becomes sufficient to pull the ions apart.   If this were so there should be no current and no electrolysis till the electromotive force reaches a certain amount, and then suddenly a very great current, and something like an explosive discharge of gas ; for many molecules would be in the very same state of tension and all would give way at once.

When the electrolytic decomposition of water was first observed, as it was (by Nicholson and Carlisle) immediately after the publication of Volta's first description of the pile, the great difficulty felt by every one was that the hydrogen and the oxygen came off at different places which might be far apart.   Grotthuss's theory no doubt

---

* Brugnatelli, Gilbert's Annalen, xxiii. p. 202 (1806).
† Schönbein, Pogg. xlvi. p. 109 ; xlvii. p. 101 (1839).

explained this, but after the proof of a cause of polarisation given by
Schönbein, and the accumulating evidence that Ohm's law applies to
electrolytic as well as to metallic conduction, no one could hold or
defend Grotthuss's theory, although it was retained as a sort of
makeshift until someone could think of something better. The
something better was produced by Clausius in 1857.[*] Clausius was
one of the eminent physicists to whom we owe the kinetic theory of
gases, and his theory of electrolysis is derived from an application to
solutions of the ideas involved in this kinetic theory. He supposes
that the molecules of the electrolyte move through the solution as
the molecules of a gas move, that they collide with one another as
the gas molecules do, and that it must happen that here and there
ions get separated, and remain separated for a time, cation again
uniting with anion when two of them meet under favourable conditions. There will thus always be some detached ions moving
about just as molecules do. They will not always be the same ions
that are thus detached, and a very small proportion of such loose
ions will suffice to explain the phenomena. These loose ions retain
in their separate condition the charges of electricity which they had
when united, the cations being positively and the anions negatively
charged. This is assumed to be the state of matters in any solution
of an electrolyte. If now into such a solution we place two electrodes
with any, however small, difference of potential, the cathode, being
negative, will exercise an attraction upon the positively charged
cations, and the positive anode will exercise a similar attraction on
the negatively charged anions, and thus the loose ions, which before
the introduction of the electrodes moved about in the liquid with no
definite preferred direction, will on the whole, now that the electrodes
have been introduced, move preferably, the cations towards the
cathode, and the anions towards the anode, and those which are near
the electrodes will be drawn to them and discharge their electric
charge. This theory seems therefore to explain the phenomena.
The essential difference between it and all previous theories is that
Clausius does not attribute the decomposition to the current or to the
attraction of the electrodes ; what the attraction of the electrodes does
is to separate the ions already disengaged from one another, and this
the smallest electromotive force can do. The theory is so far
adequate, but is it admissible? Can we suppose that hydrogen and
chlorine atoms can move uncombined through the solution? It is to
be noted that while Clausius does not give any opinion as to the
proportion of loose ions to the total ions in any case, he assumes that
this proportion increases as the temperature rises, on account of the
greater briskness of the movements of the particles, and points out
that this is in accordance with the fact that electrolytes conduct
better as the temperature is higher. But he says, " to explain the
conduction of the electricity it is sufficient that in the encounters of

---

[*] Clausius, Pogg. ci. p. 338 (1857).

the molecules an exchange of ions should take place here and there, and perhaps comparatively rarely."

In this connection we may look at the views expressed by Williamson in his paper on the theory of etherification.* He says, " we are thus forced to admit that in an aggregate of molecules of any compound there is an exchange constantly going on between the elements which are contained in it. For instance, a drop of hydrochloric acid being supposed to be made up of a great number of molecules of the composition ClH, the proposition at which we have just arrived would lead us to believe that each atom of hydrogen does not remain quietly in juxtaposition with the atom of chlorine with which it first united, but, on the contrary, is constantly changing places with other atoms of hydrogen, or what is the same thing, changing chlorine." Williamson founded this opinion on the observed facts of double decomposition. He made no application of this view to the case of electrolysis, and indeed does not explicitly mention the temporary detachment of the atoms during the process of exchange; this is wholly due to Clausius, who arrived at his views as to the exchanges going on in a solution in a way quite different from that followed by Williamson, and quite independently. It was not then known how closely double decomposition and electrolysis are connected. We may perhaps get a clearer idea of Clausius's theory by imagining the phenomenon to take place on a scale such that we could see the individual ions. Let us then imagine a large field with a large number of men in it, each mounted on a horse. We shall further suppose that all the men are exactly alike and that all the horses are exactly alike. They are moving at random, most of them at about the same rate but a few of them faster, a very few of them considerably faster, a few of them slower, a very few of them considerably slower, than the average. They move in straight lines until they meet an obstacle which makes them deviate. This obstacle will often be another man and horse. The collision will give both a shake, and will sometimes dismount one or both of the riders. When this happens each will look for a horse, and as all the horses are exactly alike the horse such a dismounted man finds and mounts will not always be the one he came down from. But in any case there will be always in the field some men without horses and some horses without men. And the quicker the average pace the larger will be the proportion of dismounted men and riderless horses to the total number of men and horses. And this not only because there will be more and, as a rule, more violent collisions, but also because the dismounted men will have more difficulty in catching horses, although to keep up the analogy of the ions we must suppose the horses to be as anxious to be caught as the men are to catch them. If it does not make my allegory too grotesque we might suppose places with attractions for men and for horses respectively, to correspond

---

* Williamson, Chem. Soc. Journ. iv. p. 111 (1852).

to the electrodes, so that a man looking for a horse would on the whole rather go in the direction of lunch than away from it, and if he got near the refreshment room before he found a horse, he would look in there.   An objection was made to Clausius's theory that the same thing which he supposed to happen in solution, say of hydrochloric acid, ought also to happen in the gas.   We are not yet in a position to discuss this point with much prospect of obtaining a perfectly satisfactory explanation of the difficulty, although some progress towards an intelligible theory has been made, but at the risk of being tedious, I may indicate that my allegory may show us that we need not despair of finding in due time an answer.   Let us suppose that in the field there are not only men and horses but also a large number of other moving objects, let us say, by way of example, cows.   It seems plain that whether the presence of the cows would increase the chance of a man being dismounted or not, it would sensibly interfere with his chance of catching a horse if he were.   And it will be admitted that the nature and size of these other moving objects must exercise an influence on the proportion of horseless men and riderless horses to the total number.   But these other moving objects represent the molecules of the solvent, so that we need not be surprised when we find that the electrolytic conductivity is affected by the nature of the solvent, and that where there is no solvent the conductivity is very small or even nothing.

A very important question was left only partially answered by Faraday.   It is, What substances are electrolytes?   Faraday considered the water in dilute acid as the electrolyte, and the acid as a substance having the power of increasing the conductivity of the water.   When a solution of sulphate of copper was electrolysed, he supposed that the water was primarily decomposed and that the metallic copper was a secondary product reduced by the nascent hydrogen.   He says,[*] " I have experimented on many bodies, with a view to determine whether the results were primary or secondary.   I have been surprised to find how many of them, in ordinary cases, are of the latter class, and how frequently water is the only body electrolysed in instances where other substances have been supposed to give way."   From our present point of view many of us would rather say that the direct electrolysis of water very rarely occurs, except to a very small extent.

In 1839 Daniell began a series of ingeniously devised and skilfully executed experiments with the view of determining, in the case of salt solutions, whether it is the salt or the water which is primarily electrolysed.   The results appeared in two letters from Daniell to Faraday in 1839 [†] and 1840,[‡] and in a paper by Daniell and W. A. Miller in 1844,[§] all published in the Transactions of the Royal

---

* Faraday, Experimental Researches in Electricity, par. 751 (1834).
† Daniell, Phil. Trans. 1839, p. 97.
‡ Op. cit. 1840, p. 209.
§ Daniell and Miller, Phil. Trans. 1844, p. 1.

Society. The purpose of these investigations was attained, and it was completely proved that in reference to their behaviour as electrolytes there was no difference between say potassium chloride and potassium nitrate, except that in the latter some ammonia was formed at the cathode by the reducing action of the nascent hydrogen, and it was clearly shown that from an electrolytic point of view all the oxygen acids and their salts fell into line with hydrochloric acid and the chlorides, and that NH_4 was electrolytically perfectly analogous to K. There is, however, an interest in these papers beyond this important result. In the earlier part of the work the authors measured the amount of electrolysis not only by "the amount of ions disengaged at either or both electrodes by the primary action of the current or the secondary action of the elements," but also tried to obtain a check to this way of measuring, by using a diaphragm in the electrolytic cell, and analysing the contents of the two parts of the cell, the one on the anode side and other on the cathode side of the diaphragm. This check was "founded on the hypothesis that the voltaic decomposition of an electrolyte is not only effected by the disengagement of its anion and cation at their respective electrodes, but by the equivalent transfer of each to the electrodes, so that the measure of the quantity of matter translated to either side of the diaphragm might be taken as the measure of the electrolysis." They soon found that this hypothesis was unfit to give any such measurement, and in the paper of 1844 state that their results show that the hypothesis of equivalent transfer of the ions, "although generally received, is itself destitute of foundation."

The non-equivalent transfer of the ions, incidentally observed by Daniell and Miller, and imperfectly measured by them in a few cases, was made the subject of a long and elaborate series of experiments by Hittorf. The work extended over six years from 1853 to 1859 * and is a monument of patient labour and of happy adaptation of means to a clearly perceived end. The importance of the work was not at first recognised by either physicists or chemists, indeed its meaning was scarcely understood. I shall try to put before you as shortly as I can an outline of the ideas involved in the work, and of the most important conclusions arrived at by Hittorf. As the anions and the cations are separated at their respective electrodes in equivalent quantity, that is, in the case where the valency of anion and cation is the same, in equal numbers, it never occurred to any one to doubt that they travelled towards the electrodes at the same rate, until Daniell and Miller showed that this hypothesis is erroneous. To follow their reasoning and that of Hittorf we may take an imaginary case, and suppose an electrolyte **MX** with its cation **M** and its anion X of such character that these ions when separated at the electrodes can be removed from the solution completely and at

---

* Hittorf, Pogg. lxxxix. p. 177 (1853); xcviii. p. 1 (1856); ciii. p. 1 (1858); cvi. pp. 337 and 513 (1859). Arch. Néerland. (II.) vi. p. 671 (1901).

once, and that the electrolysis is carried on in a vessel provided with two compartments, one containing the cathode and the other the anode, such that whatever happens at an electrode shall affect only the contents of the compartment containing that electrode, and so arranged that the liquid contained in each compartment can be completely removed from it and analysed. Now, let us first suppose MX to be such that its ions travel at the same rate. In the time then in which one M has entered the cathode compartment one X has left it. There is at this moment an excess of two M's in this compartment, these are deposited at the cathode, and now the concentration of the solution in this compartment is diminished by one MX. Similarly at the anode during the same time one X has entered and one M has left, two X's have been deposited and the solution has lost one MX. In this case, then, where the two sets of ions travel at the same rate, the loss of solute is the same at the two electrodes. Let us now suppose an extreme case in which one of the sets of ions (say the cations) does not travel at all. In the time in which one X leaves the cathode compartment no M enters it, the excess of one M is deposited, and the solution here has lost one MX. At the anode one X has entered and no M has left, the X is deposited, and the solution here has lost no MX. Again, take the case that the anions travel twice as fast as the cations. Here in the time in which one M enters the cathode compartment two X's leave it, the excess of three M's is deposited, and the solution has lost two MX's. At the anode during the same time one M has left and two X's have entered, the three X's have been deposited and the solution has lost one MX. Of course it will be seen that the excess of one kind in a compartment consists not only of what enters it, but also of the excess resulting from the departure of the other kind. Without taking any more cases we at once see that the speed of the cation is to that of the anion as the loss of solute at the anode is to that at the cathode. This non-equivalent transfer has sometimes been described in another way. It has been said that the ions go at the same rate, but that at the same time the solute as a whole is being moved towards one of the electrodes. But this really is the same thing. If we imagine two processions walking with the same length of step and the same number of steps a minute in opposite directions on such a moving platform as that in the Paris Exhibition, we might no doubt say that the two *walked* at the same rate, they could not be said to *travel* at the same rate. Hittorf's way of putting it is not only the simpler way, it is the only way that agrees with what has since been made out as to the rate of movement of the ions.

Hittorf's work had to wait long for recognition, but we now know its great importance, not only on account of the large number of accurate measurements, but also because of the general conclusions he drew from them. He deduced from the transfer numbers conclusions as to the nature of the solute, showing, for instance, that solution of stannic chloride electrolyses as hydrochloric acid, the stannic chloride being completely hydrolysed. He also showed that such double salts as $KCN,AgCN$; $2KCl,PtCl_4$; and $KCl,AuCl_3$, have potassium for their only cation, the silver, the platinum and the gold forming part of the anion. He also showed that $2KI,CdI_2$ behaves as a single salt with K as cation when the concentration is great, but as two salts with cadmium as well as potassium as cation in dilute solution. In these and in similar cases, Hittorf made a valuable contribution to the theory of double salts. But perhaps the most striking generalisation is that contained in the words "electrolytes are salts," and his very instructive comparison of the readiness with which a substance enters into double decomposition and the readiness with which it can be electrolysed. With the fairness to his predecessors which is characteristic of him, he quotes an almost forgotten statement of Gay-Lussac to something like the same effect.

Ladies and Gentlemen,—I wish here to tell you that within the last three weeks, Professor Hittorf entered on the fifty-first year of his professorship. The officials of the Royal Institution have authorised me to ask our Chairman, Lord Kelvin, to send your congratulations to Professor Hittorf on his jubilee.

We now come to another turning point in the development of the theory of electrolysis, inseparably associated with the name of Kohlrausch.[*] It is to Kohlrausch and to those who worked with him that we owe the methods for the accurate determination of the conductivity of electrolytes. I need not give a description of the apparatus. It is now used in every laboratory, and by means of it a series of observations of the conductivity of an electrolyte can be made at different concentrations in a very short time. An early result of Kohlrausch's investigation was his discovery that "all acids which have been examined in strong solutions show, for a definite proportion of water, a maximum of conductivity," and he shows that for many other electrolytes there is a solution which conducts better than one either a little more or a little less concentrated. Thus the maxima of conductivity of the following acids are at the following percentages: HNO_3 29·7 per cent.; HCl 18·3 per cent.; H_2SO_4 30·4 per cent.; $HC_2H_3O_2$ 16·6 per cent. "The maximal acetic acid conducts at least 38,000 times better than concentrated acetic acid." In connection with this he says, "we do not know one single liquid, which at ordinary temperature is, by itself, a good electrolytic conductor." He refers the trace of conductivity in H_2SO_4 to the dissociation into water and SO_3 observed by Marignac and by Pfaundler, and observes that as up to the present time we know only

---

[*] Kohlrausch and Nippoldt, Pogg. cxxxviii. p. 280 (1869); Kohlrausch, Pogg. Jubelband, p. 290 (1874); Kohlrausch and Grotrian, Pogg. cliv. pp. 1 and 215 (1875); Grotrian, clvii. p. 130 (1876); Kohlrausch, clix. p. 233 (1876). Göttinger Nachrichten, 1876, p. 213. Wied. vi. p. 1 (1879); xi. p. 653 (1880).

mixtures which at ordinary temperature conduct well, the supposition is not unnatural that it is mixture that makes electrolytes good conductors. And again, if what has been said is correct, we must, in order to have good conduction, protect the wandering constituents from frequent meeting with one another, and this service is performed by the solvent, which makes it possible for the ions to get over a part of their journey—and so much larger a part the more solvent there is—without re-forming molecules. It is this suggestion which I ventured a little while ago to put into an allegorical form.

In order to compare the conductivity of one electrolyte with that of another, it is necessary that we choose comparable quantities of the two, and there is no difficulty in seeing that such comparable quantities are those decomposed by the same current of electricity—that is to say, the electro-chemical equivalents of the electrolytes. Accordingly, instead of expressing the concentration in percentage of the solute, Kohlrausch uses "molecular numbers." The molecular number of a solution is the quantity, in grams, of the solute contained in a litre of the solution divided by the equivalent of the solute. Dividing the conductivity of a solution by its molecular number gives its molecular conductivity. It will be seen that "molecular" is not used here in its ordinary chemical sense, but as the meaning is quite distinctly stated no confusion need arise. Kohlrausch showed that the molecular conductivity increases as the solution becomes more dilute, and with extreme dilution approaches a constant value.

I now show an experiment to illustrate this.

The apparatus * consists of an electrolytic cell in the form of a tall rectangular trough, the back and front being broad plates of glass, while the sides are composed of narrow strips of wood completely lined with silver foil. The bottom of the cell is made of non-conducting material. The two sheets of silver serve as electrodes, being connected to binding screws by means of external wires. The cell is introduced into a battery circuit along with a galvanometer of low resistance. If the cell be filled with pure water there is scarcely an appreciable current transmitted. On removing the water and pouring in 20 cc. of a 4-normal silver nitrate solution, so as to cover the bottom to a depth of a few millimetres, a current passes as indicated by the galvanometer. If pure water be now added in successive portions and the solution stirred after each addition, an increase in the strength of the current is observed, the increase being greatest after the first dilution, and becoming less with each succeeding dilution, so that a maximum is approached. In this experiment the distance between the electrodes is constant, and the area of the electrodes and of the cross-section of the conducting solution are proportional to the volume of the solution, and the quantity of the

---

* From a paper by Noyes and Blanchard, in the Zeitschrift für physikalische Chemie, **xxxvi.** p. 9 (1901).

salt is constant; therefore any change in the strength of the current means a corresponding change in the molecular conductivity of the dissolved salt. The molecular conductivity, therefore, increases with the dilution, and asymptotically approaches a maximum.

I cannot here enter into a description of the great experimental difficulties connected with the determination of the conductivity of extremely dilute solutions, but I may refer to one of them, namely the small but variable conductivity of the water used in preparing the solutions. There seems now to be no doubt that water is in itself an electrolyte. But the purest water that has been obtained has a conductivity of only about 10^{-10} as compared with that of mercury as unit. The minutest traces of salts greatly increase the conductivity, so that ordinary distilled water has a conductivity of 3×10^{-10} or more. With solutions of moderate dilution the variation of this very small quantity is of little consequence, but with extremely dilute solutions the conductivity to be measured is of the same order as that of the water.

For our present purpose the most important conclusion drawn by Kohlrausch from his observations is his law of the independent rate of motion of the ions in dilute solutions. The rate of motion of any ion towards the electrode depends on the gradient of potential. But Kohlrausch shows that the rate of motion of each ion in dilute solution is proportional to a number, the same whatever be the other ion of the electrolyte. Thus the rate at which the cation K moves towards the cathode in dilute solution, is the same in solutions of KCl, KNO_3, $KC_2H_3O_2$, etc. Kohlrausch gives these numbers for six cations and ten anions. The results calculated from these numbers agree well with the observed conductivities.

Methods have been devised for directly observing and measuring the rate at which ions travel. In this connection I may specially mention the names of Oliver Lodge, Whetham, and Masson. These measurements agree very well with the rates calculated by Kohlrausch.

I now show an experiment indicating a way in which such measurements can be made.

The apparatus * consists of a glass U-tube, with a long stopcock-funnel connected to the lower part of it. The tube is nearly half filled with a dilute (about 0·03 per cent.) solution of potassium nitrate, and then about the same quantity of a solution of potassium permanganate, of the same conductivity as the other solution, is slowly introduced by means of the funnel. The permanganate solution is loaded with urea (a non-electrolyte) so as to make it denser than the nitrate solution; the permanganate solution now lies in the lower part of the U-tube with a sharp interface between it and the nitrate solution above it in each limb of the tube. If now we connect

---

* Experiment from a paper by Nernst. in the Zeitschrift für Elektrochemie, iii. p. 308 (1897).

the electrodes, which were preliminarily inserted into the upper parts of the two limbs of the tube, with a battery with high difference of potential a current will pass, and a transference of ions will take place, cations (K) towards the cathode and anions (NO_3 and MnO_4) towards the anode, and the column of pink colour will rise in the limb containing the anode and fall by an equal amount in the other. By this means an approximation can be made to the rate of travel of the ions.

We now come to a new chapter beginning with 1887; but before entering on it we must turn aside for a little to a subject which does not at first sight seem to have a very close relation with the matter we have in hand. The subject is that of what may be called the osmotic phenomena. These are all connected with the concentration or with the dilution of solutions. They all involve the idea of the work done in concentrating a solution. We need not discuss the theory of these phenomena, we are interested in them now only as they give us methods of ascertaining the molecular concentration of a solution. In 1883 * Raoult showed that in the case of a great many substances, equimolecular solutions (with the same solvent) have the same freezing-point. In 1886 † he showed that equi-molecular solutions with the same volatile solvent have the same boiling-point. Molecular is here used in its ordinary chemical sense. These discoveries were eagerly taken up by chemists as promising an important addition to the means at their disposal for determining the molecular weights of substances. Convenient arrangements for applying the methods were devised by Beckmann, ‡ and soon came into use in nearly every laboratory. They were almost exclusively used for the determination of the molecular weight of organic substances, and have been found trustworthy in such cases. When, however, van't Hoff § in his study of the theory of solutions concluded from theoretical considerations that the depression of the freezing-point and the rise of the boiling-point are proportional to osmotic pressure in the case of dilute solutions, the observations made by Raoult and others furnished a number of facts ready for testing the theory. He found that, while in many cases the osmotic pressure calculated from his formula $PV = RT$ agreed, within the limits of experimental error, with the value calculated from the observation, there were a very considerable number where the observed value differed from that given by the formula. He accordingly modified the formula by the introduction of a factor i, so as to make it $PV = iRT$. This factor i

---

* Raoult, Compt. rend. xciv. p. 1517; xcv. pp. 187 and 1030 (1882); xcvi. p. 1653; xcvii. p. 941 (1883).

† Raoult, Compt. rend. ciii. p. 1125 (1886); civ. pp. 976 and 1430; cv. p. 857 (1887). Zeitschrift f. physik. Chemie, ii. p. 353 (1888).

‡ Beckmann, Zeitschrift f. physik. Chemie, ii. pp. 638 and 715 (1888); iv. p. 532 (1889); viii. p. 223 (1891).

§ Van't Hoff, Zeitschrift f. physik. Chemie, i. pp. 500–508 (1887).

is unity in the cases where observation by Raoult's method gives results agreeing with the formula $PV = RT$, in other cases it is greater or less than unity, and indicates the extent of the disagreement. Arrhenius, to whom Van't Hoff showed these numbers, pointed out that all the substances which had i greater than unity were electrolytes, and that the deviation had to do with their splitting up into ions. Arrhenius * had before this time (1887) been working at the subject of electrolysis and of the relation between the readiness with which substances undergo electrolysis and the readiness with which they enter into chemical reactions. He had been looking for an explanation of the fact that the conductivity of a solution of an electrolyte is not proportional to its concentration, and had come to the conclusion that this must depend on some of the molecules of the solute being "active," that is, taking part in the conduction, while others were inactive, behaving like molecules of a non-electrolyte, and that the proportion of active molecules increases with dilution.

Van't Hoff's factor i enabled Arrhenius to give precision to these ideas, and in 1887† he formulated the theory that the "active" molecules were those which were split into ions. It was now possible to calculate i in two ways and compare the results. Arrhenius gives a list of eighty-four substances, for which there existed at that time data for such calculations, and calculating the value of i as deduced on his new theory from the conductivity, compares it with the value of i derived from freezing-point observations in each of the eighty-four substances. The agreement does not at first sight strike one as very close, but there are several circumstances which have to be considered in judging them. The whole mass of published observations was taken, the limits of probable error are very different in different cases, and the freezing-point measurements were all made at temperatures a little below 0°, while the conductivity measurements were made at 18°. The comparison was made, not as a demonstration of the theory, but rather as a preliminary trial with such materials as were at hand. The real testing of the theory necessarily came later. So I think we may agree with Arrhenius that, considering all the circumstances, the agreement is not unsatisfactory, except in the case of nine of the substances, and that most of these nine cases are liable to suspicion on other grounds. In 1887, almost at the time when Arrhenius published the paper of which I have just been speaking, Planck‡ discussed the subjects of the diminution of the vapour pressure and the lowering of the freezing-point in dilute salt solutions from the thermo-dynamic point of view, and starting from the principle of the increase of entropy, deduced formulæ connecting these quantities with the molecular

---

* Arrhenius, Bihang till kongl. Svenska vetensk. Akad. Handlingar, 1884 Nos. 13 and 14.

† Arrhenius, Zeitschrift f. physik. Chemie, i. p 631 (1887).

‡ Planck, Wied. xxxii. p. 495 (1887).

weight. He says, in conclusion, " This formula claims exact numerical validity. It gives for most substances a greater molecular number than that usually assumed, i.e. a partial or complete chemical decomposition of the substance in the solution. Even if the consequences of this proposition should require an essential modification of the generally prevailing views as to the constitution of solutions, I do not know any fact which shows it to be untenable. Indeed, many observations in other departments (the proportionally strong affinities of dilute solutions, which remind one of the properties of the nascent state, the easy decomposability by the weakest galvanic current, the phenomena of internal friction), are directly in favour of the view that in all dilute solutions a more or less complete decomposition of the molecules of the dissolved substance takes place. Besides, this conception adapts itself well to the opinions developed by L. Meyer, W. Ostwald and S. Arrhenius on the state of the molecules of dissolved substances, as it only goes a step further and fixes numerically the degree of the decomposition."

An objection was taken to Planck's argument. It was said that as his formula contains the ratio of the molecular numbers of the solute and of the solvent, it could not be inferred that that of the solute is greater than its formula leads to, for it might be that the molecular number of the solvent is less than that indicated by its formula. Planck's answer was immediate and obvious. In any expression in which the molecular number of the solvent appears, there also appears as a factor the molecular weight. For instance, in the formula for the depression of the freezing-point the molecular number of the solvent is multiplied by the latent heat of one molecule of the solvent, and similarly in other cases. So that it makes no difference what molecular weight we assume for the solvent, and the use of its molecular number is merely a convenient way of expressing its quantity.

This increase in the number of the molecules, or splitting into ions, was called " electrolytic dissociation." It will be seen that it is what Lodge in 1885, in speaking of Clausius's theory, called dissociation. But while it has some obvious resemblances to the dissociation of a gas, there are very striking differences between the cases, and perhaps some of the difficulties in the way of the acceptance of the theory may have arisen from the use of the same word for two things differing so much. We need not, however, discuss the name, but it is well to look for a little at the essentially different nature of the things. This essential distinction consists in the products of the electrolytical dissociation being charged, the one set with positive, the other set with negative, electricity, so that, while in the body of the solution they can move about independently, they cannot be separated by diffusion as the products of the dissociation of a gas can. It is true that the quicker moving ions can, to a small extent, forerun the slower moving ions, and diffuse a little further into pure water or into a more dilute solution, as is shown by the fact that when two solutions of the same electrolyte of different concentration are in contact there is a difference of electric potential between them, but they cannot be separated to any weighable extent in this way. In order to separate from one another two gases uniformly mixed, a certain calculable amount of work has to be done, so that after a gas has been dissociated and wholly or partially converted into a mixture of the two gaseous products, some work has still to be done to get them separately. So it is also in the case of electrolytic dissociation ; but while in the former case the decomposition work is the main thing, and the separation work very small, in the latter it is quite the other way. Here the heat of dissociation, that is the work spent in decomposing the electrolyte into its ions, is small (indeed sometimes negative), while the work to be done to *separate* the ions is always very much greater. Indeed we may quite correctly say that in most highly dissociated solution of hydrochloric acid the hydrogen and the chlorine are still very firmly united, not indeed atom to atom, but each atom of the one kind to all the atoms of the other kind within a certain distance from it. A man does not lose his money when he takes it out of his pocket and puts it into a bank. He does indeed lose his relation to the individual gold and silver coins, and does not know and does not care where these particular pieces of metal are, but he is interested in knowing that they or their like are at his command, and the same sort of work will be required to impoverish him whether his money is in the bank or in his pocket. (I assume, of course, that the bank of our present imagination cannot become insolvent.)

I have said that the test of the theory would come later. It has been going on since 1887, and if time would allow I could give you many cases in which deductions from the theory have been found to agree with close quantitative accuracy with experimental observations. I shall mention only the first, still among the most important, namely Ostwald's determination of the affinity constants, and his application of Guldberg and Waage's principle to the ions. I could also give you instances in which there have been discrepancies, or apparent discrepancies, and show how in some of these cases the difficulties have been cleared up. The history of this theory has in fact so far been that of every useful theory, for it is in this way only that a theory does its work. I shall select two points for illustration, not because they are more important than others, but because I can illustrate them by means of experiments which do not occupy much time, and can be made visible in a large room. The first has reference to the question, What are the ions in the case of a dibasic acid? As HNO_3 gives as its ions H and NO_3 so we might expect H_2SO_4 to give 2H and SO_4. But we find that until the dilution has advanced to a considerable extent the ions of sulphuric acid are mainly H and HSO_4. This is quite in harmony with the chemical action of H_2SO_4, for, as every chemist knows, at moderate temperatures we have the action $H_2SO_4 + NaCl = HCl + NaHSO_4$ and the temperature has to be raised in order to get the action $NaHSO_4 + NaCl$

$= HCl + Na_2SO_4$. In the first of these experiments we take as the electrolyte a concentrated solution of potassium hydrogen sulphate $KHSO_4$. This gives the ions K and HSO_4. The latter go to the anode and there, on being discharged, form persulphuric acid, or its ions, and potassium persulphate $K_2S_2O_8$, being sparingly soluble, crystallises out. This is the method by means of which Dr. Marshall discovered the persulphates. The next experiment will illustrate the formation and discharge at the anode of the anion SO_4. We have here dilute sulphuric acid with which is mixed a little manganous sulphate $MnSO_4$. The ion SO_4 when discharged, adds itself to $2MnSO_4$ and forms manganic sulphate $Mn_2(SO_4)_3$, recognised by its red colour. This, even in acid solution, is quickly hydrolysed, giving insoluble manganic hydrate.

The other point I wish to illustrate is the application of Guldberg and Waage's principle to ions. Without entering into any general discussion of this question, I shall merely say that theory leads to the result that the addition of a soluble acetate to a solution of acetic acid diminishes the concentration of H ions, and so makes the solution less effectively acid. This was experimentally proved by Arrhenius in 1890,[*] by measuring the rate at which cane-sugar is inverted by acetic acid alone, and with varying quantities of sodium acetate added to it. But as such an experiment cannot be made visible to a large number of spectators at once, I thought of a way of showing the same thing, which, while not capable of the same degree of accuracy, would prove the principle qualitatively. I have here a solution of ferrous acetate to which I have added enough acetic acid to prevent the precipitation of ferrous sulphide on the addition of sulphuretted hydrogen. I add sulphuretted hydrogen, of course no precipitate is formed. I now add a solution of sodium acetate mixed with rather more than three equivalents of acetic acid, so as to make it plain that the effect is not due to the formation of an acid acetate, and you see that we have at once a precipitate of ferrous sulphide. To show that the addition of the water has not produced the result, I add to another portion of the same solution as much water, and you see that no precipitation takes place.

I have not spoken of non-aqueous solutions. At the rise of the dissociation theory, these were generally supposed to be non-conductors, but many of them have now been examined both by scientific workers in the old world, and very specially by our colleagues on the other side of the Atlantic, and been found to conduct electrolytically. It seems likely that these investigations will throw much light on the influence of the solvent on the conductivity of the dissolved salt. Particularly interesting is the relation, indicated in some cases, between the specific inductive capacity of a solvent and the dissociation of the dissolved salt. But this is one of the questions not yet ripe for treatment in a discourse such as this.

---

[*] Arrhenius, Zeitschrift f. physik. Chemie, v. p. 1 (1890).

I had also thought of saying something as to the atomic character of electricity, and the compounds of electricity with what we may venture to call the other chemical elements, and had even some idea of poaching on Lord Kelvin's domain of "Aepinus atomised," but time has saved me from this.

I have been describing the history of the theory of electrolysis from the time of Faraday, in such a way as is possible within the limits of an hour. I have necessarily omitted mention of many active, able and successful workers, and I cannot in every case justify the omission except by referring to the time limit. I have as far as I could explained the evidence which we have for the theories described, but I have not intended to argue for or against the essential truth of them. I have sometimes been asked in reference to the theory of electrolytic dissociation, Do you really believe it to be true? My answer to that question is, I believe it to be an eminently useful theory. It has led to a great deal of most valuable experimental work. It has enabled us to group together things that without its help seemed very little connected. It has led to the discussion of problems that could scarcely, without its suggestion, have occurred to any one. It does not seem to be exhausted, and I look forward to much good to be got from it yet, and therefore I am willing to take it as a guide. But I do not look on it as an infallible guide; we cannot expect, we do not need, an infallible guide in physical science. A long life may be anticipated for this theory; if that be so, we may be sure that it will undergo modifications, for if it is to act, it will be acted on.

Nothing but good can come from the fullest discussion, either of the theoretical basis or of the experimental evidence for or against a theory. No great principle in science or in law can be satisfactorily settled without full argument by competent advocates on both sides, and the eager hunt for evidence by those who attack and by those who defend, will lead to a more complete investigation of the whole field than would be attained without such—shall we call it partisan —interest.

[A. C. B.]

GENERAL MONTHLY MEETING,

Monday, February 3, 1902.

SIR JAMES CRICHTON-BROWNE, M.D. LL.D. F.R.S., Treasurer and
Vice-President, in the Chair.

William Asch, Esq.
The Rt. Hon. Lord Iveagh, K.P. LL.D.
Rev. Samuel James Norman, M.A.
Hunter Finlay Tod, Esq. F.R.C.S.
Lieut.-Col. R. de Villamil,
George Westinghouse, Esq.
James Martin White, Esq. J.P.

were elected Members of the Royal Institution.

It was announced from the Chair that the following valuable
relics of Michael Faraday, bequeathed by the late Mr. Thomas J. F.
Deacon, of Newcastle-upon-Tyne, to the Royal Institution of Great
Britain, had been received :—

Medals of Silver and Bronze (20)⎱
Two Foreign Orders   ⎰ In a small mahogany box.
Faraday's Book of Portraits and Autographs.
A Daguerrotype of a Consultation of Faraday with Professor
Daniell.
Framed Drawing in Colours of the Laboratory of the Royal
Institution.
Manuscript Book intituled "A Class Book for the Reception
of Mental Exercises, instituted July 1818."

The Special Thanks of the Members were returned for the following Donations to the Fund for the Promotion of Experimental Research
at Low Temperatures :—

Sir Frederick Bramwell, Bart., F.R.S.   £100
Dr. Frank McClean, F.R.S.   . · .   £50

The following Resolution, passed by the Managers, was read and
adopted :—.

Resolved, That the Managers of the Royal Institution desire to record their
sense of the loss sustained by the Institution in the decease of Mr. Hugh Leonard,
F.S.A. M. INST. C.E.

Becoming a Member of the Royal Institution in 1881, he was elected a
Visitor in 1894 and a Manager in 1898. He was a regular attendant at the
Meetings, and was well known to many Members. He took a keen interest in

all matters concerning the welfare of the Royal Institution, and was a generous donor to the Research Fund for the prosecution of experimental inquiry. His personality and his services to the Institution will be long remembered.

The PRESENTS received since the last Meeting were laid on the table, and the thanks of the Members returned for the same, viz. :—

FROM

The Secretary of State for India—Progress Report of the Archæological Survey of Western India for 1901. 8vo.
Archæological Survey Memoirs, Vol. XXVI. Part 1. 4to. 1901.

The Governor-General of India—Memoirs, Vol. XXXII. Parts 2, 3. 8vo. 1901.

Accademia dei Lincei, Reale, Roma—Classe di Scienze Fisiche, Matematiche e Naturali. Atti, Serie Quinta : Rendiconti. 2° Semestre, Vol. X. Fasc. 10, 11. 8vo.

American Academy of Arts and Sciences—Proceedings, Vol. XXXVII. Nos. 1-5. 8vo. 1901.

Asiatic Society, Royal—Journal for Jan. 1902. 8vo.

Astronomical Society, Royal—Monthly Notices, Vol. LXII. Nos. 1, 2. 8vo. 1901.

Bankers, Institute of—Journal, Vol. XXIII. Part 1. 8vo. 1901.

Balliere, Tindall and Cox, Messrs. (the Publishers)—The Composition of Dutch Butter. By Dr. J. J. L. Van Ryn. 8vo. 1902.

Batavia Observatory—Rainfall in East Indian Archipelago for 1900. 8vo. 1901.

Belgium Royal Academy of Sciences—Mem. cour. et autres mém. Tomes LVIII.-LX. 8vo. 1901.

Berlin Academy of Sciences—Sitzungsberichte, 1901, Nos. 39-53. 8vo. 1901.

Boston Public Library—Annual List of Books, 1900-1901. 8vo. 1902.
Monthly Bulletin for Dec. 8vo. 1901.

British Architects, Royal Institute of—Journal, 3rd Series, Vol. IX. Nos. 3-7. 4to. 1902.

British Astronomical Association—Journal, Vol. XII. No. 2. 8vo. 1901.
Memoirs, Vol. XI. Parts 9, 10. 8vo. 1901.

Camera Club—Journal for Dec. 1901 and Jan. 1902. 8vo.

Canada, Geological Survey of—Catalogue of Canadian Birds, Part 1. 8vo. 1900.

Chemical Industry, Society of—Journal, Vol. XX. Nos. 11, 12; Vol. XXI. No. 1. 8vo. 1901.

Chemical Society—Proceedings, Nos. 243, 244. 8vo. 1901.
Journal for Dec. 1901 and Jan. 1902. 8vo.

Church, Prof. A. H. F.R.S. M.R.I.—Food Grains of India, Supplement, 1901. 8vo.
The Chemistry of Paints and Painting. 3rd ed. 8vo. 1901.

Cole, G. W. Esq. (the Author)—Bermuda and the 'Challenger' Expedition. 8vo. 1901.

Cracovie, l'Académie des Sciences—Bulletin, 1901, Nos. 7, 8. 8vo. 1901.

Dublin Society, Royal—Proceedings, Vol. IX. Parts 3, 4. 8vo. 1900-1901.
Transactions, Vol. VII. No. 13. 4to. 1901.

Editors—Aeronautical Journal for Dec. 1901 and Jan. 1902. 8vo.
American Journal of Science for Dec. 1901 and Jan. 1902. 8vo.
Analyst for Dec. 1901 and Jan. 1902. 8vo.
Anthony's Photographic Bulletin for Dec. 1901 and Jan. 1902. 8vo.
Astrophysical Journal for Dec. 1901 and Jan. 1902.
Athenæum for Dec. 1901 and Jan. 1902. 4to.
Author for Dec. 1901 and Jan. 1902. 8vo.

Editors—continued.
Brewers' Journal for Dec. 1901 and Jan. 1902. 8vo.
Chemical News for Dec. 1901 and Jan. 1902. 4to.
Chemist and Druggist for Dec. 1901 and Jan. 1902. 8vo.
Electrical Engineer for Dec. 1901 and Jan. 1902. fol.
Electrical Review for Dec. 1901 and Jan. 1902. 8vo.
Electrical Times for Dec. 1901 and Jan. 1902. 4to.
Electricity for Dec. 1901 and Jan. 1902. 8vo.
Electro-Chemist and Metallurgist for Dec. 1901 and Jan. 1902. 8vo.
Engineer for Dec. 1901 and Jan. 1902. fol.
Engineering for Dec. 1901 and Jan. 1902. fol.
Homœopathic Review for Dec. 1901 and Jan. 1902. 8vo.
Horological Journal for Dec. 1901 and Jan. 1902. 8vo.
Invention for Dec. 1901 and Jan. 1902.
Journal of the British Dental Association for Dec. 1901 and Jan. 1902. 8vo.
Journal of Medical Research for Dec. 1901. 8vo.
Journal of Physical Chemistry for Dec. 1901. 8vo.
Journal of State Medicine for Dec. 1901 and Jan. 1902. 8vo.
Law Journal for Dec. 1901 and Jan. 1902. 8vo.
London Technical Education Gazette for Dec. 1901 and Jan. 1902.
Machinery Market for Dec. 1901 and Jan. 1902. 8vo.
Motor Car Journal for Dec. 1901 and Jan. 1902. 8vo.
Nature for Dec. 1901 and Jan. 1902. 4to.
New Church Magazine for Dec. 1901 and Jan. 1902. 8vo.
Nuovo Cimento for Dec. 1901 and Jan. 1902. 8vo.
Pharmaceutical Journal for Dec. 1901 and Jan. 1902. 8vo.
Photographic News for Dec. 1901 and Jan. 1902. 8vo.
Physical Review for Dec. 1901 and Jan. 1902. 8vo.
Popular Science Monthly for Dec. 1901 and Jan. 1902. 8vo.
Public Health Engineer for Dec. 1901 and Jan. 1902. 8vo.
Science Abstracts for Dec. 1901 and Jan. 1902. 8vo.
Terrestrial Magnetism for Dec. 1901. 8vo.
Travel for Dec. 1901 and Jan. 1902. 8vo.
Tropical Agriculturist for Dec. 1901 and Jan. 1902. 8vo.
Zoophilist for Dec. 1901 and Jan. 1902. 4to.

Electrical Engineers, Institution of—Journal, Vol. XXX. No. 153. 8vo. 1901.
Franklin Institute—Journal for Dec. 1901 and Jan. 1902. 8vo.
Frick, John, Esq. (the Author)—On Liquid Air and its Applications. 8vo. 1901.
Geographical Society, Royal—Geographical Journal for Dec. 1901 and Jan. 1902. 8vo.
Gladstone, Dr. J. H. F.R.S. M.R.I.—Bulletin de la Société Internationale des Electriciens. 8vo. 1901.
Harlem, Société Hollandaise des Sciences—Archives Néerlandaises, Série II. Tome IV. Livr. 45. 8vo. 1901.
Heck, O. Esq.—Physiologie. 8vo. 1901.
Historical Society, Royal—Transactions, N.S. Vol. XV. 8vo. 1901.
Holmes-Forbes, A. W. Esq. M.A. M.R.I. (the Author)—The Book Wonderful, or the Bible for the People. 8vo. 1901.
Horticultural Society, Royal—Journal, Vol. XXVI. Parts 2, 3. 8vo. 1901.
Imperial Institute—Imperial Institute Journal for Dec. 1901 and Jan. 1902.
Iron and Steel Institute—Journal, 1901, No. 2. 8vo. 1901-1902.
Johns Hopkins University—University Studies, Series XIX. Nos. 4-9. 8vo. 1901.
American Chemical Journal for Dec. 1901. 8vo.
University Circulars, 154. 8vo. 1901.
Kansas Academy of Science—Transactions, Vol. XVII. 8vo. 1901.
Kansas University—Bulletin, Vol. II. No. 6. 8vo. 1901.

Leighton, John, Esq. M.R.I.—Journal of the Ex-Libris Society for Nov.-Dec. 1901. 8vo.

(The Author)—Tubular Transit. 8vo. 1902.

Liverpool, Literary and Philosophical Society—Proceedings, No. LV. 8vo. 1901.

Madrid, Royal Academy of Sciences—Memorias, Tomo XIV. Fasciculo 1, Atlas. 4to. 1890-1901.

Manchester Literary and Philosophical Society—Memoirs and Proceedings, Vol. XLVI. Part 1. 8vo. 1900-1901.

Massachusetts Institute of Technology—Technology Quarterly, Vol. XIV. No. 3. 8vo. 1901.

Mechanical Engineers, Institution of—Proceedings, 1901, No. 3. 8vo. 1901.

Mensbrugghe, Prof. Van der (the Author)—Une Triple Alliance Naturelle. 8vo. 1901.

Meteorological Society, Royal—Meteorological Record, Vol. XXI. No. 81. 8vo. 1901.

Meteorological Office—Hourly Means for 1898. 4to. 1901.

Mexico, Sociedad Cientifica—Memorias, Tomo XV. Nos. 11, 12; Tomo XVI. No. 1. 8vo. 1901.

Munich, Royal Bavarian Academy of Sciences—Abhandlungen, Band XX. Abt. 2. 4to. 1901.

Navy League—Navy League Journal for Dec. 1901. 8vo.

New South Wales, Agent-General for—Picturesque New South Wales. 8vo. 1901.

Odontological Society—Transactions, Vol XXXIV. No. 2. 8vo. 1901.

Onnes, Prof. Dr. H. K.—Communications, Nos. 72-74. 8vo. 1901.

Paris, Société Française de Physique—Séances, 1901, Fasc. 2. 8vo. 1901.

Pharmaceutical Society of Great Britain—Journal for Dec. 1901 and Jan. 1902. 8vo.

Philadelphia, Academy of Natural Sciences—Proceedings, Vol. LIII. Part 2. 8vo. 1901.

Photographic Society, Royal—Photographic Journal for Nov.-Dec. 1901 and Jan. 1902. 8vo.

Physical Society—Proceedings, Vol. XVII. Part 7. 8vo. 1901.

Rayleigh, Lord, D.C.L. F.R.S. M.R.I. (the Author)—Scientific Papers, Vol. III. 4to. 1902.

Royal Society of Edinburgh—Proceedings, Vol. XXIII. No. 5. 8vo. 1901.

Royal Society of London—Philosophical Transactions, A, Nos. 298, 299; B, Nos. 204, 205. 4to. 1901.

Proceedings, Nos. 451-454. 8vo. 1901-1902.

Salvioni, Prof. Enrico (the Author)—Sulla Volatilezzazione del Muschio. 8vo. 1901.

Pressione atmosferica. 8vo. 1901.
Un nuovo igiometro. 8vo. 1901.
Misura di masse. 8vo. 1901.

Schneider, S. Esq. (the Author)—Die Deutsche Baghdad-Bahn. 8vo. 1900.

Scottish Microscopical Society—Proceedings, Vol. III. No. 2. 8vo. 1901.

Selborne Society—Nature Notes for Dec. 1901 and Jan. 1902. 8vo.

Smithsonian Institution—Annual Report, 1900. 8vo. 1901.

Society of Arts—Journal for Dec. 1901 and Jan. 1902. 8vo.

Statistical Society, Royal—Journal, Vol. LXIV. Part 4. 8vo. 1901.

Tacchini, Prof. P. Hon. Mem. R.I. (the Author)—Memorie della Società degli Spettroscopisti Italiani, Vol. XXX. Disp. 12. 4to. 1901.

Teyler Museum, Harlem—Archives, Série II. Vol. VII. Part 4. 8vo. 1901.

United Service Institution, Royal—Journal for Dec. 1901. 8vo.

United States Army, Surgeon-General's Office—Index Catalogue to the Library. 2nd Ser. Vol. VI. 4to. 1901.

United States Department of Agriculture—Monthly Weather Review for Sept. 1901. 4to.

    Studies on Bread and Bread Making at the University of Minnesota. 8vo. 1901.

Verein zur Beförderung des Gewerbfleisses in Preussen—Verhandlungen, 1901. Heft 10. 8vo.

Vienna, Geological Institute, Imperial—Verhandlungen, 1901, Nos. 11–14. 8vo. Jahrbuch, Band LI. Heft 1. 8vo. 1901.

Washington Academy of Sciences—Proceedings, Vol. III. pp. 541–612. 8vo 1901

WEEKLY EVENING MEETING,

Friday, February 7, 1902.

The Right Hon. Sir JAMES STIRLING, M.A. LL.D. F.R.S.,
Vice-President, in the Chair.

PROFESSOR E. RAY LANKESTER, M.A. LL.D. F.R.S., Director of the
Natural History Department of the British Museum.

*The New Mammal from Central Africa, and its Relation to other
Giraffe-like Animals.*

PROFESSOR RAY LANKESTER began by pointing out how frequently
the observation *Ex Africa semper aliquid novi* had been exemplified
during the last fifty or sixty years in regard to discoveries in geography and natural history. He went on to tell how Sir Harry
Johnston had fallen in with a party of the dwarfs that inhabit the
Semliki forest, who were being taken to be shown at the Paris
Exhibition by a German speculator, and, thinking that a visit to
Paris would not be specially good for them, had conducted them
back to their own district. Travelling with them, he took the
opportunity of asking them about the animals of the forest, in
particular inquiring about an animal which had been heard of by
Sir H. M. Stanley. In this way he was informed about the Okapi,
and obtained from the dwarfs some pieces of its skin, which on
being sent home to England were not unnaturally supposed to be
of a zebra, probably a new species. Later, Mr. Eriksson, the officer
in charge of one of the posts of the Congo Independent State, sent
him a skin and two skulls of the Okapi. These he despatched to
England, at the same time identifying the animal as related to the
giraffe.

Professor Lankester then showed pictures and photographs of
the animal as reconstructed at the Natural History Museum from
this skin and one of the skulls, pointing out the peculiarities of
its markings, the arrangement of the stripes on its legs, the
absence of secondary hoofs and of horns on its head, etc. The
skull was very remarkable for the development of the hinder part,
and for a swelling over and a little behind the eye, which suggested
that there the horn would grow if there was one. This, however,
was not the place where the horns occurred in the giraffe, as was
seen by a comparison of their skulls with that of the Okapi. The
teeth of the larger Okapi skull in the Museum were in the same

condition as those of a giraffe about two-thirds grown, and this, and other facts, pointed to the conclusion that the skull belonged to an animal at that stage of development. An important fact was, that the canine teeth in the lower jaw of the Okapi were bifoliate, and the circumstance that the giraffe was the only other known animal in which this peculiarity was to be found was strong evidence of their affinity. The teeth in the skull at the Museum were only milk-teeth, but the bifoliation was detected in the permanent teeth concealed in the bone below. The Helladotherium, remains of which had been found in Greece, and to which Sir Harry Johnston had compared the Okapi, differed too much to justify naturalists in placing them in the same genus, though doubtless they were nearly related. The Okapi had a black line down the wrist, not quite in the middle, and round the actual wrist was a ring of black. Very few antelopes had anything of the kind, but it was found in some goats. Such stripings were difficult to account for, our knowledge of elusive optics, of bright markings which rendered animals invisible even in broad daylight to a person looking straight at them, being as yet insufficient to explain them. In conclusion, the lecturer discussed the types of horn-structure found in different animals, with reference to the question whether the Okapi was primitive in not having horns, or whether his ancestors had had them and lost them. The facts, he said, scarcely admitted of this question being answered definitely either way, but he himself was inclined to think that the ancestors of the Okapi had never had horns of any size, and that the Okapi and the giraffe stood near the beginning of horn development in the Ruminant Ungulates.

WEEKLY EVENING MEETING,

Friday, February 14, 1902.

Sir William de W. Abney, K.C.B. D.C.L. F.R.S.,
Vice-President, in the Chair.

Major P. A. MacMahon, R.A. D.Sc. F.R.S.

Magic Squares and other Problems on a Chess-Board.

The construction of magic squares is an amusement of great antiquity; we hear of them being constructed in India and in China
before the Christian era, whilst they appear to have been introduced
into Europe by Moschopulus, who flourished at Constantinople early
in the fifteenth century. On the diagram you see a simple example
of a magic square, one celebrated as being drawn by Albert Dürer in
his picture of "Melancholy," painted about the year 1500 (Fig. 1).

1	15	14	4
12	6	7	9
8	10	11	5
13	3	2	16

Fig. 1.

It is one of the fourth order, involving
16 compartments or cells. In describing
such squares, the horizontal lines of
cells are called "rows," the vertical lines
"columns," and the oblique lines going
from corner to corner "diagonals." In
the 16 compartments are placed the first
16 numbers, 1, 2, 3, . . . 16, and the
magic property consists in this, that the
numbers are placed in such wise that
the sum of the numbers in every row,
column, and diagonal is the same, viz. in
this case, 34.

It is probable that magic squares
were so called because the properties
they possessed seemed to be extraordinary and wonderful; they
were, indeed, regarded with superstitious reverence, and employed
as talismans. Cornelius Agrippa constructed magic squares of orders
3, 4, 5, 6, 7, 8, 9, and associated them with the seven heavenly
bodies, Saturn, Jupiter, Mars, the Sun, Venus, Mercury, and the
Moon. A magic square engraved on a silver plate was regarded as
a charm against the plague, and to this day such charms are worn
in the East.

However, what was at first merely a practice of magicians and
talisman makers has now for a long time become a serious study for
mathematicians. Not that they have imagined that it would lead
them to anything of solid advantage, but because the theory of such

squares was seen to be fraught with difficulty, and it was considered possible that some new properties of numbers might be discovered which mathematicians could turn to account. This has, in fact, proved to be the case; for from a certain point of view the subject has been found to be algebraical rather than arithmetical, and to be intimately connected with great departments of science, such as the "infinitesimal calculus," "the calculus of operations," and the "theory of groups."

In the next diagram (Fig. 2) I show you a magic square of order 5, the sum of the numbers in each row, column, and diagonal being 65. This number 65 is obtained by multi- plying 25, the number of cells, by the next higher number, 26, and then divid- ing by twice the order of the square, viz. 10. A similar rule applies in the case of a magic square of any order. The formation of these squares has a fascination for many persons, and, as a consequence, a large amount of inge- nuity has been expended in forming particular examples, and in discovering general principles of formation. As an example of the amount of labour that some have expended on this matter, it may be mentioned that in 1693 Frénicle,

17	24	1	8	15
23	5	7	14	16
4	6	13	20	22
10	12	19	21	3
11	18	25	2	9

FIG. 2.

a Frenchman, published a work of more than 500 pages upon magic squares. In this work he showed that 880 magic squares of the fourth order could be constructed, and in an appendix he gave the actual diagrams of the whole of them. The number of magic squares of the order 5 has not been exactly determined, but it has been shown that the number certainly exceeds 60,000.

As a consequence it is not very difficult to compose particular specimens, and, for the most part, the fascinated individuals to whom I have alluded have devoted their energies to the discovery of principles of formation. Of such principles I will give a few, re- marking that the cases of squares of uneven order, 1, 3, 5, . . . are more simple than those of even order, 4, 6, . . . and that no magic square of order 2 exists at all. The simplest of all methods for an uneven order is shown in the diagram (Fig. 3), where certain addi- tional cells are added to the square, the numbers written as shown in natural order diagonally, and then the numbers which are outside the square projected into the empty compartments according to an easily understood law. The second method is associated with the name of De la Loubère, though it is stated that he learnt it during a visit to Siam in 1687. The number 1 (see Fig. 2) is placed in the middle cell of the top row, and the successive numbers placed in their natural order in a diagonal line sloping upwards to the right, subject to the laws :—

(1) When the top row is reached, the next number is written at the bottom of the next column.

(2) When the right-hand column is reached, the next number is written on the left of the row above.

(3) When it is impossible to proceed according to the above rules, the number is placed in the cell immediately below the last number written.

If we commence by writing the number 1 in any cell except that above indicated a square is reached which is magic in regard to rows and columns, but not in regard to diagonals.

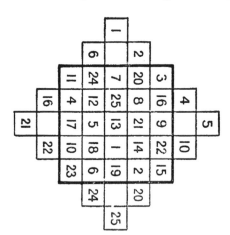

7	20	3	11	24
13	21	9	17	5
19	2	15	23	6
25	8	16	4	12
1	14	22	10	18

FIG. 3.   FIG. 4.

Subsequent writers have shown that starting with the left-hand bottom cell and using the move of the knight instead of that of the bishop, the general principal of De la Loubère will also lead to a magic square (Fig. 4). The next method is that of De la Hire, and dates from 1705. Two subsidiary squares are constructed as shown, the one involving five numbers, 1, 2, 3, 4, 5, and the other five numbers 0, 5, 10, 15, 20. When these squares are properly formed and a third square constructed by adding together the numbers in corresponding cells, this third square is magic (Fig. 5). Time does not permit me to enter into the exact method of forming the subsidiary squares, and I will merely mention that each of them possesses a particular property, viz. only five different numbers are involved, and all five appear in each column and in each row; in other words, no row and no column contains two numbers of the same kind, but no diagonal property is necessarily involved. Such squares are of a great scientific importance, and have been termed by Euler and subsequent writers "Latin squares," for a reason that will presently appear. From a scientific point of view, the chief interest of all arrangements such as I consider this evening lies, not in their actual formation, but in the enumeration of all possible ways of forming them, and in this respect very little has been hitherto achieved by mathematicians. No person living knows in how many ways it is possible to form a magic square of any order exceeding 4. The fact is, that before we can attempt to enumerate magic squares we must see our way to solve problems of a far more simple character. For example, before we can enumerate the squares that can be formed by De la Hire's method we must take a first step by finding out how many Latin squares can be formed of the different orders. For the order 5 the question is, "In how many ways can five different objects be placed in the cells so that each column and each row contains each object?" It may occur to some here this evening that

3	4	1	5	2
2	3	4	1	5
5	2	3	4	1
1	5	2	3	4
4	1	5	2	3

15	0	20	5	10
0	20	5	10	15
20	5	10	15	0
5	10	15	0	20
10	15	0	20	5

18	4	21	10	12
2	23	9	11	20
25	7	13	19	1
6	15	17	3	21
14	16	5	22	8

FIG 5.

such a discussion might be interesting or curious, but could not possibly be of any scientific value. But such is not the case. A department of mathematics that is universally acknowledged to be of fundamental importance is the "theory of groups." Operations of this theory and those connected with logical and other algebras possess what is termed a "multiplication table," which denotes the laws to which the operations are subject. In Fig. 6 you see such a table of order 6 slightly modified from Burnside's 'Treatise on the Theory of Groups'; it is, as you see, a Latin square, and the chief problem that awaits solution is the enumeration of such tables; the questions are not parallel because *all* Latin squares do not give rise to tables in the theory of groups; but still, we must walk before we can run, and a step in the right direction is the enumeration of *all* Latin squares. When I call to mind that the theory of groups has an important bearing upon many branches of physical science, notably upon dynamics, I consider that I have made good my point.

I now concentrate attention on these Latin squares, and observe that the theory of the enumeration has nothing to do with the particular numbers that occupy the compartments; the only essential is that the numbers shall be different one from another. My attention was first called to the subject of the Latin square by a work of the renowned mathematician Euler, written in 1782, 'Recherches sur une nouvelle espèce de Quarrés Magiques.' I may say that Euler seems to have been the first to grasp the necessity of considering squares possessing what may be termed a magical property of a far less recondite character than that possessed by the magic squares

I	A	B	C	D	E
A	B	I	D	E	C
B	I	A	E	C	D
C	E	D	I	B	A
D	C	E	A	I	B
E	D	C	B	A	I

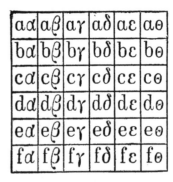

 Fig. 6.   Fig. 7.

of the ancients, and, as we shall see presently, he might have gone a step further in the same direction with advantage and have commenced with arrangements of a more simple character than that of the Latin square, with arrangements, in fact, which present no difficulties of enumeration, but which supply the key to the unlocking of the secrets of which we are in search. He commences by remarking that a curious problem had been exercising the wits of many persons. He describes it as follows: There are 36 officers of six different ranks drawn from six different regiments, and the problem is to arrange them in a square of order 6, one officer in each compartment, in such wise that in each row, as well as in each column, there appears an officer of each rank and also an officer of each regiment. Of a single regiment we have, suppose, a colonel, lieutenant-colonel, major, captain, first lieutenant and second lieutenant, and similarly for five other regiments, so that there are in all 36 officers who must be so placed that in each row and in each column each rank is represented, and also each regiment. Euler denotes the six regiments by the Latin letters *a*, *b*, *c*, *d*, *e*, *f*, and the six ranks by the Greek letters *α*, *β*, *γ*, *δ*, *ε*, *θ*, and observes that the character of an officer is determined by a combination of two letters, the one Latin and the other Greek; there are 36 such combinations, and the problem consists in placing these combinations in the 36 compartments in such wise that every row and every column contains the 6 Latin letters and also the 6 Greek letters (Fig. 7). Euler found no solution of this problem in the case of a square of order 6, and since Euler's time no one has succeeded either in finding a solution or in proving that no solution exists. Anyone interested, has, therefore, this question before him at the present moment, and I recommend it to anyone present who desires an exercise of his wits and a trial of his patience and ingenuity. It is easy to prove that when the square is of order 2, viz. the case of 4 officers of two different ranks drawn from two different regiments, there is no solution; Euler gave his opinion to the effect that no solution is possible whenever the order of the square is two greater than a multiple of four. In other simple cases he obtained solutions; for example, for the order 3, the problem of 9 officers of three different ranks drawn from three different regiments, it is easy to discover the solution shown in the diagram (Fig. 8), and, as demonstrated by Euler, whenever one solution has been constructed there is a simple process by which a certain number of others can be derived from it. Now, if you look at that diagram and suppose the Greek letters obliterated, you will see that the Latin letters are arranged so that each of the letters occurs in each row and in each column, the magical property mentioned above,

Fig. 8.

and for this reason Euler termed such arrangements Latin squares and stated that the first step in the solution of the problem is to enumerate the Latin squares of a given order. As showing the intimate connection between the Græco-Latin square of Euler and ordinary magic squares, it should be noticed that the method of De la Hire, by employing Latin and Greek letters for the elements in his two subsidiary Latin squares, gives rise immediately to the Græco-Latin square of Euler. Euler says in regard to the problem of the Latin square, " The complete enumeration of the Latin squares of a given order is a very important question, but seems to me of extreme difficulty, the more so as all known methods of the doctrine of combinations appear to give us no help," and again, " the enumeration appears to be beyond the bounds of possibility when the order exceeds 5." Moreover, Cayley, in 1890, that is 108 years later, gave a *résumé* of what had been done in the matter, but did not see his way to a solution of the question. Under these circumstances, you will see how futile it is to expect a solution of the magic-square problem when the far simpler question of the Latin square has for so long proved such a tough nut to crack. The problem of the Latin square has eventually been completely solved, and in order to lead you up gradually to an understanding of the method that has proved successful, I ask you to look at the Latin square of order 5 that you see in the diagram (Fig. 9). The first row of letters can be written in any order, but not so the second row, for each column when the second row is written must contain two different letters. We must, therefore, be able to solve the comparatively simple question of the number of possible arrangements of the first two rows. For a given order of the letters in the first row, in how many ways can we write the letters in the second row so that each column contains a pair of different letters? This is a famous question, of which the solution is well known; it is known to mathematicians as the " problème des rencontres." It may be stated in a variety of ways; one of the most interesting is as follows : A person writes a number of letters and addresses the corresponding envelopes ; if he now puts the letters at random into the envelopes, what is the probability that not a single letter is in the right envelope?

a	b	c	d	e
b	d	e	a	c
c	e	d	b	a
d	c	a	e	b
e	a	b	c	d

Fig. 9.

Passing on to the problem of determining the number of ways of arranging the first three rows so that each column contains three different letters, it may be stated that up to 1898 no solution of it had been given ; while it is obvious that as the number of the rows is increased the resulting problems will be of enhanced difficulty. A particular case of the three-row problem had, however, been considered under the title "problème des ménages" and a solution obtained. It may be stated as follows :—

A given number of married ladies take their seats at a round table in given positions ; in how many ways can their husbands be seated so that each is between two ladies, but not next to his own wife? For order 5, that is five ladies, the question comes to this : Write down 5 letters and underneath them the same letters shifted one place to the left ; in how many ways can the third row be written so that each column contains three different letters? This particular case of the three-row problem for any order presents no real difficulty. The results are that in the cases of 3, 4, 5, 6 . . . married couples there are 1, 2, 13, 80, etc. ways.

Since the year 1890, the problem of the Latin square has been completely solved by an entirely new method, which has also proved successful in solving similar questions of a far more recondite character, and I am here this evening to attempt to give you some notion of the method and some account of the series of problems to which that method has been found to be applicable.

There is, as viewed mathematically, a fundamental difference between arithmetic and algebra ; the former may be regarded as an algebra in which the numerical magnitudes under consideration are restricted to be integers ; the two branches contemplate discontinuous and continuous magnitude respectively. Similarly, in geometry we have the continuous theory, which contemplates figures generated by points moving from one place to another and in doing so passing over an infinite succession of points, tracing a line in a plane or in space, and also a discontinuous theory, in which the position of a point varies suddenly, *per saltum*, and we are not concerned with any continuously varying motion or position. The present problems are concerned sometimes with this discontinuous geometry and sometimes with an additional discontinuity in regard to numerical magnitude, and the object is to count and not to measure. Far removed as these questions are, apparently, from the subject-matter of a calculus of infinitely small quantities and the variation of quantities by infinitesimal increments, my purpose is to show that they are intimately connected with them and that success

is a necessary consequence of the relationship. I must first take you to a much simpler problem than that of the Latin square, to one which in a variety of ways is very easy of solution, but which happens to be perhaps the simplest illustration of the method. In the game of chess a castle can move either horizontally or vertically, and it is easy to place 8 castles on the board so that no piece can be taken by any other piece. One such arrangement is shown in Fig. 10. The condition is simply that one castle must be in each row and

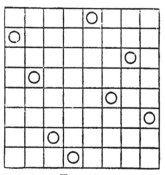

Fig. 10.

also in each column. Every such arrangement is a diagrammatic representation of a certain mathematical process performed upon a certain algebraical function. For consider the process of differentiating x^8 ; it may be performed as follows : Write down x^8 as the product of 8 x's,

$$x \ x \ x \ x \ x \ x \ x \ x,$$

and now substitute unity for x in all possible ways and add the results; the substitution can take place in eight different ways, and the addition results in $8 x^7$, which will be recognised as the differential coefficient. Observe that the process of differentiation is thus broken up into eight minor processes, each of which may be diagrammatically represented on the first row of the chess-board by a unit placed in the compartment corresponding to the particular x for which unity has been substituted. If we now perform differentiation a second time, we may take the results of the above minor processes and in each of them again substitute unity for x in all possible ways ; since in each the substitution can take place in seven different ways, it is seen that we can regard the process of differentiating twice as composed of $8 \times 7 = 56$ minor processes, each of which can be diagrammatically represented by two units, one in each of the first two rows of the chess-board, in positions corresponding to the substitutions of unity for x that have been carried out. Proceeding in this manner in regular order up to the eighth differentiation, we find that the whole process of differentiating x^8 eight times in succession can be decomposed into $8 \times 7 \times 6 \times 5 \times 4 \times 3 \times 2 \times 1 = 40,320$ minor processes, each of which is denoted by a diagram which slight reflection shows is a solution of the castle problem (Fig. 11). There

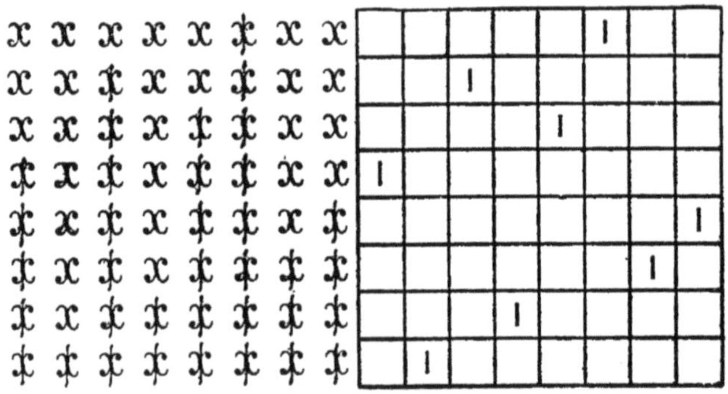

Fig. 11.

are, in fact, no more solutions, and the whole series of 40,320 diagrams constitutes a picture in detail of the differentiations. Simple differentiations of integral powers thus yield the enumerative solutions of the castle problem on chess-boards of any size.

We have here a clue to a method for the investigation of these chess-board problems; it is the grain of mustard seed which has grown up into a tree of vigorous growth, throwing out branches and roots in all sorts of unexpected directions. The above illustrations of differentiation gave birth to the idea that it might be possible to design pairs of mathematical processes and functions which would yield the solution of chess-board problems of a more difficult character. Two plans of operation present themselves. In the first place we may take up a particular question, the Latin square for instance, and attempt to design, on the one hand, a process, and, on the other hand, a function the combination of which will lead to the series of diagrams. In the second place, we may have no particular problem in view, but simply start by designing a process and a function, and examine the properties of the series of diagrams to which the combination leads. The first of these plans is the more difficult, but was actually accomplished in the case of the Latin square and some other questions; but the second plan, which is the proper method of investigation, met with great success, and the Latin square was one of its first victims, a solution of a more elegant nature being obtained than that which had resulted from the first plan of operations. There is such an extensive choice of processes and functions that many solutions are obtainable of any particular problem. I will now give you an idea of a solution of the Latin square, which is not the most elegant that has been found, but which is the most suitable to explain to an audience. Suppose we have five collections of objects, each collection containing the same five different objects, *a, b, c, d, e* (Fig. 12). I suppose the objects distributed amongst five different

$$(abcde) \quad (abcde) \quad (abcde) \quad (abcde) \quad (abcde)$$
$$(.bcde) \quad (a.cde) \quad (ab.de) \quad (abc.e) \quad (abcd.)$$
$$(..cde) \quad (a.c.e) \quad (ab.d.) \quad (.bc.e) \quad (ab.d.)$$
$$(...de) \quad (a.c..) \quad (ab...) \quad (..c.e) \quad (.b.d.)$$
$$(....e) \quad (a....) \quad (.b...) \quad (..c..) \quad (...d.)$$

FIG. 12.

persons in the following manner: The first person takes one object from each collection, so as to obtain each of the five objects; he can do this in 120 different ways; we will suppose that he takes *a* from the first, *b* from the second, *c* from the third, *d* from the fourth, *e* from the fifth; the collections then become as you see in Fig. 12, second row. Now suppose the second man to advance with the intention of taking one object from each collection and obtaining each of the five objects: he has not the same liberty of choice as had the first, because he cannot take *a* from the first collection or *b* from the second, etc. However, he has a good choice in his selection, and we will suppose him to take *b* from the first collection, *d* from the second, *e* from the third, *a* from the fourth, *c* from the fifth. The collections then become as you see in the third row. The third man who has the same task finds his choice more restricted, but he elects to take *c* from the first, *e* from the second, *d* from the third, *b* from the fourth, and *a* from the fifth. The fourth man finds he can take *d, c, a, e, b,* and this leaves *e, a, b, c, d* for the last man. If we plot the selections that have been made by the five men, we find the Latin square shown in Fig. 9.

Every division of the objects that can be made on this plan gives rise to a Latin square, and all possible distributions give rise to the whole of the Latin squares. Now it happens that a mathematical process exists (connected with algebraical symmetric functions) that acts towards a function representing the five collections in exactly the same way as I have supposed the men to act, and when the process is performed five times in succession, an integer results which denotes exactly the number of Latin squares of order 5 that can be constructed. Moreover, *en route* the "problème des rencontres" and the problems connected with any definite number of rows of the square are also solved.

I will now mention some questions of a more difficult character that are readily sclved by the method. In the "problème des ménages" you will recollect that the condition was that no man must sit next to his wife. If the condition be that there must be at least four (or any even number) persons between him and his wife, the

$a\ a\ b\ c$	$a\ a\ b\ b$	$a\ a\ a\ b$
$a\ b\ c\ a$	$a\ b\ a\ b$	$a\ a\ b\ a$
$b\ c\ a\ a$	$b\ b\ a\ a$	$a\ b\ a\ a$
$c\ a\ a\ b$	$b\ a\ b\ a$	$b\ a\ a\ a$

Fig. 13.

question is just as easily solved. Latin squares where the letters are not all different in each row and column are easily counted. Illustrations of these are shown in Fig. 13. One of these extended to order 8 gives the solution of the problem of placing 16 castles on a chess-board, 8 black and 8 white, so that no castle can take another of its own colour.

Theoretically, the Græco-Latin squares of Euler can be counted, but I am bound to say that the most laborious calculations are necessary to arrive at a numerical result, or even to establish that in certain cases the number sought is zero.

abcd		ef	g
e	abc	dg	f
f	deg	ab	c
g	f	c	abde

Fig. 14.

Next consider a square of any size and any number of different letters, each of which must appear in each row and in each column, while there is no restriction as to the number that may appear in any one compartment. In this case the result is very simple: suppose the square of order 4 (Fig. 14), and that there are seven different letters that must appear in each row and column; the number of arrangements is $(4!)^7$, viz., 4 the order of the square, must be multiplied by each lower number, and the number thus reached multiplied seven times by itself.

Finally, if there be given for each row and for each column a different assemblage of letters and no restriction be placed upon the contents of any compartment, the number of squares in which all these conditions are satisfied can be counted. This, of course, is a far more recondite question than that of the Latin square, and cannot be attacked at all by any other method.

I now pass to certain purely numerical problems. Suppose we have a square lattice of any size and are told that numbers are to be placed in the compartments in such wise that the sums of the numbers in the different rows and columns are to have any given values the same or different. This very general question, hitherto regarded as unassailable, is solved quite easily. The solution is not more difficult when the lattice is rectangular instead of square and when any desired limitation is imposed upon the magnitude of the numbers.

Up to this point, the solutions obtained depend upon processes of the differential calculus. A whole series of other problems, similar in general character, but in one respect essentially different, arises from the processes of the calculus of finite differences. Into these time does not permit me to enter. In the case of magic squares as generally understood, the method brought forward marks a distinct advance in connection with De la Hire's method of formation by means of a pair of Latin squares, but apart from this a great difficulty is involved in the condition that no two numbers must be the same. Still, a statement can be made as to a succession of mathematical processes which result in a number which enumerates the magic squares of a given order. In any cases except those of the first few orders, the processes involve an absolutely prohibitive amount of labour, so that it cannot yet be said that a practical solution of the question has been obtained.

Scientifically speaking, it is the assignment of the processes and not the actual performance of them that is interesting; it is the method involved rather than the results flowing from the method that is attractive; it is the connecting link between two, to all appearance, widely separated departments of mathematics that it has been fascinating to forge and to strengthen. Of all the subjects that for hundreds of years past have from time to time engaged the attention of mathematicians, perhaps the most isolated has been the subject of these chess-board arrangements. This isolation does not, I believe, any longer exist. The whole series of diagrams formed according to any given laws must be regarded as a pictorial representation, in greatest detail, of the manner in which a certain process is performed. We have to exercise our wits to discover what this process is. To say and to establish that problems of the general nature of the magic

square are intimately connected with the infinitesimal calculus and
the calculus of finite differences is to sum the matter up. Much,
however, remains to be done. The present method is not able to deal
with diagonal properties, or with arrangements which depend upon
the knight's move. The subject is only in its infancy at present.
More workers are required who will, without doubt, introduce new
ideas and obtain results far transcending those we are in possession

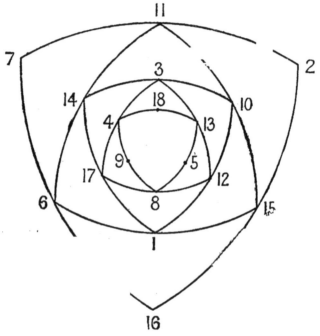

$$
\begin{aligned}
19 =\ & 7 + 12 = 14 +\ \ 5 =\ \ 4 + 15 \\
 =\ & 6 + 13 = 17 +\ \ 2 =\ \ 9 + 10 \\
 =\ & 16 +\ \ 3 =\ \ 1 + 18 =\ \ 8 + 11 \\
38 =\ & 7 + 11 + 14 +\ \ 6 = 11 +\ \ 2 + 15 + 10 = 15 + 16 +\ \ 6 +\ \ 1 \\
 =\ & 11 + 10 +\ \ 3 + 14 = 10 + 15 +\ \ 1 + 12 =\ \ 1 +\ \ 6 + 14 + 17 \\
 =\ & 14 +\ \ 3 +\ \ 4 + 17 =\ \ 3 + 10 + 12 + 13 = 12 +\ \ 1 + 17 +\ \ 8 \\
 =\ & 3 + 13 + 18 +\ \ 4 = 13 + 12 +\ \ 8 +\ \ 5 =\ \ 8 + 17 +\ \ 4 +\ \ 9 \\
57 =\ & 7 + 14 +\ \ 4 +\ \ 5 + 12 + 15 =\ \ 6 + 17 +\ \ 9 + 13 + 10 +\ \ 2 = 16 +\ \ 1 +\ \ 8 + 1\ +\ \ 3 + 11 \\
 =\ & 7 + 11 +\ \ 2 + 15 + 16 +\ \ 6 = 11 + 10 + 15 +\ \ 1 +\ \ 6 + 14 = 14 +\ \ 3 + 10 + 18 +\ \ 1 + 17 \\
 =\ & 3 + 13 + 12 +\ \ 8 + 17 +\ \ 4 =\ \ 4 + 18 + 13 +\ \ 5 +\ \ 8 +\ \ 9 \\
 =\ & 9 +\ \ 4 +\ \ 3 + 10 + 15 + 16 = 18 + 13 + 12 +\ \ 1 +\ \ 6 +\ \ 7 =\ \ \ \ +\ \ 8 + 17 + 14 + 11 + \\
 =\ & 9 +\ \ 8 + 12 + 10 + 11 +\ \ 7 = 18 +\ \ 4 + 17 +\ \ 1 + 15 +\ \ 2 =\ \ 5 + 13 +\ \ 3 + 14 +\ \ 6 + 16
\end{aligned}
$$

Fig. 15.

of now. The latest work has shown that the method is applicable to
boards of triangular and trapezoidal shapes, and also to solid boards
in three dimensions, so that the remote ground occupied by magic
and Nasik cubes will soon be invaded.

In conclusion, I bring before you an interesting example of magic
arrangement that I found whilst engaged in rummaging amongst the

books and documents of the old Mathematical Society of Spitalfields (1717–1845) for the purpose of extracting something which might interest or amuse, if it might not instruct, the audience I addressed in Section A of the British Association for the Advancement of Science at Glasgow last autumn.   It is an arrangement for the first eighteen numbers on five connected triangles; the magical property consists in the circumstance that the numbers 19, 38 and 57 appear as sums in a variety of ways.   The number 19 appears nine times, 38 twelve times and 57 fourteen times (Fig. 15).

I should say that I feel conscious that I have not been able to introduce the subject of my lecture without occasional and, perhaps, in the circumstances, unavoidable obscurity.   For the rest, I have felt somewhat doubtful as to the interest I might arouse in these problems, but the managers honoured me by inviting me to display to you some of the chips from a pure mathematician's workshop, and I felt no hesitation in accepting.

WEEKLY EVENING MEETING,

Friday, February 21, 1902.

His Grace The Duke of Northumberland, K.G. D.C.L. F.R.S.,
President, in the Chair.

W. Duddell, Esq.

Musical and Talking Electric Arcs.

It is almost exactly a century since the discovery, and first exhibition
of the electric arc in the lecture theatre of the Royal Institution by
Sir Humphry Davy. During these hundred years attention has
been chiefly centred on the light given out by the arc and on its
practical utilisation. This evening, however, I will treat more particularly of the sounds it can produce. It is only of late years that
any interest has been taken in these sounds, and this interest has
chiefly originated from the work of Mrs. Ayrton, who classified and
described the noises which arcs produce under different conditions of
working.

The sounds emitted by electric arcs can be divided into two
classes—those produced spontaneously, and those due to outside causes
acting on the arc itself or on the circuit supplying it. In either case
the sound waves seem due to the rapid variations in the volume of
the vapour column which exists between the two electrodes. If the
current through any direct current arc be increased or decreased
slowly, and if the vapour column, or better still, its image projected
on a screen be examined, it will be at once apparent that an increase
in the current flowing through the arc is accompanied by an increase
in the size of the vapour column, and vice versa. Let the current be
periodically increased and decreased, and the vapour column will also
expand and contract periodically, and if the variations in the current
and the consequent periodic expansion and contraction of the vapour
occurs with sufficient frequency, audible sounds are produced.
This can easily be demonstrated either by causing part of the direct
current which flows through the arc to pass through and operate an
electrically driven tuning-fork, or by adding to the direct current
flowing through the arc an alternating current.

The sensibility of the arc to small variations in the current supplying it is really surprising; for instance, the small fluctuations in
the current supplied by a direct current dynamo, caused by the commutator segments passing under the brushes, can generally be distinctly heard in any arcs supplied by it. A Wehnelt or other
powerful interruptor connected to the street mains will cause any

arcs fed from these mains to give out the characteristic noises, even though considerable distances intervene between the points of connection to the mains.

In some experiments on the resistance of the arc, in the course of which a small alternating current of high frequency was added to a steady direct current, I found that if even as small an alternating current as $\frac{1}{1000}$ ampere was passed through an arc taking 10 amperes direct current, a distinctly audible note was produced by the arc at frequencies ranging from a few hundreds up to 8000 periods per second; and further experiments showed that if the added alternating current was only $\frac{1}{20}$ to $\frac{1}{10}$ ampere, the note did not become inaudible until frequencies as high as 30,000 periods per second were reached.[*]

This great sensibility of the arc to small and rapid changes in the current through it at once suggests that the arc might be used as a telephone receiver. This was first tried experimentally by Herr Simon,[†] who caused the current from an ordinary microphone to flow round one coil of a mutual induction, or transformer, the second coil of which was placed in series with an arc. The variations in current flowing through the microphone due to speaking into it, induce corresponding currents in this latter coil and through the arc, causing the arc to give out sounds similar to those made near the microphone. By this means the arc may be caused to sing, whistle, or even talk distinctly.

There are a large number of other ways of superposing the microphone current or the direct current through the arc, the microphone may be shunted by an inductive or non-inductive resistance and placed directly in series with the arc; or the microphone may be connected in series with a battery and highly inductive coil, the terminals of the latter coil being connected direct to the arc terminals through a condenser, or the microphone and battery may be connected to the primary of a transformer the secondary of which is connected in series with a condenser as a shunt to the arc. I have usually found the last-mentioned arrangement give the best results.[‡]

Any good microphone will work, though the resulting speech from the arc varies greatly according to the microphone and the battery power used. A "Hunningscone" or "Delville" with two accumulators will, if carefully adjusted, give loud and distinct speech.

As the sound-waves are produced by the variations in the volume of the vapour column, it is natural that the loudness will depend on its size. In order to obtain loudness it is therefore necessary to use a long flaming arc and a fairly large current; such arcs can be best produced by using cored carbons or by introducing impurities, such

---

[*] Journal of the Institution of Electrical Engineers, 1901, vol. 30, p. 238.
[†] Annalen der Physik und der Chemie, 1898, vol. 64, p. 233.
[‡] Journal of the Institution of Electrical Engineers, 1901, vol. 30, p. 239; Elektrotechnische Zeitschrift, 1901, pp. 197 and 510.

as the salts of the metals of the alkaline earths, into the arc. An easy way to do this is to use a glass rod as a core to one of the electrodes, the sodium from the glass making it easy to burn an arc an inch or two long.

Not only is the vapour column of the arc so sensitive to any changes that may take place in the current that it can follow and reproduce all the numerous vibrations which constitute articulate speech, but I have also found that the intensity of the light given out by the arc is also capable of following very rapid small changes in the current. Thus, if the direct current through the arc be increased and decreased rapidly even over a small range, the light emitted will increase and decrease in a similar manner.*

The light given out by one of the talking arcs varies in intensity in a manner corresponding to the vibrations which form the words the arc is saying, although the eye does not appreciate the fact, so that if the light were allowed to fall on any apparatus or substance which was sensitive to such rapid changes in the intensity of its illumination and which could reproduce such variation of its illumination as corresponding mechanical movements, then it would be possible to reconvert the variation of the light into sound waves again. One such substance is the metal selenium, which, when suitably prepared, has the property that its electrical resistance depends on whether the light falling on it is strong or weak; in a strong light its resistance is much lower than in a weak light.

An arrangement of selenium and electrodes to show the effect of light on the resistance of selenium is generally called a selenium cell.

On this basis Herr Simon † has founded a method of light telephony: in his method the light given out by a talking arc at the transmitting station is collected by a lens or reflector and focussed either directly or by means of other lens and reflectors on a selenium cell placed at the distant receiving station. This selenium cell is connected in series with a battery and a telephone receiver in which the transmitted speech is received. The action of the arrangement is as follows; talking to the microphone at the transmitting station causes the currents through it to vary, increasing and decreasing the current in the talking arc; each increase of the arc current is accompanied by an increase in the light given out, which, by means of the reflectors and lens, increases the illumination of the selenium cell at the distant receiving station. The increase of illumination causes the cell's resistance to fall and allows the battery to send a larger current through it and the telephone receiver, so that every increase of current through the talking arc produces a corresponding increase in the current through the telephone receiver. Thus the variations of current due to the microphone are transmitted along the beam of

---

* Journal of the Institution of Electrical Engineers, 1901, vol. 30, p. 236.
† Elektrotechnische Zeitschrift, 1901, pp. 197 and 513.

light as variations in its intensity, and are faithfully reproduced in the telephone receiver, conveying sounds and speech just as if connecting wires had been established between the two stations.

Transmission of speech by this method has, I believe, been already accomplished over distances of several kilometers by Herr Ruhmer.

A further development of the uses of the talking arc, originated by Herr Ruhmer,[*] is the production of a new phonograph founded on similar principles. A permanent record of the variations of the light emitted is obtained by photographing on a long rapidly moving film the light given out by a talking arc so as to obtain, after development, a band of deposit, the changes of density along the length of which are a record of the variations in the intensity of the light emitted by the talking arc. If such a film be subsequently caused to travel at the same speed as during the recording, and if a beam of light be passed through it on to a selenium cell, the intensity of the light beam and the illumination of the cell will vary, and speech will be reproduced in a telephone receiver connected in series with the cell as before. This new phonograph adds one more to the methods, mechanical, magnetic, and chemical, by which sound waves can be permanently recorded, and it should give us much information as to the nature of sounds owing to the ease with which the record can be visually examined.

The second part of my subject treats of quite a different property of the electric arc, a property by means of which a direct current can be automatically converted into an alternating current of almost any frequency. This property is the instability of the arc under certain conditions.

Before proceeding further I will recall the properties of the discharge (or charge) of a condenser, or Leyden jar, through an inductive circuit. Lord Kelvin proved that if the resistance of the inductive circuit be below a certain critical value (depending on the capacity of the jar and self-induction of the circuit) then the discharge current would not flow continuously round the circuit in one direction till the jar was discharged, but would first flow round the circuit in one direction and then in the other, executing a series of swings or oscillations like a pendulum, the maximum value of the current growing less and less each swing, until the oscillations finally died away and the jar was discharged. If the resistance of the circuit was so small that it could be neglected, Lord Kelvin also showed that the periodic time, or time taken for the current to rise from zero to a maximum in one direction, then die away, reach a maximum in the opposite direction and again come to zero, was given by the expression

$$\text{Periodic time} = 2\pi \sqrt{\text{L.F.}}$$

* *Elektrotechnische Zeitschrift*, 1901, p. 198.

where F is the capacity of the condenser or Leyden jar and L is the self-induction of the circuit. So that the frequency of the oscillations, or the number of complete swings which can take place in one second, is the reciprocal of the above or

$$\text{Frequency} = \frac{1}{2\pi\sqrt{\text{L.F.}}}.$$

Such a circuit in which the discharge of the condenser is oscillatory I will refer to, in what follows, as an oscillatory circuit, and the above periodic time as the periodic time of the circuit.

If an unstable arc is made to form part of an oscillatory circuit then any oscillations in this circuit, instead of dying away as they would do were the arc not present, will persist indefinitely forming an alternating current, the energy required being automatically converted by the unstable arc into a suitable form to maintain them.

The electrical instability that is necessary for this purpose, is that if the current A through the arc be increased by any small quantity δA then the potential difference V between its terminals must decrease by an amount δV so that the ratio $\frac{\delta V}{\delta A}$ will be a negative quantity. Messrs. Frith and Rodgers [*] determined the value of the ratio $\frac{\delta V}{\delta A}$ for a large number of arcs between carbon electrodes, and they found that whereas the ratio is always *positive* when both *cored* carbons are used, it is *negative* for *solid* carbons.

If a perfectly steady current direct arc between *solid* carbon electrodes be shunted with an oscillatory circuit consisting of a suitable condenser and self-induction in series, then alternating currents of constant amplitude will be maintained in the circuit composed of the condenser, the self-induction and the arc, these alternating currents will cause the current through the arc to vary, and the arc will give out a musical note. I have therefore called it the Musical Arc. This will not occur if an arc between *cored* carbons be used, because such an arc has not the requisite instability, or, if the resistance of the inductive coil and connections exceeds a certain critical value. In a previous paper I have shown [†] that the necessary conditions in order to convert direct into alternating current by means of the arc are:—

(1) $\frac{\delta V}{\delta A}$, must be negative.

(2) $\frac{\delta V}{\delta A}$, must be numerically greater than the resistance of the

---

[*] Proceedings of the Physical Society, 1896, vol. 14, p. 307.
[†] Proceedings of the Institution of Electrical Engineers, 1901, vol. 30, p. 262.

shunt circuit (exclusive of the condenser, which should of course be very high).

Besides resistance, any other dissipation of energy such as hysteresis or eddy-currents, may, if sufficient, cause the arc to fail to produce its note, and for this reason the inductive coils used are preferably without iron cores.

The pitch of the note depends on the frequency of the alternating currents in the shunt circuit. If the resistance of the circuit is small, as it should be for the best results, the frequency of these currents may be calculated from Lord Kelvin's formula mentioned above

$$\text{Frequency} = \frac{1}{2\,\pi\,\sqrt{\text{L.F.}}}$$

where L is the self-induction and F the capacity in the oscillatory circuit shunting the arc.

In order to show that this formula is really obeyed, a constant self-induction was taken and a series of condensers calculated to produce the eight notes of an octave which were connected to a keyboard; to produce a second octave $\frac{1}{4}$ of the self-induction was used in conjunction with the same condensers, with this arrangement popular airs were played on the arc.

In view of the great sensitiveness of the ear to small differences from true musical intervals, the fact that the notes produced by this set of calculated condensers was even approximately in tune, indicates that Lord Kelvin's law was closely obeyed.

Chords could not be played on the above-mentioned key-board, as owing to the same inductive coil being used for each note, the depressing of two or more keys simply put the condensers in parallel and produced a lower note. Several experiments were made to see if it were possible to make one and the same arc give out two notes at the same time by shunting it with two distinct oscillatory circuits. When the periodic times of the two oscillatory circuits had a simple ratio such as the octave, the arc would emit both notes together and form a chord, but the condition seemed to be unstable, the chord often changing to a sound which was neither one note nor the other.

To obtain loud results, several arcs in series should be used all shunted by the one oscillatory circuit. A very convenient arrangement for working on 200-volt mains is to use four arcs in two parallel circuits, each circuit consisting of two arcs and a steadying resistance in series; by this means the four arcs, though supplied with power in two parallel circuits from the mains, will be metallically connected in series, and may be thus all four shunted by one oscillatory circuit.

A method of still further increasing the sound, first shown me by Professor Slaby, is to shunt the main condenser by a condenser consisting of alternate loose sheets of paper and tin-foil. In this case a loud note due to the alternating current is produced by the loose paper condenser as well as the note by the arcs.

It must not be forgotten that the arcs are ordinary *direct* current arcs between *solid* carbon electrodes which are taking energy from the direct current supply and converting it automatically without any moving machinery into energy in the form of alternating currents in the oscillatory circuit shunting the arcs. The frequency of this alternating current can be easily varied over a very wide range. I have already mentioned the range of two octaves which gave alternating currents having frequencies from 545 to about 2200 periods per second, and higher frequencies up to a limit of about 10,000 periods per second were easily obtained. No lower limit has yet been found, frequencies down to 170 having been used, or a range of just under six octaves. In order to produce the lowest frequency, a self-induction of about $0 \cdot 15$ henry and a capacity of six microfarads was used; under these circumstances it was found that although the arcs were only supplied from 200-volt direct-current mains, the alternating voltage between the terminals of the condenser sometimes reached nearly 2000 volts.

When the musical arc was first set up at the Central Technical College, where all the experiments were originally devised, an interesting observation in connection with the sensibility of the arc as a telephone receiver was made. It was noticed that an arc in Sir W. de W. Abney's laboratory played tunes, and this was found to be due to the fact that my musical arc and the arc in Sir W. de W. Abney's laboratory were supplied from the same street mains. Although his arc was not specially adjusted to a sensitive condition, and in spite of the two arcs being in separate buildings, 400 yards apart in a straight line, and at a considerably greater distance if measured along the mains, yet the sensibility of the arc was such that whenever tunes were played on the musical arc, they could be distinctly heard in Sir W. de W. Abney's laboratory.

The conversion of direct current into alternating current by means of the musical arc, besides being of scientific interest, forms a very convenient means of obtaining alternating currents, with which a large number of well-known experiments may be easily performed.[*] It has, however, two further advantages, namely, that the frequency can be easily adjusted to any required value, and that the frequency is at once self-evident from the note the arc is giving out.

The comparison of the capacity condensers and the self-induction of coils can be easily made by comparing the pitch of the notes produced when they are used in the circuit shunting the arc. The effect of closed secondaries and cores on the apparent self-induction of coils is also very striking. The Elihu Thomson experiment of the repulsion of a ring by a coil carrying an alternating current obtained from the musical arc can be easily performed.

If the primary of an induction coil (the contact maker being put

---

[*] W. Peukert, Elektrotechnische Zeitschrift, 1901, p. 467.

out of action) be used as the self-induction in the shunt circuit, then a high frequency flame discharge can be obtained from the secondary. Using a spark gap and Tesla coil in the ordinary way and supplying it from the secondary of the above induction coil all the ordinary Tesla phenomena can be produced. This experiment is of scientific interest, in that, without the use of a high frequency alternator or of any mechanically moving parts, low voltage direct current is converted in two steps into high voltage and very high frequency Tesla currents. Each step consists of the use of an electrically unstable conductor, the arc in one case the spark in the second, in conjunction with a circuit in which oscillations can take place. The first step converts the harmless direct current into dangerous high potential current ; the second step of the transformation, by still further raising the voltage and the frequency, again renders it harmless to human beings.

Magnetic space telegraphy can be very well demonstrated by using the self-induction coil in the oscillatory circuit as the transmitter. The effect of tuning the receiving circuit to the same note as the transmitter can be well illustrated. For instance, if a small incandescent lamp be used as the indicator in the receiving circuit, it is easy to so tune it to correspond with one note in the octave that the lamp will light up brightly when this note is sounded, will only just glow red if the next note higher or lower is depressed, and will not light at all for any other note ; so that if a scale or tune be played on the musical arc, the lamp in the receiving circuit will light up brightly every time the note occurs to which the receiving circuit is tuned. It would be quite easy to tune a separate receiving circuit to each of the notes of the octave and so form a system of syntonised wireless telegraphy on a small scale.

There is no doubt that the commercial success of wireless telegraphy will largely depend on the perfection of the syntony which can be maintained between the transmitting and receiving stations, as the more perfect the syntony the larger the number of messages which will be able to be transmitted simultaneously, and the longer the distance over which they will be able to be transmitted. So long as the transmitter only sends out trains of waves of decreasing amplitude at regular intervals, as at present, no very perfect syntony can be hoped for. As soon, however, as the transmitter can be supplied continuously with high voltage, high frequency currents of constant amplitude and thereby caused to send waves of a definite constant frequency and amplitude, so soon will very accurate syntony become possible and a great advance in wireless telegraphy and transmission of energy be made. May not the principle of the musical arc render this possible in the future ?

W. D.

WEEKLY EVENING MEETING.

Friday, February 28, 1902.

GEORGE MATTHEY, Esq., F.R.S., Vice-President,
in the Chair.

Professor HENRY A. MIERS, M.A. D.Sc. F.R.S.

Gold Mining in Klondike.

MY visit to Klondike took place at the end of August 1901, at a very
interesting time, and under most favourable conditions.   The journey
was made at the invitation of the Hon. Clifford Sifton, Canadian
Minister of the Interior, and in company with Professor Coleman of
Toronto.   Mr. Sifton provided us with a most efficient military escort
in the person of Captain Strickland of the North-West Mounted
Police, who possesses an intimate knowledge of the far North-
Western Territory and its inhabitants.   Consequently, we were able
in a short visit to see a great deal and to become acquainted with
the leading features of the mining industry.

The time was particularly interesting, because in 1901 the conditions of life were still to a large extent those of a young mining
camp, but were undergoing rapid transformation into the social and
political conditions of an organised and civilised community.

Access to Klondike is now practically confined to a single route
available for the ordinary traveller.   It is true that a considerable
amount of merchandise is taken in by the sea and river route which
consists of a voyage (from Seattle) of 2700 miles by sea to St. Michaels
at the mouth of the Yukon, followed by a voyage of 1370 miles up
that mighty river to Dawson City—a total of about 4000 miles.   But
the ordinary passenger route is the following:—a sea voyage of
900 miles, up the quiet waters that lie between the islands and the
mainland, from Vancouver to the little American port of Skagway;
a journey of 112 miles across the coast range, by the newly constructed White Pass and Yukon Railway, to the little town of White
Horse on the Yukon; and a voyage of 450 miles down the Yukon
from White Horse to Dawson City in a stern-wheel steamer:—a
journey of only 1460 miles in all.   By this route the sea voyage
occupies from three to five days, the railway journey about twelve
hours, and the river voyage from two to three days.

The railway climbs the icy precipices of the coast range, and
crosses into Canadian territory at the summit of the famous White
Pass which earned so unenviable a reputation as the scene of disasters

and deaths during the great rush into the country in the winter of 1897; and the traveller is brought down to the banks of the Yukon just low enough to escape the terrible White Horse Rapids, where also so many lives were lost in those early days. The voyage hence to Dawson is a quick one, with a stream whose average current is 5 miles an hour.

Dawson City, which in 1898 was only a collection of huts on a frozen mud swamp situated at the point where the Klondike River enters the Yukon, is now a town of about 10,000 inhabitants, consisting, it is true, of wooden buildings and chiefly of log cabins, but possessing hotels, clubs, theatres, saw-mills, large stores, electric light, telephones, power works and all the resources of modern civilisation. New government buildings were rising at the time of my visit, and the town wore an aspect of considerable and prosperous activity.

It is interesting to watch the life of this remarkable city, situated 1500 miles from Vancouver and close upon the Arctic circle : upon the plank sidewalks are slouch-batted, long-booted miners who throng dance-halls and saloons and pay from pouches of gold-dust; busy merchants, traders, and storekeepers of all nationalities ; well-dressed ladies and children ; military men, surveyors, engineers and lawyers ; while in the dusty roadway are to be seen men riding long-tailed horses with Mexican saddles, driving pack-mules laden with boxes, or urging yelping teams of dogs with the cry "mush, mush."

Popular accounts of this country generally represent it in its winter dress of snow, and relate tales of the rigorous severities of the Arctic frost. At the time when I was there Dawson was enjoying the mild and equable climate which prevails in the summer months, when the temperature may even rise to 90° F.; no snow was visible save that which clothed the serrated peaks of the northern Rockies, and their majestic chain was only to be seen from the sum mit of Moosehide Mountain above Dawson, and at a distance of about 40 miles to the north. The inhabitants had begun to grow potatoes, cabbages, lettuce and other vegetables, and a considerable market garden was being laid out on the left bank of the Yukon. And yet there is one remarkable feature of the country which prevents the traveller from ever forgetting that he is close upon the Arctic circle : if a hole be dug only 3 feet deep at any spot in the boggy ground, it will be found permanently frozen at that shallow depth ; even in Dawson itself the log cabins rest upon a foundation of ice which never thaws.

There is also a striking feature of life in Dawson which ever reminds the visitor that he is in a mining camp. He will have to pay 1s. or 2s. for a bootblack or a barber, 2s. for a glass of cow's milk, 6s. for three boiled eggs or a mutton chop, 30s. for a bottle of claret, perhaps 20l. a day for the hire of a rig and team of horses. The rent of a log cabin is about 120l., and a sense of economic insecurity is inspired by the fact that the rent of a house is about half its value, and that 60 per cent. is an ordinary rate of interest for loans. Labourers' wages are 2*l.* a day. Although almost any-thing required may be purchased in Dawson, all goods have been imported at great expense into a country which of itself has pro-duced nothing but gold and wood. The freight-rates on the White Pass route are about 6 cents a pound, and 23*l.* a ton may be paid from Vancouver to Dawson.

The mining camp is situated to the south-east of Dawson at a distance of about 13 miles. The productive area is about 30 miles square, and is bounded on the north by the Klondike River, on the west by the Yukon River, and on the south by the Indian River. The district is a gently undulating upland or plateau, attaining an average height of nearly 3000 feet above the Yukon, and intersected by deep flat-bottomed valleys which radiate from its central and highest point, a rounded hill named the Dome. The valleys are separated by hog-backed ridges; the whole district is fairly thickly wooded with spruce and poplar except on the summits of the ridges; the bottoms of the valleys are occupied by flat marshy bogs, and the streams are rarely more than 10 feet broad. The bog, which is from 5 to 10 feet thick, is frozen at a short depth below the surface and keeps the underlying gravel, which may be from 10 to 30 feet thick, permanently frozen right down to the bed-rock.

The principal streams are known as creeks; the short steep tribu-taries which flow into them as " gulches "; and the streamlets which feed these as " pups." The most important creeks are from 7 to 10 miles in length, and are productive over perhaps half their course, so that there may be about 50 miles of richly productive gravel in the district. I was informed that one stretch of $3\frac{1}{2}$ miles on El Dorado Creek produced no less than 6,000,000*l.* of gold.

Recently constructed government roads lead from Dawson to the camp and connect the various creeks; they were being completed at the time of our visit, and were still very rough or almost impassable at some points among the creeks; numerous rudely built but fairly comfortable " road-houses " afford lodging to the traveller. A small town of log cabins, known as Grand Forks, has sprung up at the junction of the two most famous creeks, Bonanza and El Dorado, and is inhabited by perhaps 1200 miners and others. There is another small town named Cariboo on Dominion Creek.

The creeks no longer present the dreary appearance of bog and forest, which made them look so unpromising to the early prospectors; the hill-sides have been largely stripped of their timber, and the valley bottoms are in many parts the scene of active mining opera-tions and rendered unsightly by machinery; the mining is also carried on upon the hill-sides at a height of 300 or 400 feet, where numerous adits penetrate the white gravel, and are marked by long heaps of tailings which descend from them towards the creek.

What little is known of the geology of the Klondike district can be stated in a few words. The auriferous area is occupied by Palæozoic schists, which may be roughly distinguished as grey or green chlorite-schist and mica-schist, and a light-coloured or white sericite-schist. These are bounded on the north—upon the right bank of the Klondike River—by a mass of diabase and serpentine, which constitutes the Moosehide mountain ; and on the south—on the left bank of the Indian River—by a series of quartzitic slates, schists, and crystalline limestones.

The auriferous creeks are entirely situated in the micaceous schists, which constitute the bed-rock everywhere. Mr. McConnel, the government geologist, regards these schists as having originated from quartz-porphyry and other eruptive rocks, but they have been much crushed and altered and entirely re-crystallised from their original condition. They are intersected by numerous bands and bosses of more recent eruptive rocks—quartz-porphyry, rhyolite, augite-andesite, diorite, basalt, etc.—and also by numerous quartz veins. In the northern and north-western portions of the area occupied by these Klondike schists, are both broad and narrow bands of a black graphitic schist, which can sometimes be traced across the valleys.

The veins and stringers of quartz which are so frequent throughout the district have for the most part a very barren appearance, but they are sometimes mineralised to a small extent and contain a little iron pyrites, argentiferous galena, and—very rarely—gold.

Up to the present, however, the gold has been exclusively won from the gravels in the valleys, and not from the quartz veins.

The gravels are mainly of two sorts :—(1) those which constitute the floors of the present valleys and have been laid down by the present streams ; (2) those which cover terraces upon the sides of the valleys and represent old valley gravels which have been cut through by the present streams.

The gold mining was at first carried on entirely in the lower gravels, and it was in them that the precious metal was first discovered. These are sandy gravels consisting of pebbles of quartz and schist—in fact, they are made up of the same materials as the bed-rock, and contain nothing that might not have been derived from the breaking up of the rocks of the district. There is no reason to believe that they were derived from any other source, and some of the pebbles are so lightly-rounded that they have clearly not travelled far. Among the minerals which I have seen from these gravels are hæmatite, rutile, pyrites, graphite, cyanite, garnet, cassiterite, epidote and tourmaline ; also barytes and mispickel.

The gold is very unevenly distributed in the gravel. The richest patches of pay-gravel seem to occur about half-way down the valley. In the wider portions of the valleys the pay-streak may be sometimes on one side and sometimes on the other, following no doubt the former course of the stream.

The valley gravel is generally from 10 to 30 feet thick, and is overlaid by from 5 to 15 feet of frozen bog, locally known as "muck."

The hill-side gravel is a very remarkable deposit; it consists almost entirely of boulders and pebbles of quartz and of sericite-schist, and, when it is exposed to view, presents the appearance of a uniform white ledge running horizontally along the hill-side at a height of about 700 feet above the level of the Klondike River. It sometimes attains a depth of 120 feet, and may be as much as half a mile wide. The pebbles are to a large extent sub-angular and less worn than those of the valley gravels.

The early miners of course confined their attention to the valleys, and the discovery of these rich deposits upon the hill-sides excited great surprise; they now rival the valley gravel in importance. The deposit is locally known as "white-wash" or the "White Channel."

The origin of the White Channel is shrouded in mystery; it was at first supposed to be a glacial deposit; but there are no striations or other signs of glacial action, and it is now the opinion of the local geologists that it was laid down by the sudden inrush of tumultuous streams acting over a small area. The materials have clearly not been transported far, and the gold is even more nuggety and less worn than that of the valleys.

One is naturally led to inquire whether all the gold of the lower gravels was not brought down by streams cutting through the White Channel which occupied the bed of the valleys when they were broad and shallow, so that the White Channel may be the real source of the gold. This view is supported by the fact just mentioned, that the gold of the valleys is more worn than that of the hill-side, also by the fact that the valleys are richer in their central portions, which must have been covered by the White Channel, than in the upper parts which are above the level of that deposit.

Still there is no reason to doubt that the White Channel itself is of local origin: its materials are those of the district, and have not travelled far. (There are a few gravels in this area which consist of pebbles foreign to the district, and they are not auriferous.) The White Channel itself follows the present valley courses. On the whole, therefore, although the origin of this peculiar deposit is obscure, there can be no doubt, in my opinion, that the conclusion forced upon us by a glance at the map is correct, and that the gold has been derived from the limited area intersected by the auriferous creeks which radiate from the Dome.

Some of the gold adheres to quartz, which exactly resembles that of the veins in the adjoining schists; and it is fairly certain that the metal came from quartz veins in the Klondike schists.

On the other hand it is certainly most remarkable that so little auriferous quartz has been found; at the time of my visit hundreds of quartz claims had been staked, but very few had been shown to contain any gold whatever; neither do the quartz boulders of the White Channel appear to be auriferous, or even mineralised. And yet it can hardly be doubted that where the valley gravels are rich in gold above their intersection with the White Channel the metal must have been derived from quartz veins in the schists.

In one instance I found direct evidence bearing on this question. In Victoria Gulch, a streamlet which descends into El Dorado Creek on its left bank high up the valley, have been found small flat crystals of gold of peculiar shape known as " spinel twins." In visiting a quartz vein at the head of Victoria Gulch (near the summit of the divide between El Dorado and Bonanza creeks), which had been lately opened and found to contain visible gold, I noticed precisely similar crystals. Here, at any rate, there can be little doubt that the gold in the creek has been derived from quartz veins in the schist.

No crushing has yet been carried on in Klondike; the gold has been entirely won by washing the gravels.

The chief difficulties of Klondike mining are due to the permanently frozen ground, which has led to certain peculiarities in the methods adopted. Every yard of gravel which is sluiced must first be thawed, either by artificial means or by exposing it to the rays of the summer sun after stripping off the overlying muck: for it is impossible to work the frozen ground with pick or spade, or even with dynamite.

Until recently shafts were sunk or tunnels were driven by laboriously thawing the ground with hot stones or wood fires; and I saw both methods in operation during my visit. The latter process—fire-setting, as it is called—is, in fact, quite frequent. A layer of dry wood is piled up against the face of the gravel, blanketed behind by a layer of green wood, ignited, and allowed to burn itself out; twelve hours of burning would thaw out little more than one foot in depth; and the process is then repeated.

Upon the larger properties this method has been entirely replaced by " steam-thawing." In this four to six-foot lengths of iron piping, tipped with steel nozzles, are inserted into the gravel, and steam is forced through them at a pressure of about 120 lb. These pipes are known as " points "; one point is inserted to about each square yard, and is driven in gradually by taps from a hammer; each point will thaw from two to five cubic yards of gravel. As contrasted with fire-setting, the steam-thawing obviates the suffocating fumes of burning wood, or the danger of thawing the frozen roof in underground workings.

A recent innovation which I saw coming into operation was thawing with water by means of the pulsometer pump, which seems to be economical since the water can be used over and over again; this process seems likely to come into more general use.

The gravel is raised from the shaft in buckets by windlass or steam-hoist, and in winter is dumped on to heaps for summer work, or in summer may be emptied straight into the sluice-boxes.

Much ingenuity has been exercised in the construction of self-dumping hoists, in which, by a single rope, the bucket is raised from the bottom of the shaft, caught by a travelling clutch, carried along a horizontal rope to the dump-heap or the sluice-boxes, where it is automatically tilted. All this is done by an engine in charge of one man, and saves much labour.

As regards the washing of the gravel, the old-fashioned hand-rockers are still to be seen in operation upon many of the gulches, but have been for the most part replaced by sluice-boxes. These are long wooden troughs made in 12-foot lengths and about 10 inches broad; the bottom is lined with wooden riffles, consisting generally of longitudinal bars (Pole riffles), but sometimes of transverse bars (Hungarian riffles), by which the gold and heavy minerals are caught. Sometimes Auger riffles—planks with circular holes—are employed; mercury is seldom used. A sluice-head of 75 miners' inches, i.e. 112 cubic feet of water per minute, is usual, and a fall of about 8 inches in the 12-foot box length.

Water, which is very scarce in the district, and must be used economically, is conducted to the sluice-boxes by long wooden flumes, which are themselves a serious expense on account of the cost of wood (about $110 a thousand feet). Some of these flumes are half-a-mile in length; in a wide valley, where the pay-streak is on the opposite side from the stream, it is necessary to raise it by centrifugal pumps to a height of 30 or 40 feet, and to convey it across the valley by a long flume.

In the final " wash-up," by which the gold-dust is extracted from the sluice-boxes, the riffles are taken out, and a copious stream of water is sent down, which carries away the fine gravel and leaves the gold, and a heavy black sand which accompanies it. This black sand consists mainly of magnetic ore, and it is removed partly by a magnet and partly by shaking with the hand and blowing with the mouth in a small metal tray.

The mining operations on the creeks and upon the hill-sides are somewhat different. On the creek a shaft is sunk down to bed-rock, four lateral tunnels are driven from the shaft along the surface of the bed-rock, and opened out in a fan-like manner to the limits of the claim. The outermost portions are worked out first, and as the excavation is carried back to the shaft, the roof and overlying muck are allowed to cave in and settle down on to the bed-rock. Timbering is thus entirely avoided. This absence of timbering in the Klondike shafts and tunnels is one of the most striking features in the mining; the frozen ground requires no support—it never thaws—and chambers as much as 100 feet square are covered by an icy roof which never breaks down.

The operations in the creeks are carried on upon a very considerable scale, and there is a large amount of machinery in the country; upon some groups of claims the work is not carried on by sinking and drifting, but is more of the nature of open quarrying. Night work is prosecuted by electric light or the acetylene lamp.

Formerly the raising of the gravel and its storage in dumps were mainly carried on in the winter, and the sluicing in the summer, and enormous winter dumps were accumulated for summer work. This last year the greater part of both has been carried on simultaneously during the summer, and it seems likely that the winter work will become less usual, and may even be abandoned altogether.

The mining upon the hill-side claims bears no resemblance in appearance to ordinary placer mining. Horizontal tunnels are driven into the White Channel from the face of the hill, and shafts are sunk into it from the surface in a manner that more nearly resembles the working in ordinary metalliferous lodes. In one such mine which I visited, a horizontal tunnel 700 feet in length had been driven into the gravel, and at right angles to this, and at intervals of about 60 feet, lateral tunnels were being driven to a distance of 70 feet on either side; and there were 200 feet of pay gravel above the tunnel; the men were working with pick and shovel at the end of the long tunnels, and cars of rock were being wheeled along a tramway to the head of the long wooden shoot which carried the gravel down to the creek.

In these high-bench claims, as they are called, two great difficulties are encountered: (1) the difficulty of disposing of the tailings, which cannot be allowed to slide down upon the creek claims but must be artificially banked up; and (2) the difficulty of obtaining water at this height. Both have co-operated to prevent hydraulicking, which would otherwise be the obvious way of working gravels situated upon a steep hill-side.

On Hunker Creek, however, Mr. Johanson, who owns both creek and hill-side claims, has, with much enterprise and at very great expense, introduced hydraulicking upon a considerable scale. The water was here derived from a reservoir in the creek, and was raised · by a 140 horse-power engine to a height of 260 feet, the level of the hill-side workings, and then to an additional height of 40 feet to an elevated tank, which gives a total fall of about 60 feet available for hydraulicking. The water was conducted through a 10-inch pipe and 6-inch hose terminating in a 2½-inch nozzle. About 1200 Canadian gallons a minute could be delivered. This enabled one man to wash out no less than 8 cubic yards per hour, and the gravel was washed straight into the sluice-boxes without the necessity for intermediate labour.

If ever water becomes more abundant and accessible in the district, there can be no doubt that hydraulicking will be largely employed.

On Bonanza Creek I witnessed another novelty in the first operations of a new dredging plant which had just been introduced, having been formerly employed on an auriferous sand bar upon the Lewes River. It is very possible that dredging will prove to be an efficient and economical way of working over some of the old claims in the creeks which have only been treated by the cruder methods of the earlier minors.

There are other introductions which were new at the time of my visit; although no crushing of quartz had then been effected, a small Tremaine mill had just been erected on the banks of the Klondike in the immediate neighbourhood of Dawson, and it is to be hoped that we shall soon hear of promising results among the quartz discoveries. An auriferous conglomerate found in considerable deposits on the Indian River was attracting much attention, chiefly on account of its superficial resemblance to the South African banket.

In default of resources other than gold, the prosperity of Klondike in the immediate future appears to me to depend mainly upon the extent to which in the creeks water can be more economically and bountifully supplied, labour and the necessaries of life more cheaply obtained, and communication be made more easy, so that it may be possible to work low-grade gravel at a profit. There is much auriferous material which it does not at present pay to touch. The Government is giving every encouragement, and in Mr. Ross the Territory has a strong governor; roads have been constructed; the royalty has been reduced to 5 per cent., and on all claims $5000 of gold are exempt. The necessary charges are only $10 for a miner's license, $15 for recording a claim, $50 for surveying, $15 for renewal, and an owner is only required to put $500 worth of work on to his claim each year.

But the cost of water, wood, labour, and materials is almost prohibitive; the standard of living is high, although there is, I think, a constantly increasing number of steady and thrifty men coming into the country and replacing the more gambling element of the early camp.

The ultimate prosperity of the country depends largely, I think, upon the extent to which auriferous quartz may be discovered, and other resources developed.

But it is certain that Dawson has come with the intention of staying, and that the country is very far from played out. Not only is there a considerable quantity of ground yet to be worked in the Klondike creeks, but it must be remembered that much of the vast Yukon territory is auriferous, and that attention has only been distracted from other localities by the extraordinary wealth of the Klondike area.

Now that the district possesses a large town, inhabited throughout the year, now that communication is being facilitated, that freight rates are being lowered, and that the population is increasing, it ought to be possible to open up districts that could never have been attempted under the more adverse conditions of two or three years ago.

Coal is being mined at Cliff Creek, 55 miles below Dawson, and at Five Fingers, about 200 miles above Dawson; placer copper exists in large quantities on the White River; copper ores have been found near White Horse; the Atlin district promises well; horses, cattle, and sheep will shortly be supported in the country itself, and vegetables and other produce will be raised.

It only remains to be seen whether the cost of production can be so far diminished that this far North-Western Territory will be able to compete with other regions which are more favourably situated.

That the inhabitants have the necessary enterprise and energy I know from what I have seen of them. It is, in fact, most interesting to note how in this isolated country native grit and intelligence have brought the best men to the front. One naturally associates the element of luck with placer mining, and no doubt many fortunes were made and lost by sudden strokes of chance. But in no mining district have previous experience and knowledge been of less avail. The conditions were so strange, that the old and experienced miners sometimes made the worst mistakes, and the men who succeeded were those who were sufficiently alert and intelligent to adapt themselves to the new conditions. One finds among the leading miners—men who have come from all places and from all classes of society— men who two or three years ago were workmen, hotel clerks, store assistants, or farmers.

I cannot conclude without a word of tribute to the magnificent work which has been done by the Canadian North-West Mounted Police, and the excellent way in which the inhabitants have settled down under their rule. A mere handful of this fine military force have sufficed to introduce law and order into the country, from the time of the great rush in 1897. The perfect quiet which prevails, and has always prevailed, on the Canadian side of the frontier, contrasts most favourably with the lawless scenes that took place till recently at Skagway and other places in American territory. Even the disorderly population which migrated into Klondike seemed to lose its character on Canadian territory, and the six-shooter was no more seen.

I doubt whether any better example exists of the manner in which, in any part of the world, the finest features of British character will prevail—the firmness, good temper, and love of order which are the dominant characteristics of our most successful colonists.

[H. A. M.]

GENERAL MONTHLY MEETING,

Monday, March 3, 1902.

SIR JAMES CRICHTON-BROWNE, M.D. LL.D. F.R.S., Treasurer and Vice-President, in the Chair.

Sir William Agnew, Bart.
Aubrey St. John Clerke, Esq.
Miss Agnes M. Clerke,
Hilder Daw, Esq.
John F. W. Deacon, Esq. M.A.
Mrs. Deacon,
Edgar Figgess, Esq.
Maurice Fitzmaurice, Esq. M.Inst.C.E.
F. Montefiore Guedalla, Esq.
John Harold, Esq. M.B. B.Ch.
Francis Legge, Esq.
H. Lewkowitsch, M.D.
Miss A. H. Little,
William Frederick Preedy, Esq.
A. Mayo Robson, Esq. F.R.C.S.
M. M. Samuel, Esq.
F. A. Smith, Esq.
John James Torre, Esq. J.P.
Clarens Tweedale, Esq.
Major-General James Waterhouse,
Philip Watts, Esq. F.R.S.

were elected Members of the Royal Institution.

It was announced from the Chair that His Royal Highness the Prince of Wales had graciously consented to become Vice-Patron of the Royal Institution of Great Britain.

The Special Thanks of the Members were returned to " An Old Member " for a Donation of £50 to the Fund for the Promotion of Experimental Research at Low Temperatures.

The PRESENTS received since the last Meeting were laid on the table, and the thanks of the Members returned for the same, viz. :—

FROM

The Secretary of State for India—General Report on Public Instruction in Bengal, 1900-1901. fol. 1901.

Accademia dei Lincei, Reale, Roma—Classe di Scienze Fisiche, Matematiche e Naturali. Atti, Serie Quinta: Rendiconti. 1º Semestre, Vol. XI. Fasc. 3. 8vo.

Astronomical Society, Royal—Monthly Notices, Vol. LXII. No. 3. 8vo. 1901.

Bankers, Institute of—Journal, Vol. XXIII. Part 2. 8vo. 1901.

Boston Public Library—Annual List of Books, 1900–1901. 8vo. 1902.
Monthly Bulletin for Feb. 1902. 8vo.

British Architects, Royal Institute of—Journal, 3rd Series, Vol. IX. No. 8. 4to. 1902.

British Astronomical Association—Journal, Vol. XII. Nos. 3, 4. 8vo. 1901.
Memoirs, Vol. X. Parts 1, 2. 8vo. 1901.

Buenos Ayres, City—Monthly Bulletin of Municipal Statistics, Dec. 1901. 4to.

Buenos Ayres, Museo Nacionale—Comunicaciones, Tom. I. No. 10, 1901. 8vo.

Camera Club—Journal for Feb. 1902. 8vo.

Chemical Industry, Society of—Journal, Vol. XXI. Nos. 2, 3. 8vo. 1902.

Chemical Society—Proceedings, No. 247. 8vo. 1902.
Journal for Feb. 1902. 8vo.

Editors—Aeronautical Journal for Feb. 1902. 8vo.
American Journal of Science for Feb. 1902. 8vo.
Analyst for Feb. and March 1902. 8vo.
Anthony's Photographic Bulletin for Feb. 1902. 8vo.
Astrophysical Journal for Feb. 1902.
Athenæum for Feb. 1902. 4to.
Author for Feb. 1902. 8vo.
Brewers' Journal for Feb. 1902. 8vo.
Chemical News for Feb. 1902. 4to.
Chemist and Druggist for Feb. 1902. 8vo.
Electrical Engineer for Feb. 1902. fol.
Electrical Review for Feb. 1902. 8vo.
Electrical Times for Feb. 1902. 4to.
Electricity for Feb. 1902. 8vo.
Engineer for Feb. 1902. fol.
Engineering for Feb. 1902. fol.
Homœopathic Review for Feb. 1902. 8vo.
Horological Journal for Feb. and March, 1902. 8vo.
Invention for Feb. 1902.
Journal of the British Dental Association for Feb. 1902. 8vo.
Journal of Medical Research for Jan. 1902. 8vo.
Journal of State Medicine for Feb. 1902. 8vo.
Law Journal for Feb. 1902. 8vo.
London Technical Education Gazette for Feb. 1902.
Machinery Market for Feb. 1902. 8vo.
Motor Car Journal for Feb. 1902. 8vo.
Nature for Feb. 1902. 4to.
New Church Magazine for Feb. 1902. 8vo.
Nuovo Cimento for Feb. 1902. 8vo.
Pharmaceutical Journal for Feb. 1902. 8vo.
Photographic News for Feb. 1902. 8vo.
Physical Review for Feb. 1902. 8vo.
Popular Science Monthly for Feb. 1902.
Public Health Engineer for Feb. 1902. 8vo.
Science Abstracts for Feb. 1902. 8vo.
Travel for Feb. 1902. 8vo.
Zoophilist for Feb. 1902. 4to.

Electrical Engineers, Institution of—Journal, Vol. XXXI. No. 154. 8vo. 1901.

Florence, Reale Accademia di—Atti, 4th Serie, Vol. XXIV. Disp. 3, 4. 8vo. 1902.

Geographical Society, Royal—Geographical Journal for Feb. and March, 1902. 8vo.

Imperial Institute—Imperial Institute Journal for Feb. and March, 1902.
Johns Hopkins University—University Circular, No. 155. 8vo. 1901.
Leighton, John, Esq. M.R.I.—Journal of the Ex-Libris Society for Jan. 1902. 8vo.
Madras Government Museum—Bulletin, Vol. IV. No. 2: Anthropology. 8vo. 1901.
Mechanical Engineers, Institution of—Proceedings, 1901, No. 4. 8vo. 1901.
Meteorological Society, Royal—Meteorological Record, Vol. XXI. No. 82. 8vo. 1901.
Microscopical Society, Royal—Journal, 1902, Part 1. 8vo.
Navy League—Navy League Journal for Jan.-March, 1902. 8vo.
New York Academy of Sciences—Annals, Vol. XIV. Part 1. 8vo. 1901.
Odontological Society—Transactions, Vol. XXXIV. No. 3. 8vo. 1901.
Photographic Society, Royal—Photographic Journal for Feb. 1902. 8vo.
Rome, Ministry of Public Works—Giornale del Genio Civile, Nov.-Dec. 1901. 8vo.
Royal Society of London—Philosophical Transactions, A, Nos. 301-303. 4to. 1902.
    Proceedings, No. 455. 8vo. 1902.
Selborne Society—Nature Notes for Feb. 1902. 8vo.
Society of Arts—Journal for Feb. 1902. 8vo.
Tacchini, Prof. P. Hon. Mem. R.I. (the Author)—Memorie della Società degli Spettroscopisti Italiani, Vol. XXXI. Disp. 1. 4to. 1902.
Thurston, Prof. Robert H. Hon. Mem. R.I.—List of Books and Papers, 1902. 8vo.
United Service Institution, Royal—Journal for Feb. 1902. 8vo.
United States Department of Agriculture—Monthly Weather Review for Nov. 1901. 4to.
    Contributions from the U.S. National Herbarium, Vol. VII.: Plants used by the Indians of Mendocino County, California. By V. K. Chesnut. 8vo. 1902.
    Nutrition Investigations among Fruitarians and Chinese, 1899-1901. 8vo. 1901.
Verein zur Beförderung des Gewerbfleisses in Preussen—Verhandlungen, 1902. Heft 2. 8vo.

FIG. 1.

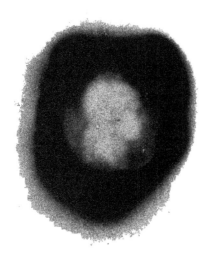

FIG. 2.

WEEKLY EVENING MEETING,

Friday, March 7, 1902.

Sir William Crookes, F.R.S., Honorary Secretary
and Vice-President, in the Chair.

Professor H. Becquerel, D.C.L. Ph.D., Membre de l'Académie des
Sciences, Paris, *Hon. Mem. R.I.*

Sur la Radio-activité de la Matière.

La propriété que possèdent certains corps d'émettre un rayonnement
invisible et pénétrant, était inconnue il y a six ans. Le mouvement
d'idées que suscitèrent les expériences de Röntgen conduisit à re-
chercher si la matière n'émettrait pas de semblables radiations, et l'on
pensa d'abord au phénomène de la phosphorescence qui réalisait un
mode connu de transformation et d'émission d'énergie. Cette idée
ne devait pas s'appliquer au phénomène qui nous occupe, mais elle
fut cependant féconde. Elle conduisit à choisir parmi les corps
phosphorescents, les sels d'uranium dont la constitution optique est
remarquable par la série harmonique des bandes de leurs spectres
d'absorption et de phosphorescence. C'est en expérimentant avec ces
corps que j'ai vu, en 1896, le phénomène nouveau dont je me propose
de vous entretenir.

Voici, reproduites par la méthode de Lippmann, les lamelles de
sulfate double d'uranium et de potassium qui ont servi aux premières
expériences. Après avoir posé une de ces lamelles sur le papier noir
qui enveloppait une plaque photographique, et l'avoir laissée ainsi
pendant quelques heures, j'ai observé, en développant la plaque, que
le sel d'uranium avait émis des rayons actifs, traversant le papier
noir, ainsi que divers écrans interposés entre le corps rayonnant et la
plaque (lamelles minces de verre, d'aluminium, de cuivre).

Je ne tardai pas à reconnaître que le phénomène était indépendant
de la phosphorescence, et même de toute excitation de nature connue,
telle qu'une excitation lumineuse ou électrique, ou une variation très
notable de température.

On était donc en présence d'un phénomène spontané, d'un ordre
nouveau. L'absence de cause excitatrice connue sur un produit préparé depuis plusieurs années dans le laboratoire, permettait de penser
que le phénomène eût été le même à quelque moment qu'on l'eût
observé; il devait donc paraître permanent, c'est à dire qu'il ne
devait pas manifester un affaiblissement appréciable pendant un temps
très long. C'est en effet ce que j'ai pu vérifier depuis six ans. Je
mets sous vos yeux la première épreuve qui m'a révélé la spontanéité
du rayonnement; celui-ci a traversé le papier noir qui enveloppait la

plaque, et une lame mince de cuivre en forme de croix (Fig. 1). Voici encore la radiographie faite à la même époque, d'une médaille en aluminium (Fig. 2); l'absorption inégale par les différentes épaisseurs du métal a fait apparaître l'effigie.

Dès la première observation, j'ai reconnu que le rayonnement nouveau déchargeait à distance dans l'air, les corps électrisés, phénomène qui donne une seconde méthode pour étudier ces rayons; la méthode photographique est surtout qualitative, l'électromètre fournit des éléments numériques de comparaison.

Au cours de ces premières constatations, je fus détourné de la voie dans laquelle les expériences ultérieurs devaient me ramener, par plusieurs faits dont le principal est le suivant : Ayant protégé une plaque photographique par une plaque d'aluminium de 2 mm. d'épaisseur, et ayant disposé sur l'aluminium divers échantillons de poudres phosphorescentes, reposant sur des lamelles de verre et recouvertes de petits tubes en forme de cloche, comme le montre la figure ci-contre (Fig. 3), l'épreuve obtenue au bout de 48 heures de pose, et que je mets sous vos yeux (Fig. 4), donna des silhouettes des lamelles de verre telles qu'elles eussent été produites par la réfraction et la réflexion totale de rayons identiques à ceux de la lumière, mais qui auraient traversé les 2 mm. d'aluminium. Cette épreuve est unique : je n'ai pu la reproduire ni obtenir d'action avec le même échantillon de sulfure de calcium, ni avec aucune autre préparation phosphorescente. À la même époque M. Niewenglowski avait obtenu une impression avec du sulfure de calcium, et M. Troost avec de la blende hexagonale. J'ignore encore la cause de l'activité de ces produits, et de sa disparition. Ces faits et quelques autres m'avaient conduit à penser que le nouveau rayonnement pouvait être un mouvement transversal de l'éther analogue à la lumière; l'absence de réfraction et un grand nombre d'autres expériences me firent abandonner cette hypothèse.

Dans cette même année 1896, je reconnus que tous les sels d'uranium émettent des radiations de même nature et que la propriété radiante est une propriété atomique liée à l'élément uranium; les mesures électriques me montrèrent que l'uranium métallique était environ trois fois et demie plus actif pour ioniser l'air, que ne l'est le sulfate double d'uranium et de potassium. La même méthode permit d'étudier le rôle des gaz dans le décharge, de reconnaître qu'une sphère d'uranium électrisée conserve sa charge dans le vide, tandis que, dans l'air, elle se décharge. La vitesse de la chute du potentiel est sensiblement proportionnelle au potentiel si celui-ci est de quelques volts; elle devient constante et indépendante du potentiel pour les potentiels très élevés. Le gaz rendu conducteur par le rayonnement conserve cette propriété pendant quelques instants. Entre deux conducteurs maintenus à des potentiels constants le rayonnement établit, dans l'air, un courant continu.

Ces expériences ont été reprises et variées en 1897 par Lord Kelvin, puis par MM. Beattie et S. de Smolan. En 1897 Mr.

FIG. 3.

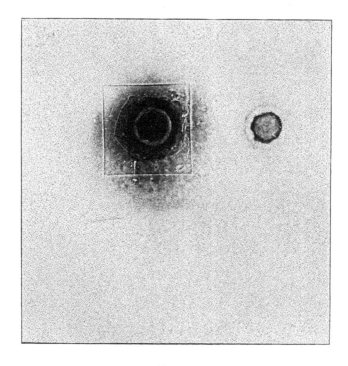

FIG. 4.

Rutherford montra comment les phénomènes dus à la conductibilité communiquée aux gaz par l'uranium, et l'existence d'un maximum dans le courant produit, peuvent s'expliquer dans l'hypothèse de l'ionisation à laquelle les beaux travaux de Mr. J. J. Thomson ont donné tant d'autorité.

En 1898, M. Schmidt et Mme. Curie observèrent séparément que le thorium a des propriétés analogues à celles de l'uranium, propriétés qui furent étudiées en particulier par Mr. Owens et par Mr. Rutherford. Mme. Curie, ayant mesuré l'activité ionisante d'un grand nombre de minéraux contenant de l'uranium ou du thorium, signala ce fait remarquable que plusieurs minerais étaient plus actifs que l'uranium métallique. M. et Mme. Curie en conclurent qu'il devait exister dans le minerai un corps plus actif que l'uranium ; et ils entreprirent de l'isoler. Traitant alors l'un des plus actifs de ces minerais, la pechblende de Joachimsthal, ils en séparèrent d'abord du bismuth actif auquel ils donnèrent le nom de Polonium, puis peu après du baryum très actif, contenant un élément nouveau, le Radium.

Ces produits se préparent par des précipitations fractionnées pour lesquelles on est guidé par les indications de l'electromètre ; l'activité des produits obtenus a dépassé 100,000 fois celle de l'uranium. Vers la même époque M. Giesel est parvenu à préparer des substances très actives, et, en 1900, M. Debierne a annoncé l'existence d'un nouvel élément, l'Actinium, sur lequel on a jusqu'ici peu de renseignements. De ces diverses préparations le radium seul est caractérisé comme élément nouveau ; il possède un spectre d'émission formé de lignes qui n'appartiennent à aucun autre corps connu, et le poids atomique des sels de baryum radifères augmente avec leur teneur en radium.

Le rayonnement de l'uranium était trop faible pour exciter le phosphorescence des corps. M. et Mme. Curie observèrent ce phénomène avec les rayons du radium ; bien plus les sels de radium se rendent lumineux eux-mêmes ; leur luminosité, comme leur rayonnement, est spontanée. Le rayonnement du radium produit des actions chimiques diverses, colore le verre, transforme l'oxygène en ozone, le phosphore blanc en phosphore rouge ; il ionise non seulement le gaz, mais encore les liquides (pétroles, air liquide) et les solides isolants, tels que la paraffine, en développant dans ce dernier corps une conductibilité résiduelle qui dure fort longtemps après que le rayonnement a cessé d'agir. Il provoque sur les tissus organiques des brûlures profondes analogues à celles que font les rayons X.

L'échantillon de radium que M. Curie m'a prêté pour cette conférence, me permet de vous montrer quelques-uns de ces phénomènes: ionisation de l'air, luminosité, phosphorescence.

J'ai constaté par l'épreuve photographique que je montre ici (Fig. 5), que le rayonnement du polonium ne traversait pas une mince feuille de papier noir qui formait un petit cylindre fermé par de l'aluminium ou du mica, et au fond duquel était la matière pulvérulente ; le rayonnement du radium traverse facilement cette enveloppe : nous verrons qu'il existe entre les deux rayonnements, des différences encore plus profondes.

Le rayonnement du radium redonne à certains cristaux et au verre la propriété d'être phosphorescent par la chaleur, quand ces corps l'ont perdue par une élevation de température préalable.

Les phénomènes d'absorption étudiés soit par la photographie, ou par la phosphorescence, ou par l'ionisation de l'air, avaient montré l'hétérogénéité du faisceau de radiations émises. Une observation nouvelle vint élargir le champ de ces recherches.

Vers la fin de l'année 1899, M. Giesel, puis MM. Meyer et Schweidler observèrent que le rayonnement de préparations actives était dévié par un champ magnétique, comme le sont les rayons cathodiques. De mon côté, à la même époque, sans avoir eu connaissance de ces expériences, je faisais la même observation avec le rayonnement du radium. On peut opérer de la manière suivante : Sur une plaque photographique enveloppée de papier noir, et placée horizontalement entre deux pôles d'aimant, on dépose une petite caisse en papier contenant quelques grains de matière active : le rayonnement est tout entier rejeté sur la plaque, d'un seul côté.

Presque aussitôt, je reconnus que les rayons du polonium ne sont pas déviés, et, par suite, qu'il existe deux espèces de rayons, les uns déviables et les autres non déviables. M. et Mme. Curie firent une étude électrique qui leur montra la présence simultanée des deux espèces de rayons dans le rayonnement du radium, leur inégale perméabilité variable avec la distance des écrans absorbants. L'épreuve photographique ci-jointe (Fig. 6) montre les deux espèces de rayons avec le radium ; j'ai reconnu récemment que la radiation du thorium comprenait les deux espèces de rayons, et que l'uranium émettait uniquement des rayons déviables, sous la réserve de l'existence de rayons non déviables beaucoup moins actifs. Il existe en effet une troisième espèce de rayons, qui ne sont pas déviables mais sont extrèmement pénétrants ; ils ont été plus particulièrement mis en évidence par M. Villard.

Ainsi le rayonnement des corps radio-actifs comprend trois espèces de rayons : des rayons déviables par un champ magnétique qui paraissent identiques aux rayons cathodiques ; des rayons non déviables de deux sortes, les uns très absorbables, les autres qui ressemblent à des rayons X très pénétrants. L'uranium émet surtout la première espèce, le polonium n'émet que la seconde, et le radium émet les trois à la fois.

Revenons aux rayons déviables. On peut leur appliquer la théorie matérielle édifiée par Sir W. Crookes et Mr. J. J. Thomson, et l'on en vérifie les conséquences avec la plus grande facilité. Dans un champ magnétique uniforme les trajectoires perpendiculaires au champ sont des circonférences de rayon ρ qui ramènent le rayonnement au point d'émission. Pour une émission oblique faisant l'angle a avec le champ les trajectoires sont des hélices qui s'enroulent sur des cylindres de rayon $\rho \sin a$. En plaçant sur une plaque photo-

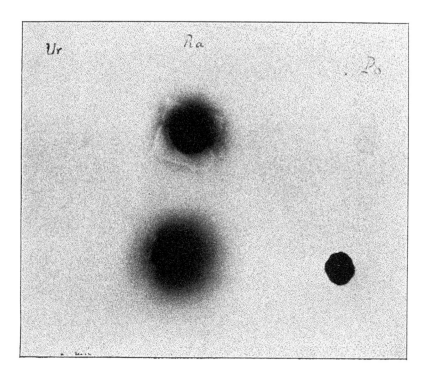

FIG. 5.

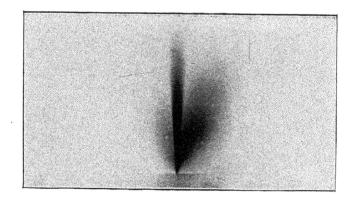

FIG. 6.

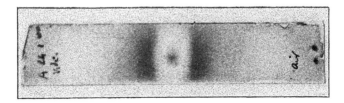

Fig. 7.

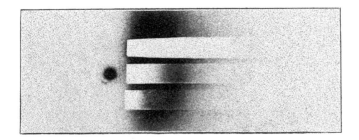

Fig. 8.

Fig 9.

graphique horizontale parallèle au champ uniforme une petite cuve en plomb contenant quelques grains de baryum radifère et formant une source de très petite diamètre, le rayonnement est ramené sur la plaque et l'impressionne d'un seul côté ; un faisceau de rayons simples émis dans le plan normal à la plaque et parallèle au champ, doit figurer théoriquement un arc d'ellipse dont les axes sont dans le rapport des nombres 2 et π. L'épreuve ci-contre (Fig. 7) montre ces arcs théoriques, obtenus en renversant le sens du champ, l'un dans l'air, l'autre dans le vide, sur une plaque enveloppé de papier noir ; l'intensité du champ magnétique était environ 4000 unités C.G.S.

Si l'on n'enveloppe pas la plaque photographique, et si l'on dispose sur celle-ci diverses bandes de papier ou de métal, formant écrans, on observe dans l'impression du rayonnement dispersé par le champ magnétique, des sortes de spectres d'absorption. À chaque trajectoire de courbure différente, correspondent des rayons de vitesses différentes qui ont des pouvoirs de pénétration différents.

Voici un exemple de ces épreuves, obtenu dans un champ de 1740 unités C.G.S. environ ; les écrans sont une bande de papier noir, une bande d'aluminium de 0,10 mm. d'épaisseur, et une bande de platine de 0,03 mm. d'épaisseur (Fig. 8). Pour avoir un spectre pur tel qu'en chaque point de la plaque il n'arrive qu'un faisceau dont les trajectoires aient la même courbure, on doit assujetir le rayonnement issu de la source ponctuelle à passer par un trou étroit. Le résultat est le même que le précédent comme le montre la figure ci-contre (Fig. 9). Celle-ci fait encore voir une impression très intense, due à des rayons secondaires provoquées par les rayons qui étaient arrêtés par une gouttière de plomb recouvrant la source, et dans laquelle était pratiquée l'ouverture qui donnait passage au spectre pur.

L'absorption varie avec la distance des écrans à la source radiante, et des rayons qui sont arrêtés par un écran placé sur la plaque peuvent traverser ce même écran lorsque celui-ci est interposé sur leur trajet, près de la source.

Ces expériences laissaient peu de doutes sur l'identité du rayonnement déviable et des rayons cathodiques. Cependant, il était nécessaire de démontrer qu'ils transportent des charges d'électricité négative, et qu'ils sont déviés par un champ électrique.

M. et Mme. Curie, dans une très belle expérience, ont montré que les rayons du radium chargent négativement les corps que reçoivent le rayonnement, et que la source se charge elle-même positivement. Pour cette double expérience il est nécessaire que tous les conducteurs et la source elle-même soient enveloppés complètement de matières isolantes telles que la paraffine. Pour la préparation active étudiée, la charge était de $4 \cdot 10^{-13}$ unités C.G.S. par centimètre carré de surface radiante et par seconde.

De mon côté je suis parvenu à mettre en évidence et à mesurer la **déviation électrostatique** en projetant sur une plaque photographique l'ombre déviée d'un écran plan perpendiculaire au champ.

L'un des appareils est figuré ci-contre (Fig. 10) ainsi que l'une des épreuves obtenues (Fig. 11), dans laquelle sur les deux moitiés d'une même plaque apparaissent les ombres déviées correspondant au renversement du champ électrique, dont l'intensité était environ $1,02.10^{12}$.

L'hypothèse balistique attribue ces phénomènes à l'existence de masses matérielles transportant des charges d'électricité négative avec une vitesse considérable. Soit m la masse matériel d'une particule, e sa charge, v sa vitesse. On sait que dans un champ magnétique d'intensité H le rayon de courbure ρ de la trajectoire circulaire est donné par la relation

$$H\rho = \frac{m}{e}v.$$

La valeur numérique du produit $H\rho$ peut servir à caractériser la nature de chaque rayon simple. D'autre part, dans un champ électrique d'intensité F le paramètre de la trajectoire parabolique est

$$\frac{m}{e}\frac{v^2}{F}.$$

La connaissance de ces deux grandeurs donne $\frac{m}{e}$ et v. Pour une valeur de $H\rho = 1600$ j'ai obtenu approximativement $v = 1,6.10^{10}$ et $\frac{e}{m} = 10^7$. Ces nombres sont tout à fait de l'ordre de grandeur de ceux auxquels conduisent les mesures faites avec les rayons cathodiques, les considérations théoriques relatives à l'expérience de Zeeman et les déterminations de M. Lenard sur l'émission provoquée par les rayons ultraviolets.

Des nombres ci-dessus on déduit que par le fait du rayonnement déviable considéré il s'échapperait par chaque centimètre carré de surface radiante 1,2 mgr. matière en un milliard d'années.

En étendant ces mesures à des rayons de diverses natures et bien déterminés, on doit reconnaître si le rapport $\frac{e}{m}$ est constant ou variable d'un rayon à un autre, et si ceux-ci ne diffèrent pas uniquement par leur vitesse : je n'ai pas terminé les expériences que j'avais entreprises pour résoudre cette question fondamentale, mais récemment M. Kaufmann s'est proposé de l'élucider. Il a combiné à angle droit l'action magnétique et l'action électrique ; malheureusement l'expérience qui est difficile, ne lui a donné qu'une seule plaque bonne à mesurer. Pour des valeurs de $H\rho$ comprises entre 1800 et 4600, il a trouvé que le rapport $\frac{e}{m}$ variait de $1,3.10^7$ à $0,6.10^7$, et la vitesse v de $2,3.10^{10}$ à $2,8.10^{10}$.

La constatation d'une variation régulière dans le rapport calculé $\frac{e}{m}$ a une importance théorique considérable : si ce rapport était

Fig. 10.

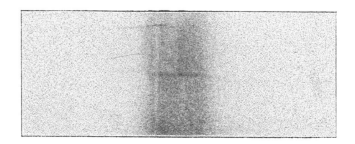

Fig. 11.

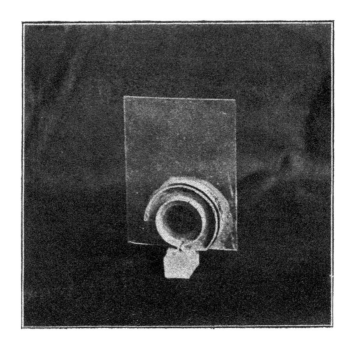

FIG. 12.

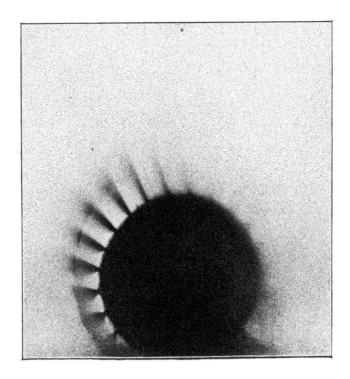

FIG. 13.

constant comme cela semblait résulter d'un grand nombre de mesures, on devrait en conclure que les rayons peu déviables pour lesquels le produit Hρ dépasse 5000, ont des vitesses notablement supérieures à celle de la lumière.

D'autre part, des considérations théoriques ont fait penser que la vitesse ne pouvait dépasser celle de la propagation des perturbations electro-magnétiques, c'est à dire la vitesse de la lumière, et l'on a été conduit à considérer les masses mobiles dans un champ magnétique comme douées d'une inertie particulière fonction de la vitesse. Dans ces conditions, la masse calculée devrait être, en partie au moins, apparente, et grandir indéfiniment à mesure que la vitesse réelle s'approche de celle de la lumière. Les nombres publiés par M. Kaufmann sont conformes à cette hypothèse.

Une autre conséquence de cette manière de voir, serait qu'il devrait y avoir continuité entre les rayons déviables et ceux qui ne le sont pas, car le rayon de courbure des trajectoires devient infini en même temps que la masse apparente.

L'épreuve photographique déjà mentionné (Fig. 6), ainsi que l'une des épreuves suivantes (Fig. 13), manifestent au contraire une discontinuité bien nette, et, dans la seconde épreuve en question, la pose a été assez prolongée pour que l'impression des rayons les moins actifs tels que les rayons pénétrants non déviés, soit nettement visible.

Cette épreuve a été obtenue en employant le dispositif suivant : Dans le champ magnétique uniforme, d'un aimant permanent on dispose normalement au champ une plaque photographique, puis on applique sur celle-ci des écrans formés de lames de plomb fixées sur une lame de verre comme l'indique la figure ci-contre (Fig. 12). Ces écrans sont percés d'ouvertures en forme de fentes plus ou moins profondes, normales à la plaque, et destinées à limiter des faisceaux étroits ; sur le trajet de ces faisceaux on peut disposer par le même procédé des écrans, tels que des lames d'aluminium. Au-dessous de la plaque, en regard d'une fente étroite pratiquée dans une lame de plomb, on place un petit bloc de plomb contenant une rainure profonde normale à la plaque, et dans laquelle on met la matière radiante. On a ainsi une source linéaire étroite, normale à la plaque, et de quelques millimètres de longueur. La rainure est recouverte d'une lame mince d'aluminium pour arrêter les rayons lumineux.

L'impression représente une section faite normalement au champ, du faisceau dont une partie est déviée. Chaque faisceau, correspondant à une vitesse déterminée, donne une impression sensiblement circulaire comme si le trajectoire entière était marquée sur la plaque. Dans ces épreuves, l'intérieur des cylindres crénelés formant les écrans est très fortement impressionné par l'émission secondaire du plomb. La première épreuve (Fig. 13) montre que par chaque ouverture il passe une infinité de rayons constituant des portions de spectres purs. Ceux-ci rencontrent une lame d'aluminium de 0,1 mm. d'épaisseur, et la traversent sans déviation, mais inégalement bien. Les rayons peu déviés sont pénétrants, et excitent des rayons secondaires à leur sortie de l'aluminium.   Les rayons très déviables sont arrêtés, et font naître aux points frappés un rayonnement secondaire intense.   Une seule des deux catégories de rayons non déviables apparaît sous la forme de deux lignes fines à l'opposé de la source ; ce sont les rayons très pénétrants.   Les autres avaient été arrêtés près de la source.

La deuxième épreuve (Fig. 14) montre des faisceaux simples obtenus par une double série d'ouvertures.   Par l'une d'elles il peut parfois passer deux trajectoires distinctes.   Cette épreuve montre la transmission de rayons simples au travers de 'l'aluminium, et les effets secondaires qu'ils provoquent.

Le même procédé a permis de constater que les rayons secondaires étaient eux-mêmes déviables par le champ magnétique, dans le même sens que les rayons excitateurs.

Le rayonnement du radium comprend une partie très pénétrante formée des rayons les moins déviables et de rayons non déviables dont les propriétés semblent les mêmes que celles des rayons de Röntgen.   Ces rayons pénétrants ne sont que très peu absorbés, et par suite leur action sur une plaque photographique ou sur l'air est très faible, de sorte que l'on ne peut, par les méthodes précédentes, avoir une idée exacte de leur intensité.   Si l'on interpose sur leur trajet un écran très absorbant, ils le traversent partiellement, mais ils s'y transforment en partie en rayons plus absorbables.   Cette transformation rappelle celle de la fluorescence, et par suite de l'action secondaire, l'effet, immédiatement derrière l'écran, est plus fort que si celui-ci n'existait pas.   Une plaque photographique recevant le rayonnement filtré au travers d'une épaisseur de plomb de 1 cm. s'impressionne plus sous une plaque de plomb de 1 mm. d'épaisseur que dans les régions non recouverte par cet écran.   L'épreuve ci-contre (Fig. 15) montre l'effet du rayonnement qui sort des parois d'une cuve de plomb après avoir traversé 5 à 12 millimètres de métal.

Ces phénomènes secondaires peuvent rendre compte en partie, des apparences d'ombres portées que donnent les bords de tous les écrans plus ou moins transparents placés sur les plaques photographiques.

Tous les faits qui viennent d'être exposés sont exclusivement relatifs au rayonnement obscur qui traverse les corps opaques, les métaux, le verre, le mica ; mais il existe un phénomène tout différent, dont les effets sont arrêtés par le verre et par le mica, et sont comparables à ceux que produirait une vapeur d'une nature particulière. Ce phénomène fut découvert en 1899, simultanément par Mr. Rutherford et par M. et Mme. Curie.

Mr. Rutherford, en étudiant le rayonnement du thorium, vit qu'à côté de la radiation ordinaire, il y avait un effet produit par une " émanation," comparable à une sorte de vapeur ionisant l'air.   Cette émanation se dépose sur les corps, principalement sur les corps électrisés négativement, et les rend momentanément radio-actifs. Mr. Rutherford fit sur ce phénomène de très interessantes mesures.

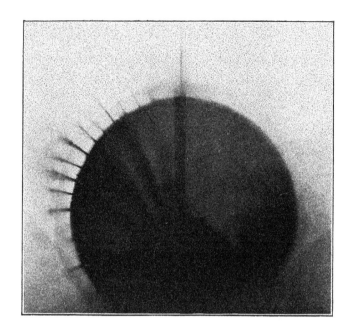

FIG. 14.

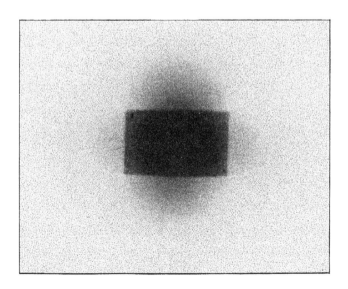

FIG. 15.

ʹEn même temps M. et Mme. Curie découvraient que sous l'influence du radium les corps devenaient temporairement radio-actifs. Ce n'est pas l'effet secondaire décrit plus haut, c'est un phénomène persistant qui disparaît assez lentement à partir du moment où l'action du radium a cessé. M. Curie a appelé ce phé-nomène " la radio-activité induite," et il en fit un étude très complète ; il a reconnu que le phénomène se produit avec une grande intensité dans un espace clos, que l'activité induite est la même sur tous les corps, et indépendante de la pression à l'intérieur de l'enceinte, mais que l'activation ne se produit pas si on maintient constamment le vide en enlevant les gaz produits ; les solutions des sels de radium pro-duisent le phénomène avec plus d'intensité que les sels solides. Les liquides, l'eau de cristallisation extraite des sels actifs, ou l'eau séparée d'une solution active par une membrane semi-perméable en celluloïde, deviennent fortement radio-actifs. Il en est de même des gaz. Ces corps activés produisent les mêmes effets que le radium ; ils émettent un rayonnement pénétrant qui traverse les enveloppes de verre qui les contiennent, et rendent celles-ci lumineuses. L'activité induite se propage de proche en proche dans le gaz d'une enceinte fermée, même au travers de tubes capillaires et de fissures imper-ceptibles ; les corps s'activent d'autant plus que le volume de gaz qui est en regard de leur surface est plus considérable. Les corps phos-phorescents deviennent lumineux en s'activant.

Dans un récent travail, MM. Elster et Geitel ont reconnu que l'air atmosphérique présentait des propriétés analogues à celles des gaz activés, et M. Geitel a pu recueillir sur des fils électrisés néga-tivement, des traces de produits radio-actifs. La cause de cette radio-activité est un problème d'un haut intérêt.

Enfin, il est un mode remarquable d'induction qui est de nature à inspirer les plus grandes réserves dans les conclusions que l'on peut formuler relativement à la présence d'éléments nouveaux dans les préparations radio-actives. Toute substance inactive que l'on introduit dans une dissolution d'un sel d'uranium ou de radium, et que l'on sépare ensuite par précipitation, est devenue radio-active, et perd lentement cette radio-activité. Ce fait a d'abord été observé par M. Curie et par M. Giesel qui a ainsi activé du bismuth. Avec l'uranium, une trace de baryum précipitée à l'état de sulfate devient notablement plus active que l'uranium ; le baryum ainsi activé n'émet, comme l'uranium, que des rayons déviables.

Après cette précipitation, le sel d'uranium ramené à l'état solide est moins actif qu'auparavant ; on peut même accentuer cet affaiblis-sement par des opérations successives, mais les produits reprennent, peu à peu, spontanément, leur activité première.

L'affaiblissement temporaire de l'activité à la suite d'une dis-solution est un fait général pour les sels d'uranium et de radium.

Avec les sels d'actinium M. Debierne a communiqué au baryum une activité très grande. Le baryum activé peut se séparer du baryum inactif ; il se fractionne comme le chlorure de baryum radifère, les parties les plus actives étant les moins solubles dans l'eau et l'acide chlorhydrique. M. Debierne a obtenu ainsi un produit mille fois plus actif que l'uranium. Le baryum activé se comporte donc comme un faux radium, mais il diffère du radium véritable par l'absence de spectre et parce qu'il perd son activité avec le temps.

Parmi les préparations radio-actives, un grand nombre pourraient être des corps activés. Tel est le cas du polonium qui est vraisemblablement du bismuth activé.

L'uranium et le radium sont caractérisés par leur spectres d'émission, et par la stabilité de leur radio-activité. L'accroissement spontané que l'on observe sur les sels déposés des dissolutions, pourrait trouver une explication dans un phénomène d'auto-induction des molécules actives sur les molécules inactives qui leur sont associées.

L'origine de l'énergie rayonnée par les corps radio-actifs est toujours un énigme. Dans l'hypothèse matérielle, il ne paraît pas invraisemblable d'assimiler le phénomène à l'évaporation d'un corps odorant, de comparer l'émanation à une sorte de gaz dont les molécules auraient des masses de l'ordre de grandeur de celles des ions électrolytiques, et d'identifier le rayonnement à des rayons cathodiques provenant de la dislocation de ces ions et donnant en même temps une émission de rayons X. On imputerait ainsi la dépense d'énergie à la dissipation de la matière active. Bien que cette hypothèse rende à peu près compte de la plupart des faits, il n'existe aucune expérience précise qui lui donne une sanction.

Je ne puis m'étendre plus longuement sur ce sujet dont j'ai résumé très incomplètement l'état actuel, en insistant sur la partie physique qui est plus spécialement de mon domaine, et bien que la partie chimique ait donné lieu à des travaux du plus haut intérêt.

Ces questions ont fait naître des aperçus nouveaux sur les transformations de la matière. En dehors des conditions exceptionelles dans lesquelles elles permettent d'étudier les rayons cathodiques, elles ont soulevé et soulèvent chaque jour des problèmes nouveaux, dont le premier et le plus mystérieux est la spontanéité du rayonnement.

[H. B.]

WEEKLY EVENING MEETING,

Friday, March 21, 1902.

George Matthey, Esq. F.R.S. Vice-President, in the Chair.

Geheimrath Professor Otto N. Witt, Ph.D. F.C.S., of Berlin.

Recent Developments in Colouring-Matters.

The love of colour is innate in the human mind, and this alone, if nothing else, would be sufficient to account for the interest with which the coal-tar colour-industry has met from its beginning. In England especially its progress has been watched with great attention, and only two days ago the Vice-Patron of this Institution, His Royal Highness the Prince of Wales, has shown by some remarks, made in his Opening Address of the New Technical Institute at Bushey, that the interest taken in this subject has in no way abated.

Artificial colouring-matters have formed so often the subject of more or less popular lectures, and this subject has been treated with such ability by eminent scientists, that it becomes difficult to show this domain of chemistry in a light new and interesting to an audience such as I have to-day the honour to address. Many years ago you have seen in this room the early achievements of the newly-created industry, marvellous for their beauty and brilliancy. Later on the progress of this industry has been duly recorded. More recently still it has become the custom in this country to view colour-making not so much from its chemical or industrial side, as from the standpoint of the national economist, who contemplates the values produced by industrial enterprise, and investigates the reasons why these values should be unevenly distributed amongst the different nations, striving side by side for progress and engaged in friendly, yet none the less eager competition.

I may say at once, that I have no intention to treat my subject from either of these points of view. I take it for granted, that everybody is acquainted with the marvellous variety and brilliancy of artificial dye-stuffs, and I am too much of a chemist and too little of an economist to offer any original or valuable view about that side of the question which I have just mentioned. But I shall make an attempt to trace in this lecture the influence of the development of theoretical chemistry on the progress of the colour industry. If in so doing, I should refer now and then to theoretical points without being able to explain them in detail, I hope to be forgiven.

In beginning this lecture allow me briefly to refer to the history of it.

When I received from Sir William Crookes the flattering invitation to speak before you this evening, my thoughts naturally wandered back to some recollections in connection with this Institution. I remembered vividly several brilliant lectures to which I had the privilege of listening in this room, where the spirits of Davy and Faraday, of Graham and Huxley, of Würtz and of my immortal friend A. W. von Hofmann seem still to be hovering. I felt loth to raise my own voice in such hallowed precincts. But then I also remembered an almost forgotten episode in my own life, which I ask your permission to tell.

I remembered, that almost exactly five-and-twenty years before receiving this invitation, I, then a very young chemist, had read before the Chemical Society of London, a paper containing a then somewhat daring speculation on the connection of the constitution of colouring-matters with their properties, a paper which the Publication Committee refused to print. A lively discussion followed, which was wound up by some encouraging remarks from the president, the late Mr. De la Rue. He said, that he hoped, that this speculative paper would prove useful in clearing up the complicated domain of colouring-matters, and that perhaps on some future occasion I should be in a position to place before the world, in a *Royal Institution lecture*, the results which had been obtained by its help.

This strange reminiscence, coupled with the curious fact that Mr. De la Rue's prophetic words were fulfilled just when the period commonly assigned to a Jubilee had elapsed, gave me the courage to accept Sir William's kind invitation. For though I have done comparatively little towards the increase of our knowledge of colouring-matters, the five-and-twenty years past have sufficed to shed a brilliant light on what Mr. De la Rue could then justly call a very imperfectly known domain of chemistry, and innumerable facts brought to light during this period by a whole army of assiduous workers are now by common assent being classified under a theory which is neither more nor less than the suggestions contained in that rejected paper of mine, which I had fortunately published in another journal.

I may add, that I have been guilty in later times of another theory, which refers to the domain of dyeing, and which has still many opponents. This theory is the direct outcome of the theory of colouring-matters, and may be illustrated by some simple, yet striking experiments, some of which I intend to show you.

A fundamental question in the chemistry of dye-stuffs, and one not at all easy to answer, is this: "What is a dye-stuff?" Clearly it is something totally different from a substance only endowed with the power of selective absorption of light, a power which causes it to appear coloured. We know now that there are more substances in creation which possess this power than bodies which lack it. In this very room we have learned, that the air itself, through which the solar rays penetrate on to the surface of the earth, is blue and not colourless, as we used to think. But even if we leave out of the question such faintly coloured substances as air and water, if we restrict our consideration to compounds endowed with a very intense power of selective absorption and at the same time soluble in the water which we employ for preparing our dye-baths, we do not arrive yet at the true definition of the dye-stuff. Cupric salts, soluble chromates and many other intensely-coloured bodies are no dye-stuffs, as may be easily shown by experiment. Yet these compounds penetrate into the interior of textile fibres which are immersed into their solution. They must do so, according to the laws of Osmose so ably expounded by Thomas Graham, because they are crystalloids and the fibres are invariably colloids.

We know now that the laws of Osmose are identical with the laws governing solution, and that crystalloids are able to wander into the interior of colloids because they are soluble in their substance. Osmotic processes may be observed between two liquids which cannot be mixed with each other, just as well as between a liquid and a colloid immersed into it. Consequently we are justified in assuming that the same powers are at work in both cases.

In my first experiment (Exp. I.) * I intend to show you that a crystalloid, dissolved in some liquid such as water, and brought into contact with another liquid, not miscible with the first, such as ether, will either remain indifferent to the ether altogether, or it will leave the water and wander into the ether, or it will be distributed according to a certain ratio between the two solvents. In this latter case we have reason to believe that a constant interchange of molecules takes place between the two solutions. Clearly, this will only happen if there exists no great difference in the solubilities of the crystalloid in water and in ether. In that case the water will continually abstract nearly as many molecules of the crystalloid from the ether as the latter will take up from the water, and thus an equilibrium will be reached. If, on the other side, there is a great dissimilarity in the solubility of the crystalloid in the two solvents, then this process of mutual interchange will become so one-sided that it practically amounts to the absorption of the whole of the crystalloid by one of the solvents.

My second experiment (Exp. II.) † is a more striking illustration of these fundamental facts. If we mix together two coloured solutions, one an aqueous one of a substance much more soluble in ether than in water, and the other an ethereal one of a substance more soluble in water than in ether, then the two solutions, on shaking, change colour, and their shades are reversed.

---

* Details of experiment: An aqueous solution of magenta does not yield its colouring-matter to ether; indophenol, on the contrary, is entirely taken up by ether. The dye-stuff, which is partly taken up by ether, is also a member of the indophenol group, the constitution of which is not yet fully established.

† The ethereal solution used contained magenta acetate, whilst the aqueous one was prepared with trichloro-indophenol.

According to my theory, the process of dyeing, considered so problematical by many experts in this ancient and useful art, is strictly analogous with this wandering of molecules governed by the laws of solution, which we can so easily observe and control in operating with two non-miscible liquid solvents.

In my next experiment (Exp. III.) * we see that a dye-stuff wanders from the bath on to the fibre in much the same way as it wandered from water into ether. And if the fibre be previously dyed with a colouring matter little soluble in its substance, then this may be expelled and replaced by another of greater solubility. (Exp. IV.) *

We see now that, in order to become a dye-stuff, a substance must not only be so intensely coloured that it can communicate its own shade to colourless substances holding it in solution; it must not only be soluble in water or any other liquid suitable for preparing a dye-bath; but it must also be soluble, and even much more soluble than in water, in the colloid, which forms the substance of the textile fibre. The finished dyed fabric is nothing more nor less than a solid solution of the dye-stuff in the substance of the fibre, unless there are secondary chemical influences, such as that of the mordants, at work, which change the solution into a suspension by precipitating the dye-stuff after its immigration into the fibre.

This peculiar combination of solubilities is very rarely met with amongst the coloured substances of an anorganic nature. In the vast domain of organic compounds of the aliphatic series we meet with very few dye-stuffs, because its members are mostly colourless, or but very faintly coloured. In the aromatic series, on the contrary, the power of selective absorption of light is so very frequent, that it would be very curious indeed if just that combination of solubilities, which is the making of the dye-stuff, were not of common occurrence. Taking as a basis the universally admitted axiom, that the physical properties of every compound are direct functions of its molecular constitution, we may easily believe that that peculiar combination of solubilities which I have shown to be the characteristic feature of the dye-stuff, would be the result of certain general conditions fulfilled in the constitution of many members of the aromatic group. My theory, proposed five-and-twenty years ago, was nothing else than an attempt to ascertain these general conditions by investigating the constitutional peculiarities of all those dye-stuffs the constitution of which had been fully established in those days.

I have no intention to tax your patience by explaining in detail the results of that old investigation. It will be sufficient to summarise them by saying that in the molecule of every colouring matter, the constitution of which has been ascertained to this day

---

* In Exp. III. wool was dyed with erythrosine in the ordinary way, whilst in Exp IV. a cotton cloth, previously dyed with patent blue, was treated in a bath of Congo red.

(and there are many thousands of them), certain atomic constellations have been observed which seem to be essential, and of which always two must be present. One of these constellations is a group of atoms, which is the cause of the selective absorption of light. This group of atoms I call a *chromophore*. The number of atomic groups endowed with chromophoric properties amounts at present to about two dozen, and is being constantly increased by the progress of chemical research. All the chromophores, however, have that in common, that they are unable to exert their influence unless they are helped by the presence of another group of atoms, which I call the *auxochromic* group. Very few auxochromic groups are known, and they belong to those which occur most frequently in the whole domain of organic chemistry—the amino group in its various forms, the hydroxyl group occurring in all the phenols, the sulpho- and the carboxyl group. None of these will cause a substance to become a dye-stuff unless this substance also contain a chromophore, but the latter is equally helpless if deprived of the assistance of the auxochromic group. Thus we meet in the molecular world that condition of the necessity of mutual help and assistance between two heterogeneous forms, which we can also trace in Sociology, a fact the establishment of which will no doubt be greeted with satisfaction by the ladies in this audience.

Our ideas on the nature and constitution of those groups which may act as chromophores have of course undergone many changes. Undoubtedly there must exist a law which governs the formation of chromophoric groups, but so far this law has not been definitely established. Some progress has, however, been made towards this end. At first the chromophores which we had gradually collected formed rather a motley crowd, and seemed to have no points in common. At present chemists working in this domain are inclined to attribute a quinoid structure to the great majority of colouring-matters. If this view be correct, then all these substances would be derivatives, not of benzene and its congeners, but of hydrocarbons containing two hydrogen atoms more in their molecule, derived from dihydrobenzene as a prototype. As sometimes it is almost impossible to decide in favour of one view or the other, the convenient hypothesis of tautomerism was resorted to, but in some cases we have been able to establish definitely the quinoid formula. Such is the case with the large and brilliantly-coloured group of dye-stuffs called phtaleines, which, according to modern view, must be considered as quinoid derivatives of benzoylbenzoic acid. The experiments which lead to this conclusion are so striking, that I cannot refrain from producing one of them, which has never been shown yet, though the time at my disposal does not allow its exhaustive discussion from a theoretical point of view. If we dissolve the well-known phenolphtaleine in anhydrous ether containing some ammonia, the solution is perfectly colourless, but if we add ordinary water to this solution (Exp. V.), it assumes a beautiful red coloration. This peculiar fact that water alone is sufficient to cause the formation of this colour is perfectly incomprehensible if the old views on the constitution of phtaleines, which are still given in the majority of text-books, be adhered to, but it is exactly what we might expect to happen if we assume that the ammonium salt of phenolphtaleine possesses a cycloid constitution in its ethereal solution, and that it is isomerised into the quinonoid form by the addition of water.*

Thus our knowledge of the chemical causes of the physical properties of colouring-matters is continuously developing. Quite lately we have even begun to form definite views about the connection of the chemical constitution of aromatic substances with that peculiar form of selective absorption of light which we call fluorescence, and which has formed, from the physical point of view, the subject of the masterly investigations of Sir Gabriel Stokes. The phenomenon of fluorescence is very frequently met with in dye-stuffs, and in the raw materials used for their manufacture. It can be exhibited in a very striking way with the help of electricity, either by allowing an arrow of electric light to penetrate into the solution of a fluorescent substance or by working a Geissler tube of suitable shape submerged in such a solution (Exp. VI.). The fact that the fluorescence of many substances is chiefly caused by the ultra-violet light, I shall try to demonstrate by the following, somewhat delicate, experiment: I have here, submerged in a solution of eosine, a Geissler tube, the lower part of which is ground out of a piece of rock-crystal, whilst the upper half is made of glass. When the electric current passes this tube the fluorescence round the quartz part of it is stronger than that in the neighbourhood of the glass, because the latter absorbs a good deal of ultra-violet light, whilst the quartz is almost free from such absorption. (Exp. VII.)

An immense amount of patient work has been accomplished by many chemists in the hope of establishing definite views on the constitution of the azo-colours, that group of dye-stuffs the introduction of which into the colour-industry was the direct consequence of our early efforts to cast off empiricism, and to conduct our search for new colouring-matters according to definite scientific principles. Simple and transparent as the constitution of azo-colours appears to be if viewed superficially, yet it offers some problems of extraordinary

---

* The isomerism of the two forms of the ammonium salt of phenolphtaleine is best explained by their constitutional formulæ:

Cycloid Form.   Quinonoid Form.

difficulty, which have not been solved so far. But fortunately these difficulties have in no way interfered with the technical development of this family of dye-stuffs, which has been for a whole quarter of a century one continued and unparalleled series of successes. The process for producing these dye-stuffs is of the greatest simplicity. It consists in pouring together (Exp. VIII.) cold aqueous solutions or suspensions of diazo-compounds and phenols or amines. The dye-stuff is formed at once in a state of absolute purity, and with a yield absolutely theoretical; it need only be collected and dried to form a saleable product. No wonder, then, that these dye-stuffs gradually became the leading ones, and to a great extent superseded the old empirical products which were concocted in many compli- cated operations, with yields very far from satisfactory. As the number of diazo-compounds and of phenols and amines at our dis- posal is very large, the number of dye-stuffs which may thus be prepared is quite extraordinary; it has been computed, according to the rules of permutation. 3,159,000 different individual dye-stuffs have thus been proved to be at present easily accessible to our in- dustry. Of these at least 25,000 form the subject of German patent specifications and of corresponding specifications in England, France, the United States, and other countries. Over five hundred are regularly manufactured on the larger scale.

The prolific nature of the azo-colour-reaction explains the fact, that in this group we can choose, much better than in any other, substances possessing that ratio of solubilities in water and in the colloid substance of the various textile fibres, which we require. We can produce, quite at will, azo-dye-stuffs which dye wool or silk or cotton, which dye slowly or quickly, which will stand soap or acid or alkali. This possibility of adjusting the chemical properties of dye-stuffs with an almost mechanical precision has been the cause of one of the greatest successes of the colour-industry, the introduction of what is now known under the name of "substantive dye-stuffs," an expression which means dye-stuffs that will dye cotton and other vegetable fibres from a simple aqueous dye-bath without the use of any mordant. The difference of the solvent power of cellulose and of water is for the vast majority of dye-stuffs so small, that the pro- cess of dyeing vegetable fibres with these ordinary colouring-matters can only be compared to that case of the joint action of ether and water on some substance soluble in both these solvents, where an almost equal division of this substance takes place between the two solvents. Such cases exist, as you saw in the first experiment. It is amongst the azo-dyes that we have found compounds which are so much more soluble in cellulose than in water, that they readily leave their aqueous solution and take up their abode in the fibre. And we have not only found these dye-stuffs but also the law which governs this most valuable abnormal solubility: it appears in all azo-colours, which are prepared with diazo-compounds derived from symmetrical para-diamines. A novel and extremely fertile field for a systematic search for new dye-stuffs was thus opened, a field which has occupied hundreds of busy workers for many years, many of whom carried home a rich reward.

But whilst this field bore its rich harvest, others were by no means neglected. The search for dye-stuffs, which will dye cotton without a mordant, could not make us forget that just those colouring-matters which imperatively demand the use of mordants are those which from times immemorial have been used in preference for the production of fast and lasting shades. The brilliant synthesis of alizarine by Graebe and Liebermann, which made the world ring with admiration early in the seventies, had given us ample proof that the old and to this day not wholly forgotten axiom, that there are two kinds of dyes: natural ones, which are fast, and artificial ones, which are fugitive, was a preconceived idea, totally devoid of any scientific foundation. The enormous financial success of the alizarine industry formed a tempting invitation to search for other dye-stuffs, which, similar to alizarine, would be endowed with the power of forming almost indestructible lakes with mordants of a sesquioxydic nature. Here too, like everywhere in science, we have marched for some time on the paths of empiricism, but here too logical deduction has come to our aid in disclosing the laws which govern the formation of lakes. In this case it is not (as in the substantive azo-dyes) the carbonic nucleus which determines the *physical* properties (viz. the ratio of solubilities) of the dye-stuff, but it is the peculiar position of the auxochromic groups contained in the molecule, which governs its chemical properties. We know now, that a dye-stuff must contain, in order to be able to form lakes with sesquioxydic mordants, two hydroxyl groups in juxtaposition. If this condition be fulfilled, the dye-stuff will dye in the same way and with equal fastness as alizarine, even if it be no derivative of anthracene, like the early alizarine dyes; and if these two hydroxyl groups or a suitable equivalent for them be missing it will lack all power of dyeing mordants, even though derived from anthracene. With this law once established the synthesis of mordant-dyestuffs became a very easy matter, and to-day there is hardly a group of colouring-matters in which there are not some members possessed of this peculiarity and owing it to the same uniform cause. Still the group of the oxyketones, to which alizarine itself belongs, remains the true home of mordant-dyes, but this group has grown to-day into a very numerous and varied one. Mordant-dyes of every shade are to be found in it, and cotton is no longer the only fibre to which such dyes are applied. It is a fact worthy of notice, that amongst the many dyes of this class which we now possess and the constitution of which is fully established there is not a small number, the molecule of which contains three, four, five or even six hydroxyl groups. Yet this increase of auxochromic groups does not influence in the least the behaviour of these dyes to mordants, this is only governed by the two hydroxyl groups in ortho-position, and any other such group introduced into the molecule only changes the shade, not the characteristic chemical properties.

A greater variety still than by the achievements of modern synthetical work will come into this group of mordant-dyes by the progress of the elucidation of the constitution of the natural dye-stuffs occurring in roots, barks and woods. A good many of them are still unsolved mysteries, but there can be no doubt that they owe, like alizarine, purpurine and the other madder dye-stuffs, their property of dyeing metallic mordants to the presence of hydroxyl groups in ortho-position in their molecule.

A very large and varied group of colouring-matters, which for a long time resisted all attempts at unravelling their constitution, are the Saffranines, Eurhodines, Oxazines, Thionines, Indulines and other allied groups. They are now completely understood, and have been recognised as the amino- and oxy-derivatives of certain peculiar substances such as the azines and azonium-bases, the molecule of which possesses a ring-structure. Here no longer carbon atoms only form the closed chain, but nitrogen, oxygen and even sulphur atoms participate in its structure and bring about the peculiar properties of the compounds. When this fact was at first ascertained it seemed sufficient for the explanation of the behaviour of such compounds as dyes. It was only somewhat later on that we recognised that in these classes of dye-stuffs especially a quinonoid structure is essential.

The greatest and most brilliant success of the chemistry of dye-stuffs is however the industrial synthesis of indigo. This offers so many points of general interest, that I am sure to meet with your approval if I refer to it in some detail.

The indigo problem is one of the oldest problems of chemistry. When Baeyer took it up more than thirty years ago he found the ground well prepared by others who had worked before him. But his is the merit of having completely elucidated the constitution of this extrordinary product of nature. He and others have also shown various methods for the synthesis or artificial production of indigo. In the laboratories artificial indigo has been known for the last twenty years.

But in this case the scientific synthesis of a natural product proved to be by no means identical with the industrial one. Industrial methods can only enter into competition with nature if they work more economically than nature does. In the case of indigo there seemed to be little hope for fulfilling this condition. The most enthusiastic admirers of the modern synthetical industry could not help seeing that all evidence in our hands went against the probability of the practical synthesis of indigo, and just those who understood most of these things could least of all close their eyes to that fact. It could not be denied that every possible synthesis of indigo, those known as well as those which might still be expected, had to start from some aromatic derivative of benzene, containing one carbonic and one nitrogenous side-chain in ortho-position. Of all the products at our disposal which fulfil that condition, ortho-nitrotoluene is the most easily accessible. Now, taking it for granted that indigo could be prepared regularly and with good yields from orthonitrotoluene, there still remained that difficulty, that all the toluene produced in the world, even if we suppose that all the other uses to which this hydrocarbon is put at present could be suppressed, would not suffice for the production of the world's consumption of indigo.

If, under such circumstances, the industry of artificial dye-stuffs continued to work at the indigo problem, it did so more for the general interest attached to it, and with a view to securing some of the finer applications of indigo in printing, than in the hope of being able to compete with the natural product in the great consumption of vat dyeing. If, on the other hand, the indigo planters in the far East showed but small apprehension of the danger of which they were occasionally warned, we cannot blame them for it; they had, no doubt, taken the advice of competent people, and these had told them what was correct according to the knowledge of the time.

The final result has shown all the calculations of experts to be wrong, but in such a way that they, too, can surely not be blamed for the error they committed.

The process by which indigo is at present manufactured on a colossal scale by the Badische Anilin- und Soda- Fabrik in Ludwigs-hafen on the Rhine, is based on Heumann's synthesis of this most important dye-stuff, which consists in submitting phenylglycine to a fusion with caustic alkali. Phenylglycine is prepared by the action of monochloracetic acid upon aniline. The yield of indigo obtained is a poor one, but it can be very much improved if, instead of phenyl-glycine, we take its orthocarbonic acid. In this we have again the presence of a nitrogenous and a carbonic side-chain in ortho-position. To prepare this acid we should have to start, according to the ordinary rules, from toluene, transforming it by a succession of operations. Thus we come again to toluene as a starting-point, and to the difficulty already explained.

There is, however, one somewhat abnormal process of preparing the same compound from phtalic acid. It consists in converting this into phtalimide, and treating the latter with sodium hypo-chlorite. By a somewhat complicated reaction, the nature of which need not be explained, one of the carboxyl groups of the phtalic acid is replaced by the amido group, anthranilic acid is formed, and this, if treated with monochloracetic acid, yields phenylglycine-carbonic acid, which has proved so important for the manufacture of indigo. Now phtalic acid is prepared by a powerful oxidation of naphtha-lene, and naphthalene again is that constituent of coal-tar which is present in by far the largest quantity.

It is true that the process for transforming naphthalene into phtalic acid, which was the only one known at the time when all these facts were first recognised, gave very bad yields, and was at the same time costly. The whole indigo problem stood thus reduced to the problem of transforming naphthalene cheaply and economically into phtalic acid. This has been accomplished by the Badische Anilin- und Soda-Fabrik by heating naphthalene with fuming sulphuric acid in the presence of mercury salts. Torrents of sulphur dioxide escape, and the whole process can only be carried out properly if the means be given to convert this gas again into fuming sulphuric acid, which may be used again for treating fresh quantities of the hydrocarbon. The new sulphuric acid process of the Badische Anilin- und Soda-Fabrik has thus been of paramount importance for the working out of the indigo problem.

Some of the older synthetical methods of producing indigo are so easy and rapid that they can easily be shown as a lecture experiment. If, for instance, we add a caustic potash solution to a solution of orthonitrobenzoic aldehyde in acetone, indigo is formed at once and settles out in dark-blue crystalline flakes (Exp. IX.). The synthesis now in practical use is a little more delicate in its execution, but there are certain modifications of it which are rapid enough to be shown in a lecture experiment (Exp. X.).

The action of the alkali on the phenylglycine-carbonic acid does not at once produce indigo, a colourless derivative of the dye, indoxylcarbonic acid, or rather its potash salt, is formed at first, but if we dissolve this in hot water and introduce a current of air, it is at once and with a quantitative yield transformed into indigo which settles out in the shape of a crystalline deposit of infinitely fine division. This is collected in filter-presses and delivered into commerce in the shape of a paste or a powder.

The industrial synthesis of indigo is extremely interesting, because it is a triumph not wholly due to chemical science. Science has shown the way to success, but it was quite unable to clear away the difficulties arising from practical and economic considerations. Here the representatives of our great industry had to advance independently and on paths for which theoretical knowledge could not serve them as a guide. Unlimited praise and admiration is certainly due to them for the masterly way in which they grappled with colossal difficulties and for the courage with which they staked millions on the realisation of one great idea.

At the same time we cannot help feeling some regret for the indigo planters in the far East, who, after enjoying more than a century of easy prosperity, see now that more serious times are in store for them. They see the day coming when the indigo plantations will disappear, in the same way in which the madder fields of Avignon have vanished. But we are consoled by the knowledge that, especially for India, the time has already come, which has been so vividly described by Sir William Crookes, in one of his addresses to the British Association, as the future in store, sooner or later, for all humanity, the time when bread begins to be scarce. It seems to me that any one who, by bringing about some great commercial revolution such as we have seen in the indigo trade, causes land in India to become free for the growing of rice and other cereals, renders a great service to large numbers of poor natives, and need therefore not be blamed for lessening to some extent the prosperity of a class of people who have had an unusually good opportunity of accumulating wealth in the past.

It is a strange fact, that with indigo the history of the fight of the madder root against artificial alizarine is almost literally repeated in spite of the great difference of original conditions in the two cases. Madder was a product containing at its best only 4 per cent. of actual colouring matter; the rest was useless fibre and obnoxious impurities which greatly hampered the dyer in his work. Alizarine, entering into competition with this natural product, was, on the contrary, the colouring matter in a pure state, and therefore not only cheaper but also much easier in its application. Indigo, such as we receive it from India and Java, is a manufactured article, the best qualities of which contain 59, 60, or even 70 per cent. of pure dye-stuff, besides impurities which have always been considered as perfectly harmless. Thus the artificial product did not seem to have much scope for improvement as far as the quality came into consideration. Here again we have committed a mistake. We know now that the impurities are not harmless, and that the blues dyed with artificial indigo are quite as superior in brightness and purity of shade to those obtained with natural indigo, as alizarine reds were to madder reds. This has, however, not always proved to be an advantage for the manufacturers of artificial indigo. The world does not ask for bright indigo shades, and a good many prejudices in that respect had to be overcome before artificial indigo was admitted as a legitimate substitute for the natural product in some of its most important applications. Yet a simple consideration will show that it is always easy to deteriorate the brilliancy of a dyed shade, whereas no art of the dyer will suffice to produce brilliant shades on textile fabrics with dye-stuffs that carry their share of dirty admixtures within them.

In its application to the fibre, indigo is perhaps the most remarkable of all dye-stuffs, for it is the principal representative of that extraordinary class of colouring-matters which must be applied by the vat process. This process, which consists in first reducing the dye-stuff into a leuco-compound before applying it to the fibre, on which the original colouring-matter is formed again by the action of the oxygen of the air, seems to have nothing in common with the ordinary dyeing processes. If, however, we consider it more closely, we come to the conclusion that vat colours are a class of dye-stuffs in which the functions of dyeing and of selective absorption of light are distributed on two different forms of the substance, one of which contains two atoms of hydrogen more in its molecule than the other.

This theory is supported to some extent by the fact that what we are pleased to call leucocompounds, are in the majority of cases by no means colourless. Indigo-white itself is not white but yellow in its alkaline solution which we call a vat. Other vat-dyes have leucocompounds which are even more strongly coloured. Thus the leucocompound of indanthrene, a beautiful new colouring-matter, is blue like indanthrene itself; flavanthrene, a yellow dye-stuff, which has not yet left the laboratory of its inventor, has a blue leucocompound. One may say, that with all vat-colours the real dye-stuff is the leucocompound which is afterwards, when once fixed on the fibre, transformed into a pigment by the oxidising influence of the air.

In 1825 Faraday discovered, in this very house, benzene; the original specimen, prepared by his own hands, is before you. We look upon it reverently, like on a sacred relic bequeathed to us by a master-mind. But what a development has sprung from this first attempt to unravel the mysteries of the aromatic series! Our science as well as our industry have been revolutionised by the investigation of the derivatives of benzene, and the world has been embellished by the gay and brilliant dyes of which it is the mother-substance. The study of the chemistry of these dye-stuffs has become a domain of science which, for variety and fascination, can hardly be surpassed by any other. The deeper we penetrate into it the more it proves an inexhaustible mine of the most subtle scientific thought, yet one which never loses touch with practical life; it is interesting alike to the philosophical mind that wishes to revel in the wonderful perfection and order of nature, and to the philanthropic spirit which rejoices in seeing many thousands of hands occupied and princely fortunes produced by the utilisation of what was only a short time ago a refuse and an encumbrance. It teaches a lesson even to those who are not attached to science and apt to consider it as a kind of pastime for people who lack ability for practical life. For they cannot help seeing that, in this case, the most intricate science has led to something eminently practical, commensurate to a standard which, though unknown to the Bureau International des Poids et Mesures, is to some people the only reliable one, viz. the one of £ *s. d.*

I am afraid that the high praise which I feel justified in bestowing on what has been the favourite pursuit of my life is not fully substantiated by the contents of this lecture. The subject which I had to treat is so vast, that all I have been able to say is nothing but a sketch or a programme of what would require a long series of lectures if full justice were to be done to it. My one excuse for attempting to sketch, in the short space of one hour, so vast a subject, is the place in which I had the honour to speak: An audience that has been addressed more than once by the pioneers of the chemistry of dye-stuffs, by Faraday, Hofmann, William Perkin, and others, could, from one of the Epigones, not have looked for more, than a few notes and additions.

[O. N. W.]

GENERAL MONTHLY MEETING

Monday, April 7, 1902.

Sir James Crichton-Browne, M.D. LL.D. F.R.S. Treasurer and
Vice-President, in the Chair.

Walter Baily, Esq. M.A.
Mark Barr, Esq.
Cecil West Darley, Esq. M. Inst. C.E.
W. E. L. Gaine, Esq.
Carl Hentschel, Esq.
James Charles Inglis, Esq. M. Inst. C.E.
George Northcroft, Esq.
H. C. Plimmer, Esq. F.L.S. F.R.M.S.
F. E. Robertson, Esq. C.I.E. M. Inst. C.E.
George O. Wilson, Esq. M.A.

were elected Members of the Royal Institution.

The Special Thanks of the Members were returned to Dr. Francis
Elgar, F.R S. for a Donation of £50 to the Fund for the Promotion
of Experimental Research at Low Temperatures.

The Presents received since the last Meeting were laid on the
table, and the thanks of the Members returned for the same, viz :—

FROM

The Lords of the Admiralty—Nautical Almanac for 1905. 8vo.
American Academy of Arts and Sciences—Proceedings, Vol. XXXVII. Parts 6-10.
8vo. 1901.
American Geographical Society—Bulletin, Vol. XXXIV. No. 1. 8vo. 1902.
Accademia dei Lincei, Reale, Roma—Classe di Scienze Fisiche, Matematiche e
Naturali. Atti, Serie Quinta : Rendiconti. 1º Semestre, Vol. XI. Fasc. 4, 5.
8vo.
Classe di Scienze Morali, Storiche, Serie Quinta, Vol. X. Fasc. 11, 12. 8vo.
Astronomical Society, Royal—Monthly Notices, Vol. LXII. No. 3. 8vo. 1901.
Berlin Chemical Society—August Wilhelm von Hofman. Ein Lebensbild. 8vo.
1902.
Boston Public Library—Monthly Bulletin for March, 1902. 8vo.
Bristol Museum—Report of the Museum Committee, 1901. 8vo. 1902.
British Architects, Royal Institute of—Journal, 3rd Series, Vol. IX. No. 10. 4to.
1902.
British Astronomical Association—Journal, Vol. XII. No. 5. 8vo. 1901.
British South Africa Co.—Southern Rhodesia : Information for Intending Settlers.
8vo. 1902.
Buenos Ayres, City—Monthly Bulletin of Municipal Statistics, Jan. 1902. 4to.
Camera Club—Journal for March, 1902. 8vo.

Chemical Industry, Society of—Journal, Vol. XXI. Nos. 4–6. 8vo. 1902.
Chemical Society—Proceedings, Nos. 248, 249. 8vo. 1902.
  Journal for April, 1902. 8vo.
Cracovie, l'Académie des Sciences—Bulletin, Classes des Sciences Mathématiques
  et Naturelles, 1901, No. 9; 1902, No. 1. 8vo.
  Bulletin, Classe de Philologie, 1901, No. 10; 1902, No. 1. 8vo.
Crawford and Balcarres, The Earl of, K.T. M.P. F.R.S. M.R.I.—Bibliotheca
  Lindesiana: Collections and Notes, No. 6.
Dax—Société de Borda—Bulletin, 1901, Part 3. 8vo.
Editors—American Journal of Science for March, 1902. 8vo.
  Anthony's Photographic Bulletin for March, 1902. 8vo.
  Astrophysical Journal for March, 1902.
  Athenæum for March, 1902. 4to.
  Author for March, 1902. 8vo.
  Brewers' Journal for March, 1902. 8vo.
  Chemical News for March, 1902. 4to.
  Chemist and Druggist for March, 1902. 8vo.
  Electrical Engineer for March, 1902. fol.
  Electrical Review for March, 1902. 8vo.
  Electrical Times for March, 1902. 4to.
  Electricity for March, 1902. 8vo.
  Engineer for March, 1902. fol.
  Engineering for March, 1902. fol.
  Homœopathic Review for March–April, 1902. 8vo.
  Horological Journal for April, 1902. 8vo.
  Invention for March, 1902.
  Journal of the British Dental Association for March, 1902. 8vo.
  Journal of Medical Research for March, 1902. 8vo.
  Journal of State Medicine for March, 1902. 8vo.
  Law Journal for March, 1902. 8vo.
  London Technical Education Gazette for March, 1902.
  Machinery Market for March, 1902. 8vo.
  Motor Car Journal for March, 1902. 8vo.
  Nature for March, 1902. 4to.
  New Church Magazine for March, 1902. 8vo.
  Nuovo Cimento for March, 1902. 8vo.
  Pharmaceutical Journal for March, 1902. 8vo.
  Photographic News for March, 1902. 8vo.
  Physical Review for March, 1902. 8vo.
  Popular Science Monthly for March–April, 1902.
  Public Health Engineer for March, 1902. 8vo.
  Science Abstracts for March, 1902. 8vo.
  Terrestrial Magnetism for March, 1902. 8vo.
  Travel for March, 1902. 8vo.
  Zoophilist for March, 1902. 4to.
Edwards, J. Hall, Esq. (the Author)—Bullets and their Billets. Experiences with
  the X Rays in South Africa. 12mo. 1902.
Electrical Engineers, Institution of—Journal, Vol. XXXI. No. 155. 8vo. 1902.
Franklin Institute—Journal, Vol. CLIII. No. 3. 8vo. 1902.
Geneva, Société de Physique—Mémoires, Vol. XXXIV. Fasc. 1. 4to. 1902.
Geographical Society, Royal—Geographical Journal for April, 1902. 8vo.
Imperial Institute—Imperial Institute Journal for April, 1902.
Johns Hopkins University—University Circular, No. 156. 8vo. 1902.
Manchester Geological Society—Transactions, Vol. XXVII. Parts 8, 9. 8vo.
  1902.
Manchester Literary and Philosophical Society—Memoirs and Proceedings,
  Vol. XLVI. Parts 3, 4. 8vo. 1902.
Massachusetts Institute of Technology—Technology Quarterly, Vol. XIV. No 4.
  8vo. 1902.

Meteorological Society, Royal—Meteorological Record, Vol. XXI. No. 82. 8vo. 1901.
Mitchell & Co. Messrs. C. (the Publishers)—Newspaper Press Directory, 1902. 8vo.
Munich, Royal Bavarian Academy of Sciences—Sitzungsberichte, 1901, Heft IV. 8vo.
Navy League—Navy League Journal for April, 1902. 8vo.
New South Wales, Agent-General for—The Year Book of New South Wales. 8vo. 1902.
New Zealand, Registrar-General of—New Zealand Official Year Book, 1901. 8vo. 1901.
Odontological Society—Transactions, Vol. XXXIV. No. 4. 8vo. 1902.
Paris, Société Française de Physique—Séances de Pâques, 1902. 8vo.
Photographic Society, Royal—Photographic Journal for March, 1902. 8vo.
Rochechouart, La Société les Amis des Sciences et Arts—Bulletin, Tome XI. Nos. 2, 3. 8vo. 1901.
Royal Engineers, Corps of—Professional Papers, Vol. XXVII. 8vo. 1901.
Royal Society of Edinburgh—Proceedings, Vol. XXIII. No. 6. 8vo. 1902.
　Transactions, Vol. XL. Part 1. 4to. 1901.
Royal Society of London—Philosophical Transactions, A, Nos. 304-306. 4to. 1902.
　Proceedings, Nos. 456, 457. 8vo. 1902.
　Reports to Malaria Committee, Sixth Series. 8vo. 1902.
Scottish Meteorological Society—Journal, Third Series, No. XVII. 8vo. 1902.
Selborne Society—Nature Notes for March and April, 1902. 8vo.
Smithsonian Institution—Miscellaneous Collections, Vol. XLIII. The Smithsonian Institution, Vol. II. 1835-1899. 8vo. 1901.
Society of Arts—Journal for March, 1902. 8vo.
Statistical Society, Royal—Journal, Vol. LXV. Part 1. 8vo. 1902.
Tacchini, Prof. P. Hon. Mem. R.I. (the Author)—Memorie della Società degli Spettroscopisti Italiani, Vol. XXXI. Disp. 2. 4to. 1902.
Toulouse, La Société Archéologique du Midi de la France—Bulletin, No. 28. 8vo. 1901.
United Service Institution, Royal—Journal for March, 1902. 8vo.
United States Department of Agriculture—Experiment Station Record, Vol. XIII. Nos. 5, 6. 8vo. 1902.
University College of South Wales—The Calendar, 1901-2. 8vo. 1902.
Verein zur Beförderung des Gewerbfleisses in Preussen—Verhandlungen, 1902. Heft 3. 8vo.
Vienna, Imperial Geological Institute—Jahrbuch, 1901, Heft 2. 8vo. 1902.
　Abhandlungen, Band XVII. Heft 5. 4to. 1901.
　Verhandlungen, 1901, Nos. 17, 18. 8vo.
Washington, Academy of Sciences—Proceedings, Vol. IV. pp. 1-116. 8vo. 1902.
　Memoirs, Vol. VIII. 1898.
Western Society of Engineers—Journal, Vol. VII. No. 1. 8vo. 1902.
Winbolt, F. J. Esq. (the Author)—Frithiof The Bold. 8vo. 1902.
Wright & Co. Messrs J. (the Publishers)—The Medical Annual, 1902. 8vo.
Yorkshire Archæological Society—Journal, Part 64. 8vo. 1902.

WEEKLY EVENING MEETING,

Friday, April 18, 1902.

HIS GRACE THE DUKE OF NORTHUMBERLAND, K.G. D.C.L. F.R.S.,
President, in the Chair.

THE RIGHT HON. SIR JOHN H. A. MACDONALD, K.C.B. LL.D.
F.R.SS.(L. & E.), M.I.E.E.,
Lord Justice-Clerk of Scotland.

Auto-Cars.

WHEN the governing body of this venerable and honoured institution
asks that a lecture be delivered upon the Auto-Car, and when such
an assemblage is gathered together as I see before me to listen, it is
manifest that motor traction has become a practical factor in land
locomotion. No doubt very considerable differences of opinion exist
in the community about it, and I suppose those differences are fully
represented in this room. There are the keen enthusiasts, of whom
a few are here; there are the people who are favourable, but holding back—looking for what they call developments—which I find
generally means a first-rate engine and an extremely handsome
carriage at a "Cheap Jack" price. Then we have the coldly indifferent, who will not give their countenance to automobilism until
it has got a little stronger in influence than it is now. And lastly,
there are the uncompromising opponents, who hate the motor of the
road with a hatred as cordial as that which their grandfathers and
great-grandfathers bestowed upon the motor of the iron railway.

But I do not think that any one fails now to recognise that automobilism is certain to progress, for pleasure, for social convenience, for
business travelling and for commerce. It has been bitterly opposed
in the past, and no doubt it will be bitterly opposed still; but nothing
will stop its progress, because it has become quite manifest to those
who have looked into the matter with care that it is going to confer
very great benefits upon the community—in some cases benefits which
can hardly be measured. I cannot enter upon these to-night, and I
shall only say—and I think the contention cannot be disputed—
that at the present moment in this case, as in so very many cases in
the past, those who are most keenly opposed to motor traction on the
road are the very people to whom it will bring the greatest benefit.
The squire and the farmer, the local tradesman and the country inn-
keeper, these form the classes who put every obstacle they can in the
way of motor traction. They are so short-sighted as not to see that
it means for them a revival of the road such as seemed absolutely
and for ever impossible since the days when the iron horse swept the
traffic from the highways of the country. Now we have reached a

time when the main lines of railway are becoming congested, and
there is a perpetual outcry from day to day against the tyranny of
railway rates.   Soon we shall have fast, cheap, convenient, long-
distance traffic upon the roads, and it is the landed proprietor, the
tiller of the soil, the merchant in small places, the tradesman, and
above all the roadside inn-keeper who will ·reap benefit from the
development of motor traction.   Whatever their feelings have been,
these feelings will change as the feelings of many people changed in
regard to the establishment of railways in the country.

I know that there are certain districts where there is a determined
set made against the auto-car.   I know also, for certain, that if that
set is successful, it will have the same effect as befel the City of
Sheffield when, in the early days of railways, the inhabitants of that
town did their very utmost, and successfully, to prevent any railway
being brought near them, because they said it would ruin their town.
For more than a generation Sheffield was not upon any practical main
road to the North from London, to its own great loss.

I have had a considerable amount of anxious thought as to
how I should treat this subject to-night.   As the man in difficulties generally does, I consulted my friends.   I asked one friend,
a man of great experience, who knows this place well, how the
subject should be treated.   He told me that many experts would
be present, and therefore I should avoid being elementary.   I consulted another; he said, " whatever you do be elementary."   I asked
a third friend, in whom I have great confidence and trust; he said,
"In medio tutissimus ibis," which, for the benefit of the ladies
who may be unacquainted with the Latin tongue, may be translated: "keep in the middle of the road."   I went to yet a fourth
friend, in whom also I have very great faith, and who also knows the
subject of the auto-car very well, and he said: " My dear fellow, you
must take it for granted there is not one person in ten in your
audience that knows anything about the auto-car, except that it has
not got horses in front of it, and that by law it is obliged to make a
series of ' toots.' "   I accordingly resolved that I would be more or
less elementary, and give my illustrations in the simplest form.   I
wish particularly to give such illustrations as the ladies will be able
to understand, because I think the auto-car is going to be a great
favourite with the ladies, who, if they are strong enough, have plenty
of nerve for driving upon the road.   My experience with ladies is
this, that they first say they don't want to get on to a motor car at
all, but whenever they do mount, they always want to go faster, and
always ask you to take them out again.

Now there is no doubt that motor driving upon the road is
not by any means a new thing.   In the early part of last century,
between 1820 and 1830, at the time when railways were beginning
to find their place in this country, there were far-seeing and inventive men who produced carriages driven by mechanical power upon
the road, which were eminently successful; and the names of such
men as Gurney and Hancock in England, and Nasmyth and Scott

Russell in Scotland, will go down to history as having been the pioneers who had to hack their way forward through the thickets of prejudice of ignorant and biassed minds. They were eminently successful for that time. They had to work with materials very inferior to those which are open to inventive men now. Steam was the only power they could use, and the steam engine had not at that time become the highly developed magnificent instrument which it is now. Nevertheless, they ran coaches in England and in Scotland over considerable distances. There is a gentleman present in this room to-night who has travelled as a passenger upon one of these coaches, which I will show you upon the screen. [Numerous lantern illustrations of early motor coaches were here thrown upon the screen.]

It is a remarkable fact that for fifty years after that period motor traction upon the road was only attempted experimentally. The railways opposed motor traction upon the roads bitterly; the squires opposed it bitterly; the farmers opposed it bitterly; the inn-keepers opposed it bitterly. Stones of large size were laid across the road in front of the carriages; ruts were dug across the roads, into which the inventor and his vehicle might go down; and prohibitive tolls, varying from 1*l.* 8*s.* to 2*l.* 10*s.*, were exacted at bars where 4*s.* or 5*s.* was the charge for a four-horse coach. The result was that, notwithstanding the great enterprise of the promoters and the great success of the coaches, they were crushed out of existence, with the exception of certain experimental carriages, and that attractive object the traction engine.

All the legislation of the British Parliament was made for the traction engine only, notwithstanding the fact that a Select Committee sat in the House of Commons in the year 1832 and unanimously reported that "The substitution of inanimate for animal power in draught on common roads is one of the most important improvements in the means of internal communication ever introduced. Its practicability they consider to have been fully established." The report stated that "They are perfectly safe for passengers, that they are not (or need not be if properly constructed) nuisances to the public, that they will become a speedier and cheaper mode of conveyance than carriages drawn by horses; that they will cause less wear of road than carriages drawn by horses." Notwithstanding this report, a signally favourable one, prejudice prevailed, and motor traffic was stopped for fifty years, and so we have been tied down to the traction engine—that iron elephant of the road—which has settled for us our laws up to the year 1896.

By an invention of twenty-two or twenty-three years ago, road traction became common on the Continent, and, while hundreds of carriages were there careering about gaily, here no one could take out a mechanically-driven vehicle, even if no bigger than a tricycle or a governess tub-cart, without having more men in attendance upon it than are required for a 90-ton express upon the Great Northern or the London and North Western Railway. Three men were required, one of whom was ordered to go before it carrying a red flag to wave in front of the noses of the horses to prevent them from being nervous. [Illustrations on screen.]

That state of things, of course, could not continue. It was ruinous to us as a matter of trade. If you will look at this diagram to the left here, you will see the extent to which it is still holding us down commercially. No one in this country could afford

IMPORTS OF FOREIGN AUTO-CARS, 1902.

January	..	..	..	..	..	..	£43,855
February	..	..	..	..	..	..	52,682
March	..	..	..	..	..	..	70,513
							£167,050

At the rate per annum of £668,200.

to expend capital to build motor cars until the Act of 1896 was passed, while makers on the Continent were establishing a very large business. Last January the imports of foreign motor cars were valued at 43,855*l.*; in February they were 52,682*l.*; and in March, 70,513*l.* —thus increasing month by month, as they will do during the rest of the summer.* The total for the three months is 167,050*l.*, which is at the rate per annum of 668,200*l.* It will give you an idea of it if I express it in this way : we are sending out of the country every day 2120*l.* of money—more than the salary for a whole year of the President of the Board of Trade himself. Such a state of things could not continue ; and it is to the great credit of Mr. Shaw Lefevre, and succeeding him Mr. Henry Chaplin, himself a lover of horses and an agriculturist, that in 1896 a reasonable Bill was passed, which everyone now thinks must be modified—that is to say extended, with proper safeguards.

The motor most chiefly in use for road traction is, what I shall call, a spirit-driven engine. It resembles the gas engine, with which we are familiar both on a small and on a large scale. You all know what happens when a housemaid goes into a room where there has been an escape of gas, carrying a naked light. The housemaid and the room suffer severely. Still more awful accidents happen in coal pits, when men digging into the bowels of the earth let free gas which, mixing with the air and being set fire to, causes terrible explosions, destructive of life. It is like the old story of the fisherman in the Arabian Nights, who found a bottle sealed up with the seal of Solomon. When he opened it a vapour came out, which formed itself into a gigantic genius, who immediately announced his intention to destroy his deliverer.

Till fifty or sixty years ago it had not occurred to anyone that it

---

* Since the date of this discourse, the rate per annum has increased to 1,586,688*l*, if taken on the month of August, in which the net imports amounted to 132,224*l.*, or, in round figures, 4265*l.* daily.

was possible to bring gas and air together in small quantities, and instead of having enormous destructive genii to deal with, to have a number of bottle imps which you could let out one after another and make each do serviceable duty.   The gas engine was a device for letting a small quantity of gas and a small quantity of air meet together and explode in a confined space, so that by a succession of small explosions sufficient power was developed to drive an engine. I will explain to you how the gas engine works.   It somewhat resembles a steam engine in general appearance, but the action is quite different.   I have prepared this model with a cylinder of transparent glass for the purpose of demonstrating it.   The first thing we have to do is to get our cylinder filled with a mixture of gas and air, whether coal gas or vapour of petrol or alcohol.   In order to make it clear to you, I attach a piece of blue paper to the piston, which, when drawn up and showing through the glass cylinder, symbolises the mixture of gas and air.   When the piston is drawn up and you have sucked into the cylinder a mixture of gas and air, the next thing you have got to do is to force the piston down again to squeeze and press that mixture close together ; then it is ignited, by means which will be afterwards described, and the piston again ascends, but this time it is driven up by the explosion, while the cylinder is filled by the products of the combustion, which now I represent by a brown paper.   These products of combustion have, by a fourth motion of the piston, to be excluded and driven out again.   You suck the mixture of gas and air in ; you then press it ; it is then expanded and the piston driven up by force of explosion ; and then the piston descends again, pushes out the exhausted gas, and it is ready to begin its work of sucking up gas and air again ; and so it goes on in a succession of strokes.

Now you will, I think, ask me three questions, which will occur to everyone.   First, how do you get your first two strokes when you have nothing to do it with, as the explosion does not take place until the third time the piston moves ?   Next you will ask when the piston is driven up why does not it stay there ?   When driven up by a strong explosion it would be expected to stop there or knock the top of the cylinder off.   The third question is, if gas is being admitted into this space, how is it that when the explosion takes place, the fire is not communicated to all the fuel carried, so burning it up at once, and causing a conflagration.

I will explain these things to you very shortly.   In the first place, one of the difficulties of the gas engine is that it has to be started by some other power before it can do its work.   To start a steam engine, steam is let in first at one end of the cylinder and then at the other, and so pushes the piston from one end to the other alternately.   But in this spirit engine something has to be done first, mechanically, in order to give the gas the opportunity of being sucked in along with the air and compressed so as to make an effective explosion.   I daresay some of you have seen people on the road, whose engines have stopped, busy with a sort of barrel-organ handle turning it round vigorously. That is to induce suction and compression to get their engine to start. There was a driver in Glasgow busy turning his handle, as a working man came up, who was a great connoisseur in music. He stood for a moment, put his hand in his pocket, and took out a penny. After watching a few turns, he said, "Man, have ye no forgotten to put a barrel in?" Presently there was a loud rum— rum—rum—rum—suggestive of the beginning of the overture to 'William Tell.' But no music followed upon that, and the man remarked, "I think I'll just keep my penny till ye get her inside right."

The second question was, why the piston when forced up did not stay jammed at the end of the cylinder. It would do so, but for the use of a fly-wheel. I asked one of my lady friends the other day if she knew what a fly-wheel was. She thought it had something to do with Santos Dumont; she evidently knew nothing about it. The fly-wheel is based upon a very simple principle, that if you are going to move anything, you cannot move it without putting some energy into it. If a ball is thrown along the ground it rolls until the energy imparted to it is exhausted, and it comes to a rest. A light ball cannot go far—a ping-pong ball thrown would not hurt a lady, but a golf ball driven by a club, or a cricket ball thrown by hand or driven by a bat, might hurt severely. Here are two wheels of the same size on one axle [model exhibited], and although I set them spinning together, so far as I can honestly do it putting the same amount of force on to both wheels, you notice that one has stopped while the other is still revolving. The only difference between them is that one has a tyre upon it and the other has not. The one with the tyre is of course a little heavier, and therefore takes more force out of my hand than the other to start it at the same speed, and so it runs on longer. That is a simple illustration of the fly-wheel, and here is the simplest and perhaps the oldest of all illustrations of the fly-wheel—a spinning-wheel [model exhibited]. A spinning-wheel is always made with a heavy outside rim, which ensures equable motion, and it will run when the hand has to be removed to attend to the thread. That is the principle of the fly-wheel. Here is a model of one of these spirit engines with a heavy fly-wheel; when the fly-wheel is first turned round, the energy taken up by it keeps the piston moving, causing it, through its crank, to change its direction instead of being jammed against the end of the cylinder, and so making the piston continue to move.

The next point is, how is the explosion obtained?

There are two modes by which we fire our mixture in the motor car. One is by means of a small platinum tube which goes through the side of the cylinder. This tube is inserted in the space in which the mixture is being pressed at the end of the cylinder; it is kept white hot by a lamp outside, and as the compressed air and gas are driven against it, it ignites them and the piston moves. A preferable mode, in my opinion, and one that is used more frequently, is by electricity, with the aid of an induction coil and a commutator—into the details of which I do not mean to enter to-night—by which a small spark is passed at the right moment between two platinum points in the inside of the cylinder and so explodes the mixture. [Induction coil and commutators were shown.]

These are the general principles upon which the gas engine works. What remained to be discovered before it could be applied to road engines was some means of making a suitable gas on the road. Ordinary gas you could not use, because you could only carry it in a carriage by having it compressed into strong and very powerful cylinders, which were necessarily very heavy. If any accident should happen, and if by any chance your compressed gas got out, very disastrous results would follow. But the difficulty was solved by using the vapour of a volatile oil which could be safely carried and made gaseous from time to time as required. Hence " petrol," which is the essence of " petroleum," and extremely inflammable, is generally used. But alcohol or vapour of benzoline can be used effectively. This is passed through what is called a carburettor ; a certain regulated quantity being brought from your tank in order that you may first get the gaseous vapour which is to be mixed with the air. There are various ways in which this is accomplished.

It is obvious, as I said before, that a third question will occur to you—is it not very dangerous to have an explosion when you are carrying a quantity of a substance giving off an inflammable gas? That difficulty is got over by the vapour from the essence being passed through small holes or through a piece of very fine wire gauze, and we then have absolute certainty that no flame will go back past that point to the store of inflammable liquid. A piece of wire gauze as you see has this effect, that a flame does not pass through. It is really a succession of very small tubes, so small that the combustion cannot be maintained, and the flame cannot establish itself beyond the gauze. That is the origin of the old Davy lamp for the safety of miners. [Lamp shown.] A lamp covered with gauze in this way can never set fire to an explosive mixture. In the suction-stroke of the piston vapour is drawn off from the carburettor, much as a lady's scent sprinkler draws off and sprays out scent when the india-rubber bulb is pinched, the only difference being, that in the one case the air is forced past and in the other it is sucked past the orifice.

The carburettor having done this duty, the gas is let into the cylinder, mixed with the air, and compressed, and we get our explosion and our motion, and, having got that, we have to deal with it according to reason, because you see you cannot possibly have an effective explosion except when the cylinder is closed up. Accordingly we have to use valves, so that when the explosion comes nothing goes out at the entrance at which the gas and air come in. The best illustration I can give is that of the ordinary bellows. There is a little hole in the bellows and a little piece of wood working loose when you open the bellows but closing tight by the compression of the air when you shut them. It is exactly the same thing with a petrol engine, except that we assist the valve to close by providing it with a spring to make it move more quickly.

The next thing is the means by which we keep this engine cool. The engine is running as high as 700, 800 or even 1200 or more revolutions every minute, and these constant explosions which are taking place every second time the piston moves, necessarily cause an enormous amount of heat. In the case of very small motors we can have working along with the machine a small fanner to keep it cool, but with large motors this is quite inefficient. We have to use a series of pipes with gills upon them: water is kept circulating in these pipes either by means of a pump or other means, so that the heat taken up from the cylinder is drawn off, and the water returns and again draws off the heat.

The next thing is how we apply speed to get our running, because, as your engine is going round and round at 700 or 800 or more revolutions a minute, the speed must be brought down, so as to apply the power effectively to the wheels at a much slower rate, and to vary the speed as required. [Illustrations on screen of methods regulating speed by various sizes of pulleys, gear wheels, chains and sprockets, etc.]

Then we come to the tyres which we put upon our wheels. In the days of the " Enterprise " coach there were no rubber tyres. It was rough work, and required very powerfully-built carriages. Here is a specimen of the latest tyre for the motor car, if you want very comfortable running and are prepared to risk puncture every now and again. There are other specimens here, which can be examined afterwards, of the different modes in which the pneumatic tyre is made up. Here are specimens of the solid tyre, one of them being composed of hard rubber outside, and softer material for the inner part. [Specimens of tyre exhibited.]

There is a necessity for having very powerful brakes upon these carriages. Here is one which closes round a drum upon the axle. These carriages can, in fact, be pulled up in an extraordinarily short distance; if people are willing to sacrifice a tyre upon some occasions, they can be stopped almost instantly. [Specimens of brakes exhibited.]

That is a description generally of what I call a spirit engine. I have used illustrations more particularly of Herr Daimler's engine. He is dead, but he was the great pioneer in this direction. I have no wish in this lecture to praise any person's particular work, or any individuals; but everyone can, without partiality, give a tribute to the memory of Daimler as one of the great inventors and manufacturers of our time. It was he who practically adapted the gas engine to the use of petrol vapour, which at present is nearly universally found suitable for the smaller machines, such as tricycles and bicycles. We very often see, indeed, an engine supported between two wheels, and there is one which has the whole apparatus lodged inside the spokes of the hind wheel. You will appreciate the ingenuity required to fix on such a small machine as a tricycle or bicycle, an engine of 2, $2\frac{1}{2}$ or $2\frac{3}{4}$ horse-power to run at as much as 1200 or more revolutions a minute. Such vehicles have a great future before them. There are other auto-mobile carriages in which a heavier oil is used, applying heat to the oil, and thus bringing it into a vaporised condition. This has not made much progress as yet, but its promoters are persevering, and, if crowned with success, may find a field of usefulness in places where so volatile a liquid as petrol cannot be carried in large quantities— as, for instance, in very hot climates—without great danger either in consequence of the heat or from the proximity of other things.

Steam cars have become highly successful, but of these I need not speak at much length. For the benefit of the ladies I shall only say, that while petrol engines resemble " fives " or rackets, where the ball, corresponding to the piston, is struck from only one end, and from its own rebound given back to be struck again, the steam engine is like ping-pong or lawn tennis, in which the ball is struck and forced from one end to the other alternately. In the steam engine, as the piston arrives at one end of the cylinder the steam is shut off from the other end, and let in at the former, and thus the piston is driven backwards and forwards. This mode has its advantages and disadvan-tages, which it is impossible now to enter into, as the time is limited.

Another class of automobile is moved by electric power—a luxurious class of auto-car, if I may say so, and there is a great field before it. Its advantages are ease of starting and control, absence of complication, and noiselessness and cleanliness. Its disadvantages at present are its dead weight, and the impossibility, when its power is exhausted, of renewing it except at places where you can recharge from accumulators, as you can only run 60 or 70 miles without coming to the end of your power. For going about town, making calls or shopping, or visiting places of amusement, or taking a short drive, nothing can possibly be better. I have no time to explain to you what a storage battery is, beyond stating that it is a box in which there is sufficient energy stored up to do certain work when you want it.

Allow me a few words in conclusion upon the question of how the motor car will affect the general community. It will undoubtedly facilitate the residence in the country of those requiring to do their work in towns. Parts of the country, and those of the most pic-turesque and healthy, are unapproachable by rail or tramway, both because of the gradients and the width of the roads, and the present paucity of population. The tendency of suburban dwellers to cluster around railway stations, or along tramcar routes, will be modified. The range in distance from town will be increased, and the choice in rural scenery will be enlarged.

There is a limit to human endurance of daily tramcar runs if the distance is great. And what a relief it will be to many, who now travel daily in and out by rail, to be able to do so by road if the time occupied is no more than at present. In many cases the journey will occupy less time. At present paterfamilias is compelled to gobble a hasty breakfast and make off, often not at a proper post-prandial speed, sometimes in hail, rain or snow, to reach a station half-a-mile or more from his house. When he arrives there, he finds that, instead of being nearly too late, the train does not come in till long after its proper time, and when it does take him up, it stops time after time at signals, and gets further and further still behind its time. Then on arrival he has still perhaps a mile to go by omnibus before he can get within some hundred yards of his office. All this would be changed, and countless citizens who cannot afford to set up a horse and stable, or who would not have the space, could find accommodation for their voiturettes in their warehouse, or put them up elsewhere for a trifle, till the time for the return journey.

One who is a little better off can take a servant with him, and may send back the car to enable my lady to do her shopping or make her calls, coming back for him in the afternoon, doing its 50 or 60 miles in one day, which no horse could do, or if it could, would certainly not be fit for work on the day following.

The car can be applied to purposes such as sawing, or watering, pumping, thrashing, and other useful works. As regards the business uses of the motor vehicle, will it not be a boon to the occupiers of land to be able to send their milk and butter, and field produce, long distances of 30 or even 40 miles into the large towns, either by their own vehicles or by contract, without the intervention of the middleman, and the arbitrary rates and delays of the railway companies? The time occupied and the expense incurred would be less, for the carting to a station, the loading on to railway trucks, the waiting for train despatch, the unloading on to another cart, are all complications bad for the goods and tending to pile up expense.

Further, the farmer who is many miles from a railway station would be able to compete on fair terms with others which now he cannot do, and therefore his landlord gets a smaller rent, and he is somewhat limited in his choice of means of making a profit. To the manufacturer, the merchant and the town tradesman, will it not be a great advantage to have a range for delivery of their goods and for taking orders by their own vehicles practically trebled, as will certainly be the case. Their vans can go a round of 50 miles, or even more, whereas now the range is limited by the necessity of baiting and watering, and by the capacity of the horse for work. Between Manchester and Liverpool, within a very short time, enormous quantities of goods will be carried, and quite as fast as they could be conveyed by railway.

The wayside inn-keeper, who has been a faded and decayed person ever since the railways took possession of our road traffic, will be benefited. I have seen some, in places which a short time ago were never visited, greatly rejoicing when three or four cars came in, from the occupants of which they were able to draw a good profit. There is now an excellent outlook for the country inn-keeper.

A little higher in the social scale the doctor, whether in town or country, will have his capacity for making rounds greatly increased. He will have no need to go out of his way to change his horse, and will not find himself crippled in his work by his nag having been over-fatigued by too long a round on the previous day, or chilled by standing in evil weather. When suddenly summoned he can be on the road in two minutes, and will not have to choose between being too late in an emergency, or cruelly treating a wearied animal. A physician or surgeon, summoned in consultation to an urgent case, will not have to consult the A.B.C., only to find that perhaps he cannot even start for two hours, and that when he does, his train, even if punctual, will, after an hour-and-a-half of travelling, only land him three or four miles from his destination. On such distances as 30 miles or so, he could outstrip the railway without going at any dangerous speed upon the road. It may give you some idea of the difference between what may be done now and what used to be possible to refer to this diagram.

This diagram shows roughly the number of places of considerable

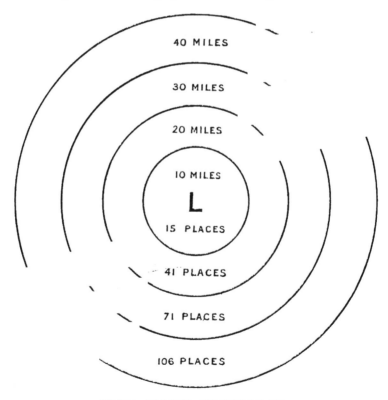

TOTAL PLACES BY HORSE 56.

TOTAL PLACES BY AUTOCAR 233.

population within radii of different lengths from the centre of this city. There is no difficulty in seeing at once that, while by horse haulage 15 places might be visited, or 41 places with great difficulty, an auto-car could make a visit to several places within the inner areas, and could visit any of the 71 places in the 30-mile radius, or of the 160 places in the 60-mile radius. These places include Oxford, Aylesbury, Buckingham, Northampton, Bedford, Cambridge, Bury St. Edmunds, Ipswich, Harwich, Colchester, Margate, Ramsgate, Dover, Folkestone, Hastings, Eastbourne and Brighton, Chichester, Portsmouth, Southampton and Winchester. In such journeys the charms of country can be enjoyed as they cannot be from a railway line. Who knows the beauties of Kent or Surrey, or Herts, who never sees these counties except out of the window of a railway carriage, even when, as I am told sometimes happens, the speed gives plenty of time for admiring the scenery?

Remember this too, that there is considerable economy in not having to pay anything when not using your auto-car. If you go for a month or six weeks to the Continent, instead of spending money in feeding a horse you will have that money to spend abroad. Also the gain to be made is not limited to those who enjoy the use of the vehicle or those who profit by it in their business. In removing the congestion of the streets, or in clearing them of pollution, the gain would be very great. If, when a continuous line of vehicles is crawling from Charing Cross to the Bank every horse disappeared, as by the wave of a fairy's wand, and each vehicle moved automatically, the time occupied in travelling between the two places would be diminished by more than one-half; every check to allow traffic to cross the line of vehicles would occupy only half of the time it does now, and any vehicle that could pass another would do so much more easily, there being no horse to draw out into the roadway before you can make any progress. A string of carriages at a Drawing Room or the Opera would occupy only half the distance, and double the number of carriages would be cleared at the same time, both in arriving and departing, thereby saving a great deal of grumbling at the ladies by the gentlemen, and *vice versa*. All over the busy parts of the great town the tendency to congestion would be diminished.

Then there is the question of pollution. How many hundreds of tons of deleterious matter are blown about or caked into the face of our insanitary wood pavement and ground into fine dust, to be blown into our noses, mouths and eyes? And how much of it is carried by fashionable skirts into the houses of Belgravia or Mayfair, to be thrown up next morning by the housemaid's broom? Can anything be imagined worse in a great city, where the air is much more injurious than in the country, contaminated as it is by other unavoidable impurities? Ask any medical man whether the sick rate, ay, and the death rate, would not go down if the pollution resulting from the use of animal haulage could be sensibly diminished. Since

I wrote these words Sir Henry Thompson has said, "The sanitary condition of our great city would be considerably improved, since the streets would no longer be, as they are now, the daily receptacles of many tons of manure, which becomes not merely offensive but prejudicial to health. Its components soak into the wood pavement and pollute the surrounding air by evaporation, and in hot and windy weather with dust also, which is inhaled by the inhabitants. This is a cause of deranged health to many. It is second in degree, but the same in kind, as that occasioned by intramural interment, long ago forbidden."

Finally, Ladies and Gentlemen, one thing, and one thing only, I will ask, and it is, that you will not refuse to realise that the advent of the motor vehicle upon the road is a thing to be looked upon in a broad and not in a narrow spirit. It is certain that it will affect the whole community in many directions—national prosperity, social and commercial convenience and economy. Is it to be a benefit to the whole of the public taken in the aggregate? This is a question which must be considered, and with a serious and unbiassed mind. It will be so considered by you, I feel sure, and when it is I am certain of your judgment. It will be, that although this new mode of traction has its drawbacks, as every great change in this world must have, the advantages which it promises to give do so immeasurably outweigh them, that it is the duty of a good citizen—subject, of course to reasonable legislative control—to give it a fair and a free field.

[J. M.]

WEEKLY EVENING MEETING,

Friday, April 25, 1902.

Sir James Crichton-Browne, M.D. LL.D. F.R.S., Treasurer and Vice-President, in the Chair.

James Mackenzie Davidson, Esq. M.B. C.M. *M.R.I.*

X-Rays and Localisation.

At the end of the year 1895 Professor Röntgen announced that he had discovered a new kind of rays coming from a Crookes' Tube, which he called the X-rays. This discovery created a widespread interest in all parts of the world and among all classes of people, and the reason is not far to seek, because the fact that these rays have the power of passing through our bodies and casting shadows of our bones upon a photographic plate or a fluorescent screen, at once gave the discovery an intensely human interest.

It is my privilege to-night to tell you something about these rays and their application in surgical practice. At the risk of being too elementary, I nevertheless purpose beginning with a short historical sketch of the scientific discoveries and developments which led up to their discovery.

We will begin with the electric spark. Two types of instruments are employed to produce disruptive electric discharges ; the Wimshurst or Holtz Influence Machine which, by merely turning a handle produces electricity at a high pressure. The other instrument is an induction coil, and it is the latter which I will use to-night in carrying out my experiments.

The Induction coil consists of a coil of insulated thick wire bound round a bundle of soft iron wires, such as I show you here. This is called the primary. It is introduced into the interior of a large coil, consisting of a great length of very fine insulated copper wire ; this is called the secondary. In this instrument here to-night there are about nine miles of fine copper wire. The action of this instrument is based upon the discovery made in this place by that great philosopher and experimentalist, Faraday. If a current of electricity be passed through the primary coil at the moment of starting it, there is a current induced in the secondary coil in a reverse direction ; and when the current is broken in the primary, there is induced a powerful reverse current in the secondary ; so that at each make and break of the current in the primary we have two currents correspondingly induced in the secondary in opposite directions, but at a much higher potential.

The current induced in the secondary when the current is broken in the primary, is about fifty times more powerful than the current produced at the make in the primary. In this way if the terminals of the secondary are widely separated only one of the induced currents has sufficient energy to spark across, so that the secondary discharge is thus rendered unidirectional. A commutator introduced in the primary circuit enables the operator to make either end of the secondary a plus or a minus pole at will.

Induction coils are usually supplied with a spring hammer break tipped with platinum, which automatically makes and breaks the current in the primary, but to-night I shall use instead, a simple form of mercury break. This consists of a tank or small vessel containing mercury, into which one end of the primary of the coil dips. A small electric motor is placed in an inclined position, so that its spindle with a small copper blade attached at right angles to it dips down obliquely into the vessel containing the mercury. The battery has then one of its poles attached to the other end of the primary coil, while the other pole of the battery is attached by means of a spring to the spindle of the motor. It is so placed that as the motor rotates the blade at its end dips into the mercury, and in this way makes good contact and the current of the primary is made; as it rotates it comes out of the mercury and so the current in the primary is suddenly broken. To avoid arcing, the mercury is covered to a considerable depth with paraffin or alcohol. In this way we can control the number of breaks which we can produce, and if the motor is running at the same speed, the same number of makes and breaks can be repeated in the same period of time. The speed of the motor is regulated by a small adjustable resistance. The higher the voltage, the greater speed is required in the motor.

In order to increase the efficiency of the coil a condenser is always introduced as a shunt to the primary across the break terminals. A condenser consists of sheets of tin foil with sheets of paraffin paper between in such a manner that the odd numbers of the sheets of tin foil are all joined together and the even numbers are all joined together, and this arrangement allows the sheets of tin foil to be oppositely charged, and allows the break in the primary with ordinary speeds to be more sudden. Now when I start the break and turn on the current you see the spark leaping across between the terminals. Sparks of this kind were much studied after the invention of the first electric machine about 200 years ago, which produced them, and I will now show you some peculiarities of these sparks.

Once a spark has passed through the air and broken it down, it so alters the molecules that it makes it easier for another spark to follow in its wake. Instead of the knobs we take two vertical wires parallel to each other and connect them with the terminals of the coil, and turning on the current you see the spark begins at the bottom of the wires and each succeeding spark takes place higher up than the previous one ; in this way, you have as it were, a ladder of flame. The reason is that the decomposed parts of air by reason of the heat produced by the spark, rises a little and forms an easier path for the succeeding spark.   The conducting power of the air is so greatly increased that the spark will sometimes prefer to follow the path of this altered air in preference to metal, which is such an excellent conductor of electricity.

It is only natural that experimentalists should have wondered how this electric discharge would behave if it was allowed to pass through a vessel from which the air had been pumped out.   Faraday, I think, was the first to call attention to the peculiar appearance of the electrical discharge in rarefied air.   In the 13th Series of his Experimental Researches on Electricity, 1838, he observes :—

" I will now notice a very remarkable instance in the luminous discharge, accompanied by a negative glow, which may perhaps be correctly traced hereafter into discharges of much higher intensity. Two brass rods $0\cdot3$ of an inch in diameter, entering a glass globe on opposite sides, had their ends brought in contact, and the air about them very much rarefied.   A discharge of electricity from the machine was then made through them, and whilst that was continued the ends were separated from each other.   At the moment of separation a continuous glow came over the end of the negative rod, the positive termination remaining quite dark.   As the distance was increased, a purple stream or haze appeared on the end of the positive rod, and proceeded directly outwards towards the negative rod, elongating as the interval was enlarged, but never joining the negative glow, there being always a dark space between.   This space of about $\frac{1}{16}$ or $\frac{1}{20}$ inch was apparently invariable in its extent and its positive relation to the negative rod, nor did the negative glow vary.   Whether the negative end were inductric or inducteous, the same effect was produced.   It was strange to see the positive purple haze diminish or lengthen as the ends were separated, and yet this dark space and the negative glow remained unaltered."

You may look upon this as the first step towards the discovery of X-rays.

De la Rue, Müller, Gassiot, Spottiswoode and Geissler, all did excellent work in studying electric discharges through vacuum tubes. After the discovery of the mercury pump by Dr. Herman Sprengel the degree of exhaustion of these tubes was rendered far more perfect.   Professor Hittorf had done some very fine work in Germany with these highly exhausted tubes, and then Sir William Crookes, in this country, did his remarkable researches between 1874 and 1875. By carrying exhaustion to a higher degree than had ever been previously attained, he came to the conclusion that the particles of air remaining in these highly exhausted tubes behaved so differently from an ordinary gas, that he considered it to have attained a fourth condition of matter, which he called *radiant matter*, so that we could no longer only speak of solids, liquids and gases, but had to include radiant matter.   The gas in the tube assumes this condition when the vacuum is about a millionth of an atmosphere ; in other words, if a tube is broken and air is allowed to rush in through the opening, there is about a million times more air in it at the ordinary pressure than was the case before the opening was made. When we pass a discharge through such a tube from the negative or cathode side, a stream of particles starts, and anything upon which they impinge is made to glow. This cathode ray has the capacity of producing great heat; it will make platinum red hot, and even melt it; it will cause substances to fluoresce and phosphoresce; it will only travel in straight lines, it will not turn round corners, and it is deflected by a magnet; it will cast shadows of any solid substances placed in its path.

All these observations had been made by Crookes, and necessarily created much interest. Dr. Philip Lenard, in 1894, endeavoured to bring the cathode rays outside the tube into the open, and as glass was opaque to them, he endeavoured to make an opening for them by introducing a small window of thin aluminium foil opposite the cathode. He succeeded, for the rays which passed through his aluminium window produced fluorescence. They affected a photographic plate. He found that they would pass through thin sheets of aluminium, and even of copper; he also found them to affect a photographic plate through metal.

Professor Röntgen of Würtzburg took up the subject where Lenard left it. When I had the pleasure of an interview with Professor Röntgen, shortly after his discovery, I asked him what intention he had in his mind when he commenced the research. He said that he was looking for any invisible rays that might be produced from a Crookes' tube. The next question I asked him was, why had he used a screen of Barium platino-cyanide. He replied that he had used this substance because, by its brilliant fluorescence it was capable of revealing the invisible rays of the spectrum. He then told me that he had covered the Crookes' tube with black paper, so that when it was excited by the electric current no visible light was seen, but he suddenly observed that a piece of cardboard, which had been covered with crystals of Barium platino-cyanide, glowed brilliantly. I asked him what he thought when he saw this. His reply was, "I did not think; I experimented." He showed me the original screen with which this great discovery was made, and I suggested that such an historical screen should be carefully protected in a glass case, and I trust that my suggestion has been carried out.

He speedily found that various substances interposed between a Crookes' tube and a fluorescent screen allowed the rays to pass through in varying degrees, and doubtless he was immensely surprised and interested to observe that when he interposed his hand, his bones inside his flesh were revealed on the screen. One of the earliest experiments which he tried was to take a photograph through a deal door between two of his laboratories. He placed the tube on one side of the door and affixed to it a strip of platinum, and at a corresponding part on the other side of the door he placed a photo-

graphic plate. He then excited the tube by the electric current, and
then developed the photographic plate. He found the image of his
strip of platinum, as he expected, but he also found a second strip
which he could not explain. On examining the door he found that
the unexpected strip came to be opposite the beading. It was a pine
door, varnished but unpainted. At first he wondered whether the
additional thickness of the beading would account for the unexpected
strip, but he found wood so transparent, that this view was untenable.
He then removed the beading from the door. The explanation was
then obvious. Some white lead had been used to cement the strip to
the door, and it was the white lead which by its density had caused
the shadow on the photographic plate.

It is remarkable that Professor Röntgen, in his original communication concerning the X-rays, described almost all the properties
which they are known to possess. Very little, indeed, has been added
during these seven years. The chief advance has been in the methods
of their application to surgical and medical work.

I will now explain how the X-rays take origin in a Crookes' tube.
This rough model illustrates the path of the cathodal stream by
means of these cords. A curved cathode, similar in shape to an
ordinary reflector, is made of aluminium. The cathode rays emerge
from it normal to its surface and converge towards a point, then
proceed for a short distance in a straight line, and then gradually
diverge—*wherever this cathodal stream impinges upon solid matter X-rays
take their origin.* From what has been previously said, it would
naturally occur to you that X-rays must have been produced long
before they were discovered. This is so. Looking backward, we
now know that Crookes in his researches must have been producing
X-rays from his tubes. Lenard, no doubt, had X-rays intermingled
with the cathode rays which he had so laboriously got outside his
tubes. In the early tubes which Röntgen used, the cathode stream
was allowed to impinge upon the bottom of the glass tube. The consequence was that the X-rays took origin from a more or less large
surface, and consequently the photographs produced by a tube of this
kind were necessarily rather blurred; but Professor Jackson, of
King's College, London, suggested putting in the interior of the tube
a metal anode, inclined as you see in this modern tube, and in this
way the converging cathode stream from the concave cathode impinged
upon the platinum target, giving rise to X-rays richly from a comparatively small point or surface. This model served to illustrate
what happens. The slide now thrown upon the screen illustrates the
importance of the proper distance of the anode from the cathode. It
should be so placed as to meet the cathode stream at its narrowest
part. The impact thus gives rise to a rich production of X-rays from
a point, and sharp shadows on the fluorescent screen or photographic
plate are thereby produced.

A matter of some practical importance is that all parts of the
tube which fluoresce green give off X-rays to some extent. And

while a sharp shadow of an object is being cast on a photographic plate by the point on the anode, a blurred image is being produced from the large and diffuse area of the bulb. These rays from the glass may be called secondary, and photographically are about six to ten times weaker than the primary rays.

A strip of lead ½ inch wide and too thick to let any X-rays pass through, was applied closely to the tube—it cast a shadow—no primary X-rays could reach the plate within the shadow; but the two halves of the fluorescing glass supplied diffuse X-rays, and the slide shows two shadows of a single wire interposed.

To cut off the superfluous rays which fog and blur the true X-ray image I surround the tube with non-conducting opaque material—red lead, red lead and plaster, or place it in a box lined with a mixture of white and red lead with a small hole in the lid to act as a diaphragm.

The properties of X-rays may be very shortly summed up as follows:—they proceed in straight lines from their point of origin on the anode to the destination, be it a photographic plate or a fluorescent screen. They are invisible. They cannot be refracted, and they cannot be reflected, except to a small extent, and then only a scattered reflection, just as light is reflected from white paper. They cause certain substances to fluoresce. (This screen was kindly made for me by Messrs. Johnson and Matthey.) They darken a photographic plate; they pass through substances in the order of their atomic weights, the lighter and less dense the substance, the more readily do they pass through, and therefore all organic substances are comparatively transparent, while metals are more or less opaque according to the thickness; they also have the power of discharging electrified bodies. The tube when worked by a coil with a rapid interrupter is illuminated continuously, and the eye gets the impression that the emission of X-rays is continuous, just as the emission of light from a candle, but the impulse of the coil is intermittent, the impact of the cathodal ray is intermittent, and the output of the X-rays is intermittent. If we reduce the speed of the break the light in the tube is seen to flicker, if the speed of the break is increased the light apparently becomes continuous. This is due to a limitation in the human eye. If an impression of light is given to the eye, however transient, it lasts for one-tenth of a second, and if the impressions are repeated oftener than ten times a second, the impression upon the eye is that of a continuous illumination. The intermittence of the highest rate of speed is shown by the rapid rotation of a small fluorescent screen. This experiment shows that the light produced in the fluorescent screen is also transient. This luminescence is in marked contrast to the lasting luminescence which is termed phosphorescence.

During Lord Rayleigh's recent lectures here on Saturday afternoons on recent Electric Developments, I was much interested in his method of producing a very sudden break in the primary of the coil by dividing a wire by a bullet fired from a pistol. He showed that the sudden break gave the full-length spark in the coil without using a condenser. I at once tried the effect of this discharge through an X-ray tube and found that a very brilliant illumination of the tube took place sufficient to take a photograph (show photograph on the screen). It then occurred to me that if I interposed the revolver nozzle with the wire in front of it, between the tube and the photographic plate, that possibly a flash from the X-ray tube would be quick enough to cast a shadow of the bullet in its flight on the photographic plate immediately after the division of the wire had taken place. This I accomplished (show slide on the screen). The shadow of the bullet is blurred showing that the X-rays flash had lasted too long, and this in marked contrast to the beautifully sharp photographs taken of bullets by Professor Boyd by means of the discharge Leyden jar. I repeated this experiment with a revolver with a muzzle velocity of 800 feet per second, and found in two independent negatives that the track of the bullet was approximately the same length and had begun at the same distance from the broken wire. With a Mauser pistol, the muzzle velocity being 1400 feet a second, I failed to get an image of the bullet within 10 inches of the stretched copper strip, but got several blurred images of this copper strip, which on this occasion I used instead of wire; the vibrations produced in the broken strip giving rise to many shadows according to its varying positions. The results of the experiments go to show that with the revolver which had an initial velocity of 800 feet a second wherever oscillations may have been produced in the coil, the duration of the X-ray flash was about $\frac{1}{2000}$ of a second. I will now show you the flash in the tube interposing the fluorescent screen by dividing this wire by the pistol, which I now show you. I advise you to close your ears so that the noise may not interfere with your observing the flash.

We now pass on to the action that X-rays have upon the photographic film. Röntgen rays while producing changes in photographic plates in many respects similar to light yet differ in some important respects. For example, I have here six films. In the dark-room they were placed in a bag superimposed like a pack of cards. I put my fingers upon the films and gave a short exposure to the X-rays. The films were numbered, the top one numbered one and so on, and the last one numbered six. On developing these films all in the same way, hardly any difference could be detected, the first and last were almost equally good, so that the X-rays do only partial work in producing a photograph, although the first film produced a firm strong negative, yet sufficient rays pass through and do equally good work on the films below.

There is one advantage in this that in cases where one is doubtful of the exposure, two or more films can be placed together and if the exposure be correct two negatives are obtained, and if the exposure has been too short by accurately superimposing the films a picture of sufficient density could be obtained. But there is another and more striking difference between the action of light upon a photographic film and X-rays. Before saying more upon this matter I will take a photograph by means of X-rays now. In this paper bag there is a plate 20 inches by 24 inches. It is laid upon the table and upon it is placed a metal design. The tube is now placed above it, and, excited by the coil, produces X-rays. The metal stops the passage of the X-rays reaching the films, but all the part not covered by the metal receives the X-rays, because the paper envelope though being quite opaque to light is transparent to the Röntgen rays. The usual proceeding after the exposure is made is to take the plate into the usual dark-room and develope it in the usual way, but I think it would be more interesting to you if I developed this plate before you in this room. I therefore, now that the exposure is complete, take it out of its coverings, and it is now being exposed to this brilliant electric light, and no doubt many suppose that with this excessive exposure to ordinary light that the X-ray picture will be fogged out and spoilt. But I hope that this will not be the case. I now put it in the developing dish and allow the developer to flow over it in the usual way, and if this experiment is as successful as the rehearsal ones have been, I think you will see the photograph gradually appear. The picture you see is the reverse of the usual X-ray negative; in the ordinary negative the part of the film that has been protected by an opaque substance appears naturally white because the silver salts are dissolved out by the fixing solution; after development in this case the metal part is black and the unprotected film is comparatively light. As far as my scanty leisure will allow I have carried out a few experiments testing the effect of X-rays and light combined upon various kinds of photographic plates, and the explanation of what you have witnessed is given by a slide which I will now show upon the screen. It is a remarkable fact that if a photographic plate be exposed to ordinary light for a considerable time and then exposed to X-rays for a definite time, that the effect is entirely different to what occurs if the order is reversed, that is to say, if a plate is previously exposed to X-rays for a definite time and then to ordinary light.

The next slide is from a negative where the exposures to X-rays and to light were made in strips at right angles to each other. You will notice the reversal produced by the X-rays throughout, the combined action of X-rays and then light making the film lighter than the separate action of either. It seems to me that these curious effects are well worth a systematic investigation. [Slide; it came up ordinarily, and then suddenly reversed and remained as now shown.]

I am pleased to be able to express my thanks to Dr. Findley, for his help in carrying out some of these photographic experiments.

LOCALISATION.

I must now leave the physical side, and proceed at once to tell you something as to the surgical application of these rays in locating bullets, etc., in the human body. From what we have seen this evening, you will readily understand that an X-ray photograph is simply a shadow of the object interposed, and the appearance of a single photograph, however realistic, gives no reliable or accurate guide as to the actual relative position of the parts. An X-ray photograph is simply a central projection. It soon became necessary, when the use of X-rays became more general in surgical practice, that some means should be adopted to give reliable data for the position of bullets, needles, etc., in the body. Various plans have been suggested, but the one which I shall describe to-night is the one with which I am most familiar, and one which gives results as accurate as possible. The method may be shortly described as follows:— The anode of the Crookes' tube is placed vertically above a point where two stretched wires intersect each other at right angles. The vertical distance is measured and noted down. The tube is attached to a holder which slides along a horizontal bar, and this bar is carefully arranged parallel to one of the wires. The tube is then displaced 3 centimetres to one side of the zero, a scale on the horizontal bar enabling this to be done correctly. The part of the patient to be photographed is now placed upon the cross wires, which are usually stretched over a space covered by stretched calf-skin. The wires may be brushed over with some aniline dye, so that the mark of the cross wires may be left upon the patient's skin. The photographic plate is now placed beneath the cross wires, and pressed tightly against the parchment. One exposure is made. The photographic plate is then removed and a fresh one put in its place; the tube is then brought back to zero and moved 3 centimetres the other side, and a second photograph is taken. These two photographs will show a parallactic displacement, as they have been taken from two points 6 centimetres apart, approximately the distance between our two eyes; they will not be exactly alike, and if, for example, we were locating a needle in the hand, the different position of the needle in each photograph in relation to the cross wires can be at once shown by taking a tracing from each negative, with a sheet of celluloid with a cross marked upon it. The celluloid is put upon the photographic film, and the cross upon it is brought into register, that is, superimposed over the shadow of the crossed wires in the negative. A tracing is then made of the needle, and the same is done with the other negative, and it will then be found that the needle images occupy different positions. To interpret this I use an apparatus called a cross-thread localiser, which enables a graphic reconstruction of the conditions in which these negatives were taken.

The cross-thread localiser consists of a large sheet of glass with two lines scratched upon it at right angles to each other, and crossing at the middle point. A small T-piece of metal with three notches in it is supported on a vertical rod, and can slide up and down. It is so arranged that the three notches are in a line parallel with one of the diamond scratches, and three centimetres separate the centre notch from the adjacent ones on either side. The tracing from the negatives on the celluloid is placed upon the stage with the crosses superimposed. The metal support, with the notches, is then fixed so as to be the same height from the stage as the anode was from the cross-wires. The two lateral notches now represent the two positions to which the anode of the Crookes' tube was respectively displaced, and as X-rays travel in straight lines without deviation, fine threads can be used to trace their path. Therefore if a thread, attached to a needle, is placed upon the end of the tracing of one of the needles, and a thread through the other notch is placed upon its own shadow at a corresponding point of the needle, the threads will intersect, and the point where they intersect represents the actual position in space of that end of the needle. Now the distance of this point of intersection from the photographic plate is the actual depth of that point below the surface of the patient's skin next the photographic plate. If a perpendicular plane be raised upon each of the horizontal wires, and the vertical distance from each taken, we get three measurements which enable us, from the markings of the cross on the patient's hand, to fix a point upon his skin beneath which, at a known depth, the foreign body is certain to be found. In geometrical language, we get three co-ordinates of the point in relation to three planes at right angles to each other, and as the relation of the patient's body to these planes is also known, it is a simple matter to decide the actual position of the needle or other foreign body; more than that, the size and shape of the foreign body can be ascertained. But these two photographs, while enabling this information to be derived from them of ordinary triangulation, are really stereoscopic, and when viewed in a Wheatstone's stereoscope by an observer possessing binocular vision, the eyes will triangulate the negatives, as it were. The negatives should give a combined image in perfect stereoscopic relief, so that by this method not only can actual measurements and data be given to the surgeon, but a view of the parts be also afforded. As an indication of how reliable this method is, I will mention that I have applied it in detecting very minute particles in the eyeball and orbit in a great number of cases, and this proves of signal service in eye surgery. I have detected and located in the eyeball particles of glass, steel, brass, iron, copper, silver and lead. The method for localisation is based upon the same principles.

In warfare as well as in civil life the X-rays when properly used are of invaluable service, and I consider that the stereoscopic method is of the highest value in giving the maximum and accurate information that could possibly be afforded by X-rays. Several examples of stereoscopic skiagrams are exhibited in the Library, and I am glad to state that there is also exhibited there an instrument for obtaining the stereoscopic image directly on the fluorescent screen.

On two occasions at the Soirée of the Royal Society I have exhibited apparatus for giving the stereoscopic picture on the fluorescent screen, but, while interesting, they were too cumbersome for practical use. The problem to be solved mechanically was quite definite. Two tubes had to be placed opposite the observer's two eyes, the distance between the anodes preferably should be the same distance apart, namely about 6 centimetres, as the distance between the observer's two eyes. A fluorescent screen interposed midway had to be illuminated by one tube, and the shadow of whatever object was cast upon the screen had only to be observed by one eye, the other being eclipsed, then the next tube had to be illuminated and then the other eye only had to see the shadow produced by it. If this alternating arrangement could be repeated quicker than ten times a second the impression of the two shadows becomes continuous in each eye, the consequence being that the combined mental image stands out in striking stereoscopic relief. This diagram shows the principal features of the instrument which I am exhibiting in the Library. I have to thank my friend Dr. Muirhead for the help he has given me in the construction and designing of this instrument. A previous one was made for me by Mr. Crawley, representing Messrs. Muirhead, but which depended for its action upon two synchronous motors, but the latter instrument with the oscillating eye-piece is likely to supersede the former.

The practical value of getting a clear and brilliant stereoscopic image upon the screen is immense, for not only are the parts seen in their correct position, but when the apparatus is properly arranged, it is possible for the observer, with a metal probe, needle or forceps, to touch any object he desires which he sees in the stereoscopic image.

[J. M. D.]

ANNUAL MEETING,

Thursday, May 1, 1902.

SIR JAMES CRICHTON-BROWNE, M.D. LL.D. F.R.S., Treasurer and Vice-President, in the Chair.

The Annual Report of the Committee of Visitors for the year 1901, testifying to the continued prosperity and efficient management of the Institution, was read and adopted, and the Report on the Davy Faraday Research Laboratory of the Royal Institution, which accompanied it, was also read.

Forty-four new Members were elected in 1901.

Sixty-three Lectures and Seventeen Evening Discourses were delivered in 1901.

The Books and Pamphlets presented in 1901 amounted to about 253 volumes, making, with 722 volumes (including Periodicals bound) purchased by the Managers, a total of 975 volumes added to the Library in the year.

Thanks were voted to the President, Treasurer, and the Honorary Secretary, to the Committees of Managers and Visitors, and to the Professors, for their valuable services to the Institution during the past year.

The following Gentlemen were unanimously elected as Officers for the ensuing year:

PRESIDENT—The Duke of Northumberland, K.G. D.C.L. F.R.S.
TREASURER—Sir James Crichton-Browne, M.D. LL.D. F.R.S.
SECRETARY—Sir William Crookes, F.R.S.

MANAGERS.	VISITORS.
The Right Hon. Lord Alverstone, G.C.M.G. M.A. LL.D.	Henry E. Armstrong, Esq. Ph.D. LL.D. F.R.S.
Sir James Blyth, Bart. J.P.	Charles Edward Beevor, M.D. F.R.C.P.
Sir Frederick Bramwell, Bart. D.C.L. LL.D. F.R.S. M.Inst.C.E.	John B. Broun-Morison, Esq. J.P. D.L. F.S.A.
Thomas Buzzard, M.D. F.R.C.P.	Francis Elgar, Esq. LL.D. F.R.S. M.Inst.C.E.
Donald William Charles Hood, C.V.O. M.D. F.R.C.P.	Francis Gaskell, Esq. M.A. F.G.S.
Sir Francis Henry Laking, K.C.V.O. M.D.	James Dundas Grant, M.D. F.R.C.S.
George Matthey, Esq. F.R.S.	Lord Greenock, D.L. J.P.
Ludwig Mond, Esq. Ph.D. F.R.S.	Maures Horner, Esq. J.P. F.R.A.S.
Hugo W. Muller, Esq. Ph.D. LL.D. F.R.S.	Sir Henry Irving, Litt.D. LL.D.
Edward Pollock, Esq. F.R.C.S.	Wilson Noble, Esq. M.A.
Sir Owen Roberts, M.A. D.C.L. F.S.A.	Winter Randall Pidgeon, Esq. M.A.
Sir Felix Semon, C.V.O. M.D. F.R.C.P.	Arthur Rigg, Esq.
The Right Hon. Sir James Stirling, M.A. LL.D. F.R.S.	William Stevens Squire, Esq. Ph.D. F.C.S.
John Isaac Thornycroft, Esq. LL.D. F.R.S. M.Inst.C.E.	Harold Swithinbank, Esq. J.P. F.R.G.S.
James Wimshurst, Esq. F.R.S.	Charles Wightman, Esq.

WEEKLY EVENING MEETING,

Friday, May 2, 1902.

Sir William Crookes, F.R.S., Honorary Secretary and
Vice-President, in the Chair.

A. E. Tutton, Esq. B.Sc. F.R.S. F.C.S.

Experimental Researches on the Constitution of Crystals.

Experimental work in connection with the study of crystals offers
attractions of a more than usually fascinating kind. For, in the first
place, crystals themselves are such wonderfully beautiful objects;
they are, indeed, unquestionably the most beautiful of all the inanimate productions of Nature. It is even doubtful whether we are
right in considering them inanimate, for the force of crystallisation,
though it may lie dormant for thousands of years, is ever ready,
when a suitable environment offers, to re-assert itself. In the second
place, the study of crystals involves the investigation of that most
exquisite of all the forms of energy, light—that extraordinary effect
of wave-motion in the ethereal all-pervading medium, which we now
believe to be due to an exceedingly rapid, periodic, oscillating change
in the electrical condition of the atoms and molecules of the incandescent light-giving source.

We will allow the beam of light from an electric lantern to fall
on this cluster of diamonds, carbon in its most exquisite crystalline
form. The diamonds are arranged, as you see, in the shape of a
crown, and were generously lent for this lecture by Mr. Streeter.
You observe a magnificent play of light waves rippling towards
you in every variety of colour and scintillation. The object of
the experiment is that you may distinguish the two distinct types
of emanation, namely, exterior reflections of white light from the
facets and coloured rays due to penetration of the light into the
interior structure of the diamonds, and subsequent refraction and
reflection outwards again. The former kind you observe best in
the innumerable images in white light of the carbon points of the
lanterns, reflected on the ceiling and screen.

It is by the determination of the direction of the exterior reflections that we are enabled to ascertain the wonderfully regular angular
relations of the various faces of the crystals; and it is by the study
of the light which has penetrated that we gain the best information
concerning the internal structure.

We shall consider this evening some of the results of a study of
the crystals of certain series of definitely related chemical salts.

There has been a great want of correlated work of this kind, having for its definite object the elucidation of the relationship between the chemical composition of a crystallised substance and the peculiar type of symmetry which it exhibits.

The salts which have, up to the present, been studied, are the sulphates and selenates of the alkali metals potassium, rubidium and cæsium, which crystallise in the rhombic system of symmetry; and thirty members of the well-known monoclinic series of double sulphates and selenates, containing one of the three alkali-metals just mentioned, another metal of the magnesium or iron type, and six molecules of water of crystallisation.

The research had two main objects. First, to discover the effect of replacing one alkali-metal by another; and second, the effect of replacing the lighter element, sulphur, contained in the acid radicle of the salt, by the heavier element, selenium. The three metals, potassium, rubidium and cæsium, belong strictly to the same family of chemical elements, according to the now famous periodic classification of Newlands and Mendeleeff. They are the most electropositive metals known, and the weights of their atoms are related in a most interesting manner, that of rubidium being the mean of the atomic weights of potassium and cæsium. It was, therefore, to be expected that any difference of form or properties brought about by the replacement of any one of these metals by another would be as great as could ever be produced by a change of this kind, and it might be hoped would be adequately great to enable a true idea of its character to be obtained. The very fact that no differences in the angles of the crystals of so-called isomorphous salts had, up to the commencement of this work, been detected with certainty, shows how very small, at most, such differences must be. It will also be evident that for such a research only the most perfectly formed and homogeneous of crystals must be employed. It is a primary essential that the faces of the crystals shall yield perfect images of the signal-slit of the goniometer, and in order that this may be so they must be absolutely plane surfaces. Indeed, if one may be forgiven a slight equivocation in the spelling of a word, the ladies present may be interested to hear that the beauty of crystals lies in the planeness of their faces.

The mode of measuring the angles between crystal faces on the goniometer, and at the same time the difficulty offered by imperfect faces, may be illustrated with the aid of this large crystal of quartz. It is mounted upon one of the actual goniometers employed in the work, but the size of the crystal is enormously greater than those actually used in the research, which rarely exceeded the size of a pin's head. Such small crystals are much more free from distortion than larger ones.

You now see on the screen the image of the goniometer signal-slit reflected from one of the faces of the quartz crystal. This image is an irreproachable one, the face reflecting it being truly plane. Its narrow part is capable of accurate adjustment to a vertical cross-wire. On rotating the crystal so as to bring the image from the next face, of the particular zone which has been previously adjusted, into the field, you observe that this also is a very good image, but much weaker than the other. This is simply because it is derived from a much smaller face, but the face is quite plane. The next image you see is a bad one, being not only a multiple image, but distorted. Such an image would be quite useless for our purpose. Thus, on rotating round the whole zone—for it is one of the geometrical properties of crystals that the faces lie in zones—we find that some images are good and some are bad. In the case of very small crystals it generally happens that the greater number of images afforded by the faces are either one thing or the other, and crystals must be selected yielding the maximum of good images.

The first result of importance that has been brought to light is that, with regard to each series of salts, there is an interesting relationship between the general exterior character—or habit, as the crystallographer terms it—of any triplet of salts containing potassium, rubidium and cæsium respectively.

We will see on the screen, for instance, the configurations of the three double salts, potassium zinc sulphate, rubidium zinc sulphate and cæsium zinc sulphate. The same planes are present in all, but very differently developed. In the potassium salt the shape is determined by the large development of the prism zone (p faces), and the large flat end faces of the basal plane, c. The cæsium salt, on the other hand, exhibits a prismatic habit formed by the faces of the clinodome, q, and the basal plane c is reduced to a strip. Intermediate between these two types comes the rubidium salt, and this is the case with the rubidium salt of every one of the sixteen triplets examined.

Hence the habit clearly follows the order of the atomic weights of the alkali metals.

The next general result arrived at is, that small angular differences between the faces have been established, but they rarely amount to a degree in magnitude. And what is of even more interest is, that the faces of every rubidium salt are inclined to each other at intermediate angles to those of the potassium and cæsium salts.

This important fact may be illustrated by the vacuum-tube model on the table, which has been specially constructed to emphasise the point. The three crystal outlines represent sections through one of the principal planes of the alkaline sulphates or selenates. The outer one represents the outline of cæsium sulphate or selenate, the middle one that of the rubidium salt, and the inner one a section of the potassium salt. The intermediate inclination of the domal faces will be clearly apparent. This is only one of some sixty angles which have been examined, and all show the same beautiful progression in the order of the atomic weights.

In the case of the principal angle of the monoclinic series, which determines the inclination of the inclined axis characteristic of that system of symmetry, the change of angle is directly proportional to the change in atomic weight.

Before concluding what may be said about the exterior morphology, reference may be made to one other fact, which follows from that just referred to and from the results of careful determinations of the specific gravity of the salts. The latter enables one to arrive at the molecular volume, which in the case of salts belonging to the same series does actually represent the relative volume of the chemical molecules. The size of the molecules of the rubidium salt was in every case found to be intermediate between the sizes of the molecules of the potassium and cæsium salts. By combining these molecular volumes with the lengths of the crystallographical axes as afforded by the angular measurements, it has been possible to determine the relative dimensions, in the three directions of space, of each chemical molecule.

This was preceded by a proof, which has been confirmed by Dr. Fock, of Berlin, from an entirely different point of view, that the unit of the crystal structure in these salts was identical with the chemical molecule, and was not an aggregate of such.

The result has been to show that the sizes, or at any rate the distances apart from centre to centre, of the chemical molecules in the three rectangular directions of space are intermediate in the case of every rubidium salt. Thus the directional dimensions of the molecules, like the angles of the crystal structure, follow the order of the atomic weights of the alkali metals.

This fact is clearly exhibited in exaggerated fashion by the model, and it will be observed that the intermediate position of the rubidium salt is somewhat nearer to that of the potassium salt than to that of the cæsium salt.

In turning now to the consideration of the optical characters of the crystals of the series of salts in question, it will be necessary to remember two main facts. The first is, that the symmetry of the rhombic system, in which the sulphates and selenates of the alkalies crystallise, determines that the velocity of light transmission shall be different along the three morphological axes of the crystals, and that two of these directions shall be those of maximum and minimum velocity respectively. In other words, the ellipsoid which, as is well known, in general represents the velocity of light transmitted through a crystalline medium, is one whose three rectangular axes are of unequal lengths, but which is fixed in direction, for these axes are identical in direction with the three morphological axes. The second is, that in the case of the monoclinic double sulphates and selenates the symmetry of the system demands that only one of the three rectangular axes of the optical ellipsoid shall be identical with a morphological axis, the latter being the one symmetry axis of the system. The ellipsoid may therefore be regarded as free to rotate about this axis, for the inclination of the other two rectangular axes of the ellipsoid, which lie in the plane of symmetry, may be any whatsoever with respect to the two inclined morphological axes which lie in that plane.

Within the three models of sections of our rhombic sulphate crystals you now see three illuminated ellipses. They represent the sections of the three optical ellipsoids; but in this case the inner one corresponds to the cæsium salt, and represents the ellipsoidal line to which light waves emanating from the imaginary centre of the crystal would penetrate in a given interval of time. The middle one shows the distance to which they would penetrate if the crystal were one of rubidium sulphate or selenate; and any point on the outer one, which is much nearer to the middle one than the outer one is, represents the position at which the light waves would arrive in the same interval of time if the crystal were one of potassium sulphate or selenate.

The velocity with which light travels through the crystals of the three salts is thus observed to vary in the same manner as does the atomic weight of the alkali metal present in the salt.

In the case of the monoclinic double sulphates and selenates, we have a further phenomenon of even greater interest exhibited. For not only does the velocity along the three rectangular axes of the ellipsoid vary with the atomic weight, but the possible rotation of the whole ellipsoid about the symmetry axis is found to actually occur; and to occur, moreover, to a very considerable extent, sometimes amounting to as much as 20°.

Further, most interesting of all, the rotation varies in a perfectly regular manner throughout all the triplets, in accordance with the atomic weight of the alkali metal present in the salt. This may be demonstrated to you by means of the lantern slide which you now see projected on the screen. You observe the outline of a crystal, arranged so that the screen represents the symmetry plane, and within it an ellipsoid. The latter is at present arranged as it is usually situated in any potassium salt of the series, its major axis being inclined so that its top is somewhat to the left of the vertical morphological axis. The ellipsoid will now be rotated to the approximate position, further on the left of the vertical axis, which it occupies in any rubidium salt of the series, by means of a simple mechanical device; and again it is rotated much further, to about the situation which it takes up in any cæsium salt.

In this phenomenon the accelerating progression according to the atomic weight of the alkali metal is beautifully exhibited.

It may now interest you to learn how these results have been obtained. We have to determine, first the position of the optical ellipsoid, and next its dimensions. That is to say, we have to determine the velocity of light transmission in all the various directions in the crystal. Determinations of the refractive index in different directions will afford us the necessary data, and if we can discover the directions of maximum and minimum velocity we shall obtain all that we require if we make the determinations of refractive index for these two directions and for a third direction at right angles to the plane containing them. For these three directions will be those of the three axes of the ellipsoid.

It will next be demonstrated to you how we discover the positions of the axes of the optical ellipsoid.

It is scarcely necessary to introduce to a Royal Institution audience the magnificent pair of Nicol prisms which formerly belonged to the late Mr. Spottiswoode. The analysing Nicol is at present arranged with its vibrating direction parallel to that of the polarising prism, and to the horizontal cross-wire which you see appearing on the screen, so that light passes to the screen.

There is also focussed the outline of a section of a monoclinic crystal, placed between the two Nicols, and which you may take as representing a crystal of one of our double sulphates or selenates; for it behaves precisely similarly, and is immensely larger than any crystal that could be obtained of one of those salts. We now rotate the analyser so that its vibrating direction is at right angles to that of the polariser, when you see the crystal section brilliantly coloured on the dark field.

If, however, we rotate the crystal section in its own plane, you observe that the coloured light becomes weaker, until, at a certain position, it is altogether extinguished. Further rotation causes light to again appear, and if we complete the circle of rotation we shall find that the crystal becomes four times dark and four times light. That is, there are two directions, at right angles to each other, in which extinction occurs. These two directions are those of two axes of the optical ellipsoid. If the crystal is so prepared as to show on its edge traces of one or two natural faces, and if our rotating stage is divided, it is quite easy to determine the angle between any given trace of a face and either of the extinction directions. In the case of monoclinic crystals the section is cut parallel to the symmetry plane, and the reference edge will usually be a trace of one of the faces in the primary zone perpendicular to the symmetry plane. That is the case with the two faces whose traces you see are left after the grinding of the large section you observe on the screen. A much more refined method of measuring the angle between the extinction direction and the basal plane, based on this principle, was employed in the research.

We will next illustrate the determination of the three refractive indices, corresponding to the three velocities along the three axes of the ellipsoid. There is here a large 60°-prism of a crystal which behaves similarly to the salts we are discussing, and which has been specially cut and polished for this lecture by Mr. Hilger. The refracting edge is parallel to an axis of the ellipsoid. Employing it in the usual manner to throw a spectrum on the screen, we observe that instead of a single spectrum, as when we use a prism of glass, we have two spectra produced, differing considerably in their dispersion. Moreover, on placing a Nicol prism in the path of the light, we see that these spectra consist of polarised light, for one extinguishes when the Nicol is arranged with its vibration direction at 0°, and the other when the Nicol is at 90°. One of the spectra, in fact, is formed by light vibrating parallel to the refracting edge, and therefore to that axis of the ellipsoid which runs parallel with it, and the other by light vibrating at right angles to that direction. If, in cutting our prism, we make this perpendicular direction to coincide with a second axis of the ellipsoid, the two spectra, when set to their minimum deviation, will, provided we know also the angle of the prism, at once afford us the data for computing the refractive indices corresponding to the two axes of the ellipsoid in question. A second prism, similarly cut so as to have vibration directions parallel to one of these axes and to the third axis, will afford us the third refractive index as well as a repetition of one of the first two.

You observe that the two spectra are considerably separated on the screen. If the two axial directions are those of maximum and minimum velocity, the amount of separation is a measure of the double refraction. Now it has been observed that the amount of the double refraction also varies progressively according to the atomic weight, and this fact may be illustrated realistically by means of a carefully made lantern slide, reproducing the actual positions, as seen in the spectrometer, of the two spectra afforded by a triplet of salts. The triplet chosen consists of potassium zinc sulphate, rubidium zinc sulphate and cæsium zinc sulphate. The spectra are represented by the images of the spectrometer slit for two wavelengths of light, those corresponding to the red and greenish-blue hydrogen lines. The spectra were produced by three prisms of the respective salts, of precisely the same angle, near 60°. You see that the two upper spectra, representing the potassium salt, are the furthest apart, and that the separation diminishes for the rubidium salt, and it becomes so much less in the case of the cæsium salt that the two lower spectra formed by this salt partially overlap.

This relative behaviour as regards double refraction is quite general throughout all the four series of salts.

The point can be further illustrated by the curves which you now see projected on the screen. The two outer curves represent the maximum and minimum refraction, and the inner one the intermediate refraction along the third rectangular axis of the ellipsoid. You observe that the two outer curves converge towards each other as the atomic weight of the alkali metal increases.

This leads in four particular cases to perhaps the most interesting of all the results derived from the work. We will illustrate it by the case of cæsium magnesium selenate. The curves which you now see on the screen are those for the magnesium triplet of double selenates. On arrival at the cæsium salt the convergence has actually brought the curve for the minimum refraction into contact with that for the intermediate refraction, so that we have here a crystal whose total double refraction is exceptionally small, and in which the velocities along two of the three ellipsoidal axes are approximately equal, which is characteristic of tetragonal and hexagonal crystals, but a perfect anomaly in a monoclinic crystal. Let us see, however, if the identity is a fact for all wave-lengths, as would be the case in a truly uniaxial crystal of tetragonal or hexagonal symmetry. We will reproduce for you the appearance in the spectrometer afforded by a prism so cut as to give us the two spectra corresponding to these two intersecting curves, that is, whose vibration directions are those of the two apparently equal axes of the velocity ellipsoid.

The images in three colours which you see are those corresponding to the wave-lengths of red lithium light, green thallium light and violet hydrogen light. You observe that they completely overlap, and if we only used the same magnification as we used for the other images we saw just now, you would say that they are absolutely identical images. We are attempting to faithfully reproduce the whole phenomena by the use of separate slides, shown by two equal lanterns, and when one of them is shut off, the effect is exactly as when a Nicol is introduced at 0° in the spectrometer, which extinguishes one of the spectra. When the other is shut off instead, what you observe is the same as if we rotated the Nicol to 90°, which would extinguish the second spectrum. Commencing with both lanterns on, as when no Nicol is being used, or if it is, it is arranged at 45°, and examining the images closely, we notice that the red and violet images are distinctly double, while the central green image is a truly single one. Now we cut off one lantern, and you please imagine that we are introducing a Nicol at 0°; the effect has been that the outermost image in the case of both red and violet has disappeared. Now shutting off the other lantern instead, and you please imagine we are rotating the Nicol to 90°, the innermost images disappear. All this time the green image remains fixed and single. This evidently means that for the middle part of the spectrum there is absolute identity of refraction and therefore of velocity, while for the red end there is a minute difference of refraction in one sense, and for the violet end a similar small difference in the opposite sense. The fact is, the two images, corresponding to the two ellipsoidal axes, for red are slightly separated; they approach and coalesce for green; then they pass each other and re-separate on the opposite side of each other as violet is approached.

Thus our crystal is only truly uniaxial for one wave-length of light, and this apparent anomalous refraction is merely the effect of **the operation of the rule of progression.**

You will have gathered that for the prosecution of this work it has been necessary to prepare some hundreds of 60°-prisms and parallel-sided section-plates, all accurately cut to the desired orientation with respect to the crystal faces. For this purpose it has been found necessary to have the delicate apparatus constructed which you see before you on the table, and a photograph of which is also thrown on the screen. The crystal, held in a grip-holder, is suspended from a refined apparatus which serves not only for the adjustment of a zone of the crystal's faces to the vertical axis of the instrument, as determined by the observer through the telescope of the collimator signal-slit, but also, as the movements are graduated, for its setting to any position with respect to the axis. Separate and interchangeable cutting and grinding gear are provided, and also a delicate means of varying the pressure of the crystal on the grinding disc, so that the most fragile crystals can be manipulated without danger of fracturing them.

It is not too much to say of this instrument, that without it the work described to you this evening would never have been possible.

Another original instrument which you see before you is an apparatus for producing spectrum monochromatic light of any wavelength whatsoever. For all the optical researches have to be carried out in pure monochromatic light, that is, for a series of colours of light, each of which is composed of vibrations of as nearly one wavelength as possible; for all the optical constants vary considerably for different wave-lengths of light. It is essentially a spectroscope constructed to transmit as large a proportion of the light as possible which streams from the condenser of an electric lantern; a broad spectrum is produced by a highly refractive prism, and is focussed on the back of a second slit, which permits only a selected line of the spectrum to escape, in the same manner as in the well-known apparatus of Sir Wm. Abney. By rotation of the prism the spectrum is moved over the exit slit, so as to permit any desired colour to escape, whose wave-length is known from the reading of the calibrated circle on which the prism is mounted.

The apparatus is placed before you just as it is arranged, in front of the goniometer-spectrometer, when determining refractive indices for a series of different wave-lengths.

There will now be introduced to you another means which we possess of determining the position and shape of the ellipsoid. In an ellipsoid having three unequal rectangular axes, it will be evident that if we consider the elliptical section containing the maximum and minimum axes, there must be a point somewhere on each elliptical quadrant where the radius vector will be equal to the intermediate axis. Such are the four points C on the figure projected on the screen. The two sections of the ellipsoid which contain these points and the intermediate axis will consequently be circles, and rays of light which pass through the crystal at right angles to these sections will be able to vibrate with equal velocity in all directions perpendicular to the line of propagation. Consequently in these two directions, which are termed optic axes, the crystal will exhibit no double refraction. When the Y velocity approaches nearer to the X velocity the inclination of the two circular sections to each other naturally diminishes, and the angle between the optic axes becomes proportionately less, until at length we have equality of X and Y, the ellipsoid becoming a spheroid, with a single optic axis. This is the special case which we observe in tetragonal and hexagonal crystals, such as the well-known cases of calcite and quartz.

We will now see some very beautiful phenomena in connection with these optic axes, employing for the purpose a projection polariscope, which is furnished with a special set of lenses to render the light strongly convergent while passing through the crystal. First let us investigate the phenomena exhibited by one of the uniaxial crystals we have just referred to. A section-plate cut perpendicular to the single optic axis is now between the crossed Nicols, and you see the beautiful spectrum rings and the black cross, characteristic of uniaxial crystals, which are produced.

We will next see the effect produced by a section-plate of a biaxial crystal, similar to the salts we are discussing. The plate is cut perpendicular to that axis of the ellipsoid which is the bisectrix of the acute angle between the optic axes. You observe the loci of the two optic axes are marked by hyperbolic brushes, in the present position of the section, and they are surrounded by separate rainbow-coloured rings, which in turn are surrounded by lemniscates and eventually ellipse-like curves. If we rotate the section 45°, the hyperbolæ join up to form a cross, but the loci of the optic axes remain marked by the rings. These rings are large and brilliant, to render the demonstration clear, and are afforded by a very large section-plate. But for research purposes we require the rings to be very small and the hyperbolæ very narrow, so that the measurement of their position may be very accurate.

We will put in another section-plate, much smaller because of the impossibility of obtaining larger ones, of rubidium magnesium selenate, and you see how small and sharp, although naturally fainter, the rings and hyperbolæ are.

Next let us demonstrate how we measure the angle of separation between the two optic axes. Another section-plate has been arranged on a small goniometer, so that it can be rotated in the plane of the axes. You see we are able to bring first one and then the other optic axis up to the cross wires, which you see also focussed on the screen.

If we note the reading of the goniometer circle while one optic axis is so adjusted, and then rotate until the other is in position and read the circle again, the difference between the two readings will give us the apparent angle between the axes as seen in air.

In order to arrive at the true angle within the crystal, it is necessary to cut another section perpendicular to the bisectrix of the obtuse angle between the axes, and to measure this obtuse angle. As, however, this large angle is usually invisible in air, owing to internal reflection, it is necessary that both angles shall be measured while the sections are immersed in some highly refractive liquid. A simple calculation, involving the two angles measured, enables us then to deduce the true optic axial angle within the crystal.

The result of a large number of such measurements of optic axial angles has been to show that in all cases where the interference figures are normal, the angle of the rubidium salt is intermediate in value between the angles of separation of the optic axes of the potassium and cæsium salts of the same triplet.

We will conclude the lecture by demonstrating to you the beautiful optic axial phenomena in the four abnormal cases to which reference was made when discussing the refraction phenomena, in which for a certain wave-length of light a prism of the crystal only gives one refraction image instead of two.

You see on the screen the type of interference figure which is given by rubidium sulphate. It is characteristic of the few known interesting salts which exhibit the phenomenon of crossed axial plane dispersion.

You see next the interference figure afforded by cæsium magnesium selenate. It is characterised by large dispersion, and it will be proved to you in a moment that for blue light this crystal also is apparently uniaxial, and that the figure in white light which you are now contemplating is due to the fact, that for the most luminous colours of the spectrum separation of the optic axes in the horizontal plane occurs.

It would be interesting to analyse this figure by showing you the curves afforded by the crystal in the pure monochromatic light from the spectroscopic illuminator. But although this can be done most brilliantly for one person at a time looking through the so illuminated observing instrument, there is not light enough for projection. But a series of six photographs, for six specific wave-lengths of light, have been taken with the aid of the apparatus, and you now see them on the screen, each with as exact a reproduction of the colour for which it was taken as possible. The first shows the separation for red lithium light, the second the diminished angle for yellow sodium light, the third the still smaller angle for green thallium light, the fourth the further approach towards the centre for greenish-blue hydrogen light, the fifth shows the uniaxial figure in light of wave-length 466 for which crossing occurs, and the sixth shows the separation in the perpendicular plane for violet hydrogen light.

We will also show you another six, to prove to you that not only does change of wave-length in the illuminating light provoke extraordinary changes in the optic axial angle, but that change of temperature is also provocative of remarkable changes of angle. This set represents the phenomena observed at about 80°. The angle for lithium light is now very much smaller than it was at the ordinary temperature; for red hydrogen light it is still smaller, and for sodium light the crossing has now actually arrived, while for thallium green, hydrogen greenish-blue, and the blue light for which crossing occurred at the ordinary temperature, the axes are separated at increasing angles in the perpendicular plane.

Our last experiment will illustrate the effect of change of temperature in the case of cæsium selenate. The behaviour of this salt is so similar to that of the long-known case of gypsum, that, as I can obtain a very much larger crystal of this beautiful mineral, suitable for projection purposes, I shall use it to demonstrate the phenomena. You see at the ordinary temperature nothing whatever beyond the fact that some light gets through to the screen. The optic axes are at present separated in the horizontal plane to such an extent that they are well outside the field. We are now warming the crystal, and you see colour making its appearance at the sides; now the axes themselves, surrounded by their coloured rings, are appearing. They approach the centre, they now coalesce to form the uniaxial cross and spectrum circles. They now separate in the vertical plane, and we remove the source of heat, the axes still continuing to separate vertically until the crystal and the metal frame in which it is being heated have taken up the same temperature. Now the motion stops, and the phenomena repeat themselves in the inverse order, once more coming to a cross and circles, and again separating in the horizontal plane, until finally they disappear on the margin of the field.

These beautiful cases of crossed axial plane dispersion, you will remember, are entirely due to the operation of the rule which we found to govern the progress of the double refraction, in accordance with the progress in the atomic weight of the alkali metal, so that these very exceptional cases are in reality strong proofs of the main generalisation derived from this work.

It has thus been amply demonstrated to you that the chief result to which these researches have led is, that the members of every series of salts known to the chemist, which differ by containing different elements of the same family group, exhibit perfectly regular variations in their exterior morphology and in their interior physical properties; and that these variations follow the order of the differences between the atomic weights of those interchangeable elements. Thus, it has been shown that the crystallographical properties of the elements are in line with all their other properties, chemical and physical, in exhibiting the same progressive character which is so conveniently expressed by their atomic weights.

We do not yet know why a particular series of salts chooses the specific type of symmetry which is common to its members, but we have reasonable ground for hope that further work in the direction indicated, together with a successful development of the interesting mathematical and geometrical researches now being conducted by several fellow-workers, on the possible modes of partitioning space and the types of molecular packing, will eventually lead to a solution of this important question.

During the researches described in this lecture many thousands of crystals have had to be prepared in a state of perfection and purity, many hundreds of section-plates and prisms cut and ground, and innumerable measurements with the most refined of instruments carried out. But one forgets the labour in the contemplation of knowledge truly gained, and one remembers only the delights of the way, the glorious phenomena of colour which one has enjoyed, and the exquisite beauty of the crystals themselves.

[A. E. T.]

GENERAL MONTHLY MEETING,

Monday, May 5, 1902.

His Grace The Duke of Northumberland, K.G. D.C.L. F.R.S.,
President, in the Chair.

The following Vice-Presidents for the ensuing year were
announced :—

The Rt. Hon. Lord Alverstone, G.C.M.G. LL.D.
Sir Frederick Bramwell, Bart. D.C.L. LL.D. F.R.S.
Donald W. C. Hood, C.V.O. M.D.
George Matthey, Esq. F.R.S.
Ludwig Mond, Esq. Ph.D. F.R.S.
The Rt. Hon. Sir James Stirling, M.A. LL.D.
Sir James Crichton-Browne, M.D. LL.D. F.R.S. *Treasurer.*
Sir William Crookes, F.R.S. *Honorary Secretary.*

Alfred Hillier, Esq. B.A. M.D.
Sydney Lupton, Esq. M.A.
Carl F. von Siemens, Esq.
Mrs. C. F. von Siemens,

were elected Members of the Royal Institution.

The Special Thanks of the Members were returned to Sir Thomas
Sanderson, G.C.B. for a Donation of £5 5s. to the Fund for the Promotion of Experimental Research at Low Temperatures.

The following Resolution, passed by the Managers, was read and
adopted :—

It was unanimously *Resolved,* That the Managers of the Royal Institution of
Great Britain desire to record their sense of the deep loss sustained by the Insti-
tution and the whole scientific world in the decease of Professor Alfred Cornu,
Membre de l'Institut, Officier de la Légion d'Honneur, Professeur à l'École Poly-
technique, D.C.L. of Oxford, Sc.D. of Cambridge, F.R.S. of London, Honorary
Member of the Royal Institution of Great Britain.
Professor Alfred Cornu as a brilliant investigator was early recognised in
this country. He has been invited by the Managers to deliver three Friday
Evening Discourses at the Royal Institution in 1875, 1879 and 1895, on the
" Velocity of Light " (in French), " Étude Optique de l'Élasticité," and " Phéno-
mènes Physiques des Hautes Régions de l'Atmosphère," in which he summarised
his important investigations.
He was made an Honorary Member when present on the occasion of the
Faraday Centenary of 1891, and attended in 1899 the celebration of the Cen-

tenary of the foundation of the Royal Institution of Great Britain, delivered an admirable speech at the Banquet, and afterwards on his return to Paris gave a report of the celebration to the French Academy of Sciences.

The Managers desire to offer to Madame Cornu and Family the expression of their most sincere sympathy and condolence with them in their bereavement.

The PRESENTS received since the last Meeting were laid on the table, and the thanks of the Members returned for the same, viz.:—

FROM

The Secretary of State for India—Archæological Survey of India, Vol. XX. 4to. 1901.
    The Jain Stupa. By Vincent C. Smith.
    Moghul Colour Decoration of Agra. By E. W. Smith. Part 1. 1901. 4to.
The Astronomer Royal—Greenwich Observations, 1899. 4to. 1901.
    Greenwich Spectroscopic and Photographic Results, 1899. 4to. 1900.
    Cape Meridian Observations, 1877–79. 2 vols. 8vo. 1901.
        Ditto   1896–97, 1898–99. 2 vols. 4to. 1901.
The British Museum (Natural History)—Catalogue of Fossil Fishes, Part 4. 8vo. 1901.
    Catalogue of Lepidoptera Phalænæ, Vol. III. and Plates. 8vo. 1901.
    Hand List of Birds, Vol. III. 8vo. 1901.
American Academy of Arts and Sciences—Proceedings, Vol. XXXVII. Parts 12–14. 8vo. 1901.
American Philosophical Society—Proceedings, Vol. XL. No. 167. 8vo. 1901.
Asiatic Society, Royal—Journal, April, 1902. 8vo.
Astronomical Society, Royal—Monthly Notices, Vol. LXII. Nos. 4, 5. 8vo. 1901.
Basle, Natural History Society of—Tycho Brahe, 1546–1601. Von Fr. Burck-hardt. 8vo. 1901.
    Verhandlungen, Band XIII.
Batavia, Royal Magnetical and Meteorological Observatory—Observations, Vol. XXII. Part 2. 4to. 1901.
Boston Public Library—Monthly Bulletin for April, 1902. 8vo.
British Architects, Royal Institute of—Journal, Third Series, Vol. IX. Nos. 11, 12. 4to. 1902.
British Association—Report of Seventy-first Meeting. 8vo. 1901.
British Astronomical Association—Journal, Vol. XII. No. 6. 8vo. 1902.
    Memoirs, Vol. XI. Part 1. 8vo. 1902.
Buenos Ayres, City—Monthly Bulletin of Municipal Statistics, Feb. 1902. 4to.
Cambridge Philosophical Society—Proceedings, Vol. XI. Part 5. 8vo. 1902.
Cambridge University Press, Syndics of the—Mathematical and Physical Papers of Sir George Stokes, Vol. III. 1902. 8vo.
Camera Club—Journal for April, 1902. 8vo.
Chemical Industry, Society of—Journal, Vol. XXI. Nos. 7, 8. 8vo. 1902.
Chemical Society—Proceedings, No. 250. 8vo. 1902.
    Journal for May, 1902. 8vo.
    List of Members, 1902. 8vo.
City of London, The Corporation of the—The Prince and Princess of Wales and the City of London and the Colonies. 12mo. 1902.
Committee of International Engineering Congress (Glasgow), 1901—Proceedings of Sections. 2 vols. 8vo. 1901.
    Handbook of the Industries of Glasgow and the West of Scotland. 8vo. 1901.
Comité du Cinquantenaire Scientifique de M. Berthelot—Cinquantenaire Scientifique de M. Berthelot, 24 Novembre, 1901. 4to. 1902.
Cracovie, l'Académie des Sciences—Bulletin, Classes des Sciences Mathématiques et Naturelles, 1901, No. 9; 1902, Nos. 2, 3. 8vo.
    Bulletin, Classe di Philologie, 1901, No. 10; 1902, Nos. 2, 3. 8vo.
Dewar, Professor, M.A. F.R.S. M.R.I.—Lehrbuch der Anorganischen Chemie. Von Prof. Dr. H. Erdmann. 3rd edition. 1902. 8vo.

Editors—Aeronautical Journal for April.
  American Journal of Science for April, 1902. 8vo.
  Anthony's Photographic Bulletin for April, 1902. 8vo
  Astrophysical Journal for April, 1902.
  Athenæum for April, 1902. 4to.
  Author for April, 1902. 8vo.
  Brewers' Journal for April, 1902. 8vo.
  Chemical News for April, 1902. 4to.
  Chemist and Druggist for April, 1902. 8vo.
  Electrical Engineer for April, 1902. fol.
  Electrical Review for April, 1902. 8vo.
  Electrical Times for April, 1902. 4to.
  Electricity for April, 1902. 8vo.
  Electro Chemist and Metallurgist for March, 1902. 8vo.
  Engineer for April, 1902. fol.
  Engineering for April, 1902. fol.
  Feilden's Magazine for May, 1902. 8vo.
  Invention for April, 1902.
  Journal of the British Dental Association for April, 1902. 8vo.
  Journal of Medical Research for April, 1902. 8vo.
  Journal of State Medicine for April, 1902. 8vo.
  Law Journal for April, 1902. 8vo.
  Machinery Market for April, 1902. 8vo.
  Motor Car Journal for April, 1902. 8vo.
  Nature for April, 1902. 4to.
  New Church Magazine for April, 1902. 8vo.
  Pharmaceutical Journal for April, 1902. 8vo.
  Photographic News for April, 1902. 8vo.
  Physical Review for April, 1902. 8vo.
  Public Health Engineer for April, 1902. 8vo.
  Science Abstracts for April, 1902. 8vo.
  Travel for April, 1902. 8vo.
  Zoophilist for April–May, 1902. 4to.
Electrical Engineers, Institution of—Journal, Vol. XXXI. No 156. 8vo. 1902.
  Report of the Committee on Electrical Legislation. fol. 1902.
Entomological Society—Transactions, Part 5, 1901. 8vo. 1902.
Franklin Institute—Journal, Vol. CLIII. No. 4. 8vo. 1902.
Geographical Society, Royal—Geographical Journal for May, 1902. 8vo.
Imperial Institute—Imperial Institute Journal for May, 1902.
Johns Hopkins University—University Circular, No. 157. 8vo. 1902.
  American Journal of Philology, Vol. XXII. No. 4. 8vo. 1901.
Kansas University—Bulletin, Vol. II. No. 7. 8vo. 1902.
Linnean Society—Journal: Zoology, Vol. XXVIII. No. 184; Botany, Vol. XXXV.
  No. 244. 8vo. 1902.
Mexico, Imperial Geological Institute—Mémoires, No. 15. 4to. 1901.
Microscopical Society, Royal—Journal, 1902, Part 2. 8vo.
Navy League—Navy League Journal for May, 1902. 8vo.
North of England Institute of Mining and Mechanical Engineers—Transactions,
  Vol. LI. Part 2. 8vo. 1902.
Odontological Society—Transactions, Vol. XXXIV. No. 5. 8vo. 1902.
Photographic Society, Royal—Photographic Journal for April, 1902. 8vo.
Physical Society—Proceedings, Vol. XVIII. Part 1. 8vo. 1902.
  List of Fellows. 8vo. 1902.
Quekett Microscopical Club—Journal, Series II, Vol. VIII. No. 50. 8vo. 1902.
Quesneville, Dr. G. (the Author)—Théorie Nouvelle de la Loupe. 8vo. 1902.
Rio de Janeiro, Observatory—Monthly Bulletin, April–June, 1901. 8vo.
Rome, Ministry of Public Works—Giornale del Genio Civile, Feb. 1902. 8vo.
Royal Irish Academy—Transactions, Vol. XXXI. Parts 12–14; Vol. XXXII.
  Section A, Parts 1, 2. 4to. 1901.

Royal Society of London—Philosophical Transactions, A, Nos. 307–309; B, Nos. 208, 209. 4to. 1902.
 Proceedings, No. 458. 8vo. 1902.
 Year Book. 8vo. 1902.
Sanitary Institute—Journal, Vol. XXIII. Part 1. 8vo. 1902. And Supplement.
Selborne Society—Nature Notes for May, 1902. 8vo.
Sidgreaves, Rev. W. S. J.—Stonyhurst College Observatory, Results of Meteorological and Magnetic Observations, 1901. 8vo. 1902.
Society of Arts—Journal for March, 1902. 8vo.
Tacchini, Prof. P. Hon. Mem. R.I. (the Author)—Memorie della Società degli Spettroscopisti Italiani, Vol. XXXI. Disp. 3. 4to. 1902.
Tasmania, Agent-General for—Handbook of Tasmania. 8vo. 1899.
 Review of Reviews of Australia, Feb. 1901. 8vo.
United Service Institution, Royal—Journal for April, 1902. 8vo.
United States Department of Agriculture—Monthly Weather Review for Dec. 1901 and Jan. 1902. 4to.
 Loss of Life by Lightning. 8vo. 1901.
 Annual Summary for 1901. 4to. 1902.
United States Geological Survey—Twenty-first Annual Report, 1899–1900, Parts 2–4. 4to.
Verein zur Beförderung des Gewerbfleisses in Preussen—Verhandlungen, 1902, Heft 4. 8vo.
Vienna, Imperial Geological Institute—Verhandlungen, 1902, Nos. 1, 2. 8vo.
Zurich Naturforschende Gesellschaft—Vierteljahrsschrift, Jahrg. XLIV. Heft 3, 4. 8vo. 1902.
 Neujahrsblatt, 104 Stück. 4to. 1902.

WEEKLY EVENING MEETING,

Friday, May 9, 1902.

HIS GRACE THE DUKE OF NORTHUMBERLAND, K.G. D.C.L. F.R.S.,
President, in the Chair.

PROFESSOR J. NORMAN COLLIE, Ph.D. F.R.S.

Exploration and Climbing in the Canadian Rocky Mountains.

IN the west of Canada lies a " Great Lone Land," a land with few
inhabitants, a land almost deserted, if we except a few prospectors,
trappers, and Indians, who spend their time among its mountain
fastnesses, either hunting wild animals or searching for gold and
minerals.   This land is where the Rocky Mountains have place, where
a mass of hills, valleys, snow-fields and rushing rivers divide the
prairie land of Alberta from the canyons of the west in British
Columbia.

Since, however, the Canadian Pacific Railway has bridged the
continent, these solitudes of the far west are more accessible to the
ordinary traveller, and the wild secluded valleys of the Canadian
Rocky Mountains are becoming more frequented.   The future of this
country is not hard to foretell; unfit for agricultural purposes but
full of the most beautiful scenery, it must, like Switzerland, become a
playground for those people who care for hunting, fishing, or
mountaineering, or for those who take a delight in spending a holiday
surrounded by snow, ice, and glaciers, mighty woods, beautiful lakes
and great rivers.   No doubt, like Switzerland, it will some day be
completely overrun.   At present, however, it is unspoilt; and as
there does not seem to be any likelihood of its ever being the centre
of any great manufacturing district, and as this mountain land is
vastly greater in extent than the Alps, for many years to come it will
remain the hunting-ground of those who can spend their spare time
living amidst fine scenery and in breathing pure air.

The exploration of that part of the mountains near the Canadian
Pacific Railway is going rapidly forward, the valleys are being
explored, the snow-fields and glaciers are being mapped, and many of
the highest peaks have been ascended.   In 1897, however, when I
first visited this country, little was known of the mountains that
lay over thirty miles away from the railway.   It is true that one or
two parties had cut their way through the dense woods in the chief
valleys, but the great expanses of snow-fields and many of the higher
passes and snow-peaks were then undiscovered.

In 1897, and during other journeys that I have made to this mountain land in 1898 and 1900, I have always been north of the railway line; the furthest north was in 1898, when I reached the head waters of the Athabasca and discovered the Columbia ice-fields, which in extent surpass any others at present known in the Canadian Rocky Mountains.

In 1897, in company with Professor H. B. Dixon, Mr. G. P. Baker and some American friends belonging to the Appelachian Club of Boston, I explored the Wapta ice-field that lies fifteen miles north of the railway at Field and a peak (Mount Gordon), situated in its very centre, was ascended from whose summit higher mountains to the north-west were seen. Later in the same year Mr. Baker and I made an attempt to reach these unknown mountains; men and horses were hired at Banff and provisions were taken to last for a month. That this mountain land was almost unknown is easy to explain, for the country is far from any human habitation, and often so difficult to get at, that it takes weeks of hard work battling with the rivers and woods before even the valleys are reached, which lie at the foot of these ranges of snow and ice-covered mountains. At first, one is quite unaccustomed to the leisurely method of progression of a "pack team" amidst heavy timber, swamps and thick underbrush; at first, one is alarmed to see the sturdy small Indian ponies, with perhaps one's most treasured belongings on their backs, being swept down stream or hopelessly floundering in a morass, but later all these things are taken as a matter of course. These accidents, and the delays consequent on the cutting away fallen timber, rescuing the ponies from the rivers, and finding the easiest way through the dense woods, all after a time become part of the day's work and travelling in these mountain solitudes is neither irksome nor unpleasant.

During the latter part of August and the beginning of September, Baker and I explored that part of the Rocky Mountain system lying about 60 miles north of where the Canadian Pacific Railway crosses the continental watershed at the Kicking Horse Pass.

Starting from Laggan, we went north up to the head of the Bow Valley, thence after crossing the Bow Pass, Bear Creek was descended to the Saskatchewan River. From this point Mount Sarbach was climbed in order that a survey of the new country we were in might be obtained. To the west were visible the high peaks we were in search of, together with vast expanses of snow-fields and glaciers that constituted the eastern slopes of the main chain of the Rocky Mountains. Amongst these we spent nearly a fortnight, surveying, photographing and exploring. But bad weather and lack of time forced us to return to civilisation before we could ascend Mount Forbes, the highest mountain in the district. However, although unable to get to the summit of Forbes, a magnificent view of the main chain had been obtained from high up on the ridge of Mount Freshfield, and far away further north higher mountains still were visible. We returned to civilisation by another route, first by going over the Howse Pass, the continental watershed was crossed, we then descended the Blaeberry Creek for some distance, but this valley at last becoming quite impassable for horses, owing to the fallen timber and dense forest, a new pass (the Baker Pass) had to be discovered, which led us finally down to Field on the railway. The most important results achieved by this journey were a plane table survey map by Mr. Baker and our discovery of higher peaks still further north.

On returning to England, the only literature that I could obtain dealing with the country we had visited was in a rare parliamentary report on the Palliser Expedition of forty years ago. There an account was given by Dr. Hector of a visit he made to this region. The existence of these unknown peaks further north that I had seen from the ridge of Freshfield, took me back in the following year (1898) to the Rocky Mountains: this time in the company of Mr. H. E. M. Stutfield and Mr. H. Woolley. Again I started from Laggan with men and horses, and travelling north by way of the Pipestone Pass the main Saskatchewan was reached. The north fork of the Saskatchewan was then ascended, but owing to the immense quantity of water in the river progress was very slow. Eventually a pass was reached (the Wilcox Pass) which led over to the head waters of the Athabasca River. From this spot the Athabasca peak (11,900 feet) was ascended; from its summit, a magnificent panorama of an almost unknown land was obtained. Stretching for miles to the north-westward lay an immense snow-field surrounded by the loftiest peaks I had yet seen in the Rockies; moreover some of these were those I was in search of. This glacier was the source of three of Canada's largest rivers, the Columbia, the Saskatchewan and the Athabasca. Later we climbed a peak (the Dome, 11,700 feet) which rises from near its centre, and it is a point of some interest that the snows of this peak when melted feed rivers that flow into three oceans, the Pacific, the Arctic and the Atlantic, also it is quite impossible that any other mountains can exist on the American continent of which the same can be said. The two highest peaks discovered on this journey were named Mount Columbia and Mount Alberta, and the snow-field was called the Columbia snow-field. In 1900 a third trip was made to this fascinating land, this time to explore the entirely unknown valleys on the western slopes of the range that drain into the Columbia River. With Mr. Stutfield and Mr. Spencer the Bush Valley was visited. Starting from Donald on the railway we first went down the Columbia Valley and then turned eastwards towards the mountains up the Bush Valley.

In the valley of the Columbia, down which we travelled for several days, we hardly saw the sky. The vast forest far surpassed in size anything we had seen on the other side of the range, huge pines, cotton-wood trees, firs and spruces, reaching to a height of 150 feet or more. The undergrowth too was very dense, whilst the fallen trunks of dead trees, sometimes six or eight feet in diameter, lay scattered with others of lesser size in every kind of position. Some in their fall had been arrested by others and were waiting for the

first gale to bring them crashing to the ground. Others that had lain
perhaps scores of years in the wet underbrush had decayed and rotted,
having rich masses of decomposing vegetation. There is a marvellous
fascination about these quiet shady fastnesses of the western valleys.
As one wanders day after day through this underworld cut off from
the glaring sun of noonday and the blue sky, hardly a sound breaks
the stillness, whilst all around lie piled the ruins of ancient woods.
In these western valleys the rainfall too is far greater than on the
other side of the range, hence the forests are thicker and the swamps
more dangerous; progression, therefore, up an unknown valley is often
very tedious. This we found to be the case in the Bush Valley. It
is true that our first view of this valley from the summit of a small
hill near its mouth, held out hopes that we should soon get to the
head waters and snow-peaks fifteen miles away. Stretched out at one's
feet as we looked down on the Bush Valley, was a wide and almost
level expanse of shingle. There were no canyons or defiles that
might necessitate lengthy detours up precipitous hillsides. It is
true we saw some swamps at the sides, but along the level bottom
stretched the shingle flats, seamed by innumerable streams, and the
main river which wound first to one side and then to the other. The
whole formed a veritable puzzle of interlacing channels, islands of
pebbles, stretches of swamps and lakes all hopelessly intermingled.
The first ten miles up that valley took us ten days' incessant work.
Our way was alternately through immense timber, dense thickets of
willows, through swamps, streams, small lakes, along insecure river
banks, climbing up the hillsides, jumping logs, cutting through fallen
trees and undergrowth so thick one could hardly see a yard in front
of one, splashing, fighting and worrying ahead; we had an experience
of almost everything that could delay us, and whether the woods,
the streams, or the swamps, were worst it was impossible to say. So
the days go by, and often real mountaineering is a luxury which has
to be left to the last. But we were the pioneers; now the trails are
partly made, and the way to get to the peaks is known, therefore the
expenditure of time in arriving at any particular spot can be calculated with much greater certainty. But with this gain in time comes
also the loss of the pleasure of the uncertainty of an unknown land.

However, it will be many a long year before much real change can
be made in the valleys that lie thirty or more miles from this line; also
the snow-peaks, the marvellously clear atmosphere, the woods, lakes,
and scenery will remain the same. After a long day through those
valleys of the Canadian Rocky Mountains, one will be just as able to
pitch one's tent and enjoy over the camp fire the stories of the hour,
to eat one's dinner with the mountaineer's appetite, to smoke by the
light of the smouldering logs, and to go to sleep safely surrounded
by the mysterious and dark forests. I always think that the supreme
moments of a mountaineer's existence are more often not whilst
battling with the great mountains, but afterwards when the struggle is
done and the whole story is gone over again quietly by a camp fire.

One such evening I remember in the Bush Valley, when no victory had crowned our efforts, in fact, we were returning from an attempt to reach Mount Columbia which had proved an undoubted failure; still somehow I felt that although beaten, we had been honourably beaten, we had struggled hard, but two things had failed us—time and provisions—we were retracing our steps towards civilisation. The camp that evening had been pitched on the banks of the Bush River. In the foreground, water and shingle stretched in desolate fashion westward, to where ridges of dark pine woods sloped down from dusky peaks above, sending out point after point to strengthen the forms of the middle distance; whilst beyond, far across the Columbia, the Selkirk mountains raised their snow-peaks into the calm, clear sky, a mysterious land unexplored and unknown. Through a rift in the clouds in the far west shone the setting sun, tinging the dull grey clouds overhead and the stealthily flowing river below with its many coloured fires. A faint evening breeze softly moved the upper foliage, a couple of inquisitive chipmunks were chattering near at hand, and a small stream could be heard whispering amongst the thickets down by the banks of the river.

The great gnarled trunks of pine and fir festooned with moss, fungi, and dry lichen, the dead drooping branches and the half fallen decaying trunks propped up in dreary melancholy array, caught for a moment the sunset's ruddy glow, whilst the shadows of the dense forest darkened by contrast. And as the evening gradually passed into the mysterious night, the stillness, the solitude, and the remoteness of these great woods became more evident and quite beyond description. Such evenings compensate one for many a wet dreary day spent amongst the mountains. Nature suddenly offers them to the traveller without any toil on his part. He has only, surrounded by the dark forest, to sit watching the stretch of waters and the ever-changing glory of the setting sun; then, unmindful of the worries of yesterday, or the uncertainties of to-morrow, amidst the great stillness he feels with absolute conviction one thing and one thing only—that it is good to be alive and free. Civilisation teaches us much, but when one has tasted once the freedom of the wilds a different knowledge comes. The battling with storm, rain, cold, and sometimes hunger, and the doubt of what any day may bring forth, these at least teach that life—that mere existence—is beyond all price.

[J. N. C.]

WEEKLY EVENING MEETING,

Friday, May 16, 1902.

Sir Frederick Bramwell, Bart. D.C.L. LL.D. F.R.S. M. Inst. C.E.,
Vice-President, in the Chair.

Sir Robert Ball, M.A. LL.D. D.Sc. F.R.S.,
Lowndean Professor of Astronomy and Geometry, Cambridge,
and Director of the Cambridge Observatory.

The Nebular Theory.

I STAND here to-night with a grave task before me. I am called upon to expound, so far as my powers will permit, an exceptionally great subject. How puny do all other things appear in comparison with the great Nebular Theory! Our personal affairs, the affairs of the country, the affairs of the Empire—indeed all human affairs, past, present and future—shrink to insignificance in comparison with what is revealed in that mighty chapter from the book of Nature which we hope to open.

The grand transformations through which the solar system has passed, and is even now at this very moment passing, cannot be seen by us poor creatures of a day, they might perhaps be surveyed by beings whose pulses counted centuries instead of seconds, by beings whose minutes were longer than the duration of dynasties, by beings to whom an hour was far longer than all human history.

The Sun appears constant in size and constant in lustre during the brief interval of human observation, but the Sun has not always been the same, it did not always shine as it does now, nor will it continue for ever to shine as it does at present. Our great luminary is smaller at the end of each year than it was at the beginning, the same is true through indefinitely great periods of time. In a retrospect we see the Sun ever larger and larger, there was a time uncounted millions of years ago when the Sun had ten times the diameter that it now possesses, there was a time when the materials which now form the Sun were expanded into a volume of diameter greater than the diameter of the Earth's orbit at the present moment. But even when the Sun was millions of times as big as it is now it was not heavier—there could not have been appreciably more material in it, though that material was enormously rarefied. Thus our reasoning makes us think of an epoch when the Sun was very different indeed from the globe which we know so well. It had then no Earth to cherish with warmth and gladden with light. Our globe was in those days truly "without form and void." At the time when the

Sun was swollen out into this great ball of glowing gas the materials of the Earth were in a condition utterly different from their present state. The Earth was then part of the great nebula itself from which the Sun and all the solar system has been formed.

Laplace bade us imagine a great primæval nebula or fire-mist. He reminded us how this must be incessantly radiating its heat and gradually contracting. Laplace demonstrated that it was infinitely probable such a nebula would have some movement of rotation; he pointed out the remarkable dynamical law by which the contracting object would gradually accelerate its rotation, and he showed how the inner parts would thus revolve more quickly than the outer parts. Laplace bade us see how the denser parts of the nebula at the centre gradually drew themselves together so as to form a sun. He showed in like manner how the outer parts of the nebula gradually cohered together to form the planets. By strictly dynamical reasoning Laplace thus pointed out how from an extensive diffused nebula a solar system with Sun and planets all complete, could be duly evolved in the lapse of ages. Laplace bade us see how the subsidiary systems of satellites appropriate to each planet came into being, he made it plain that these satellites would revolve around their primaries, just as the primaries revolve around the Sun, he bade us follow in imagination the progress of the whole system, from the widely diffused nebulosity—a mere stain of milky light in the sky on the one hand, to an organised system of revolving worlds on the other.

If Laplace lived now there are many objects in the heavens to which he could point in vindication of his theory. The age of photography has dawned and the photographic plates have not only illustrated in the most marvellous manner the Spiral Nebulæ discovered by Lord Rosse, but they have succeeded in disclosing many other spiral nebulæ. The photographs have even revealed on the plate beautiful spiral nebulæ invisible to any human eye, no matter what may be the telescope to which it is applied. How strikingly do the spiral nebulæ elucidate Laplace's theory. We see in the Great Spiral how the central part is condensed—doubtless in consequence of the fact that the nebulous matter is drawing itself together. We see how the future Sun may gradually become evolved, we see how planets, also originally "without form and void," gradually come into shape, drawing as they do so, their material from the same primæval fire-mist. We have in the photograph of the Great Spiral a marvellous illustration of those principles of celestial evolution which Laplace laid down for the formation of the solar system. I try to imagine the astonishment and delight with which Kant or Laplace would look on a photograph of the Great Spiral. If we sought for the best picture of the great primæval fire-mist which has evolved into the solar system, I feel confident we could not obtain anything so effective as a photograph of this wonderful object.

I am permitted to illustrate this lecture by a series of photographs which have been most kindly sent to me by Professor Hale, the Director of the Yerkes Observatory, and Professor Campbell, the Director of the Lick Observatory, and I have also pictures taken by Dr. Isaac Roberts and by Dr. W. E. Wilson. These plates suggest the wonderful variety and abundance of the nebulous contents of the heavens, and throw much light on the Nebular Theory of the solar system. It seems to follow from the researches of the late lamented Professor Keeler that enormous numbers of spiral nebulæ lie within the reach of our photographic plates. Indeed it is not too much to say that next to a fixed star itself, the spiral nebula is the most characteristic object in the heavens. The significance of this statement in connection with the Nebular Theory can hardly be overestimated. There can be little doubt that at one stage of the history of the solar system the gradually evolving nebula must have formed an object of that type which we term spiral.

There is also another most remarkable discovery of modern times which has added much weight to the arguments in favour of the Nebular Theory. If the Sun and the Earth—to confine our attention solely to those two bodies—had originated from the primæval nebula, they would bear with them, as a mark of their common origin, a striking identity in material and composition. We do not of course mean that the nebula was homogeneous all through, Nature does not like homogeneity. The nebula was evidently irregular, vague in form, dense in some places, greatly rarefied in others. We by no means assert that if we compared a sample of the nebula in one place with a sample of the same nebula taken a hundred or a thonsand million miles away from it, that the two samples would show identity of chemical composition. We need not be surprised at this, remembering that two samples of rock from the same quarry would not be identical. But we may feel confident that the elements present in the nebula will be more or less widely dispersed through it, so that if two globes are formed by concentration in different parts of the nebula, we might reasonably expect that though these two globes would not be actually identical yet that the elementary bodies which entered into their composition would be in substantial agreement. If one element, say iron, was abundant in one body, we should reasonably expect that the same element would not be absent from the other. Laplace had no means of testing this surmise, but our modern methods enable us to investigate the chemistry of the Sun, and have shown that the elements of which the Sun is composed are practically the same elements as those of which our Earth is built. Is not this a weighty piece of testimony in favour of Laplace's theory?

Laplace knew not of these photographic and spectroscopic revelations; he based his belief in the Nebular Theory mainly on a remarkable deduction from the theory of probabilities. If the evidence thus derived seemed satisfactory to Laplace one hundred years ago, this same line of evidence, strengthened as it has been by recent discoveries, is enormously more weighty now.

Laplace was able to count up about thirty instances in which movements of revolution and movements of rotation in the solar system had a common direction. The mathematical mind of Laplace drew a remarkable inference from this unanimity. Here, he said, are thirty different movements; each of them might have been either from right to left or from left to right, but, as a matter of fact, they are all unanimous. Laplace showed that there was a thousand millions to one in favour of this unanimity being the result of some common cause, and the nebular theory offered such a cause. For as the great nebula was rotating it carried round with it, of course, the masses of nebulæ which were ultimately to form the planets. As each of the planets broke away from the central mass it was found to be revolving, just as the Moon revolves round the Earth, as the mass contracted further, its motion accelerated and the planet came to turn more quickly on its axis, though still at all times revolving in the same direction. In this way the unanimity of the movements was a natural consequence of the Nebular Theory, and no other method has ever been suggested by which so remarkable a concord could have arisen. Laplace deemed that the thirty common movements were sufficient to establish his argument.

But the modern discoveries have enormously strengthened the original argument. There are now 500 objects which revolve around the Sun, and they all move in the same direction. The numerical expression for the probability of the truth of the Nebular Theory has to be correspondingly amended. The argument has been strengthened billions of times.

The attention of astronomers at this moment is directed to the marvellous nebula associated with the new star in Perseus. Photographs taken at the Yerkes Observatory show that the nebula is in a state of rapid motion. This remarkable announcement has been confirmed by a similar series of photographs taken at the Lick Observatory. Such a discovery would be of interest were it only for the circumstance that it is the first occasion on which motion in a nebula has been certainly observed. But the most important circumstance is the extraordinarily high velocity of the motion. Nothing is known of the parallax of this star, except that it is too small to be appreciable, we cannot therefore state exactly to what its velocity amounts. It seems that it must be at least twenty thousand miles a second, but it may be even ten times as much.

Whether the phenomenon is velocity of actual matter, or whether the observations may not admit of some other explanation, must for the present remain undetermined.

In a notable lecture which Huxley gave in the year 1880 on the coming of age of 'The Origin of Species,' he mentioned some great discoveries in Geology and Biology which had taken place since the publication of the immortal work. He then announced that these fresh discoveries brought such wonderful corroboration and illustration of the truth of Darwin's theory that if the famous doctrine of natural selection had not been formed to account for the facts of

Nature as they were known to Darwin when he wrote his book, the theory of natural selection would have had to be formed to explain the facts which had been brought to light in the twenty-one years which succeeded. We may, perhaps, make a similar claim on behalf of the great Nebular Theory. If that theory had not been invented to account for the phenomena which were known to Laplace, it would have had to be invented for the purpose of explaining the additional discoveries which have been made in the century which has since run its course.

[R. B.]

WEEKLY EVENING MEETING,

Friday, May 23, 1902.

Sir James Crichton-Browne, M.D. LL.D. F.R.S., Treasurer and Vice-President, in the Chair.

The Rev. Canon Ainger, M.A. LL.D.

The Ethical Element in Shakespeare.

[No Abstract.]

WEEKLY EVENING MEETING,

Friday, May 30, 1902.

Sir William Crookes, F.R.S., Honorary Secretary and
Vice-President, in the Chair.

Dr. J. A. Fleming, M.A. F.R.S. *M.R.I.*,
Professor of Electrical Engineering, University College, London.

The Electronic Theory of Electricity.*

Considerable progress has been made of late years in our knowledge
concerning the structure and relations of atoms and electricity.
Recent discoveries have moreover placed in a new light old theories
and experimental work. The remarkable investigations and deductions made from his own experiments and those of others, which
have led Professor J. J. Thomson to the conclusion that atoms can
be split up into, or can give off, smaller masses, which he calls
corpuscles, have been explained by him on many occasions.† There
seems to be good evidence that in a glass vessel exhausted to a
high vacuum, through the walls of which are sealed platinum wires,
we have a torrent of small bodies or so-called corpuscles projected
from the kathode or negative wire, when the terminals are connected
to an induction coil or electrical machine.

Twenty-five years ago Sir William Crookes explored with wonderful skill many of the effects due to electric discharge through such
high vacua, and came to the conclusion that they could only be
explained by the supposition that there was present in the tube
matter in a *fourth state*, neither solid, liquid, nor gaseous, but
' radiant matter' projected in straight lines from the surface of the
negative pole or kathode, the particles moving with immense velocity,
and all charged with negative electricity. He showed by beautiful
experiments that this radiant matter bombarded the glass walls and
produced phosphorescence, could be focussed on to metal sheets and
render them red hot, and could drive round little windmills or vanes
included in the tube. It therefore possesses the quality of inertia,

---

* The following pages do not contain a verbatim reproduction of the discourse
delivered on this occasion, but are a reprint of an article in the ' Popular Science
Monthly,' for May 1902, by the lecturer, covering substantially the same ground,
and reproduced here by kind permission of the editor, Professor J. McKeen
Cattell.

† See ' Popular Science Monthly,' vol. lix. p. 323, " On Bodies smaller than
Atoms," by Professor J. J. Thomson, F.R.S. See also by the same author a
paper in the ' Philosophical Magazine ' for December 1899, " On the Masses of
the Ions in Gases at Low Pressures."

and, in consequence of the electric charge it carries, it is virtually an electric current, and can be deflected by a magnet. The proof which has been given by Professor Thomson that this 'radiant matter' consists of corpuscles, a thousand times smaller than an atom of hydrogen in mass, and that they are shot off from the kathode with a velocity which is comparable with that of light, explains at once both their kinetic energy and also the manner in which they are able to pass through windows of aluminium, as shown by Lenard, and get into the space outside the tube. Furthermore, evidence has been put forward to show that the electric charge carried by each one of these tiny corpuscles is exactly the same as that which a hydrogen atom carries in the act of electrolysis or when it forms a hydrogen ion.

It seems tolerably clear from all the facts of electrolysis that electricity can only pass through a conducting liquid or electrolyte by being carried on atoms or groups of atoms which are called *ions*— i.e. *wanderers*. The quantity thus carried by a hydrogen atom or other monad element, such as sodium, silver or potassium, is a definite natural unit of electricity. The quantity carried by any other atom or group of atoms acting as an ion is always an exact integer multiple of this natural unit. This small indivisible quantity of electricity has been called by Dr. Johnstone Stoney an *electron* or *atom of electricity*. The artificial or conventional unit of electric quantity on the centimetre-gramme-second system, as defined by the British Association Committee on Electrical Units, is as follows :

An *electrostatic unit* of electric quantity is the charge which when placed upon a very small sphere repels another similarly charged sphere, the centres being one centimetre apart, with a mechanical force of one dyne. The *dyne* is a mechanical unit of force, and is that force which acting for one second on a mass of one gramme gives it a velocity of one centimetre per second. Hence, by the law of inverse squares the force in dynes exerted by two equal charges Q at a distance D is equal to Q^2/D^2. Two other units of electric quantity are in use. The *electromagnetic unit*, which is thirty thousand million times as great as the electrostatic unit, and the *practical unit* called the coulomb or ampère-second, which is three thousand million times the electrostatic unit. We can calculate easily the relation between the *electron* and the *coulomb* ; that is, between *Nature's unit of electricity* and the British Association unit, as follows :

If we electrolyse any electrolyte, say acidified water which yields up hydrogen at the negative electrode, we find that to evolve one cubic centimetre of hydrogen gas at 0° C. and 760 mm. we have to pass through the electrolyte a quantity of electricity equal to 8·62 coulombs. For 96,540 coulombs are required to evolve one gramme of hydrogen and 11,200 cubic centimetres at 0° C. and atmospheric pressure weigh one gramme. The number 8·62 is the quotient of 96,540 by 11,200.

Various arguments, some derived from the kinetic theory of gases, indicate that the number of molecules of hydrogen in a cubic centimetre is probably best represented by the number twenty million million million $= 2 \times 10^{19}$. Hence it follows, since there are two atoms of hydrogen in a molecule, that in electrostatic units the electric charge on a hydrogen atom or hydrogen ion is

$$\frac{96540 \times 3 \times 10^9}{11200 \times 4 \times 10^{19}} = \frac{65}{10^{11}} \text{ of a C.G.S. electrostatic unit}$$

$$= \frac{22}{10^{20}} \text{ of a coulomb.}$$

Accordingly, if the above atomic charge is called *one electron* then the conventional British Association electrostatic unit of electric quantity is equal to 1540 million electrons, and the quantity called a coulomb is nearly five million million million electrons. The electron or the electric charge carried by a hydrogen atom or ion is evidently a very important physical constant. If we electrolyse, that is decompose by electricity aqueous solutions of various salts, such as sodium chloride, zinc chloride, copper sulphate, silver nitrate, we find, in accordance with Faraday's Laws of Electrolysis, that the passage of a given quantity of electricity through these solutions decomposes them in proportional amounts such that for every 46 grammes of sodium liberated there are 65 of zinc, 63·5 of copper and 216 of silver. These masses are called chemical equivalents. Accordingly, if we imagine a number of vessels placed in a row containing these solutions and by means of platinum connecting links or plates we pass an electric current through the series, for every atom of copper or zinc carried to their respective kathodes, we shall have two atoms of silver or sodium similarly transported. Since the same quantity of electricity must pass through every vessel in the same time, it is evident that the above fact may be interpreted by assuming that whilst an atom of silver or sodium acting as an ion carries one electron, an atom of zinc or copper carries two electrons.

In the same way we may have atoms which carry three, four, five or six electrons. Thus we may interpret the facts of chemical valency and Faraday's Law of Electrolysis in terms of the electron.

We are thus confronted by the idea long ago suggested by Weber and by Von Helmholtz, that the agency we call electricity is *atomic* in *structure*, that is to say, we can only have it in amounts which are all exact multiples of a certain small unit. Electricity therefore resembles those articles of commerce like cigars, which we can buy in exact numbers, 1, 10, 50, 100, 1000, but we cannot buy half a cigar or five-sixths of a cigar. If then the law which holds good for electricity in association with atoms during electrolysis holds good generally, a very important advance has been made in establishing the fact that there is a small indivisible unit of it which can be multiplied but not divided, and every quantity of electricity, small or large, is an exact integer multiple of this unit, *the electron*.

Theories of Electricity.

Various answers have been given at different times to the question —What is electricity? It has been defined as an imponderable fluid, as a force, as a mode of motion, a form of energy, an ether strain or displacement or a molecular motion.

At one time physicists have considered it as a single entity or fluid; at others it has been pronounced to be duplex in nature, and positive and negative fluids or electricities have been hypothecated.

The state of electrification has been looked upon at one period as due to an excess or defect of a single electricity, at others as a consequence of the resolution of some neutral fluid into two components. An electrical charge on a conductor has been regarded as something given to or put upon the conductor, and also as a state of strain or displacement in the surrounding non-conductor. The intelligent but non-scientific inquirer is often disappointed when he finds no simple, and as he thinks essential, answer forthcoming to the above question, and he asks why it cannot be furnished.

We must bear in mind, however, that scientific hypotheses as to the underlying causes of phenomena are subject to the law of evolution and have their birth, maturity and decay. Theory necessarily succeeds theory, and whilst no one hypothesis justified by observations can be looked upon as expressing the whole truth, neither is any likely to be destitute of all degree of truth if it sufficiently reconciles a large number of observed facts.

The notion that we can reach an absolutely exact and ultimate explanation of any group of physical effects is a fallacious idea. We must ever be content with the best attainable sufficient hypothesis that can at any time be framed to include the whole of the observations under our notice. Hence the question—What is electricity?—no more admits of a complete and final answer to-day than does the question What is Life? Though this idea may seem discouraging, it does not follow that the trend of scientific thought is not in the right direction. We are not simply wandering round and round, chasing some elusive will-o'-the-wisp, in our pursuit after a comprehension of the structure of the universe. Each physical hypothesis serves, as it were, as a lamp to conduct us a certain stage on the journey. It illuminates a limited portion of the path, throwing a light before and behind for some distance, but it has to be discarded and exchanged at intervals because it has become exhausted and its work is done.

The construction and testing of scientific theories is therefore an important part of scientific work. The mere collection of facts or even their utilisation is not the ultimate and highest goal of scientific investigation. The aim of the most philosophic workers has always been to penetrate beneath the surface of phenomena and discover those great underlying fundamental principles on which the fabric of nature rests. From time to time a fresh endeavour has to be made to reconstruct, in the light of newly acquired knowledge, our scientific theory of any group of effects. Thus, the whole of electrical phenomena have become illuminated of late years by a theory which has been developed concerning the atomic structure of electricity, and this hypothesis is called the Electronic Theory of Electricity.

The Atomic Theory.

The opinion that matter is atomic in structure is one which has grown in strength as chemical and physical knowledge has progressed. From Democritus, who is said to have taught it in Greece, to John Dalton who gave it definiteness, and to Lord Kelvin who furnished the earliest numerical estimate of the size of atoms, in spite of adverse criticism, it has been found to be the best reconciler of very diverse and numerous observed effects. Let us consider what it really means. Suppose we take some familiar substance, such as common table salt, and divide a mass of it into the smallest grains visible to the eye. Each tiny fragment is as much entitled by all tests to be called table *salt*, or to give it the chemical name, sodic chloride, as a mountain of the material. Imagine that we continue the subdivision under a good microscope; we might finally obtain a little mass of about one hundred-thousandth of an inch in diameter, but beyond this point it would hardly be visible even under a powerful lens. We may, however, suppose the subdivision continued a hundredfold by some more delicate means until we finally arrive at a small mass of about one ten-millionth of an inch in diameter. A variety of arguments furnished by Maxwell, Boltzmann, Loschmidt, Lord Kelvin and others show that there is a high degree of probability that any further subdivision would cause the portions into which the salt is divided to be no longer identical in properties, but there would be two kinds of parts or particles, such that if all of one kind were collected together they would form a metal called sodium, and if all of the other kind were similarly picked out they would form a non-metal called chlorine. Each of these smallest portions of table salt, which if divided are no longer salt, is called a *molecule* of sodic chloride, and each of the parts into which the molecule is divisible is called an *atom*, of sodium or of chlorine. In dealing with the dimensions of these very small portions of matter an inch or a centimetre is too clumsy a unit. To express the size of an atom in fractions of an inch is worse than stating the diameter of an apple in fractions of a mile. Every one knows what is meant by a millimetre; it is nearly one twenty-fifth part of an inch. A metre is equal to a thousand millimetres. Suppose a millimetre divided into a thousand parts. Each of these is called a *micron* and denoted by the Greek letter μ. This however is still too large a unit of length for measuring the size of atoms, so we again divide the micron into a thousand parts and call each a micromillimetre or *micromil*, and denote it by the symbol $\mu\mu$. Lord Kelvin's estimate of the diameter of a molecule is that it lies between one hundredth of a micromil and two micromils, that is between $\cdot 01 \, \mu\mu$ and $2 \, \mu\mu$. This is certainly a very wide estimate, but it is the best yet to hand, and for present purposes we may take it that an atom is a small portion of matter of approximately one millionth of a millimetre or one micromil ($1 \, \mu\mu$) in diameter. On the same scale the wave-length of a ray of yellow light is about $0 \cdot 6 \, \mu$ or $600 \, \mu\mu$ that is six hundred times the size of an atom. We know nothing as yet about the relative sizes of different kinds of atoms. In the next place, as regards the number of molecules in a given space, various distinguished physicists, Maxwell, Kelvin, Boltzmann, Van der Waals and others, have given estimates for the number of molecules in a cubic centimetre of air at ordinary temperature and pressure, which vary between 10^{18} and 10^{21}, between a million billion and a thousand million billion. All we can do is to take a rough mean of these different values, and we shall consider that in one cubic centimetre of hydrogen or other gas at 0° C. and 760 mm. or freezing point and ordinary pressure there are about 2×10^{19} or twenty million million million molecules. To understand what this enormous number means we must realise that if we could pick out all the molecules in one cubic inch of air and place them side by side in a row, small as they are individually, the row would extend nearly twice the distance from the earth to the sun.

Having provided ourselves with a rough idea of the sizes and numbers of the molecules of any gas, we proceed to obtain an idea of their weight or mass. Since 11,162 cubic centimetres of hydrogen gas at 0° C and 760 mm. weigh one gramme, it follows from the above facts that each molecule of hydrogen has a mass of nearly $1/10^{23}$ of a gramme. To weigh these tiny atoms we must therefore take a unit of weight equal to one-billionth of one-billionth of a gramme and then on this scale the hydrogen molecule weighs 10 such units. We may obtain in another way an illustration of the mass, size and number of the molecules of any gas in the following manner :

First as to size. We can, in a good Whitworth measuring instrument, detect a variation in length of a metal bar equal to one millionth of an inch. This short length would be occupied by 25 molecules placed in a row close together. We can in a good microscope see a small object whose diameter is one hundred-thousandth of an inch. In a small box of this size we could pack 16 million molecules close together. The smallest weight which can be weighed on a very good chemical balance is one hundredth of a milligramme. The united weight of one million million million molecules of hydrogen would therefore just be detectable on such a balance.

Ultra-Atomic Matter.

Until a few years ago our knowledge of the divisibility of matter may be said to have ended with the chemical unit, the atom. But of late years information has been steadily accumulating which has made us acquainted with matter in a finer state of subdivision. For a long time a controversy was carried on, whether the radiation in a high vacuum tube which proceeds from the kathode was a material substance or a wave motion of some kind. But no fact yet found is inconsistent with the notion which originated with Sir William Crookes that the transfer which takes place is that of something which has the inertia quality of matter, and his term "radiant matter" is a peculiarly suitable phrase to describe the phenomena. The great advance which has since been made, by Professor J. J. Thomson and others, is that of measuring accurately the amount of bending which a stream of this radiant matter experiences under a known magnetic force, and from this deducing the ratio between the mass of the radiant particle and the electric charge carried by it. This measurement shows that if the radiant matter consists of corpuscles or particles, each of them carries a charge of one electron, but has a mass of about one-thousandth of a hydrogen atom.

The evidence therefore exists that Crookes' "radiant matter" (also called the "kathode rays") and Thomson's "corpuscles," are one and the same thing, and that these corpuscles may be described as fragments broken off from chemical atoms and possessing only a small fraction of their mass. These particles are shot off from the negative terminal or kathode of the vacuum tube with a velocity which is from one-fifth to one-third the velocity of light.

Moreover, it has been shown that when the kathode rays pass through a thin metal window in a vacuum tube and get into the space outside, thus forming Lenard's rays, they are likewise only the same or similar corpuscles in the space outside rather than inside the vacuum tube. Finally it has been proved that these electrified corpuscles are present as well in the mass of a gas through which Röntgen rays have passed, also in the mysterious radiation called Becquerel rays which proceeds from uranium and other radio-active substances, also in all flames, near all very hot bodies and in the air near certain metallic surfaces, on which ultra-violet light falls. In every case the corpuscle is charged with an electron charge of negative electricity. If a corpuscle originates as a fragment chipped off from an electrically neutral atom and is negatively charged, it follows that the remainder of the atom of matter is left positively charged.

The word "atom" therefore, as far as it signifies something which *cannot be cut,* is becoming a misnomer as applied to the chemical unit of matter, because this latter is capable of being divided into two parts of very unequal size. First, a small part which is negatively electrified and which is identically the same, no matter from what chemical atom it originates, and secondly, a much larger mass which is the remainder of the atom and is positively electrified, but which has a different nature depending on the kind of chemical atom broken up. The question has then begun to be debated whether we can distinguish between the corpuscle and the electric charge it carries, and if so in what way. In other words, can we have an unelectrified corpuscle, or is the corpuscle so identified with its electric charge that they are one and the same thing? It has been shown experimentally that an electric charge in motion is in effect an electric current, and we know that an electric current possesses something equivalent to inertia, that is, it cannot be started and stopped instantly, and it possesses energy. We call this electric inertia *inductance*, hence the question arises whether the energy of the corpuscles when in motion is solely due to the electric inductance or whether it is partly due to what may be called the ponderable inertia of the corpuscle.

This very difficult question has not yet been even approximately settled. At the present moment we have no evidence that we can separate the electron charge from the corpuscle itself. If this is the case, then the corpuscles taken together constitute for all practical purposes negative electricity, and we can no more have anything which can be called electricity apart from corpuscles than we can have momentum apart from moving matter. For this reason it is sometimes usual to speak of the corpuscle carrying its charge of one electron of negative electricity simply as *an electron*, and to drop all distinction between the electric charge and the vehicle in or on which it is conveyed.

It is remarkable that so far no one has been able to produce or find a corpuscle positively electrified. Positive electricity is only known in association with masses as large as atoms, but negative electricity is united with corpuscles or masses only a small fraction of the size of an atom. This does not prove that an atom may not include positive corpuscles or electrons, but only that so far we have not been able to isolate them.

The Electronic Theory of Electricity.

From this point of view a theory of electricity originates called the electronic theory. The principal objects of consideration in this theory are these electrons which constitute what we call electricity. An atom of matter in its neutral condition has been assumed to consist of an outer shell or envelope of negative electrons associated with some core or matrix which has an opposite electrical quality, such that if an electron is withdrawn from the atom the latter is left positively electrified.

A neutral atom *minus* an electron constitutes the natural unit of positive electricity, and the electron and the neutral atom *minus* an electron are sometimes called negative and positive ions. Deferring for a moment a further analysis of possible atomic structure we may say that with the above hypothesis in hand we have then to express our statements of electrical facts in terms of the electron as the fundamental idea.

All that can be attempted here is a very brief exposition of the success which has so far attended this effort to create a new range of electrical conceptions. Let us consider first the fundamental difference between substances in respect of electrical conductivity. In the electronic theory, what is the distinction between conductors and non-conductors? It must be remembered that on the electronic hypothesis an electric current is a movement of electrons. Hence a conductor must be a substance in which electrons free to move exist. It is considered therefore that in metals and good conductors a certain proportion of the atoms are broken up into positive and negative ions or into electrons and remainders of atoms which we may call coelectrons. There may be a constant decomposition and recomposition of atoms taking place, and any given electron so to speak flits about, now forming part of one atom and now of another and anon enjoying a free existence. It resembles a person visiting from house to house forming a unit in different households and in between being a solitary person in the street. In non-conductors, on the other hand, the electrons are much restricted in their movements, and can be displaced a little way but are pulled back again when released. The positive and negative ions or electrons and coelectrons never have the opportunity to part company very far.

The reader who is familiar with the modern doctrine of the ionization of salts in solution will see that a close similarity exists between this view of the atomic state of a metal and the chemical state of a salt in solution. The ionic theory of solution is that if some salt, say sodic chloride, is placed in water a certain proportion of the molecules of sodic chloride are dissociated into sodium and chlorine ions, that is to say, atoms possessing electric charges, and the electric conductivity of the solution is due to the mobility of these saline ions.

On the electronic theory a certain proportion of the atoms of a conductor are similarly in a state of electronization. The application of an electromotive-force to the conductor thus at once causes the electrons to begin to migrate. If we compare conductors and non-conductors we shall see that the former are mostly elementary bodies, the metals and alloys or graphitic carbon, whilst the latter are all very complex substances such as glass, ebonite, the oils, shellac, gutta-percha, etc. These last have large and complex molecules, but the good conductors have all simple molecules and small atomic volumes. The exceptions apparently are sulphur and carbon in the form of diamond. When, however, we remember that carbon and sulphur are elements very prone to polymerise and so to speak combine with themselves they may not really be an exception. The electrons may, therefore have much more difficulty in exchanging from atom to atom or in making their way between or through the molecules when these are very complex than when they are simple.

The question then may be asked why these free electrons do not all escape from the conductor. The answer is that there must be an equal quantity of electrons and coelectrons or remainders of atoms or of so-called negative and positive ions and the strong attraction between these involves the expenditure of work to separate them. The radio-active substances, such as uranium, polonium, radium, actinium and others, to which so much attention has been paid lately, do seem to have the power of emitting their corpuscles or electrons and scattering them abroad, and hence can only do this at the expense of some of their own internal molecular energy or else drawing upon the heat of surrounding bodies.

We come next to the explanation of the familiar fact of electrification by friction. Why is it that when we rub a glass rod with a bit of silk the two things are equally and oppositely electrified? To explain this on the electronic theory we have to consider the state of affairs at the surface of any substance immersed say in air. At the surface where the air and glass meet there will be an electronization of atoms which appears to result in the formation of a double layer of electrons and coelectrons or negative and positive ions. This is probably an attempt on the part of the glass and air to combine chemically together. The same state exists at the surface of the silk. When we rub these two things together these double layers are very roughly treated and are broken up. The whole lot of electrons and coelectrons or residual portions of atoms get mixed up and more or less divided up between the two surfaces. As however every negative electron has its positive coelectron, it follows that what one surface gains the other must lose. Hence in the end we may have a majority of negative ions or electrons left on the one surface and a majority of positive ions or coelectrons left on the other surface; and the glass and the silk are then electrified with equal quantities but opposite signs. Owing to the mutual repulsion of the similar electrons the charge resides wholly on the surface.

This conception of the existence of a double layer of opposite electricities or ions at the surface of contact of two substances has been put forward to account for the familiar effect of the electrification of air by falling drops of water. It has long been known that the air in the neighbourhood of waterfalls of fresh water is electrified negatively, whereas the air in the neighbourhood of splashing salt water, as at the seaside, is positively electrified, and the explanation that has been given by Professor J. J. Thomson is that this is due to the breaking up of this double layer of ions at the surface of the drop when it strikes the ground.

Atomic Valency.

At this stage it may be well to indicate that any valid theory of electricity must involve an explanation of the facts of chemical combination and chemical valency as well. At present all ideas on the structure of atoms must necessarily be purely speculative. So much advance has been made however in the development of a department of chemistry called stereo-chemistry that we need not despair of coming to know in time much about the architecture of atoms and molecules.

The way is cleared, however, for some consistent explanations if we can assume that one or more free electrons can attach themselves to a neutral atom and so give it a negative charge of electricity. We may suppose as a first assumption that in a neutral atom which is otherwise complete, there exist localities at which one or more electrons can find a permanent attachment. The atom is then no longer neutral but negatively electrified. If the atom can as it were accommodate *one* electron it is a monovalent element, if *two* it is divalent, and so on. If it cannot accommodate any at all it is an *avalent* or *non-valent* element.

Consider the case of gaseous molecules. Chemical facts teach us that the molecules of free gaseous hydrogen, oxygen or other gases contain two atoms, so that these free molecules are represented by the symbols H_2, O_2, etc. In these cases hydrogen and oxygen are so to speak combined with themselves. We can explain this by the supposition that most neutral atoms are unstable structures. In contact with each other some lose one or more electrons and an equal number gain one or more electrons. Hence in a mass say of hydrogen we have some atoms which are positively electrified and some which are negatively electrified then called atomic ions, and these ions united pair and pair form the molecules of hydrogen which may be represented by $(\overset{+}{H}, \overset{-}{H})$. Similarly for other gases. Certain neutral atoms such as those of argon are monatomic and non-valent and these appear to be unable to enter into combination either with each other or with other atoms. Accordingly, in a mass of free hydrogen there are no free electrons and all the positively charged and negatively charged H atoms are in union. Hence the gas is a non-conductor of electricity. But we can make it a conductor by heating it to a high temperature. The explanation of this is that a high temperature dissociates some of the molecules into atoms and these under the action of electric force move in opposite directions, thus creating an electric current. Thus air at ordinary temperatures is an almost perfect non-conductor, but at a white heat it conducts electricity freely.

The monovalent elements like hydrogen are those neutral atomic structures which can lose one electron or take up one electron, becoming respectively positive atomic ions and negative atomic ions. In the same way the divalent elements such as oxygen are those neutral atomic structures which can part with two electrons and take up two, and so on for trivalent, quadrivalent, etc., atoms. The work required to remove the second electron probably is very much greater than that required to remove the first. Hence in polyvalent atoms the valencies have unequal energy values.

Consider now a mass of intermingled oxygen and hydrogen consisting of neutral molecules. The state is a stable one as long as all the molecules are neutral. If, however, we dissociate a few of the hydrogen and oxygen molecules by an electric spark or by heat then there is a recombination. A positive oxygen ion unites with two negative hydrogen ions and a negative oxygen ion with two positive hydrogen ions and the result is two neutral molecules of water. This combination takes place because the union of oxygen ions with hydrogen ions to form water evolves more heat and exhausts more potential energy than the combination of oxygen with oxygen and hydrogen with hydrogen ions in equivalent quantity. The energy set free by the union of the O and H is sufficient to continue the dissociation of further gaseous molecules, so the action is explosive and is propagated throughout the mass.

There is however a broad distinction between the elements in this respect, viz.: that some atoms are prevalently electropositive and others electronegative. A metallic atom for instance is electropositive, but the atoms of non-metals are mostly electronegative. Moreover metals in the mass are electrically good conductors, whereas non-metals in the mass are non-conductors or bad conductors. This may be explained by the varying degree of force required to detach electrons from neutral atoms and conversely the varying degree of attachment of electrons for neutral atoms. Thus we may consider that the metallic atoms lose very easily one or more electrons, and also that there is a somewhat feeble attachment in their case between the neutral atom and the free electron. Hence metals in the mass are conductors because there are plenty of free electrons present in them. On the other hand, in the case of non-metallic atoms the force required to detach one or more electrons from the atom is much greater, and conversely the attachment of free electrons for the neutral atom is larger. Accordingly, in non-metals there are few free electrons, and they are therefore nonconductors. Moreover, the presence of positive and negative atomic ions causes them to link together into more or less complex molecules, and they exhibit polyvalency and act as the grouping elements in molecular complexes. This is a very characteristic quality of the elements sulphur, silicon and carbon.

Helmholtz long ago laid stress on the fact that certain physical and chemical effects could only be explained by assuming a varying attraction of electricity for matter. The same idea followed out leads to an hypothesis of chemical combination and dissociation of salts in solution. Thus a molecule of sodic chloride is the electrical union of a monovalent sodium ion or sodium atom *minus* one electron with a chlorine ion which is a chlorine atom *plus* one electron. It may be asked why in this case does not the extra electron pass over from the chlorine to the sodium ion and leave two neutral atoms. The answer is because the union between the electron and the chlorine is probably far more intimate than that between the atomic groups. These latter may revolve round their common centre of mass like a double star, but the electron which gives rise to the binding attraction may be more intimately attached to the atomic group into which it has penetrated.

Voltaic Action.

Any theory of electricity must in addition present some adequate account of such fundamental facts as voltaic action and magneto-electric induction. Let us briefly consider the former. Suppose a strip of copper attached to one of zinc and the compound bar immersed in water to which a little hydrochloric acid has been added.

All chemical knowledge seems to point to the necessity and indeed validity of the assumption that the *work* required to be done to remove an electron from a neutral atom varies with the atom. Conversely the attraction which exists between a free electron and an atom deprived of an electron also varies. Accordingly the attraction between atomic ions, that is, atoms one of which has gained and one of which has lost electrons, is different. Upon this specific attraction of an atomic ion for electrons or their relative desire to form themselves into neutral molecules depends what used to be called chemical affinity. Mr. Rutherford has shown that negative ions gave up their charges more readily to some metals than others, and most readily to the electro-positive metals. Hence a zinc atomic ion is more ready to take up electrons and again become neutral than a copper ion.

Consider then the simple voltaic couple above described. In the electrolyte we have hydrogen ions which are H atoms *minus* an electron, and chlorine ions which are chlorine atoms *plus* an electron. These are wandering about in a menstruum which consists of water molecules and hydrochloric acid molecules. Then in the metal bar we have zinc and copper divalent ions which are these atoms each *minus* two electrons, and also an equivalent number of free and mobile electrons.

If we adopt Volta's original view of contact electricity, we must assume that at the surface of contact of the metals there is some action which drives electrons across the boundary from the zinc to the copper. This may be due to the neutral copper atom having a slightly greater attraction for electrons than the neutral zinc atom. The zinc is therefore slightly electrified positively and the copper negatively. Accordingly in the electrolyte the negative chlorine ions move to the zinc and combine with positive zinc ions, forming neutral zinc chloride, two chlorine ions going to one zinc ion. The hydrogen ions therefore diffuse to the copper side and each takes up a free electron from the copper, becoming neutral hydrogen atoms and there escape.

In proportion as the zinc atomic ions are removed from the zinc bar and the corresponding free electrons from the copper, so must there be a gradual diffusion of electrons from the zinc bar to the copper bar across the metallic junction. But this constitutes the voltaic current flowing in the circuit. It is a current of negative electricity flowing from zinc to copper and equivalent to a positive current from copper to zinc. The energy of this current arises from the differential attraction of zinc and copper ions for chlorine ions, and is therefore the equivalent of the exhaustion of the chemical potential energy of the cell. Thus the electronic theory outlines for us in a simple manner the meaning of voltaic action. Even if we do not admit the existence of a metallic junction volta contact force, the theory of the cell may be based on the view that the movement of the saline ions in the electrolyte is determined by the law that that motion takes place which results in the greatest exhaustion of potential energy. Hence the chlorine ions move to the zinc and not to the copper.

In the same manner the electronic theory supplies a clue to the explanation of the production of an electric current when a conductor is moved across a magnetic field. Every electron in motion creates a magnetic force. Hence a uniform magnetic field may be considered as if due to a moving sheet of electrons. The ' cutting ' of a conductor across a magnetic field will therefore be accompanied by the same reactions as if a procession of electrons were suddenly started in it. This, however, would involve at the moment of starting a backward push on surrounding electrons, just as when a boat is set in motion by oars the boat is pushed forward and the water is pushed back. Hence there is an induced current at the moment when the field begins in the conductor. Similarly the reaction at stopping the procession would drag the surrounding electron with it. Accordingly the induced current when the field ceases is in the opposite direction to that when it begins.

The electronic theory has in the hands of other theorists such as Professors P. Drude and E. Riecke been known to be capable of rendering an account of most thermomagnetic effects on metals, contact electricity, the so-called Thomson effects in thermoelectricity, and also the Hall effect in metals when placed in a magnetic field.

Electrons and Æther.

The ultimate nature of an electron and its relation to the æther has engaged the attention of many physicists, but we may refer here more particularly to the views of Dr. J. Larmor whose investigations in this difficult subject are described in his book on ' Æther and Matter ' and also in a series of important papers in the ' Transactions ' of the Royal Society of London, entitled ' A Dynamical Theory of the Electric and Luminiferous Medium.' * Larmor starts with the assumption of an æther which is a frictionless fluid, but possesses the property of inertia; in other words, he assumes that its various parts can have motion with respect to each other and that this motion involves the association of energy with the medium. He regards the electron as a strain centre in the æther, that is as a locality from which æther strain radiates. Electrons can therefore be either posi-

* Phil. Trans. Roy. Soc., 1893, 1895, 1898.

tive or negative according to the direction of the strain, and to every positive electron there is a corresponding negative one. Atoms according to him are collocations of electrons in stable orbital motion like star clusters or systems.

An electron in motion is in fact a shifting centre of æther strain and it can be displaced through a stationary æther just as a kink or knot in a rope can be changed from place to place on the rope.

An electron in vibration creates an æther wave, but it radiates only when its velocity is being accelerated and not when it is uniform.

The type of æther which Larmor assumes as the basis of his reasoning is one which has a rotational elasticity, that is to say, the various portions of it do not resist being sheared or slid over each other, but they resist being given a rotation round any axis. Starting from these postulates and guided by the general and fundamental principle of Least Action, he has erected a consistent scheme of molecular physics in which he finds an explanation of most observed facts.

The discovery by Zeeman of the effects of a strong magnetic field in triplicating or multiplicating the lines in the spectrum of a flame placed in a magnetic field meets with an obvious explanation when we remember that the effect of a magnetic field on an electron in motion is to accelerate it always transversely to its own motion and the direction of the field. Hence it follows that a magnetic field properly situated will increase the velocity of an electron rotating in one direction and retard it if rotating in another. But a linear vibration may be resolved into the sum of two oppositely directed circular motions and accordingly a magnetic force properly applied must act on a single spectral line, which results from the vibration of an electron in such manner as to create two other lines on either side, one representing a slightly quicker and the other a slightly slower vibration.

The notion of an electron or point charge of electricity as the ultimate element in the structure of matter having been accepted, we are started on a further inquiry as to the nature of the electron itself. It is obvious that if the electron is a strain centre or singular point in the æther, then corresponding to every negative electron there must be a positive one. In other words, electrons must exist in pairs of such kind that their simultaneous presence at one point would result in the annihilation of both of them.

On the view that material atoms are built up of electrons we have to seek for a structural form of atom which shall be stable and equal to the production of effects we find to exist.

The first idea which occurs is that an atom may be a collection of electrons in static equilibrium. But it can be shown that if the electrons simply attract and repel each other at all distances according to the law of the inverse square no such structure can exist. The next idea is that the equilibrium may be dynamic rather than static, that an atom may consist of electrons, as suggested by Larmor, in orbital motion round each other, in fact that each atom is a miniature solar system.

Against this view, however, Mr. T. H. Jeans * has pointed out that an infinite number of vibrations of the electrons would be possible about each state of steady motion and hence the spectrum of a gas would be a continuous one and not a bright-line spectrum.

If we are to assume an atom to consist wholly of positive and negative electrons or point charges of electricity, Mr. Jeans has suggested that we may obtain a stable structure by postulating that the electrons, no matter whether similar or dissimilar, all repel each other at very small distances.

We might then imagine an atom to be built up of concentric shells of electrons like the coats of an onion alternately positive and negative, the outermost layer being in all cases negative. The difference between the total number of positive and negative electrons is the valency of the atom.

On this view an atom of hydrogen would consist of from 700 to 1000 positive and negative electrons arranged in concentric layers in a spherical form. The vibrations which emit light are not those of the atom as a whole but of the individual electrons which compose it.

The reason for assuming that in all cases the outermost layer of electrons is negative is that if it were not so, if some atoms had their outer layers of negative and some of positive electrons, two atoms when they collided would become entangled and totally lose their individuality. There would be no permanence. Hence our present atoms may be, so to speak, the survivors in a struggle for existence which has resulted in the survival only of all atoms which are of like sign in the outer layer of electrons. We see an instance of a similar action in the case of the like directed rotation of all the planets round the sun which is due to the operation of the law of conservation of angular momentum. As a consequence of the equality of sign of the outer layer of electrons two atoms cannot approach infinitely near to each other. They mutually repel at very small distances. This suggestion affords a possible clue to the reason why we only know at present free negative electrons; it is because we can only detach a corpuscle or electron from the outer layer of an atom. It is clear, however, that the complete law of mutual action of electrons has yet to be determined. We have also to account for gravitation, and this involves the postulate that all atomic groups of electrons without regard to sign must attract each other. Hence we need some second Newton who shall formulate for us the true law of action of these electrons which form the " foundation stones of the material universe." Facts seem to suggest that the complete mathematical expression for the law of mutual action of two electrons must show:

1. That at exceedingly small distances they must all repel each other without regard to sign.

---

* 'Mechanism of Radiation,' Proc. Phys. Soc. Lond., vol. xvii. p. 760.

2. That at greater distances positive electrons must repel positive and negative repel negative, but unlike electrons attract, with a force which varies inversely as the square of the distance.

3. Superimposed on the above there must be a resultant effect such that all atoms attract each at distances great compared with their size without regard to the relative number of positive and negative electrons which compose them, inversely as the square of the distance.

In this last condition we have the necessary postulate to account for universal gravitation in accordance with Newton's law.

It is conceivable, however, that this differential or resultant universal attraction to which gravitation is due, is only true of electrons when gathered together so as to form atoms. In order words, every atom attracts every other atom; but every electron does not attract every other electron. Universal gravitation may be an effect due to the collocation of electrons to form atoms and molecules, but not an attribute of electrons in themselves, though, if the gravitative effect is proportional to the product of the total number of electrons in each mass, the Newtonian law will be fulfilled. It has been also suggested that a sufficient source for the necessary resultant mass attraction may be found in a slight superiority of the attractive force between two opposite electrons over the repulsion between two similar electrons.

Conclusion.

In the above sketch of the electronic theory we have made no attempt to present a detailed account of discoveries in their historical order or connect them especially with their authors. The only object has been to show the evolution of the idea that electricity is atomic in structure, and thus these atoms of electricity called electrons attach themselves to material atoms and are separable from them. These detachable particles constitute as far as we yet know negative electricity. The regular free movements of electrons create what we call an electric current in a conductor, whilst their vibrations when attached to atoms are the cause of æther waves or radiation, whether actinic, luminous, or thermal. The æther can only move and be moved by electrons. Hence it is the electron which has a grip of the æther and which, by its rapid motions, creates radiation, and in turn is affected by it. We have therefore to think of an atom as a sort of planet accompanied by smaller satellites which are the electrons. Moreover the electrons are capable of an independent existence, in which case they are particles of so-called negative electricity, The atom having its proper quota of electrons is electrically neutral, but with electrons subtracted, it is a positive atomic ion, and with electrons added to it it is a negative atomic ion. It has been shown from a quantitive study of such diverse phenomena as the Zeeman effect, the conductibility produced in gases by Röntgen rays or by ultra-violet light and from the magnetic deflection of kathode rays, that in all cases where we have to deal with free moving, or vibrating electrons, the electric charge they carry is the same as that conveyed by a hydrogen atom in electrolysis.

There is good ground for the view that when a gas is made ineandescent, either by an electric discharge or in any other way, the vibrating bodies which give rise to the light waves are these electrons in association with the atom. The energy of mass movement of the atom determines temperature, but the fact that we may have light given out without heat, in short, *cold light*, becomes at once possible if it is the vibrating electric particle attached to the atom which is the cause of eye-affecting radiation or light.

Lorentz, Helmholtz, Thomson and others have shown that such a conception of atomic structure enables us to explain many electro-optic phenomena which are inexplicable on any other theory. Maxwell's theory that electric and magnetic effects are due to strains and stresses in the æther, rendered an intelligible account of electric phenomena, so to say, in empty space, and its verification by Hertz placed on a firm basis the theory that the agencies we call electric and magnetic force are affections of the æther. But the complications introduced by the presence of matter in the electric and magnetic fields presented immense difficulties which Maxwell's theory was not able to overcome.

The electronic theory of electricity, which is an expansion of an idea originally due to Weber, does not invalidate the ideas which lie at the base of Maxwell's theory, but it supplements them by a new conception, viz., that of the electron or electric particle as the thing which is moved by electric force and which in turn gives rise to magnetic force as it moves. The conception of the electron as a point or small region towards which lines of strain in the æther converge, necessitates the correlative motion of positive and negative electrons. We are then led to ask whether the atom is not merely a collocation of electrons. If so, all mechanical and material effects must be translated into the language of electricity. We ought not to seek to create mechanical explanations of electrical phenomena but rather electrical ones of mechanical effects. The inertia of matter is simply due to the inductance of the electron, and ultimately to the time element which is involved in the creation of æther strain in a new place. All the facts of electricity and magnetism are capable of being re-stated in terms of the electron idea. All chemical changes are due to the electric forces brought into existence between atoms which have gained or lost electrons. If moving electrons constitute an electric current, then electrons in rotation are the cause of magnetic effects. In optics it is capable of giving a consistent explanation of dispersion, absorption and anomalous dispersion and the relation of the index of refraction to the dielectric constant. A scientific hypothesis, with this wide embrace, which opens many closed doors and enables us to trace out the hidden connection between such various departments of physical phenomena, is one which must continue to attract investigators. Physical inquirers are at present, however, groping for guiding facts in this difficult field of investigation, but we have confidence that mathematical and experimental research will in due time bring the reward of greater light.

[J. A. F.]

GENERAL MONTHLY MEETING,

Monday, June 2, 1902.

His Grace the DUKE OF NORTHUMBERLAND, K.G. D.C.L. F.R.S.,
President, in the Chair.

John Christie, Esq.
Francis Kennedy McClean, Esq.
Sir John Denison Pender, K.C.M.G.

were elected Members of the Royal Institution.

The PRESENTS received since the last Meeting were laid on the
table, and the thanks of the Members returned for the same, viz. :—

FROM

Accademia dei Lincei, Reale, Roma—Classe di Scienze Fisiche, Matematiche e
Naturali. Atti, Serie Quinta: Rendiconti. 1o Semestre, Vol. XI. Fasc.
7-9. 8vo.
Classe di Scienze Morali, Storiche, etc. Serie Quinta, Vol. XI. Fasc. 1, 2. 8vo.
American Geographical Society—Bulletin, Vol. XXXIV. No. 2. 8vo. 1902.
Antiquaries, Society of—Archæologia, Vol LVII. Part 2. 4to. 1902.
Proceedings, Vol. XVIII. No. 11.
Astronomical Society, Royal—Monthly Notices, Vol LXII. No. 6. 8vo. 1901.
Bankers, Institute of—Journal, Vol. XXIII. Part V. 8vo. 1902.
Bartholomew's Hospital, St.—Statistical Tables for 1901. 8vo. 1902.
Belgium, Royal Academy of Sciences—Bulletin, 1901 ; 1902, 1-3. 8vo.
Annuaire, 1902. 8vo.
Mém. Cour. et des savants étrang. Tome LIX. Fasc. 1, 2. 4to. 1901.
Mém. Cour. et autres Mém. Tome LXI. 8vo. 1902.
Mémoires, Tome LIV. Fasc. 1-4. 4to. 1901.
Berlin, Academy of Sciences—Sitzungsberichte, 1902, Nos. 1-22. 8vo.
Boston Public Library—Monthly Bulletin for May, 1902. 8vo
Boston Society of Natural History—Proceedings, Vol. XXIX. Nos. 15-18 ; Vol.
XXX. Nos. 1, 2. 8vo. 1901.
Index to North American Orthoptera. By S. H. Scudder. 8vo. 1901.
British Architects, Royal Institute of—Journal, Third Series, Vol. IX. Nos. 13, 14.
4to. 1902.
British Astronomical Association—Journal, Vol. XII. No. 7. 8vo. 1901.
Brooks, H. Jamyn, Esq. (the Author)—The Elements of Mind. 8vo. 1902.
Brymner, D. Esq.—Report on Canadian Archives, 1901. 8vo.
Buenos Ayres, City—Monthly Bulletin of Municipal Statistics, March, 1902. 4to.
California, University of—Publications, 1901. 8vo.
Cambridge University Library—Report for the year 1901. 4to. 1902.
Camera Club—Journal for May, 1902. 8vo.
Canada, Geological Survey of—Contributions to Canadian Palæontology, Vol. II.
Part 2 ; Vol. IV. Part 2. 8vo. 1900-1.
Catalogue of Marine Invertebrata of Eastern Canada. 8vo. 1901.
Chemical Industry, Society of—Journal, Vol. XXI. Nos. 9, 10. 8vo. 1902.

Chemical Society—Proceedings, Nos. 251, 252. 8vo. 1902
Journal for June, 1902. 8vo.
Chicago, Field Columbian Museum—Publications:
Annual Report for 1900–1901. 8vo.
Geological Series, Vol. I. Nos. 10, 11. 8vo. 1901.
Congress, Library of—Report of the Librarian for 1901. 8vo. 1902.
Dewar, Professor, M.A. F.R.S. M.R.I.—Early Metallurgy in Europe. By W
Gowland. 4to. 1899.
Editors—American Journal of Science for May, 1902. 8vo.
Astrophysical Journal for April, 1902.
Athenæum for April, 1902. 4to.
Author for June, 1902. 8vo.
Brewers' Journal for May, 1902. 8vo.
Chemical News for May, 1902. 4to.
Chemist and Druggist for May, 1902. 8vo.
Electrical Engineer for May, 1902. fol.
Electrical Review for May, 1902. 8vo.
Electrical Times for May, 1902. 4to.
Electricity for May, 1902. 8vo.
Engineer for May, 1902. fol.
Engineering for May, 1902. fol.
Homœopathic Review for May, 1902. 8vo.
Horological Journal for June, 1902. 8vo.
Invention for May, 1902.
Journal of the British Dental Association for May, 1902. 8vo.
Journal of Medical Research for May, 1902. 8vo.
Journal of State Medicine for May, 1902. 8vo.
Law Journal for May, 1902. 8vo.
London Technical Education Gazette for April, 1902.
Machinery Market for May, 1902. 8vo.
Mois Scientifique, May, 1902. 8vo.
Motor Car Journal for May, 1902. 8vo.
Nature for May, 1902. 4to.
New Church Magazine for May, June, 1902. 8vo.
Nuovo Cimento for April, 1902. 8vo.
Pharmaceutical Journal for May, 1902. 8vo.
Photographic News for May, 1902. 8vo.
Photographic Times Bulletin for May, 1902. 8vo.
Physical Review for April, May and June, 1902. 8vo.
Public Health Engineer for May, 1902. 8vo.
Science Abstracts for May, 1902. 8vo.
Travel for May, 1902. 8vo.
Zoophilist for May, 1902. 4to.
Florence, Reale Accademia di—Atti, Vol. XXV. Disp. 1." 8vo. 1902.
Franklin Institute—Journal, Vol. CLIII. No. 5. 8vo. 1902.
Geographical Society, Royal—Geographical Journal for May, 1902. 8vo.
Geological Society—Quarterly Journal, Vol. LVIII. Part 2. 8vo. 1902.
Geological Survey of the South African Republic—Annual Report for the year
1898. 4to. 1902. (Translated from the Dutch.)
Horticultural Society, Royal—Journal, Vol. XXVI. Part 4. 8vo. 1902.
Imperial Institute—Imperial Institute Journal for April, 1902.
Johns Hopkins University—Studies, Series XIX. Nos. 10–12; Series XX. No. 1,
8vo. 1901–2.
Life-Boat Institution, Royal National—Annual Report, 1901. 8vo.
Manchester Literary and Philosophical Society—Memoirs and Proceedings,
Vol. XLVI. Part 5. 8vo. 1902.
Meteorological Society, Royal—Meteorological Record, Vol. XXI. No. 83. 8vo. 1901.
Quarterly Journal, Vol. XXVIII. No. 122. 8vo. 1902.
Hints to Meteorological Observers. By W. Marriott. 8vo. 1902.

ontana, University of—Bulletin, No. 3. Summer Birds of Flathead Lake. 8vo. 1901.

Munich, Royal Bavarian Academy of Sciences—Sitzungsberichte, 1902, Heft I. 8vo.

Navy League—Navy League Journal for May, 1902. 8vo.

Numismatic Society—Numismatic Chronicle, 1901, Parts 3, 4; 1902, Part 1. 8vo.

Odontological Society—Transactions, Vol. XXXIV. No. 6. 8vo. 1902.

Photographic Society, Royal—Photographic Journal for April, 1902. 8vo.

Queen's College, Galway—Calendar for 1901-2. 8vo. 1902.

Royal Society of Edinburgh—Proceedings, Vol. XXIV. No. 2. 8vo. 1902.

Royal Society of London—Philosophical Transactions, A, Nos. 310, 311. 4to. 1902.

Proceedings, Nos. 459, 460. 8vo. 1902.

Reports to the Evolution Committee, First Report 8vo. 1902.

Saxby & Farmer, Messrs.—Railway Safety Appliances, 1902.

Selborne Society—Nature Notes for May and June, 1902. 8vo.

"*Shakespearean,*" *A (the Author)*—The Shakespeare Anagrams as used by Ben Jonson. 8vo. 1902.

Smith, B. Leigh, Esq. M.R.I.—Transactions of the Institution of Naval Architects, Vol. XLIII. 4to. 1901.

The Scottish Geographical Magazine, Vols. XVII. and XVIII. Nos. 1-4. 8vo.

Smithsonian Institution—Sixteenth Annual Report, 1901-2.

Experiments with Ionized Air. By C. Barus. 4to. 1902.

Society of Arts—Journal for May, 1902. 8vo.

Tacchini, Prof. P. Hon. Mem. R.I. (the Author)—Memorie della Società degli Spettroscopisti Italiani, Vol. XXXI. Disp. 4. 4to. 1902.

United Service Institution, Royal—Journal for May, 1902. 8vo.

United States Department of Agriculture—Experiment Station Record, Vol. XIII. Nos. 7, 8. 8vo. 1902.

Wind Velocity on Lake Erie. 4to. 1902.

Verein zur Beförderung des Gewerbfleisses in Preussen—Verhandlungen, 1902, Heft 5. 8vo.

Vienna, Imperial Geological Institute—Verhandlungen, 1902, Nos. 3-6. 8vo.

Washington, Philosophical Society of—Bulletin, Vol. XIV. pp. 179-204. 8vo. 1902.

Western Society of Engineers—Journal, Vol. VII. No. 2. 8vo. 1902.

Zoological Society—Report of the Council for 1901. 8vo. 1902.

F G.

WEEKLY EVENING MEETING,

Friday, June 6, 1902.

H.R.H. THE PRINCE OF WALES, K.G. G.C.V.O. LL.D. F.R.S.,
Vice-Patron, in the Chair.

SIR BENJAMIN BAKER, K.C.M.G. LL.D. D.Sc. F.R.S. *M.R.I.*,
Past President Institution of Civil Engineers.

The Nile Dams and Reservoir.

MR. CECIL RHODES, last Christmas, when riding across the hot and
dusty desert between Aswân and the Nile Reservoir works, incidentally remarked that, after all, there was no climate like England's;
and as for rain, why, it did good and hurt nobody. Glancing around
at the apparently limitless desert on all sides of us, the hills and
valleys, beautiful in form, but doomed for all time to remain of
uniform burnt-brick hue, bare of trees, and of the many-coloured
growths which adorn a rainy country, one could not but reflect how
puny were the efforts of man when attempting to combat any decree
of Nature. The desert lands of Egypt will remain desert, however
many millions of pounds are expended in Nile reservoirs. All that
man can do is to extend somewhat the narrow strip of green running
along the banks of the Nile, and to render that and the other low-
lying lands more productive than they are at present with a scanty
supply of water (Fig. 1).

The Nile Reservoir at Aswân will contain over 1000 million
tons of water. This statement will probably convey little meaning
to most people; and in truth the quantity may be made to appear
either small or large at will by a judicious selection of illustrations.
Thus the absolute insignificance to Egypt of 1000 million tons of
water in a reservoir, as compared with a reasonable rainfall, will be
apparent at once when it is considered that the annual rainfall on the
area included within the four-mile cab radius from Charing Cross is
about 100 million tons, and that the rainfall on London and its
suburbs within a thirteen-mile radius would, therefore, about suffice
to fill the Nile Reservoir. On the other hand, we may, by choosing
other illustrations, restore the Nile Reservoir to the dignity of its
just position of one of the greatest engineering works of the day.
Thus the question of the water supply of London, and its prospective
population of 11¼ millions, has been prominently before the public
for some years; and many will remember what was termed the colossal
project of our member, Sir Alexander Binnie, late Engineer of the
London County Council, for constructing reservoirs in every reasonably available valley in Wales, to store up water for London, and to

supply compensation water to the Welsh rivers affected thereby. Well, the united contents of the whole of those reservoirs would be less than half that of the great Nile Reservoir. Again, the Nile Reservoir would hold more than enough water for one year's full domestic supply to every city, town, and village in the United Kingdom with its 42 million inhabitants. But possibly the best way of giving an idea of the magnitude of the work, and its utility to cultivation in a thirsty land, is by considering the volume of the water issuing from the Reservoir during the three or four summer months, when scarcity of supply prevails in the river and the needs of the cultivators are greatest. At that time the flow from the Reservoir will be equivalent to a river double the size of the Thames in mean annual flood condition. It will be recognised at once that a good many buckets would have to be set at work to bale out a river like that, and yet the scarcity of water in the Nile itself, and in the canals, during the months of April, May and June, is such that even dipping the water out of the channels in buckets has to be controlled by strict regulations. Thus, two years ago, when the Nile was below the average in summer discharge, it was decreed in Upper Egypt that the "lifting machines," which include the shadoof, or bucket-and-pole system, and the sakieh, or oxen-driven chain of buckets, should be worked not more than from five to eleven consecutive days, and stop the following nine to thirteen days, between the middle of April and the middle of July; and the order in which the different districts were to receive a supply was carefully specified, so that, as far as possible, every crop should get watered once in about three weeks. When it is remembered that a single watering of an acre of land means, where shadoofs are concerned, raising by manual power about 400 tons of water to varying heights up to 25 feet, and that four or five waterings are required to raise a summer crop, it will be seen what a vast amount of human labour is saved throughout the world by the providential circumstance that in ordinary cases water tumbles down from the clouds, and has not, as in Egypt, to be dragged up from channels and wells. Shadoof work, under average conditions, involves one man's labour for at least one hundred days for each acre of summer crop; so that even at 6*d.* per day for labour, the extra cost of cultivation due to the absence of rain would amount to 50*s.* per acre.

The great Nile Reservoir and Dam at Aswân, the Barrage at Asyût, and various supplementary works in the way of distributing canals and regulators, are designed with the object of mitigating the evils enumerated above, by supplying in summer a larger volume of water at a higher level in the canals, so that not only can more land be irrigated, but that labour in lifting water will be saved. When the International Commission, eight years ago, recommended the construction of a large reservoir somewhere in the Nile valley, I was desirous of knowing what would be the opinion of a real old-fashioned native landowner on the subject; and was introduced to one whose qualifications were considered to be of no mean order, as he was a descendant of the Prophet, very rich, and had been twice warned by the Government that he would probably be hanged if any more bodies of servants he had quarrelled with were found floating in the Nile. He was a very stout old man, and, between paroxysms of bronchial coughing, he assured me that there could be nothing in the project of a Nile reservoir, or it would have been done at least 4000 years ago. In contrast with this I may mention that, a few months ago, the most modern and enlightened of all the rulers of Egypt, the present Khedive, when visiting the Dam, said he was proud that the great work was being carried out during his reign, and that the good services rendered by his British engineers was evidenced by the London County Council coming to his Public Works Staff for their chief engineer.

The old system of irrigation, which the descendant of the Prophet looked back upon with regret, was little more than a high Nile flooding of different areas of land or basins surrounded by embankments. Less than a hundred years ago, perennial irrigation was first attempted to be introduced, by cutting deep canals to convey the water to the lands when the Nile was at its low summer level. When the Nile rose, these canals had to be blocked by temporary earthen dams, or the current would have wrought destruction. As a result, they silted up, and had to be cleared of many millions of tons of mud each year by enforced labour, much misery and extortion resulting therefrom. About half a century ago, the first serious attempt to improve matters was made by the construction of the celebrated barrage at the apex of the Delta. This work consists, in effect, of two brick arched viaducts crossing the Rosetta and Damietta branches of the Nile, having together 132 arches of 16 ft. 4 in. span, which were entirely closed by iron sluices during the summer months, thus heading up the water some 15 ft., and throwing it at a high level into the main irrigation canals below Cairo. The latter are six in number, the largest being the central canal at the apex of the Delta, which, even in the exceptionally dry time of June 1900, was carrying a volume of water one-fourth greater than the Thames in mean flood, whilst the two canals right and left of the two branches of the river carried together one-half more than the Thames, and the Ismailieh Canal, running down to the Suez Canal, though starved in supply, was still a river twice the size of the Thames at the same time of the year. At flood times the discharges of all the canals are, of course, enormously increased. It will be recognised at once, therefore, that, as in the summer months the whole flow of the Nile is arrested and thrown into the aforesaid canals, the old barrage will always remain the most important work connected with the irrigation of Egypt. It was constructed under great difficulties by French engineers, subject to the passing whims of their Oriental chiefs. About fifteen years elapsed between the commencement of the work and the closing of all the sluices, and

another twenty years before the structure was sufficiently strengthened
by British engineers to fulfil the duties for which it was originally
designed. All the difficulties arose from the nature of the foundations,
as the timber sheet piling wholly failed to prevent the substructure
from being undermined by the head of water carrying away the fine
sand and silt upon which the barrage was built. At Asyût, cast-iron
sheet piling was used, as will hereafter be described. It is impossible
to say what the cost of the old barrage has been from first to last,
but probably nearly ten times that of the recently-completed Asyût
Barrage. Forced labour was largely employed in its construction,
and at one time 12,000 soldiers, 3000 marines, 2000 labourers, and
1000 masons were at work at the old barrage.

In connection with the Nile Reservoir, subsidiary weirs have been
constructed below the old barrage to reduce the stress on that
structure. The system adopted was a novel one, reflecting great
credit on Major Brown, Inspector-General of Irrigation in Lower
Egypt. His aim was to dispense almost entirely with plant and
skilled labour; and so, without attempting to dry the bed of the
river, he made solid masonry blocks under water, by grouting
rubble dropped by natives into a movable timber caisson. Both
branches of the Nile were thus dammed in three seasons, at a cost,
including navigation locks, of about half a million sterling. Many
other subsidiary works have been and will be constructed, including
regulators, such as that on the Bahr Yusuf Canal.

ASYÛT BARRAGE.

By far the most important of the works constructed to enable the
water stored up in the great reservoir to be utilised to the greatest
advantage is the Barrage across the Nile at Asyût, about 250 miles
above Cairo, which was commenced by Sir John Aird and Co. in the
winter of 1898, and completed this spring. As already stated,
in general principle this work resembles the old barrage at the
apex of the Delta; but in details of construction there is no similarity, nor in material, as the old work is of brick and the new one
of stone.

The total length of the structure is 2750 ft., or rather more than
half a mile, and it includes 111 arched openings of 16 ft. 4 in. span,
capable of being closed by steel sluice gates 16 ft. in height. The
object of the work is to improve the present perennial irrigation of
lands in Middle Egypt and the Fayoum, and to bring an additional area
of about 300,000 acres under such irrigation, by throwing more water
at a higher level into the great Ibrahimiyah Canal, whose intake is
immediately above the Barrage (Fig. 2).

The piers and arches are founded upon a platform of masonry
87 ft. wide and 10 ft. thick, protected up and down by a continuous
and impermeable line of cast-iron grooved and tongued sheet piling,

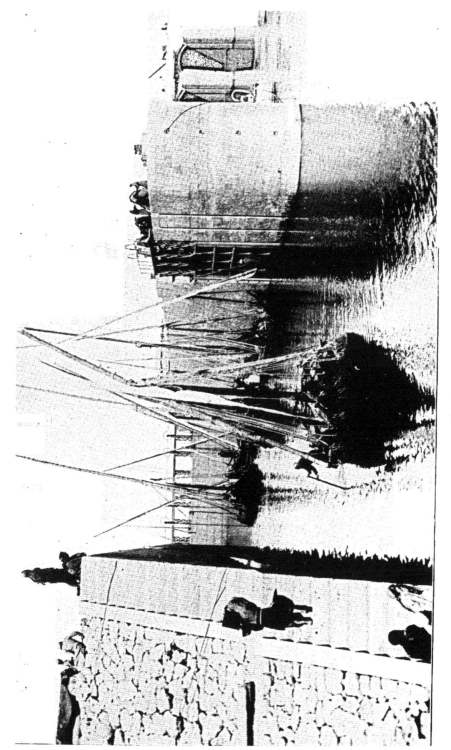

FIG. 2.—ASYÛT LOCK.

with cemented joints. This piling extends into the sand bed of the river to a depth of 23 ft. below the upper surface of the floor, and thus cuts off the water and prevents the undermining action which caused so much trouble and expense in the case of the old barrage. The height of the roadway above the floor is 41 ft., and the length of the piers up and down stream 51 ft. The river bed is protected against erosion for a width of 67 ft. up stream by stone pitching, with clay puddle underneath to check infiltration, and down stream for a similar width by stone pitching, with an inverted filter-bed underneath, so that any springs which may arise from the head of water above the sluices shall not carry sand with them from underneath the pitching.

It is easy enough to construct dams and barrages on paper, but wherever water is concerned the real difficulty and interest is in the practical execution of the works, for water never sleeps, but day and night is stealthily seeking to defeat your plans. On the Nile the conditions are very special, and in some respects advantageous. There is only one flood in the year, and within small limits the time of its occurrence can be foretold, and arrangements made accordingly. It would have been impossible to have carried out the Nile works on the system adopted had the river been subject to frequent floods. The working season for below-water work on the Nile lies practically between November and July, for nothing would be gained by starting the temporary enclosing embankments, or sudds, when the river was at a higher level than it is in November; nor would it be possible at any reasonable cost to prevent the sudds from being swept away by the flood in July. At Asyût the mode of procedure was to enclose the site of the proposed season's work by temporary dams or sudds of sandbags and earthwork, then to pump out and keep the water down by powerful centrifugal pumps, crowd on the men, excavate, drive the cast-iron sheet piling, build the masonry platform and piers, lay the aprons of puddle and pitching, and get the work some height above low Nile level before the end of June, so that the temporary dams should not require reconstruction after being swept away by the flood. The busiest months were May and June, when in the year 1900 the average daily number of men was 13,000. It is also then the hottest: the shade temperature rising to 118 degrees. To keep the water down, seventeen 12-in. centrifugal pumps, throwing enough water for the supply of a city of two million inhabitants, had to be kept going, and in a single season as many as one and a half million sandbags were used in these temporary dams. The bed of the river being of extremely mobile sand, the constant working of the pumps occasionally drew away sand from under the adjoining completed portions of the foundations, necessitating the drilling of many holes through the 10-ft. thick masonry platform, and grouting under pressure with liquid cement. About 1000 springs also burst up through the sand, each one of which required special treatment. A new regulator had to be constructed for the Ibrahimiyah Canal, with nine arches and sluices, to control the high floods and prevent damage to the canal and the works connected therewith.

Aswân.

Asyût, as already observed, is about 250 miles above Cairo. The great dam at Aswân is 600 miles above the same point. Between Asyût and Aswàn the remains of many temples exist, of far greater interest and importance than those at Philæ. The latter ruins, however, have attracted more attention in recent days, because, being situated immediately above the Dam, the filling of the Reservoir will partially flood Philæ Island during the tourist season.

It would be idle to speculate as to who first thought of constructing a reservoir in the Nile valley, or who first arrived at the conclusion that the site of the present Dam above Aswân was the best one. Mr. Willcocks, one of the ablest engineers of the Public Works Department of Egypt, who was instructed by Sir William Garstin to survey various suggested sites for a dam between Cairo and Wady Halfa, unhesitatingly decided that the Aswân site was the best, and the majority of the International Committee, who visited the sites in 1894, came to the same conclusion. This conclusion had, however, been anticipated by Sir Samuel Baker more than forty years ago, from mere inspection of the site without surveys. In suggesting a series of dams across the Nile to form reservoirs from Khartoum downwards, he wrote: " The great work might be commenced by a single dam above the first cataract at Aswân, at a spot where the river is walled in by granite hills. By raising the level of the Nile 60 ft., obstructions would be buried in the depths of the river, and sluice-gates and canals would conduct the shipping up and down stream." This single dam, proposed by Sir Samuel Baker forty years ago, is in effect the one which is now on the point of completion. Mr. Willcocks' original design consisted practically of a group of independent dams, curved on plan, and the arrangement of sluices and dimensions of the dam differed considerably from those of the executed work. There is no doubt that the single dam, $1\frac{1}{4}$ miles in length, constitutes a more imposing monumental work than a series of detached dams, and that it also offered greater facilities to a contractor for the organisation of his work and rapid construction ; and, further, the straight dam is better able to resist temperature stresses from extreme heat without cracking. Two dams across the Nile, the old barrage and the Asyût Barrage, have already been described ; and it will be hardly necessary to say, therefore, that the Aswân Dam is not a solid wall, but is pierced with sluice openings of sufficient area for the flood discharge of the river, which may amount to 15,000 tons of water per second. There are 180 such openings, mostly 23 ft. high by 6 ft. 6 in. wide ; and where subject to heavy pressure, when being moved, they are of the well-known Stoney roller pattern.

FIG. 3.—NORTH SIDE OF DAM, LOOKING WEST.

Although the preliminary studies of Mr. Willcocks and the other Government engineers occupied some four years, there was neither time nor money to sink shafts in the bed of the river, to ascertain the real character of what was called in the engineer's report "an extensive outcrop of syenite and quartz diorite clean across the valley of the Nile," giving "sound rock everywhere at a very convenient level." Unfortunately, the rock proved to be unsound in many places to a considerable depth, with schistous micaceous masses of a very friable nature, which necessitated carrying down the foundations of the dam sometimes more than 40 ft. deeper than was originally anticipated or provided for in the contract. As the thickness of the dam is nearly 100 ft. at the base, this misapprehension as to the character of the rock involved a very large increase in the contract quantity and cost of the granite masonry of the Dam. The total length of the Dam is about $1\frac{1}{4}$ miles; the maximum height from foundation, about 130 ft.; the difference of level of water above and below, 67 ft.; and the total weight of masonry over one million tons. Navigation is provided for by a "ladder" of four locks, each 260 ft. long by 32 ft. wide.

As remarked in the case of Asyût, the difficulties in dam construction are not in design, but in the carrying out of the works. It would not be too much to say that any practical man standing on the verge of one of the cataract channels, hearing and seeing the apparently irresistible torrents of foaming water thundering down, would regard the putting in of foundations to a depth of 40 ft. below the bed of the cataract in the short season available each year as an appalling undertaking. When the rotten rock in the bed was first discovered, I told Lord Cromer frankly that I could not say what the extra cost or time involved by this and other unforeseen conditions would be, and that all I could say was that, however bad the conditions, the job could be done. He replied that he must be satisfied with this assurance, and say that the Dam had to be completed whatever the time and cost. With a strong man at the head of affairs, both engineers and contractors—who often are suffering more anxiety than they care to show—are encouraged, and works, however difficult, have a habit of getting completed, and sometimes, as in the present case, in less than the original contract time.

The contract was let to Sir John Aird and Co., with Messrs. Ransomes and Rapier as sub-contractors for the steelwork, in February 1898, and they at once commenced to take possession of the site of the works, and of as much of the adjoining desert as they desired in order to construct railways, build dwellings, offices, machine shops, stores and hospitals, and provide sanitary arrangements, water supply, and the multitudinous things incidental to the transformation of a remote desert tract into a busy manufacturing town. Two months after signing the contract the permanent works were commenced, and before the end of the year thousands of native labourers and hundreds of Italian granite masons were hard at work.

On February 12, 1899, the foundation stone of the Dam was laid by H.R.H. the Duke of Connaught. Many plans were considered by the engineers and contractors for putting in the foundations of the Dam across the roaring cataract channels, and it was finally decided to form temporary rubble dams across three of the channels below the site of the great Dam, so as to break the force of the torrent and get a pond of comparatively still water up stream to work in. Stones of from 1 ton to 12 tons in weight were tipped into the cataract, and this was persevered with until finally a rubble mound appeared above the surface of the water. The first channel was successfully closed on May 17, 1899, the depth being about 30 ft. and the velocity of current nearly 15 miles an hour. In the case of another channel, the closing had to be helped by tipping in railway wagons themselves, loaded with heavy stones, and bound together with wire ropes, making a mass of about 50 tons, to resist displacement by the torrent.

These rubble dams were well tested when the high Nile ran over them ; and on work being resumed in November, after the fall of the river, water-tight sandbag dams or sudds were made around the site of the Dam foundation in the still waters above the rubble dams, and pumps were fixed to lay dry the bed of the river. This was the most exciting time in the whole stage of the operations, for no one could predict whether it would be possible to dry the bed, or whether the water would not pour through the fissured rock in altogether overwhelming volumes. Twenty-four 12-in. centrifugal pumps were provided to deal if necessary with one small channel ; but happily the sandbags and gravel and sand embankments staunched the fissures in the rock and interstices between the great boulders covering the bottom of this channel, and a couple of 12-in. pumps sufficed. The open rubble dam itself, strange to say, checked the flow sufficiently to cause a difference of nearly 10 ft. in the level of the water above and below ; but when the up-stream sand-bag dam was constructed the difference was 20 ft., so that the down-stream sandbag dam was a very small one compared with the other.

The masonry of the dam is of local granite, set in British Portland cement mortar. The interior is of rubble, set by hand, with about 40 per cent. of the bulk in cement mortar, four sand to one of cement. All the face-work is of coursed rock-faced ashlar, except the sluice linings, which are finely dressed. This was steam-crane and Italian masons' work. There was a great pressure at times to get a section completed before the inevitable rise of the Nile, and as much as 3600 tons of masonry were executed per day, chiefly at one point in the Dam. A triple line of railway, and numerous trucks and locomotives, were provided to convey the materials from quarries and stores to every part of the work. The maximum number of men employed was 11,000, of whom 1000 were European masons and other skilled men (Figs. 3, 4 and 5).

Mr. Wilfred Stokes, chief engineer and managing director of Messrs. Ransomes and Rapier, was responsible for the detailed

Fig. 4.—South Side of Dam, from West Bank.

designing and manufacture of the sluices and lock-gates; 140 of the sluices are 23 ft. high by 6 ft. 6 in. wide, and 40 of them half that height; 130 of the sluices are on the "Stoney" principle, with rollers, and the remainder move on sliding surfaces. The larger of the Stoney sluices weigh 14 tons, and are capable of being moved by hand under a head of water producing a pressure of 450 tons against the sluice.

There are five lock-gates, 32 ft. wide, and varying in height up to 60 ft. They are of an entirely different type to ordinary folding lock-gates, being hung from the top on rollers, and moving like a sliding coach-house door. This arrangement was adopted for safety, as 1000 million tons of water are stored up above the lock-gates, and each of the two upper gates is made strong enough to hold up the water, assuming the four other gates were destroyed (Fig 6).

When the river is rising, the sluices will all be open, and the red water will pass freely through, without depositing the fertilising silt. After the flood, when the water has become clear, and the discharge of the Nile has fallen to about 2000 tons per second, the gates without rollers will be closed, and then some of those with rollers'; so that between December and March the Reservoir will be gradually filled. The re-opening of the sluices will take place between May and July, according to the state of the Nile and the requirements of the crops.

Between December and May, when the Reservoir is full, the Island of Philæ will in places be slightly flooded. As the temples are founded partly on loose silt and sand, the saturation of the hitherto dry soil would cause settlement, and no doubt injury to the ruins. To obviate this risk, all the important parts, including the well-known Kiosk, or "Pharaoh's bed," have been either carried on steel girders or underpinned down to rock, or, failing that, to the present saturation level. It need hardly be said that, having regard to the shattered condition of the columns and entablatures, the friability of the stone, and the running sand foundation, the process of under-pinning was an exceptionally difficult and anxious task. There were few men to whom I would have entrusted the task, but amongst those was Mat Talbot—one of the well-known Talbots who have done such splendid service as non-commissioned officers in the army of work-men employed by contractors during the past forty years; and well has he justified his reputation at home—where his last job was the most difficult part of the Central London Railway—and the com-mendation of Dr. Ball, who had charge of the works at Philæ.

It would be invidious to single out for special acknowledgment the services of members of a staff, where all have enthusiastically done their best for the accomplishment of the great work projected and patiently persisted in against all opposition, by Lord Cromer and his trusty lieutenant, Sir William Garstin, Under Secretary of State for Public Works. The successive Director-Generals of the Reservoirs were Mr. Willcocks, Mr. Wilson, and Mr. Webb; the chief engineers at Aswân, Mr. Fitzmaurice and Mr. May, and at Asyût, Mr. Stephens. The almost unprecedented labour and anxiety of arranging all the practical contractors' details of supply of labour, materials, and execution of the work fell upon the shoulders of Mr. Blue, except as regards Asyût, where Mr. McClure relieved him of a part of his responsibility.

As regards the initial stages of the project, I may say that when the Egyptian Government informed me that they wanted the works carried out for a lump sum, and no payment to be made to the contractor until the works were completed, I felt it would be idle to invite tenders until some arrangement had been made as to finance. As in other cases of doubt and difficulty, therefore, I went to my friend, Sir Ernest Cassel, and the difficulties vanished. The way was then clear for getting offers for the work. Sir John Aird and Co. were the successful competitors, and they have completed a largely increased quantity of work in less than the contract time, to the entire satisfaction of the Egyptian Government and of every one with whom they have been associated. The same recognition is due to Messrs. Ransomes and Rapier, and their able engineer and manager, Mr. Wilfred Stokes, who was unexpectedly called upon to complete all the complicated machinery of the sluices and gates in one year under the contract time, and did it.

[B. B.]

Fig. 6.—Navigation Channel: First Lock-Gate from North.

WEEKLY EVENING MEETING,

Friday, June 13, 1902.

His Grace the Duke of Northumberland, K.G. D.C.L. F.R.S.,
President, in the Chair.

G. Marconi, Esq., M. Inst. E.E.

The Progress of Electric Space Telegraphy.

Wireless Telegraphy, or telegraphy through space without connecting wires, is a subject which at present is probably attracting more
world-wide attention than any other practical development of modern
electrical engineering. That it should be possible to actuate an instrument from a distance of hundreds or thousands of miles and oblige
it at will to reproduce audible or visible signals through the effects
of electrical oscillations transmitted to it without the aid of any continuous artificial conductor, strikes the minds of most people as being
an achievement both wonderful and mysterious. If we examine the
subject closely we may, however, come to the conclusion that, although
telegraphy through space is certainly wonderful, as are likewise all
natural and physical phenomena, yet it is certainly in no way more
wonderful than the transmission of telegrams along an ordinary telegraph wire. The light and heat waves of the sun and stars travel
to us through millions of miles of space, and sound also reaches our
ears without requiring any artificial conductor. It is not, therefore,
wonderful that man should have devised means by which he is
enabled to confine electricity conveying messages or power to a wire
and cause the effect which we call an electric current to follow all
the turns and convolutions which may exist in the wire.

We find that the first systems of telegraphy used by mankind
were truly wireless. A bonfire built on a hill by a band of aboriginal
Indians conveyed a signal wirelessly by etheric waves—in this case
light waves—to Indians on another hill, perhaps miles distant. Even
to-day there are innumerable systems of what may truly be called
wireless telegraphy in practical use. A red light at a railway crossing
conveys a signal by waves through the ether to the eye of the engine
driver. The red light is the transmitter, the eye the receiver.

The method of space telegraphy of which I intend speaking to-
night is founded on a comparatively new way of controlling and
detecting certain kinds of etheric waves, much slower in rate of
vibration than light waves, called Hertzian waves, after the scientist
who first demonstrated their existence. The mathematical and

o 2

experimental proof by Clerk Maxwell and Heinrich Hertz of the identity of light and electricity, and the knowledge of how to produce and detect certain previously unknown ether waves, made possible this new method of communication. I think I am right in saying that the importance of the discoveries of Maxwell and Hertz was realised by very few, and even, perhaps, so recently as a year ago a great number of scientific men would have hardly foreseen the advances which have been made in so brief a time in the art of space telegraphy.

The time allowed for this discourse does not permit me to describe all the various steps which have made possible the results recently obtained nor to describe the work of the numerous workers who have contributed to the advance of the subject, but I hope it may be of interest if I describe the various problems which have lately been solved, and the very interesting developments which have taken place in my own work during the last few months. I shall first briefly describe my system as used in my early experiments six years ago, and afterwards endeavour to explain the various improvements and modifications which have since been introduced into it.

The transmitter consists of a modified form of Hertzian oscillator, the main feature of which is in having one sphere of the spark discharger earthed, and the other connected to an elevated capacity area or to a comparatively vertical wire. The two spheres are also convected to the ends of the secondary winding of an induction coil or transformer. When the key is pressed the current of the battery is allowed to actuate the spark coil, which charges the spheres and the vertical wire, which, when discharging, causes a rapid succession of sparks to pass across the spark-gap. The sudden release caused by the spark discharge of the electrical strain or displacement created along certain lines of electric force through space by the charged wire causes some of the electrical energy to be thrown off in the form of a displacement wave in the ether, and, as a consequence, the vertical wire becomes a radiator of electric waves. In this connection it is interesting to remember that Lord Kelvin showed mathematically more than forty years ago the precise conditions under which such a discharge as we are considering would be oscillatory. It is easy to understand how, by pressing the key for longer or shorter intervals, it is possible to emit a long or short succession of impulses or waves which, when they influence a suitable receiver, reproduce on it a long or short effect, according to their duration, in this way reproducing the Morse or other signals transmitted from the sending station.

The receiver consists of a coherer (on the nature of which I hope to make a few further remarks later) placed in a circuit containing a local cell and a sensitive telegraph relay actuating another circuit, which works a trembler or decoherer and a recording instrument. In its normal condition the resistance of the coherer is infinite, or at least very great, and the current of the battery cannot pass through it to actuate the instruments, but when influenced by electric waves the coherer becomes a comparatively good conductor, its resistance falling to between 100 and 500 ohms.   This allows the current from the local cell to actuate the relay, which in turn causes another stronger current to work the recording instrument and also the tapper or decoherer, which is so arranged as to tap or shake the coherer, and in this way restore its sensitiveness.   The practical result is that the circuit of the recording instrument is closed for a time equal to that during which the key is pressed at the transmitting station, and in this way it is possible to obtain a graphic, acoustic or optical reproduction of the movements of the key at the sending station.   One end of the tube, or coherer, is connected to earth and the other to an insulated conductor, preferably terminating in a capacity area similar in every respect to the one employed at the transmitting station.

I first noticed that by employing similar vertical rods at both stations, it was possible to detect the effects of electric waves, and in that way convey the intelligible alphabetical signals over distances far greater than had previously been believed possible, and by means of similar arrangements distances of transmission up to about 100 miles were obtained.

It was soon, however, realised that so long as it was possible to work only two installations within what I may call their sphere of influence, a very important limit to the practical utilisation of the system was imposed.   Without some practical method of tuning the stations it would have been impossible to work a number in the vicinity of each other at the same time without interference caused by the mixing of messages.   The new methods of connection which I adopted in 1898—i.e. connecting the receiving vertical wire or aerial directly to earth instead of to the coherer, and by the introduction of a proper form of oscillation transformer in conjunction with a condenser so as to form a resonator tuned to respond best to waves given out by a given length of aerial wire—were important steps in the right direction.   I referred at length to this improvement in the discourse which I had the honour to deliver at this table on February 2, 1900.   I had, however, realised at the time that one great difficulty in the way of achieving the desired effects was caused by the action of the transmitting wire.   A straight rod in which electrical oscillations are set up forms, as is well known, a very good radiator of electrical waves.   In all what we call good radiators electrical oscillations set up by the ordinary spark-discharge method cease or are damped out very rapidly, not necessarily by resistance, but by electrical radiation removing the energy in the form of electric waves.

It is a well-known fact that when one of two tuning forks having the same period of vibration is set in motion, waves will form in the air, and the other tuning fork, if in suitable proximity, will immediately begin to vibrate in unison with the first.   In the same way a violin player, sounding a note on his instrument, will find a response from a certain wire in a piano, near by, that particular wire, out of all the wires of the piano, happening to the only one which has a period of vibration identical with that of the musical note

sounded by the violinist. Tuning-forks and violins, of course, have
to do with air waves and wireless telegraphy wth ether waves, but
the action in both cases is similar. It is very important to take into
consideration the one essential condition which must be obtained in
order that a well-marked tuning or electrical resonance may take
place. Electrical resonance, like mechanical resonance, essentially
depends upon the accumulated effect of a large number of small
impulses properly timed. Tuning can only be obtained if a sufficient
number of these timed electrical impulses reach the receiver. As
Prof. Fleming graphically puts it in one of his lectures on electrical oscillations, to " set a pendulum in vibration by puffs of air
we must not only time the puffs properly, but keep on puffing for a
considerable period." It is, therefore, clear that a dead-beat radiator
—i.e. one that does not give a train or succession of electrical
oscillations—is not suitable for tuned or syntonic space telegraphy.

As I pointed out before, a transmitter consisting of a vertical wire
discharging through a spark-gap is not a persistent oscillator. Its
electrical capacity is comparatively so small and its capability of
radiating waves so great that the oscillations which take place in it
must be considerably damped. In this case receivers or resonators
of a considerably different period or pitch will respond and be
affected by it.

Early in 1900 I obtained very good results with another arrangement in which the radiating and resonating conductors each take the
form of two concentric cylinders, the internal cylinder being earthed.
By using zinc cylinders only 7 metres high and 1·5 metres in diameter
good signals could easily be obtained between St. Catherine's Point,
Isle of Wight, and Poole, over a distance of 30 miles, these signals
not being interfered with or read by other wireless telegraph installations worked by my assistants or by the Admiralty in the immediate
vicinity. The capacity of the transmitter due to the internal conductor is so large that the energy set in motion by the spark discharge
cannot all radiate in one or two oscillations, but forms a train of
slowly-damped oscillations, which is just what is required. A simple
vertical wire may be compared with an empty teapot, which, after
being heated, would cool very rapidly, and the concentric cylinder
system with the same teapot filled with hot water, which would take
a very much longer time to cool. In the receiver the closely adjacent
cylinders which give it large electrical capacity cause it to be a
resonator possessing a very decided period of its own, and it becomes
no longer apt to respond to frequencies which differ from its own particular period of electrical oscillation, nor to be interfered with by stray
ether waves which are sometimes caused by atmospheric disturbances,
and which occasionally prove troublesome during the summer.

Another successful system of tuning or syntonising the apparatus
was the outcome of a series of experiments carried out with the discharge of condenser or Leyden jar circuits. I tried by means of
associating with the radiating wire, or capacity, a condenser circuit,

which is known to be a persistent oscillator, to set up the required number of oscillations in the radiator. An arrangement consisting of a circuit containing a condenser and a spark-gap constitutes a very persistent oscillator. Prof. Lodge has shown us how, by placing it near another similar circuit it is possible to demonstrate interesting effects of resonance by the experiment usually referred to as that of Lodge's syntonic jars. But, as Lodge points out, "a closed circuit such as this is a feeble radiator and a feeble absorber, so that it is not adapted for action at a distance." I very much doubt if it would be possible to affect an ordinary receiver at even a few hundred yards. It is, however, interesting to notice how easy it is to cause the energy contained in the circuit of this arrangement to radiate into space. It is sufficient to place near one of its sides a straight metal rod or good electrical radiator, the only other condition necessary for long-distance transmission being that the period of oscillation of the wire or rod should be equal to that of the nearly closed circuit. Stronger effects of radiation are obtained if the radiating conductor is partly bent round the circuit containing the condenser (so as to resemble the circuits of a transformer).

My first trials with this system were not successful, in consequence of the fact that I had not recognised the necessity of attempting to tune to the same period of electrical oscillations (or octaves) the two electrical circuits of the transmitting arrangement (these circuits being the circuit consisting of the condenser and primary of the transformer and the aerial or radiating conductor and secondary of the transformer). Unless this condition is fulfilled, the different periods of the two conductors create oscillations of a different frequency and phase in each circuit, with the result that the effects obtained are feeble and unsatisfactory on a tuned receiver. The syntonised transmitter is shown in Fig. 1. The period of oscillation of the vertical conductor A can be increased by introducing turns of wire, or decreased by diminishing their number, or by introducing a condenser in series with it. The condenser

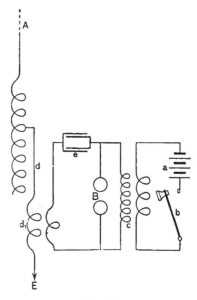

Fig. 1.

in the primary circuit is constructed in such a manner as to render it possible to vary its electrical capacity.

The receiving station arrangements are shown in Fig. 2. Here we have a vertical conductor connected to earth through the primary of

a transformer, the secondary circuit of which is joined to the coherer
or detector.   In order to make the tuning more marked, I place an
adjustable condenser across the coherer in Fig. 3.   Now, in order to
obtain best results, it is necessary that the free period of electrical
oscillation of the vertical wire primary of transformer and earth connection should be in electrical resonance with the second circuit of
the transformer, which includes the condenser.   I stated that in
order to make the tuning more marked a condenser is placed across
the coherer.   This condenser increases the capacity of the secondary
resonating circuit of the transformer, and in the case of a large series

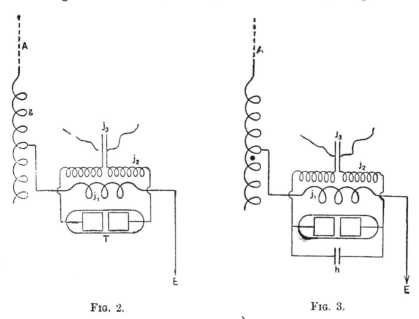

Fig. 2.   Fig. 3.

of comparatively feeble but properly timed electrical oscillations
being received, the effect of the same is summed up until the E.M.F.
at the terminals of the coherer is sufficient to break down its insulation and cause a signal to be recorded.   In order that the two systems,
transmitter and receiver, should be in tune it is necessary (if we
assume the resistance to be very small or negligible) that the product
of the capacity and inductance in all four circuits should be equal.
    It is easy to understand that if we have several stations, each
tuned to a different period of electrical vibration, and of which the
corresponding inductance and capacity at the transmitting station are
known, it will not be difficult to transmit to any one of them, without
danger of the message being picked up by the other stations for
which it is not intended.   But better than this we can connect to
the same vertical sending wire, through connections of different
inductance, several differently tuned transmitters, and to the receiving

vertical wire a number of corresponding receivers.   Different messages can be sent by each transmitter connected to the same radiating wire simultaneously and received equally simultaneously by the vertical wire connected to differently tuned receivers.   This result, which I believe was quite novel at the time, I showed to several friends of mine, including Dr. J. A. Fleming, F.R.S., nearly two years ago. Dr. Fleming made mention of the results he had seen in a letter to the London *Times* dated October 4, 1900.   I have further noticed that the tuning can be further improved by the combination of the two systems described.   In this case the cylinders are connected to the secondary of the transmitting transformer, and the receiver to a properly-tuned induction coil, and all circuits must be tuned to the same period as already described.   This arrangement is going to be further tested in long-distance experiments shortly to be undertaken between England and Canada.

The syntonic systems have not been applied generally to ships, as it has always been considered an advantage that each ship should be able, especially in case of distress, to call up any other ship or ships which may happen to be at the time within the range of its transmitter, but in the case of land stations the syntonic method has been applied in several instances where necessity demanded it.   Thus at the testing stations which maintain communication between St. Catherine's, Isle of Wight, and Poole, in Dorset, when electric waves of a certain frequency are used, no interference whatever can be caused by the working of the Admiralty installations in the vicinity.   The long-distance station at Poldhu, Cornwall, is able to transmit signals decipherable on a tuned receiver on a ship at over 1000 miles distance, while the Lloyd's wireless station at the Lizard, only 7 miles away, is not affected by the powerful waves radiated from Poldhu if tuned to a different frequency.

I am not at all prepared to say that under no possible circumstances could a wireless message transmitted between syntonic instruments be tapped or interfered with, but I wish to point out that it is now possible to work a considerable number of wireless telegraph stations simultaneously in the vicinity of each other without the messages suffering from any interference.   Of course, if a powerful transmitter, giving off waves of different frequencies, is actuated near one of the receiving stations it may prevent the reception of messages, but the ordinary systems of communication through wires may be likewise affected.   Prof. O. J. Lodge, in a report of his experiments in magnetic space telegraphy, mentions that he was able to interfere with the working of the ordinary wire telephone system in the city of Liverpool.   Sir W. H. Preece has also published results which go to show that it is possible to pick up at a distance on another circuit the conversation which may be passing through a telephone wire. About two years ago, at Cape Town, it was found impossible to work the cables landing there during certain hours when the electric tramways of the town were running, and the matter became subsequently the subject of litigation between the companies concerned. Prof. Fleming, who has witnessed the working of a great number of syntonic wireless telegraph stations, was sufficiently impressed by what he saw to make the following statement in his Cantor lectures on " Electrical Oscillations and Electric Waves," delivered before the Society of Arts in December 1900 :—" The objections as to inter- ference of stations which imperfectly informed persons are in the habit of raising with regard to Mr. Marconi's system of wireless telegraphy, as a matter of fact no longer exist."

I shall now say a few words on the subject of the detector of the electric waves, called sometimes " the electric eye," which consists of that essential part of the receiving apparatus especially affected by the electrical oscillations. In all wireless telegraph apparatus used up to quite a recent date, a detector, now called a coherer, has been employed. This detector is based on discoveries and observations made by S. A. Varley, Prof. Hughes, Colzecchi, Onesti, and especially Prof. Branly. Prof. O. J. Lodge has made large use of this apparatus, which he first named " coherer," in the very numerous experiments and studies he has carried out on the effects produced by Hertzian waves. The form of coherer I have found most trustworthy and reliable for long-distance work consists of a small glass tube about 4 cm. long, into which two metal pole-pieces are tightly fitted. They are separated from each other by a small gap, which is partly filled with a mixture of nickel and silver filings. Provided such a coherer is properly constructed, and the tapper and relay in good adjustment, it proves to be quite reliable when within range of the transmitting station. Experiments with syntonic systems have, however, shown that certain kinds of coherers can be far more advantageously employed than others. One apparently all-important condition is that the resist- ance of the coherer in its sensitive state, or after being tapped, should appear to be infinite when measured with an E.M.F. of about 1 volt. If the tapping does not entirely do away with the conductivity of the filings very poor results are obtained, which can be explained as follows. According to the systems I have described, electrical syntony between the transmitter and receiver is dependent on the proper electrical resonance of the various circuits of the transformers used in the receivers. The condenser and secondary of the transformer must not be partially short-circuited by the coherer, otherwise the oscillations cannot mount up or sum up their effect, as is essential in order to produce the difference of potential at the ends of the coherer neces- sary for breaking down its resistance; but the electrical oscillations will leak across the conductive coherer without causing it to record any signal. Of course, the condenser is short-circuited when the filings cohere under the influence of the received oscillations; but in this case the signal is already recorded, and the tapper at once restores the coherer to its non-conducting condition, and in this way restores its sensitiveness.

By using coherers containing very fine filings the necessary condition of non-conductivity when in a sensitive state is obtained. Coherers have lately been tried which will work to a certain extent satisfactorily without the necessity of employing any tapper or decoherer in connection with them. Nearly all are dependent on the use of a carbon microphonic contact or contacts which possess the curious quality of partially re-acquiring spontaneously their high resistance condition after the effect of the electrical oscillations has ceased. This enables one to obtain a far greater speed of reception than is possible by means of a mechanically-tapped coherer, the inertia of the relay and tapper which are used in connection with it being necessarily sluggish in their action. In all these self-decohering coherers a telephone which is affected by the variations of the electric current, caused by the changes in conductivity of the coherer, is used in place of the recording instrument. It has not yet been found possible, so far as I am aware, to actuate a recording instrument or a relay by means of a self-restoring coherer. The late Prof. Hughes was the first, I believe, to experiment with and receive signals on one of these coherers associated with a telephone. His experiments were carried out as early as 1879, and I regret that this pioneer work of his is not more generally known. Other self-restoring coherers were proposed by Professors Tommasina, Popoff, and others, but one which has given good results when syntonic effects were not aimed at was (according to official information communicated to me) designed by the technical personnel of the Italian navy. This coherer, at the request of the Italian Government, I tested during numerous experiments. It consists of a glass tube containing plugs of carbon or iron with between them a globule of mercury. Lieutenant Solari, who brought me this coherer, asked me to call it the "Italian Navy Coherer." Recently, however, a technical paper gave out that a signalman in the Italian navy was the inventor of the improved coherer, and I was at once accused in certain quarters of suppressing the alleged inventor's name. I therefore wrote to the Italian Minister of Marine, Admiral Morin, asking him to make an authoritative statement, to which I could refer in the course of this address, of the views of the Italian Admiralty on the matter. The head of the Italian navy was good enough to reply to me by a letter, dated the 4th inst., in which he makes the following statement, which I have translated from the original Italian:—" The coherer has been with good reason baptised with the name of ' Italian Navy Coherer,' as it must be considered fruit of the work of various individuals in the Royal Navy and not that of one." These non-tapped coherers have not been found to be sufficiently reliable for regular or commercial work. They have a way of cohering permanently when subjected to the action of strong electrical waves or atmospheric electrical disturbances, and have also an unpleasant tendency towards suspending action in the middle of a message. The fact that their electrical resistance is low and always varying, when in a sensitive state, causes them to be unsatisfactory for the reasons I have already enumerated when worked in connection with my system of syntonic wireless telegraphy.

These coherers are, however, useful if employed for temporary tests in which the complete accuracy of messages is not all-important, and when the attainment of syntonic effects is not aimed at. They are especially useful when using receiving vertical wires supported by kites or balloons, the variations of the height of the wires (and, therefore, of their capacity) caused by the wind making it extremely difficult to obtain good results on a syntonic receiver.

Coherers have long been considered as constituting almost the essential basis of electric space telegraphy, and although many other detectors of electric waves existed, none of them possessed a sensitiveness which even approached that of a coherer, and most of them were also unsuitable for the reception of telegraphic messages. With a view to producing a receiver which could be worked at a much higher speed than a coherer, I was fortunate enough to succeed in constructing a magnetic detector of electric waves, based on a principle essentially different from that of the coherer, and which I think leaves all coherers far behind in speed, facility of adjustment and efficiency when worked in tuned circuits. This detector, which I had the honour of describing in detail before the Royal Society yesterday, possesses I believe a sensitiveness which surpasses that of the best coherers. The magnetisation and demagnetisation of steel needles by the effect of electrical oscillations has long been known and was noted especially by Prof. J. Henry, Aloria, Lord Rayleigh and others. Mr. E. Rutherford also has described a magnetic detector of electric waves based on the partial demagnetisation of a small core composed of fine steel needles previously magnetised to saturation. By means of a magnetometer Mr. Rutherford succeeded in 1895 in tracing the effects of his electrical radiator up to a distance of three-quarters of a mile across Cambridge. But Mr. Rutherford's arrangement is not suitable for the reception of telegraphic messages in consequence of the fact that a careful process of re-magnetisation, which requires some time to effect, is necessary in order to restore its sensitiveness after the receipt of each impulse. Mr. Rutherford's arrangement is also considerably less sensitive than a coherer.

The detector which I am about to describe is, in my opinion, based upon the decrease of magnetic hysteresis, which takes place in iron when under certain conditions it is exposed to the effect of high frequency oscillations or Hertzian waves. As employed by me, it has been constructed in the following manner:—On a core of thin iron or steel, but preferably hard drawn iron, are wound one or two layers of thin insulated copper wire. Over this winding, insulating material is placed, and over this, again, another longer winding of thin copper wire contained in a narrow bobbin. The ends of the windings nearest the iron core are connected, one to earth and the other to an elevated conductor, or they may be connected to the secondary of a suitable receiving transformer or intensifying coil, such as are employed for syntonic wireless telegraphy. The ends of the other winding are connected to the terminals of a telephone or other suitable receiving instrument. Near the ends of the core, or in close proximity to it, is placed a horse-shoe magnet, which, by a clock-work arrangement, is so moved or revolved as to cause a slow and constant change or successive reversals in the magnetisation of the piece of iron. I have noticed that if electrical oscillations of suitable period be sent from a transmitter, rapid changes are effected in the magnetisation of the iron wires, and these changes necessarily cause induced currents in the windings, which in their turn reproduce on the telephone with great clearness and distinctness the telegraphic signals which may be sent from the transmitting station. Should the magnet be removed or its movement stopped, the receiver ceases to be perceptibly affected by the electric waves even when these are generated at very short distances from the radiator.

I have had occasion to notice that the signals audible on the telephone are weakest when the poles of the rotating magnet have just passed the core, and are increasing their distance from it, whilst they are strongest when the magnet's poles are approaching the core. Good results have also been obtained by keeping the magnet fixed, and using an endless iron rope or core of thin wires revolving on pulleys (worked by clock-work), which cause the iron to travel through the copper wire windings, in proximity to, preferably, two horse-shoe magnets with their poles close to the windings, care being taken that their poles of the same sign are adjacent. This detector has been successfully employed for some time in the reception of wireless telegraphic messages between St. Catherine's Point, Isle of Wight, and the North Haven, Poole, over a distance of 30 miles, also between Poldhu, in Cornwall, and Poole, in Dorset, over a distance of 152 miles, of which 109 are over sea and 43 over high land.

It would, no doubt, be possible to obtain signals by causing the iron core to act directly on a telephone diaphragm, and in this case the secondary winding could be omitted. This detector, as I have already stated, appears to be more sensitive and reliable than a coherer, nor does it require any of the adjustments or precautions which are necessary for the good working of the latter. It possesses a uniform and constant resistance, and, as it will work with a much lower E.M.F., the secondaries of the tuning transformers can be made to possess much less inductance, their period of oscillation being regulated by a condenser in circuit with them, which condenser may be much larger (in consequence of the smaller inductance of the circuit) than those used for the same period of oscillation in a coherer circuit, with the result that the receiving circuits can be tuned much more accurately to a particular radiator of fairly persistent electric waves. As a call, a coherer in circuit, with a relay working a bell, can always be used, and if it is found possible to make the magnetic detector record on a registering instrument (as to the possibility of which the results of recent tests have left little doubt in my mind), it may be found possible to receive wireless telegraph messages at a speed of several hundred words a minute. At present, by means of this detector, it is possible to read about 30 words a minute.

The considerations which led me to the construction of the above-described detector are the following :—It is a well-known fact-that, after any change has taken place in the magnetic force acting on a piece of iron, some time elapses before the corresponding change in the magnetic state of the iron is complete. If the applied magnetic force be caused to effect a cyclic variation, the corresponding induced magnetic variation in the iron will lag behind the changes in the applied force. To this tendency to lag behind Prof. Ewing has given the name of magnetic hysteresis. It has been shown also by Profs. Gerosa, Finzi and others, that the effect of alternating currents or high-frequency electrical oscillations acting upon iron is to reduce considerably the effects of magnetic hysteresis, causing the metal to respond readily to any influence which may tend to alter its magnetic condition. The effect of electrical oscillations probably is to bring about a momentary release of the molecules of iron from the constraint in which they are ordinarily held, diminishing their retentiveness and consequently decreasing the lag in the magnetic variation taking place in the iron. I therefore anticipated that the group of electrical waves emitted by each spark of a Hertzian radiator would, if caused to act upon a piece of iron which is being subjected at the same time to a slowly varying magnetic force, would produce sudden variations in its magnetic hysteresis, which would cause others of a sudden or jerky nature in its magnetic condition. In other words, the magnetisation of the iron, instead of slowly following the variations of the magnetic force applied, gives a sort of jump each time it is affected by the electric waves emitted by each spark of the radiator. These jerks in the magnetic condition of the iron would, I thought, cause induced currents in a coil of wire of strength sufficient to allow the signals transmitted to be detected intelligibly on a telephone, or perhaps even read on a mirror galvanometer. The results obtained go to confirm my belief that this detector can be advantageously substituted for the coherer for the purposes of long-distance space telegraphy.

During the last few years the developments in the practical applications of my system have been exceedingly rapid. Time does not allow me to give you an account of the many cases in which it has proved its usefulness, but it may be sufficient if I mention that Lloyd's have adopted the system exclusively for use at their stations at home and abroad for a period of 14 years, and that no less than 17 liners plying across the Atlantic carry permanent installations. In more than one case recorded in the daily papers the system has been of service to vessels in distress, especially in the English Channel. No less than 40 land stations (most of which are controlled by the corporation of Lloyd's) are being equipped with the system in Great Britain and Europe, and over 40 vessels in H.M. Navy carry installations. The adoption of my system in the Royal Navy has brought about a certain slight change of appearance in the rig of the ships. Some naval officers believe that this change improves the ships' appearance; others think the contrary.

The Italian Admiralty, after experimenting for some time with the self-decohering coherers to which I have referred before, have informed me officially, by a letter dated May 24th last, of its decision to equip their war vessels with the same apparatus as has been successfully employed on the transatlantic liners. On these liners commercial use is made of the system for the convenience of passengers, and as an illustration of its commercial workableness I might mention that lately the "Campania" and "Lucania" of the Cunard line have been collecting as much as 60*l.* each trip in receipts derived from passengers' wireless messages.

Nearly two years ago the facility with which communication was possible over distances of nearly 200 miles, and the improvements in syntonic methods introduced, together with the ascertained fact of the non-interference of the curvature of the earth, led me to decide to recommend the construction of a large power station in Cornwall and another one at Cape Cod, Mass., U.S.A., in order to test whether, by the employment of much greater power, it might not be possible to transmit messages across the Atlantic, and establish a trans-oceanic commercial communication which the monopoly of the Post-master-General will not apparently permit between two stations if both are situated in Great Britain. An unfortunate accident to the masts at Cape Cod seemed likely to postpone the experiments for several months, when I came to the conclusion that while the necessary repairs there were being carried out I would use a purely temporary installation in Newfoundland for the purpose of a trans-Atlantic experiment, from which I might, at any rate, be able to judge how far the arrangements in Cornwall had been conducted on right lines. Before describing the results it may be useful if I give a brief description of the nature of the apparatus used at the transmitting and receiving stations.

The transmitter at Poldhu was similar in principle to the syntonic one I have already described, but the elevated conductor at the transmitting station was much larger, and the potential to which it was charged very much in excess of any that had previously been employed, the amount of energy to be used in this transmitting station having been approximately determined by me prior to its erection. The transmitting elevated conductor consisted of 50 almost vertical naked copper wires, suspended at the top by a horizontal wire stretched between two poles, each 48 metres high, and placed 60 metres apart. These wires were separated from each other by a space of about 1 metre at the top, and, after converging together, were all connected to the transmitting instruments at the bottom. The potential to which these conductors were charged during transmission was sufficient to cause sparking between the top of the said wires and an earthed conductor across a space of 30 cm. of air. The general engineering arrangements of the electric power station erected at Poldhu for the execution of these plans and for creating the electric waves of the frequency which I desired to use were made by Dr. J. A. Fleming, F.R.S., who also devised many of the details of the appliances for producing and controlling the electric oscillations. These, together with devices introduced by me, and my special system of syntonisation of inductive circuits, have provided an electric wave-generating plant more powerful than any hitherto constructed. Mr. R. N. Vyvyan and Mr. W. S. Entwistle have also greatly assisted me in the experiments carried out with the very high tension electrical apparatus employed.

The first experiments were carried out in Newfoundland last December, and every assistance and encouragement was given me by the Newfoundland Government. As it was impossible at that time of the year to set up a permanent installation with poles, I carried out experiments with receivers joined to a vertical wire about 400 ft. long, elevated by a kite. This gave a very great deal of trouble, as in consequence of the variations of the wind constant variations in the electrical capacity of the wire were caused. My assistants in Cornwall had received instructions to send a succession of "S's," followed by a short message at a certain pre-arranged speed, every ten minutes, alternating with five minutes' rest during certain hours every day. Owing to the constant variations in the capacity of the aerial wire it was soon found out that an ordinary syntonic receiver was not suitable, although a number of doubtful signals were at one time recorded. I, therefore, tried various microphonic self-restoring coherers placed in the secondary circuit of a transformer, the signals being read on a telephone. With several of these coherers, signals were distinctly and accurately received, and only at the pre-arranged times, in many cases a succession of "S's," being heard distinctly although, probably in consequence of the weakness of the signals and the unreliability of the detector, no actual message could be deciphered. The coherers which gave the signals were one containing loose carbon filings, another, designed by myself, containing a mixture of carbon dust and cobalt filings, and thirdly, the "Italian Navy Coherer," containing a globule of mercury between two plugs. For the good results obtained I was very much indebted to two of my assistants, Mr. G. S. Kemp and Mr. P. W. Paget, who gave me very efficient aid during the tests, which the extremely severe weather prevailing in December in Newfoundland made exceedingly difficult to carry out.

The result of these tests was sufficient to convince myself and my assistants that, with permanent stations at both sides of the Atlantic, and by the employment of a little more power, messages could be sent across the ocean with the same facility as across much shorter distances. The experiments could not be continued or extended in consequence of the action which the cable company, which claims all telegraphic rights in Newfoundland, saw fit to take at the time. Having received a most generous invitation from the Government of the Dominion of Canada to continue my operations in the Dominion, it was thought undesirable to continue the experiments in Newfoundland, where I should have probably been landed into litigation with the telegraph company. I am glad to say that the Canadian Government, on the initiative of Sir Wilfred Laurier and Mr. Fielding, has shown itself most enterprising in the matter, and not only encouraged the erection of a large station in Nova Scotia, but actually granted a subsidy of 16,000*l.* towards the erection of this trans-Atlantic station, the object of which is to communicate with England from the coast of Nova Scotia. It is anticipated that the Canadian station will be ready for further tests very shortly. Another station for the same purpose is being erected on the United States coast.

Towards the end of February of this year I thought it desirable to test how far the messages transmitted by the powerful station at Poldhu could be detected on board a ship. The ship selected was the " Philadelphia," of the American line. The receiving aerial conductor was fixed to the mast, the top of which was about 60 metres above sea level. As the elevated conductor was fixed, and not floating about with a kite, as in the case of the Newfoundland experiments, very good results were obtained on an ordinary syntonic receiver, similar to those I have already described, and the signals were all recorded on tape by the ordinary Morse recorder. Readable messages on tape were received up to a distance of 1551 miles from Cornwall, and indications were received as far as 2099 miles. Most of the messages were received in the presence of the captain or the chief officer of the ship, who were good enough to sign the tapes. I have some of these tapes here, in a frame, and they can be examined at the conclusion of my discourse. It is curious to observe that signals could not be received at over 900 miles by any of the self-restoring coherers. The reason for this lies probably in the fact that the tuned receiver, when connected to a fixed aerial is more efficient. Another result of considerable scientific interest was that at distances of over 700 miles the signals transmitted during the day failed entirely, while those sent at night remained, as I have stated, quite strong up to 1551 miles, and were even decipherable up to a distance of 2099 miles. This result, which I had the honour of describing before the Royal Society yesterday afternoon, may be due to the dis-electrification of the very highly charged transmitting elevated conductor operated by the influence of daylight.

I regret time does not permit me to give you the views which have been expressed with reference to this phenomenon. I do not think, however, that the effect of daylight will be to confine the working of trans-Atlantic wireless telegraphy to the hours of darkness, as sufficient sending energy can be used during day-time, at the transmitting station, to make up for the loss of range of the signals, and therefore this business of communicating across the Atlantic will not be one of those works of darkness with which some people connected with cable companies would seem disposed to class it. It is, however, probable that had I known of this effect of light at the time of the Newfoundland experiments, and had tried receiving at night-time, the results would have been much better than those that were obtained.

The day is rapidly approaching when ships will be able to be in touch and communication with the shore across all oceans, and the quiet and isolation from the outside world which it is still possible to enjoy on board ship will, I fear, soon be things of the past. However great may be the importance of wireless telegraphy to ships and shipping, I believe it will be of even greater importance to the world if found workable and applicable over such great distances as those which divide Great Britain from her colonies and from America. Any of those who have lived in the colonies will easily appreciate what a hardship it is to have to wait, perhaps, four or five weeks before receiving an answer to a letter sent home. The cable rates are at present prohibitive to a vast majority of people. May it not, perhaps, be for wireless telegraphy to supply the want ?

I apologise for having kept you so long, but I cannot help reading you, in conclusion, a short extract from a leading article in the London *Times* of Saturday, December 21, 1901, published at the time of the Newfoundland experiments. And I do so because it expresses in language of admirable clearness the sentiments with which I myself regard this subject :—" It would probably be difficult to exaggerate the good effect of wireless telegraphy if, as Mr. Marconi and Mr. Edison evidently believe, and as the Anglo-American Company evidently fear, it can at no distant time be developed into a commercial success. The expense of telegraphy to distant countries is at present prohibitory to vast numbers of people, and even those who use it do so only in respect of matters of great urgency, or in which large money interests are at stake. The reason of the high charges must be sought, of course, in the enormous costliness of the plant, both in its original construction and in its maintenance and repair. A system of aerial telegraphy which would not require an expensive plant, and through which, therefore, messages might be sent at moderate rates, would soon become a potent agent in cementing those ties between Great Britain and the Colonies which other recent events have done so much to strengthen and even to create. A system of comparatively cheap telegraphs would do for the British Empire very much what was done by the penny post for the United Kingdom. The pathetic story of Rowland Hill, whose efforts to establish cheap postage originated in the sympathy he felt for a poor girl in a Cumberland village, who was unable to pay the sum demanded for a letter from her brother in a distant county, relates an event which in principle may be repeated to-day in many parts of the world. A cheap telegraph service would unite families, however scattered, would keep the dispersed members in close and constant touch with the old home, and would cement friendships between our own people and the Colonial nations, besides forging another link in the ties which bind this country to the United States."

[G. M.]

GENERAL MONTHLY MEETING,

Monday, July 7, 1902.

His Grace The DUKE OF NORTHUMBERLAND, K.G. D.C.L. F.R.S.,
President, in the Chair.

Mrs. Baily,
Miss S. M. Burnett,
The Right Hon. Sir Ernest Cassel, K.C.M.G.
Lady Kelvin,
Miss F. A. Musgrave,
Mrs. Otter,
Mr. Emile Schweich,

were elected Members of the Royal Institution.

The PRESENTS received since the last Meeting were laid on the
table, and the thanks of the Members returned for the same, viz.:—

FROM

The Astronomer-Royal—Report to the Board of Visitors of the Royal Observatory,
1902. 4to.
The British Museum—Catalogue of Greek Coins: Lydia. 8vo. 1901.
Catalogue of Drawings of British Artists, Vol. III. 8vo. 1901.
Catalogue of Sinhalese Printed Books. 4to. 1902.
American Academy of Arts and Sciences—Proceedings, Vol. XXXVII. Parts 15, 16.
8vo. 1901.
Accademia dei Lincei, Reale, Roma—Classe di Scienze Fisiche, Matematiche e
Naturali. Atti, Serie Quinta: Rendiconti. 1° Semestre, Vol. XI. Fasc. 10, 11.
8vo.
Astronomical Society, Royal—Monthly Notices, Vol. LXII. No 7. 8vo. 1901.
Bankers, Institute of—Journal, Vol XXIII. Part 6. 8vo. 1902.
Boston Public Library—Monthly Bulletin for June, 1902. 8vo
British Architects, Royal Institute of—Journal, Third Series, Vol. IX. Nos. 15, 16.
4to. 1902.
Memoirs, Vol. X. Part 3. 8vo. 1902.
British Astronomical Association—Journal, Vol. XII. No. 8. 8vo. 1902.
British South Africa Co.—Mining in Rhodesia. 4to. 1902.
Brothers, Arthur E. Esq.—Ten Daguerrotype Photographs taken in 1842 by
Professor Goddard.
Buenos Ayres, City—Monthly Bulletin of Municipal Statistics, April, 1902. 4to.
Camera Club—Journal for June, 1902. 8vo
Cape Town, The Colonial Secretary—Geodetic Survey of South Africa, Vol. II.
fol. 1901.
Chemical Industry, Society of—Journal, Vol. XXI. Nos. 11, 12. 8vo. 1902.
Chemical Society—Proceedings, Nos. 253, 254. 8vo. 1902.
Journal for July, 1902. 8vo.

Cracovie, L'Académie des Sciences—Bulletin, Classes des Sciences Mathématiques et Naturelles, 1902, Nos. 4, 5. 8vo.
　　Bulletin, Classe de Philologie, 1902, Nos 4, 5. 8vo.
Editors—American Journal of Science for June, 1902. 8vo.
　　Astrophysical Journal for May, 1902.
　　Athenæum for June, 1902. 4to.
　　Author for June, 1902. 8vo.
　　Brewers' Journal for June, 1902. 8vo.
　　Chemical News for June, 1902. 4to.
　　Chemist and Druggist for June, 1902. 8vo.
　　Electrical Engineer for June, 1902. fol.
　　Electrical Review for June, 1902. 8vo.
　　Electrical Times for June, 1902. 4to.
　　Electricity for June, 1902. 8vo.
　　Electro-Chemist and Metallurgist for May, 1902. 8vo.
　　Engineer for June, 1902. fol.
　　Engineering for June, 1902. fol.
　　Homœopathic Review for June, 1902. 8vo.
　　Horological Journal for July, 1902. 8vo.
　　Invention for June, 1902.
　　Journal of the British Dental Association for June, 1900. 8vo.
　　Journal of State Medicine for July, 1902. 8vo.
　　Law Journal for June, 1902. 8vo.
　　London Technical Education Gazette for June, 1902.
　　Machinery Market for June, 1902. 8vo.
　　Motor Car Journal for June, 1902. 8vo.
　　Nature for June, 1902. 4to.
　　New Church Magazine for July, 1902. 8vo.
　　Nuovo Cimento for May, 1902. 8vo.
　　Page's Magazine for July, 1902. 8vo.
　　Pharmaceutical Journal for June, 1902. 8vo.
　　Photographic News for June, 1902. 8vo.
　　Physical Review for July, 1902. 8vo.
　　Public Health Engineer for June, 1902. 8vo.
　　Science Abstracts for June, 1902. 8vo.
　　Travel for July, 1902. 8vo.
　　Zoophilist for June and July, 1902. 4to.
Electrical Engineers, Institution of—Journal, Vol. XXXI. No. 157. 8vo. 1902.
Field Columbian Museum—Publications; Anthropological Series, Vol. III. No. 2. 8vo. 1901.
Franklin Institute—Journal, Vol. CLIII. No. 6. 8vo. 1902.
Geographical Society, Royal—Geographical Journal for June, July, 1902. 8vo.
Imperial Institute—Imperial Institute Journal for June, July, 1902.
Johns Hopkins University—University Circular, No. 158. 8vo. 1902.
　　American Journal of Philology, Vol. XXIII. No. 1. 8vo. 1902.
Junior Engineers, Institution of—Transactions, Vol. XI. 8vo. 1902.
Kansas University—Bulletin, Vol. II. No. 8. 8vo. 1902.
Leighton, John, Esq. M.R.I.—Journal of the Ex-Libris Society for April, 1902. 8vo.
　　Notes on Books and Bindings.
Manchester Literary and Philosophical Society—Memoirs and Proceedings, Vol. XLVI. Part 6. 8vo. 1902.
Manchester Steam Users' Association—Boiler Explosions Acts, 1882 and 1890, Reports Nos. 1248–1319. fol. 1901.
Massachusetts Institute of Technology—Technology Quarterly, Vol. XV. No. 1. 8vo. 1902.
Mather and Crowther (the Publishers)—Practical Advertising. 8vo. 1902.
Mechanical Engineers, Institution of—Proceedings, 1901, No. 5. 8vo.
　　List of Members, 1902. 8vo.

Mersey Conservancy—Report on the State of Navigation of the River Mersey, 1901. 8vo. 1902.
Microscopical Society, Royal—Journal, 1902, Part 3. 8vo.
Navy League—Navy League Journal for June, 1902. 8vo.
  Guide to the Coronation Review. 8vo. 1902.
New York Academy of Sciences—Annals, Vol. XIV. No. 2. 8vo. 1902.
New Zealand, Registrar-General of—Statistics of the Colony of New Zealand for 1900. 4to. 1901.
North of England Institute of Mining and Mechanical Engineers—Subject Matter Index of Mining, Mechanical and Metallurgical Literature for the year 1901. 8vo. 1902.
Odontological Society—Transactions, Vol. XXXIV. No. 7. 8vo. 1902.
Onnes, Professor H. K.—Communications, Nos. 77-79. 8vo. 1902.
Paris, Société Française de Physique—Séances, 1901, Fasc. 3. 8vo. 1901.
Philadelphia Academy of Natural Sciences—Proceedings, Vol. LIII. Part 3. 8vo. 1902.
Photographic Society, Royal—Photographic Journal for May, 1902. 8vo.
Queensland, The Home Secretary of—North Queensland Ethnography Bulletin, No. 4. fol. 1902.
Rome, Ministry of Public Works—Giornale del Genio Civile. March, 1902. 8vo.
Royal Society of London—Philosophical Transactions, A, Nos. 304-606; B. No. 210. 4to. 1902.
  Proceedings, No. 461. 8vo. 1902.
Selborne Society—Nature Notes for June-July, 1902. 8vo.
Smith, B. Leigh, Esq. M.R.I.—Transactions of the Institute of Naval Architects, Vol. XIII. 4to. 1900.
  The Scottish Geographical Magazine for July, 1902. 8vo.
Society of Arts—Journal for June, 1902. 8vo.
Statistical Society, Royal—Journal, Vol. LXV. Part 2. 8vo. 1902.
Stirling, James, Esq. M.R.I.—The Geological Ages of the Gold Deposits of Victoria. By James Stirling. 8vo. 1900.
  Report on the Walhalla Gold Fields (Victoria). By H. Herman. fol. 1901.
Toronto, University of—Studies, Biological Series, No. 2. 8vo. 1902.
United Service Institution, Royal—Journal for June, 1902. 8vo.
Upsala, Royal Society of Sciences—Nova Acta, Third Series, Vol. XX. Fasc. 1. 4to. 1901.
Verein zur Beförderung des Gewerbfleisses in Preussen—Verhandlungen, 1902, Heft 6. 8vo.
Washington, Academy of Sciences—Proceedings, Vol. IV. pp. 275-292. 8vo. 1902.
Zoological Society—Proceedings, 1902, Vol. I. Part 1. 8vo.

APPARATUS, ETC.

The Badische Anilin und Soda-Fabrik—Collection of Important Dyes and Organic Products.
B. Leigh Smith, Esq. M.R.I.—Small Electric Pile given to Faraday by Volta.

GENERAL MONTHLY MEETING,

Monday, November 3, 1902.

SIR JAMES CRICHTON-BROWNE, M.D. LL.D. F.R.S., Treasurer and Vice-President, in the Chair.

Granville Hugh Baillie, Esq.
W. Deane Butcher, Esq. M.R.C.S.
Mrs. A. R. Cox.
Sir Archibald Campbell Lawrie.
Gabriel James Morrison, Esq. M. Inst. C.E.
A. B. Tubini, Esq.

were elected Members of the Royal Institution.

The Special Thanks of the Members were returned to Sir Andrew Noble, Bart. K.C.B. F.R.S., for his donation of £150, and to Dr. Ludwig Mond, Ph.D. F.R.S., for his donation of £200, to the Fund for the Promotion of Experimental Research at Low Temperatures.

The Honorary Secretary announced the decease of Dr. J. H. Gladstone, former Professor of the Royal Institution, on October 6.

Resolved, That the Managers of the Royal Institution of Great Britain desire to record their sense of the loss sustained by the Institution and by the whole scientific world in the decease of Dr. John Hall Gladstone, Ph.D. D.Sc. F.R.S. late Fullerian Professor of Chemistry.

He became a Member of the Royal Institution of Great Britain in 1854, and was elected a Visitor in 1857 and a Manager in 1860.

His first course of Chemical Lectures in the Institution was delivered in 1855, and he was subsequently, in 1874, appointed Fullerian Professor of Chemistry and Director of the Laboratory of the Royal Institution.

In addition to courses of Lectures, he gave the Christmas Lectures adapted to a Juvenile Auditory in 1874–5 and again in 1876–7, and delivered many Friday Evening Discourses on his important Discoveries. His last Discourse was given in 1898 on "The Metals used by the Great Nations of Antiquity."

He published a popular and admirable Life of Michael Faraday, and has made many contributions to scientific knowledge by his important researches on the laws of Chemical Combination and on the relations of Chemical and Optical Science.

The Managers desire to offer to the family the expression of the most sincere sympathy with them in their bereavement.

The Honorary Secretary announced the decease of Sir Frederick Abel, Bart. on September 6.

Resolved, That the Managers of the Royal Institution of Great Britain desire to record their sense of the loss sustained by the Institution and by the whole scientific world in the decease of Sir Frederick Abel, Bart. G.C.V.O. K.C.B. D.C.L. LL.D. F.R.S., late Vice-President of the Royal Institution of Great Britain, who has rendered such signal services to Science.

He became a Member of the Royal Institution of Great Britain in 1884, and was elected a Manager in 1885.

He has delivered many Friday Evening Discourses on his important investigations on the applications of chemistry to military purposes.

He always took a deep interest in the Institution and the Promotion of Scientific Research.

The Managers desire to offer to the family the expression of the most sincere sympathy with them in their bereavement.

The PRESENTS received since the last Meeting were laid on the table, and the thanks of the Members returned for the same, viz. :—

FROM

The Secretary of State for India—
  Geological Survey of India—
    General Report, 1901-2. 8vo.
    Palæontologia Indica, New Series. Vol. II. Part 1. Fol. 1902.
    The G.T. Survey of India, Vol. XVI. 4to. 1902.
Abderhalden, Dr. Emile (the Author)—Uber den Einfluss des Hohenklimas auf die Zusammensetzung des Blutes. 8vo. 1902.
Accademia dei Lincei, Reale, Roma—Classe di Scienze Fisiche, Matematiche e Naturali. Atti, Serie Quinta: Rendiconti. 1º Semestre, Vol. XI. Fasc. 12 ; 2º Semestre, Fasc. 1-7. 8vo. 1902.
    Classe di Scienze Morali, Storiche, etc., Serie Quinta, Vol. XI. Fasc. 3-6. 8vo.
American Academy of Arts and Sciences—Proceedings, Vol. XXXVII. Nos. 17-22. 8vo. 1902.
    Memoirs, Vol. XII. November. 4to. 1902.
American Philosophical Society—Proceedings, Vol. XLI. Jan.-April, 1902. 8vo.
Asiatic Society, Royal—Journal, July-Oct. 1902. 8vo.
Automobile Club—Journal for July-Oct. 4to. 1902.
Bankers, Institute of—Journal, Vol. XXIII. Parts 6, 7. 8vo. 1902.
Batavia, Meteorological Observatory—Observations, Vol. XXIII. 1900. fol. 1902.
Belgium, Royal Academy of Sciences—Bulletin, 1901 ; 1902, 4-8. 8vo.
    Mém. Cour. et des savants étrang. Tome LIX. Fasc. 3. 4to. 1902.
    Mém. Cour. et autres Mém. Tome LVI. and LXII. Fasc. 1. 8vo. 1902.
    Mémoires, Tome LIV. Fasc. 5. 4to. 1902.
Berlin, Academy of Sciences—Sitzungsberichte, 1902, Nos. 1-40. 8vo.
Booth, A. (the Author)—The Trilingual Cuneiform Inscriptions. 8vo. 1902.
Bose, Prof. J. C. (the Author)—Response in the Living and Non-Living. 8vo. 1902.
Boston Public Library—Monthly Bulletin tor July-Aug. 1902. 8vo.
    Annual Report, 1901-1902. 8vo.
Boston, Society of Medical Sciences—Journal of Medical Research for June, 1902. 8vo.
British Architects, Royal Institute of—Journal, Third Series, Vol. IX. Nos. 17-20. 4to. 1902.
    The Kalendar, 1902-1903. 8vo. 1902.
British Astronomical Association—Journal, Vol. XII. Nos. 9, 10. 8vo. 1901.
    Memoirs, Vol. X. Part 4.
    List of Members, 1902.
Buenos Ayres, City—Monthly Bulletin of Municipal Statistics, May-Aug. 1902. 4to.
Caine, Rev. Cæsar (the Editor)—The Register of St. John's Church, Garrigill, from 1699-1730. 8vo. 1901.
Cambridge Philosophical Society—Transactions, Vol. XIX. Part 2. 4to. 1902.
    Proceedings, Vol. XI. Part 6. 8vo. 1902.
Camera Club—Journal for July-October, 1902. 8vo.
Canada, Geological Survey of—Catalogue of Canadian Plants, Part VII. 8vo. 1902.
Canada, Meteorological Office—Report of the Meteorological Service for 1899. 4to. 1902.
Chemical Industry, Society of—Journal, Vol. XXI. Nos. 13-20. 8vo. 1902.

Chemical Society—Journal for Aug.–Nov. 1902. 8vo.
Chicago, Field Columbian Museum—Publications:
Zoological Series, Vol. III. No. 6. 8vo. 1902.
Geological Series, Vol. I. No. 2. 8vo. 1902.
Civil Engineers, Institution of—Minutes of Proceedings, Vols. 147–149. 8vo. 1902.
Colonial Institute, Royal—Proceedings, Vol. XXXIII. 8vo. 1902.
Comité International des Poids et Mesures—Procès-Verbaux des Séances de 1901.
Tome I. 8vo. 1902.
Travaux et Mémoires de Bureau International de Poids et Mesures. Tome
XII. 4to. 1902.
Cornwall, Royal Institution of—Proceedings, Vol. XV. Part 1. 8vo. 1902.
Cracovie, l'Académie des Sciences—Bulletins, Classe des Sciences Mathématiques
et Naturelles, 1902, No. 6. 8vo.
Classe de Philologie, 1902, No. 6. 8vo.
Dax: Société de Borda—Bulletin, 1902, Nos. 1–4. 8vo.
Despaux, A. (the Author)—Cause des Energies Attractives. 8vo. 1902.
Dewar, Prof. J. M.A. F.R.S. M.R.I.—Œuvres de Gallissard de Mariguac. Tome I.
1840–60. 4to. 1902.
Duckworth, Sir Dyce, M.R.I.—Knowledge and Wisdom in Medicine. 8vo. 1902.
Editors—Aeronautical Journal for July–Oct. 1902. 8vo.
Aeronautical World for Aug. 1902. 8vo.
American Journal of Science for July–Oct. 1902 8vo.
Astrophysical Journal for June–Oct. 1902.
Athenæum for July–Oct. 1902. 4to.
Author for Aug.–Nov. 1902. 8vo.
Board of Trade Journal for Sept. 1902. 8vo.
Brewers' Journal for Aug.–Oct. 1902. 8vo.
Chemical News for Aug.–Oct. 1902. 4to.
Chemist and Druggist for Aug.–Oct. 1902. 8vo
Electrical Engineer for Aug.–Oct. 1902. fol.
Electrical Review for Aug.–Oct. 1902. 4to.
Electrical Times for Aug.–Oct. 1902. 4to.
Electricity for Aug.–Oct. 1902. 8vo.
Engineer for Aug.–Oct. 1902. fol.
Engineering for Aug.–Oct. 1902. fol.
Homœopathic Review for Aug.–Nov. 1902. 8vo.
Horological Journal for July–Oct. 1902. 8vo.
Invention for July–Oct. 1902. fol.
Journal of the British Dental Association for July–Oct. 1902. 8vo.
Journal of State Medicine for July–Oct. 1902. 8vo.
Law Journal for July–Oct. 1902. 8vo.
London Technical Education Gazette for July–Oct. 1902.
Machinery Market for July–Oct. 1902. 8vo.
Model Engineer for July–Sept. 1902. 8vo.
Mois Scientifique for July–Oct. 1902. 8vo.
Motor Car Journal for July–Oct. 1902. 8vo.
Nature for July–Oct. 1902. 4to.
New Church Magazine for Aug.–Nov. 1902. 8vo.
Nuovo Cimento for June–Sept. 1902. 8vo.
Page's Magazine for Oct. 1902.
Pharmaceutical Journal for July–Oct., 1902. 8vo.
Photographic News for June–Sept. 1902. 8vo.
Physical Review for Aug.–Oct. 1902. 8vo.
Public Health Engineer for July–Oct. 1902. 8vo.
Science Abstracts for July–Oct. 1902. 8vo.
Travel for Aug.–Oct. 1902. 8vo.
Tea for June–July, 1902. 4to.
Zoophilist for Aug.–Oct. 1902. 4to

Electrical Engineers, Institution of—Journal, Vol. XXXI. Part 6. 8vo. 1902.
    List of Officers and Members. 8vo. 1902.
Florence, Reale Accademia dei Georgofili—Atti. Vol. XXV. Disp. 2. 8vo. 1902.
Franklin Institute—Journal, Vol. CLIV. Nos. 1-4. 8vo. 1902.
Geographical Society, Royal—Geographical Journal for Aug.-Oct. 1902. 8vo.
Geological Society—Quarterly Journal, Vol. LVIII. Part 3. 8vo. 1902.
Glasgow, Philosophical Society of—Proceedings, Vol. XXXIII. 8vo. 1902.
Hibbert, James, Esq. (the Author)—Monimenta 8vo. 1902.
Horticultural Society, Royal—Journal, Vol. XXVII. Part 1. 8vo. 1902.
Imperial Institute—Imperial Institute Journal for Aug.-Nov. 1902.
International Engineering Congress (Glasgow) 1901—The Report and Abstracts
    of Papers. 8vo. 1902.
Iron and Steel Institute—Index to Journal, Vols. XXXVI.-LVIII. 8vo. 1902.
Janet, Charles, Esq. (the Author)—Les habitations à Bou Marché dans les villes
    du moyenne importance. 8vo. 1901.
    Extrait des Mémoires de la Société Académique de l'Oise. 8vo. 1898.
    Extrait du Bulletin de la Société Zoological de France. 8vo. 1900.
    Études sur les Fourmis, les Guêpes et les Abeilles; Note 17-18. 8vo. 1898.
    Extraits des Mémoires de la Société Zoologique de France. 8vo. 1898-99.
    Essai sur la Constitution Morphologique de la tête de l'insect. 8vo. 1899.
    Notes sur les Fourmis et les Guêpes. 4to.
Johns Hopkins University—Circulars, No. 159. 1902.
    American Journal of Philology, Vol. XXIII. No. 2. 8vo. 1902.
Kyoto, Imperial University—Calendar, 2561-62 (1901-1902). 8vo.
Linnean Society—Transactions, Botany, Vol. VI. Parts 2, 3; Zoology, Vol. VIII.
    Parts 5-8. 4to. 1902.
    Journal, Botany, Vol. XXXV. No. 245; Zoology, Vol. XXVIII. No. 185. 8vo.
    1902.
London County Council (Technical Education Board)—Report on the Application
    of Science to Industry. Fol 1902.
Longe, Francis D., Esq. (the Author)—The Fiction of the Ice Age. 8vo. 1902.
Manchester Geological Society—Transactions, Vol. XXVII. Parts 10-16. 8vo.
    1902.
Marchlewski, L. Esq. (the Author)—Natural Colouring Matters. 8vo. 1902.
    Chemistry of Isatin. 8vo. 1902.
Martius, Dr. C. A., M.R.I.—Die Chemische Industrie des Neunzenten Jahrunderts
    von G. Muller. 4to 1902.
Massachusetts Institute of Technology—Technology Quarterly, Vol. XV. No. 2.
Mechanical Engineers, Institution of—Proceedings, 1902, No. 1. 8vo.
Meteorological Society, Royal—Meteorological Record, Vol. XXI. No. 84. 8vo.
    1901.
    Quarterly Journal, Vol. XXVIII. No. 123. 8vo. 1902.
Microscopical Society, Royal—Journal, Aug. and Oct. 8vo. 1902.
Middlesex Hospital—Reports for the Year 1900. 8vo. 1902.
Montpellier, Académie des Sciences—Catalogue de la Bibliothèque, Part 1. 8vo.
    1901.
Musée Teyler—Archives, Série II. Vol. VIII. Fasc I. 8vo. 1902.
Natal, Commissioner of Mines—Report on the Mining Industry of Natal for 1901.
    Folio. 1902.
Navy League—Navy League Journal for July-Oct. 1902. 8vo.
New Jersey, Geological Survey—Report for 1901. 8vo. 1902.
New South Wales, Controller of Prisons—Report on Prisons, 1901. fol. 1902.
Norfolk and Norwich Naturalists' Society—Transactions, Vol. VII. Part 3. 8vo.
    1902.
North of England Institute of Mining and Mechanical Engineers—Transactions,
    Vol. LI. Nos. 3-4. 8vo. 1902.
Odontological Society—Transactions, Vol. XXXIV. No. 8. 8vo. 1902.

Paton, Messrs. J. & J. (the Publishers)—List of Schools and Tutors. 8vo. 1902.

Philadelphia, Academy of Natural Sciences—Proceedings, Vol. LIV. Part 1. 8vo. 1902.

Photographic Society, Royal—Photographic Journal for May–Aug. 1902. 8vo.

Physical Society—Proceedings, Vol. XVIII. Part 2. 8vo. 1902.

Righi, A. Esq. (the Author)—Sui fenomeni acustici dei condensatori. 8vo. 1902.

Rio de Janeiro Observatory—Annual, 1902. 8vo.

Rochechouart, La Société les Amis des Sciences et Arts—Bulletin, Tome XI. Nos. 4–5. 8vo 1902.

Rome Ministry of Public Works—Giornale del Genio Civile for April–July. 8vo. 1902.

Royal Society of London—Philosophical Transactions, A. Nos. 312–317. 4to. 1902.

　Proceedings, Nos. 462–467. 8vo. 1902.

　Reports to the Malaria Committee, 7th Series. 8vo. 1902.

Sanitary Institute—Journal, Vol. XXIII. Part 2–3. 8vo. 1902.

　Supplement, Vol. XXIII. Part 2–3. 8vo. 1902.

Scotland, Astronomer Royal for—Annals of the Royal Observatory, Edinburgh, Vol. I. 4to. 1902.

Selborne Society—Nature Notes for Aug.–Oct. 1902. 8vo.

Smith, B. Leigh. Esq. M.R.I.—Transactions of the Institution of Naval Architects, Vol. XLIV. 4to. 1902.

　The Scottish Geographical Magazine, Vols. XVII. and XVIII. Nos. 7–10. 8vo.

Smithsonian Institution—Report of the U.S. National Museum, 1900. 8vo. 1902.

　Miscellaneous Collections, 1259, 1312-14. 8vo. 1902.

Society of Arts—Journal for July–Oct , 1902. 8vo.

Statistical Society—Journal, Vol. LXV. Part 3. 8vo. 1902.

Tacchini, Prof. P. Hon. Mem. R.I. (the Author)—Memorie della Società degli Spettroscopisti Italiani, Vol. XXXI. Disp. 5ª–8. 4to. 1902.

Tasmania, Agent-General for—Mineral Industry of Tasmania, 1901 and 1902. 8vo.

Tasmania, Royal Society of—Proceedings, 1900–1901. 8vo. 1902.

United Service Institution, Royal—Journal for July–Oct, 1902. 8vo.

United States Department of Agriculture—Eclipse Meteorology and Allied Problems.

　Bulletin Nos. 32, 109 and 117. 4to. 1902.

　Experiment Station Record, Vol. XIII. No. 9. Vol. XIV. No. 1. 8vo. 1902.

　Monthly Weather Review, March 1902. 4to.

　Report of Chief of Weather Bureau, 1900–1901, Vol. I. 4to. 1902.

United States Geological Survey—Mineral Resources of United States, 1900. 8vo. 1902.

　Bulletins, 177–190, 192, 193 and 194. 8vo.

　21st Annual Report, 1899–1900. Parts 5–7.

　Geology and Mineral Resources of Copper River District, Alaska.

　Reconnaissances in the Cape Nome and Norton Bay Regions, Alaska. 1900.

Upsal, Royal Society of Sciences—Nova Acta, Third Series, Vol. XV. Fasc. 1. 4to. 1902.

　List of Fellows, 1902. 8vo.

Verein zur Beförderung des Gewerbfleisses in Preussen—Verhandlungen, 1902. Heft 6–8. 8vo.

Victoria Institute—Journal, Vol. XXXIV. 8vo. 1902.

Vienna, Imperial Geological Institute—Verhandlungen, 1902, Nos. 7–10. 8vo.

Wadsworth, F. L. Esq. (the Author)—A New Type of Focal Plane Spectroscope.
  Theory of the Ocular Spectroscope.
  Report of Director Allegheny Observatory. 8vo. 1902.
Washington Academy of Sciences—Proceedings, Vol. IV. pp. 293-560. 8vo. 1902.
  Memoirs, Vol. XIII. (sixth memoir). 4to. 1902.
Western Society of Engineers—Journal, Vol. VII. Nos. 3 and 4. 8vo. 1902.
Williams & Norgate, Messrs. (the Publishers)—Proceedings of the Aristotelian
  Society, 1901-1902. 8vo.
Zoological Society—Transactions, Vol. XVI. Part 6. 4to. 1902.
  Proceedings, 1902, Vol. I. Part 2. 8vo. 1902.
Zurich, Naturforschende Gesellschaft—Vierteljahrsschrift, Jahrg. XLVII. Heft 1, 2.
  8vo. 1902.

WEEKLY EVENING MEETING,

Friday, March 14, 1902.

SIR FREDERICK BRAMWELL, Bart. D.C.L. LL.D. F.R.S.,
Vice-President, in the Chair.

PROFESSOR SILVANUS P. THOMPSON, B.A. D.Sc. F.R.S. *M.R.I.*

Magnetism in Transitu.

(Abstract deferred.)

GENERAL MONTHLY MEETING,

Monday, December 1, 1902.

Sir JAMES CRICHTON-BROWNE, M.D. LL.D. F.R.S., Treasurer and
Vice-President, in the Chair.

Edward Divers, M.D. B.Sc.,
Miss Amy French,
Winefred Lady Howard of Glossop,
John Herbert Whitehorn, Esq.

were elected Members of the Royal Institution.

The special thanks of the Members were returned to Mrs.
Hickman for her Donation of £21, and to Dr. Frank McClean,
F.R.S., for his Donation of £40 to the Fund for the Promotion of
Experimental Research at Low Temperatures.

The Honorary Secretary announced the decease of Sir William
Chandler Roberts-Austen, K.C.B. on November 22.

Resolved, That the Managers of the Royal Institution of Great Britain desire
to record their sense of the loss sustained by the Institution in the decease of
Sir William Chandler Roberts-Austen, K.C.B. D.C.L. D.Sc. F.R.S.

Becoming a Member of the Royal Institution of Great Britain in 1871, he
delivered many Friday Evening Discourses on his important researches on the
Properties of Metals and their Alloys, and he gave also an interesting Course of
Lectures in 1886 on "Metals as affected by small quantities of Impurity." He
was elected a Manager in 1889.

The Managers desire to offer to Lady Roberts-Austen the expression of the
most sincere sympathy with her in her bereavement.

The PRESENTS received since the last Meeting were laid on the
table, and the thanks of the Members returned for the same,
viz. :—

FROM

Allegheny Observatory—Miscellaneous Scientific Papers, New Series, Nos. 8–9·
8vo. 1902.
American Academy of Arts and Sciences—Proceedings, Vol. XXXVII. Part 23.
8vo. 1902.
American Geographical Society—Bulletin, Vol. XXXIV. No. 4. 8vo. 1902.
American Philosophical Society—Proceedings, Vol. XLI. No. 170. 8vo. 1902.
Accademia dei Lincei, Reale, Roma—Classe di Scienze Fisiche, Matematiche e
Naturali. Atti, Serie Quinta: Rendiconti. 2° Semestre, Vol. XI. Fasc. 8,
9. 8vo. 1902.
Astronomical Society, Royal—Monthly Notices, Vol. LXII. No. 9. 8vo. 1902.
Automobile Club—Journal for November, 1902. 8vo.
Backhouse, J. W. (the Author)—Publications of West Hendon House Observatory,
Sunderland. No. 2. 4to. 1902.
Bankers, Institute of—Journal, Vol. XXIII. Part 8. 8vo. 1902.

Bombay. Government of—Report on Total Solar Eclipse of January 1898, as observed at Jeur, in Western India. By K. D. Naegamoala. 4to. 1902.

Boston Public Library—Monthly Bulletin for November, 1902. 8vo.

British Architects, Royal Institute of—Journal, Third Series, Vol. X. Nos. 1, 2. 4to. 1902.

British Astronomical Association—Journal, Vol. XIII. No. 1. 8vo. 1902.

Buenos Ayres, City—Monthly Bulletin of Municipal Statistics, September, 1902. 4to.

Canada, Central Meteorological Office—Report of the Meteorological Service for 1900. 8vo. 1902.

Canada, Geological Survey of—Geological Map of Canada, Western Sheet, No. 783.

Chemical Industry, Society of—Journal, Vol. XXI No. 21. 8vo. 1902.

Chemical Society—Proceedings, Nos. 255, 256. 8vo. 1902. Journal for December, 1902. 8vo.

Clinical Society—Transactions, Vol. XXXV. 8vo. 1902.

Editors—American Journal of Science for November, 1902. 8vo.
Analyst for November, 1902. 8vo.
Astrophysical Journal for October, 1902.
Athenæum for November, 1902. 4to.
Brewers' Journal for November, 1902. 8vo.
Chemical News for November, 1902. 4to.
Chemist and Druggist for November, 1902. 8vo.
Electrical Engineer for November, 1902. fol.
Electrical Review for November, 1902. 8vo.
Electrical Times for November, 1902. 4to.
Electricity for November, 1902. 8vo.
Engineer for November, 1902. fol.
Engineering for November, 1902. fol.
Feilden's Magazine for November, 1902.
Horological Journal for December, 1902. **8vo**
Invention for November, 1902.
Ironmonger for November' 1902. 8vo.
Journal of the British Dental Association for November, 1902. 8vo.
Journal of State Medicine for November, 1902. 8vo.
Law Journal for November, 1902. 8vo.
London Technical Education Gazette for November, 1902.
Machinery Market for November, 1902. 8vo.
Model Engineer for November, 1902.
Mois Scientifique for November, 1902. 8vo.
Motor Car Journal for November, 1902. 8vo.
Motor Car World for December, 1902.
Musical Times for November, 1902. 8vo.
Nature for November, 1902. 4to.
New Church Magazine for November, 1902. 8vo.
Nuovo Cimento for October, 1902. 8vo.
Page's Magazine for November, 1902. 8vo.
Pharmaceutical Journal for Nov. 1902. 8vo.
Photographic News for Nov. 1902. 8vo.
Physical Review for Nov. 1902. 8vo.
Popular Astronomy for Nov. 1902. 8vo.
Public Health Engineer for Nov. 1902. 8vo.
Science Abstracts for Nov. 1902. 8vo.
Travel for Nov. 1902. 8vo.
Zoophilist for Nov. 1902. 4to.

Eggimann, C., & Co. (the Publishers)—Œuvres Complètes de J. C. Galissard de Marignac. Tome 1, 1840–1860. 4to. 1902.

Fleming, Prof. J. A., M.R.I. (the Author)—Waves and Ripples in Water, Air and Ether. 8vo. 1902.

Franklin Institute—Journal, CLIV. No. 5. 8vo. 1902.
Geographical Society, Royal—Geographical Journal for Nov. 1902. 8vo.
Geological Survey of the United Kingdom—Summary of Progress, 1901. 8vo. 1902.
Harlem, Société Hollandaise des Sciences—Archives Néerlandaises, Sér. II. Tome VII. Livr. 4-5. 8vo. 1902.
Herdenking van het Honderdvijftigjarig Bestaan. 8vo. 1902.
Hill, Messrs. W. E., & Sons—Antonio Stradivari: his Life and Works (1644-1737). 4to. 1902.
Huggins, Lady, M.R.I. (The Author)—The Astrolabe. 8vo. 1902.
Linnean Society—Proceedings, October, 1902. 8vo
   Journal; Botany, Vol. XXVI. Nos. 179-180. 8vo, 1902.
Manchester Geological Society—Transactions, Vol. XXVII. Part 17. 8vo. 1902.
Martius, Dr. C. A., M.R.I.—Die Arbeiterheilstatten des Landes-Versicherungs-anstalt Berlin bei Beelitz. 1902.
Massachusetts Institute of Technology—Technology Quarterly, Vol. XV. No. 3. 8vo. 1902.
Mechanical Engineers, Institution of—Proceedings, 1902, No. 2. 8vo.
Meteorological Society, Royal—Journal for October, 1902. 8vo.
   Meteorological Record, Vol. XXII. No. 85. 8vo. 1902.
Munich, Royal Bavarian Academy of Sciences—Abhandlungen, Band XXI. Abt. 3. 4to. 1902.
   Max von Pettenkofer zum Gedächtniss von Carl v. Voit. 4to. 1902.
Navy League—Navy League Journal for Nov. 1902.
New South Wales, Royal Society of—Journal and Proceedings, Vol. XXXV. 8vo. 1902.
North of England Institute of Mining and Mechanical Engineers—Annual Report, &c. 1901-1902.
Numismatic Society—Numismatic Chronicle, 1902, Part 3. 8vo.
Odontological Society—Transactions, Vol. XXXV. No. 1. 8vo. 1902.
   List of Members, 1902. 8vo.
Paris, Société Française de Physique—Séances, 1902, Fasc. 1-2. 8vo. 1902.
Photographic Society, Royal—Photographic Journal for Oct. 1902. 8vo.
Physical Society of London—Proceedings, Vol. XVIII. Part 3. 8vo. 1902.
Quekett Microscopical Club—Journal, Series 2, Vol. VIII. No. 51. 8vo. 1902.
Royal Society of London—Philosophical Transactions, A, Nos. 319, 320, 321 B, No. 322. 4to. 1902.
   Proceedings, No. 468. 8vo. 1902.
Selborne Society—Nature Notes for Nov. 1902. 8vo.
Smith, B. Leigh, Esq, M.R.I.—The Scottish Geographical Magazine for Nov. 1902. 8vo.
Society of Arts—Journal for Nov. 1902. 8vo.
Tacchini, Prof. P, Hon. Mem. R.I. (the Author)—Memorie della Società degli Spettroscopisti Italiani, Vol. XXXI. Disp. 10A. 4to. 1902.
United Service Institution, Royal—Journal for Nov. 1902. 8vo.
United States Department of Agriculture—Experiment Station Record, Vol. XIII. Nos. 10-12. 8vo. 1902.
   Monthly Weather Review, July. 4to. 1902.
   North American Fauna, No. 22. 8vo. 1902.
Verein zur Beforderung des Gewerbfleisses in Preussen—Verhandlungen 1902, Heft 9. 8vo.
Western Society of Engineers—Journal, Vol. VII. No. 5. 8vo. 1902.
Zoological Society—Proceedings, 1902, Vol. II. Part 1. 8vo.
   Transactions, Vol. XVI. Parts 5 and 7. 4to. 1902.
   Index to Proceedings, 1891-1900. 8vo. 1902.
   Catalogue of Library. 8vo. 1902.

WEEKLY EVENING MEETING,

Friday, April 11, 1902.

THE RIGHT HON. LORD KELVIN, O.M. G.C.V.O. D.C.L. LL.D. D.Sc. F.R.S., Vice-President, in the Chair.

Professor DEWAR, M.A. LL.D. D.Sc. F.R.S. *M.R.I.*

Problems of the Atmosphere.

THE present liquid ocean, neglecting everything for the moment but the water, was at a previous period of the earth's history part of the atmosphere, and its condensation has been brought about by the gradual cooling of the earth's surface. This resulting ocean is subjected to the pressure of the remaining uncondensed gases, which for the present we may regard as composed solely of nitrogen and oxygen, and as these are slightly soluble they dissolve to some extent in the fluid. The gases in solution can be taken out by distillation or by exhausting the water, and if we compare their volume with the volume of the water as steam, we should find about 1 volume of air in 60,000 volumes of steam. This would then be about the rough proportion of the relatively permanent gas to condensable gas which existed in the case of the vaporised ocean.

Now let us assume the surface of the earth gradually cooled to some 200 degrees below the freezing-point; then, after all the present ocean was frozen, and the climate became three times more intense than any arctic frost, a new ocean of liquid air would appear about thirty-five feet deep, covering the entire surface of the frozen globe.

We may now apply the same reasoning to the liquid air ocean that we formerly did to the water one, and this would lead us to anticipate that it might contain in solution some gases that may be far less condensable than the chief constituents of the fluid. In order to separate them we must imitate the method of taking the gases out of water.

If a sample of liquid air cooled to the lowest temperature that can be reached by its own evaporation was connected by a pipe to a condenser cooled in liquid hydrogen, the result would be rapid distillation, and any volatile gases present in solution would distil over with the first portions of the air, and while the nitrogen and oxygen solidified in the condenser they could be pumped off, being still gaseous at the temperature of $20°$ absolute. A diagram of the apparatus is given in my lecture entitled "Gases at the Beginning and End of the Century."[*] In this way, a gas mixture, containing, of the known gases, free hydrogen, helium and neon, has been separated from liquid air.

---

[*] Proceedings of the Royal Institution, Vol. XVI. p. 736.

It is interesting to note in passing that the relative volatilities of water and oxygen are in the same ratio as those of liquid air and hydrogen, so that the analogy between the ocean of water and that of liquid air has another suggestive parallel. The total uncondensable gas separated in this way amounts to about $\frac{1}{50000}$th of the volume of the air, which is about the same proportion as the air dissolved in water.

That free hydrogen exists in air in small amount is conclusively proved, but the actual proportion found by the process described above is very much smaller than Gautier has estimated by the combustion method. The recent experiments of Lord Rayleigh show that air does not contain more than $\frac{1}{30000}$th, so that Gautier, who estimated the hydrogen present as $\frac{1}{5000}$th, has in some way produced more hydrogen than can be extracted from air by a repetition of the same process conducted by an equally competent experimenter.

The more volatile gases, helium, hydrogen and neon, can be separated from liquid air without the use of liquid hydrogen. For this purpose the arrangement of apparatus shown in Diagram 1 is employed; the primary object being to liquefy air uncontaminated by compression pumps or the use of chemical reagents, at the pressure of the atmosphere by cooling a vessel shaped like F to some $-210°$ C. by the evaporation *in vacuo* of liquid air covering its exterior.

This is effected by connecting the tube D to a large air pump and regulating a supply of liquid air from the vessel B by means of the valve a. The external air freed from moisture by preliminary cooling on its way to the apparatus, is brought from the roof of the laboratory through glass pipes to the tube A, with its regulating stopcock. It rapidly liquefies and begins to fill up the vessel F, which has a capacity of about 130 cc., while the more volatile gases diffuse into the space above the liquid. In order to prevent their re-solution a very light and rather tight-fitting glass cylinder, shown in E, is used as a float to diminish the liquid surface. As the liquid air accumulates the float E rises in the vessel F, until finally a small amount of uncondensed gas is left at C. The external supply of liquid air from B being shut off and the stopcock A closed, the valve b connected with the tube d coming from the interior vessel F is opened. This discharges all the air which was liquefied directly to the outside of F; thereby diminishing the amount required to be supplied from the vessel B for the next operation of filling F, and the float E falls to the bottom of the vessel. During this emptying of the vessel the separated gas expands and fills the whole of F under the given conditions of temperature and pressure. Care is taken, however, to leave a layer of liquid air below the float to guard against the gaseous contents in F being taken away in part by the air pump. The valve b is now shut again and A opened so that the operation of filling may be repeated, and after five or six sequences of this kind the accumulated gas in F is drawn off into a mercury receiver G. The total volume of gas collected in this way amounted to about

PLATE I

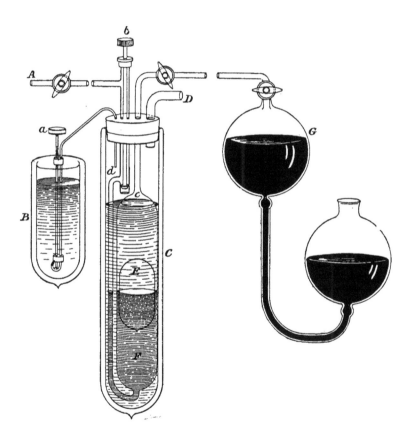

$\frac{1}{35000}$th of the volume of the air directly liquefied, and contained 38 per cent. of nitrogen, 4 per cent. of hydrogen, and 58 per cent. of mixed helium and neon. On the average some 25 cc. of gas was collected for every litre of liquid air produced, and as the apparatus works almost automatically, the method becomes a practical one for the separation of the most incondensable constituents of the atmosphere. After the removal of the hydrogen and nitrogen by sparking in the usual way with excess of oxygen over alkali, the helium and neon can be separated. This is easily done by freezing out the neon by passing the gaseous mixture through a tube cooled in liquid hydrogen; or the alternative method of spectroscopic fractionation described in a former lecture may be adopted to effect the general separation of all the constituents. The execution of the process on the larger scale will enable gases other than helium and neon to be detected and ultimately separated.

A similar sample of gas collected from air that had been passed through lead pipes and dried with strong sulphuric acid contained about 33 per cent. of hydrogen, instead of the 4 per cent. found as above, or say at the most $\frac{1}{100000}$th of free hydrogen in the original sample of air. The increased proportion of hydrogen must have originated from the contact of metal with water and acid vapours.

The total volume of the very volatile gases, helium and neon, collected in this way amounted to $\frac{1}{58000}$th of the original volume of the air. This mixed gas had a density $8\cdot7$ times that of hydrogen, so that the mixture was composed of 16 per cent. of helium and 84 per cent. of neon; if the latter be taken as having a density of 10. Thus air contains five times more neon than helium. These experiments prove that air contains as a minimum $\frac{1}{362000}$th of its volume of helium, about $\frac{1}{70000}$th of neon, and not more than $\frac{1}{100000}$th of free hydrogen. In order to make the method strictly quantitative, the correction factor for solubility of these gases under the conditions of the experiment in the liquid air would require to be known. With the use of air freed from the volatile gases by liquefaction and standing for a day or two, the method was proved to be capable of detecting the presence of $\frac{1}{50000}$th of free hydrogen in a specially prepared mixture. In a mixture containing $\frac{1}{5000}$th of free hydrogen the loss from solubility, etc., amounted to about 50 per cent.

The spectroscopic examination of these gases throws new light upon the question of the aurora and the nature of the upper air. On passing electric discharges through the tubes containing the most volatile of the atmospheric gases, they glow with a bright orange light, which is especially marked at the negative pole. The spectroscope shows that this light consists, in the visible part of the spectrum, chiefly of a succession of strong rays in the red, orange and yellow, attributed to hydrogen, helium and neon. Besides these, a vast number of rays, generally less brilliant, are distributed through the whole length of the visible spectrum. The greater part of these rays are of, as yet, unknown origin. The violet and ultra-violet part of the spectrum rivals in strength that of the red and yellow rays. As these gases probably include some of the gases that pervade inter-planetary space, search was made for the prominent nebular, coronal and auroral lines. No definite lines agreeing with the nebular spectrum could be found, but many lines occurred closely coincident with the coronal and auroral spectrum. But before discussing the spectroscopic problem, it will be necessary to consider the nature and condition of the upper air.

According to the old law of Dalton, supported by the modern dynamical theory of gases, each constituent of the atmosphere while acted upon by the force of gravity forms a separate atmosphere, completely independent, except as to temperature, of the others, and the relations between the common temperature and the pressure and altitude for each specific atmosphere can be definitely expressed.

If we assume the altitude and temperature known, then the pressure can be ascertained for the same height in the case of each of the gaseous constituents, and in this way the percentage composition of the atmosphere at that place may be deduced.

Suppose we start with a surface atmosphere having the composition of our air, only containing $\frac{2}{10000}$ths of hydrogen; then at thirty-seven miles, if a sample could be procured for analysis, we believe that it would be found to contain 12 per cent. of hydrogen and only 10 per cent. of oxygen. The carbonic acid practically disappears; and by the time we reach forty-seven miles, where the temperature is *minus* 132 degrees, assuming a gradient of 3·2 degrees per mile, the nitrogen and oxygen have so thinned out that the only constituent of the upper air which is left is hydrogen. If the gradient of temperature were doubled, the elimination of the nitrogen and oxygen would take place by the time thirty-seven miles was reached, with a temperature of *minus* 220 degrees.

The theoretical distribution of the chief components of our atmosphere, on the assumption of steady equilibrium, is graphically represented in Diagrams II. and III. In the diagrams nitrogen is represented by the red colour, oxygen by the blue, hydrogen by the yellow, argon by vermilion, and carbonic acid by black. A horizontal line drawn across the diagram at any height marked in kilometres (0·62 mile) shows the percentage by volume of the constituents at that elevation by the respective lengths of line in the colour of the individual constituents. The results of Hinrich's calculations which involve no consideration of the effects of temperature, are represented in Diagram II., and those of Ferrel, who assumes a temperature gradient of 4° per kilometre, throughout the upper air, in Diagram III. The higher the assumed temperature gradient the lower the elevation at which the nitrogen and oxygen are eliminated and the true hydrogen atmosphere begins. The elevations marked A, B, C and D in the diagram refer to the respective gradients of 4°, 3°, 2° and 1° per kilometre, and mark the end of nitrogen or the beginning of the true hydrogen atmosphere. The position A corresponds to 60 kilometres

DISTRIBUTION OF THE ATMOSPHERIC GASES

HINRICH'S FORMULÆ.

DISTRIBUTION OF THE ATMOSPHERIC GASES

FERREL'S FORMULÆ

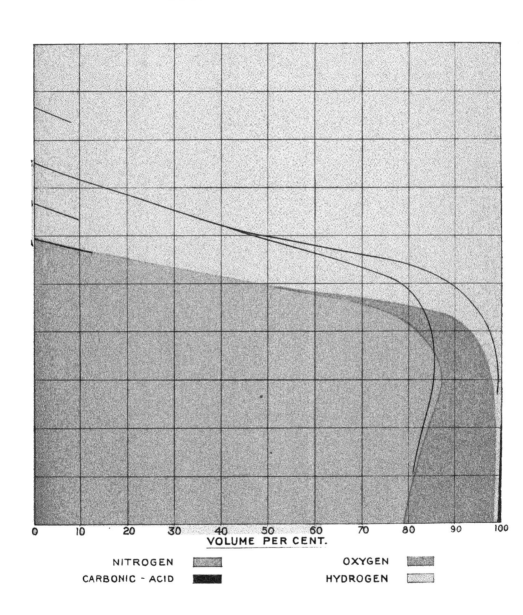

VOLUME PER CENT.

NITROGEN

OXYGEN

CARBONIC - ACID

HYDROGEN

and a temperature of $-220°$; B to 67 kilometres and $-181°$ C.; C to 76 kilometres and $-132°$; and D to 87 kilometres and $-67°$.

On any of these temperature gradient hypotheses it appears that practically above 56 miles the atmosphere would be substantially composed of hydrogen. If helium and neon had been included in the calculations they would have been found concentrated at high elevation between the regions occupied respectively by the hydrogen and the nitrogen in the diagrams. If the temperature is taken as constant, Diagram II. shows that at an elevation of some 62 miles the composition of a sample of the air, if it could be secured, would be 95·1 per cent. of hydrogen, 4·6 per cent. of nitrogen, and 0·3 per cent. of oxygen.

The permanence of the composition of the air at the highest altitudes, as deduced from the basis of the dynamical theory of gases, has been discussed by Stoney, Bryan, and others. It would appear that there is a consensus of opinion that the rate at which gases like hydrogen and helium could escape from the earth's atmosphere would be excessively slow. Considering that to compensate any such loss the same gases are being supplied by actions taking place in the crust of the earth, we may safely regard them as necessarily permanent constituents of the upper air.

The temperature at the elevations we have been discussing would not be sufficient to cause any liquefaction of the nitrogen and oxygen, on account of the pressure being so low. If we assume the mean temperature as about the boiling-point of oxygen, then a considerable amount of the carbonic acid must solidify as a mist, if the air from a lower level be cooled to this temperature; and the same result might take place with other gases of relatively small volatility which occur in air. The temperature of the upper air must be above that on the vapour pressure curve corresponding to the barometric pressure at the locality, otherwise liquid condensation must take place. In other words, the temperature must be above the dew-point of air at that place. At very high elevations, on any reasonable assumption of temperature distribution, we inevitably reach a temperature where the air would condense, just as Fourier and Poisson supposed it would, unless the temperature is arrested in some way from approaching the zero.

Both ultra-violet absorption and the prevalence of electric storms may have something to do with the maintenance of a higher mean temperature than we should anticipate, following the deductions of our assumed formulas for temperature decrements. The whole mass of the air above 40 miles is not more than $\frac{1}{700}$th part of the total mass of the atmosphere, so that any rain or snow of liquid or solid air, if it did occur, would necessarily be of a very tenuous description. In any case, the dense gases tend to accumulate in the lower strata, and the lighter ones to predominate at the higher altitudes, always assuming a steady state of equilibrium has been reached.

It must be observed, however, that a sample of air taken at an elevation of 9 miles has shown no difference in composition from that at the ground, whereas, according to our hypothesis, the oxygen ought to have been diminished to 17 per cent., and the carbonic acid should also have become much less.   This can only be explained by assuming that a large intermixture of different layers of the atmosphere is still taking place at this elevation.   This is confirmed by a study of the motions of clouds about six miles high, which reveals an average velocity of the air currents of some seventy miles an hour; such violent winds must be the means of causing the intermingling of different atmospheric strata.   Some clouds, however, during hot and thundery weather, have been seen to reach an elevation of seventeen miles, so that we have direct proof that on occasion the lower layers of atmosphere are carried to a great elevation.

The existence of an atmosphere at more than a hundred miles above the surface of the earth is revealed to us by the phenomenon of twilight and the luminosity of meteors and fireballs.   When we can take photographs of meteoric spectra, a great deal may be learnt about the composition of the upper air.   In the meantime Pickering's solitary spectrum of a meteor reveals an atmosphere of hydrogen and helium, and so far this is a corroboration of the doctrine we have been discussing.   It has long been recognised that the aurora is the result of electric discharges within the limits of the earth's atmosphere, but it was difficult to understand why its spectrum should be so entirely different from anything which could be produced artificially by electric discharges through rarefied air at the surface of the earth.   Rand Capron, in 1879, after collecting all the recorded observations, was able to enumerate no more than nine auroral rays, of which but one could with any probability be identified with rays emitted by atmospheric air under electric discharge.   Vogel attributed this want of agreement between nature and experiment, in a vague way, to difference of temperature and pressure; and Zollner thought the auroral spectrum to be one of a different order, in the sense in which the line and band spectra of nitrogen are said to be of different orders.

Such statements were merely confessions of ignorance.   But since that time observations of the spectra of auroras have been greatly multiplied, chiefly through the Swedish and Danish Polar Expeditions. The spectrum recorded on the ultra-violet side has been greatly extended by the use of photography, so that, in a recent discussion of the results, M. Henri Stassano is able to enumerate upwards of one hundred auroral rays, of which the wave-length is more or less approximately known.   Of this large number of rays he is able to identify, within the probable limits of errors of observation, about two-thirds as rays, which Professor Liveing and myself have observed to be emitted by the most volatile gases of atmospheric air unliquefiable at the temperature of liquid hydrogen.   Most of the remainder he ascribes to argon, and some might, with more probability, have been identified with krypton or xenon.

The rosy tint often seen in auroras, particularly in the streamers, appears to be due mainly to neon, of which the spectrum is remarkably rich in red and orange rays. One or two neon rays are amongst those most frequently observed, while the red ray of hydrogen and one red ray of krypton have been noticed only once. The predominance of neon is not surprising, seeing that from its relatively greater proportion in air and its low density it must tend to concentrate at higher elevations.

So large a number of probable identifications warrants the belief that we may yet be able to reproduce in our laboratories the auroral spectrum in its entirety. It is true that we have still to account for the appearance of some, and the absence of other, rays of the newly discovered gases, which in the way in which we stimulate them appear to be equally brilliant, and for the absence, with one doubtful exception, of all the rays of nitrogen. If we cannot give the reason of this, it is because we do not know the mechanism of luminescence —nor even when the particles which carry the electricity are themselves luminous, or whether they only produce stresses causing other particles which encounter them to vibrate; yet we are certain that an electric discharge in a highly rarefied mixture of gases lights one element and not another, in a way which, to our ignorance, seems capricious.

The Swedish North Polar Expedition concluded from a great number of trigonometrical measurements that the average above the ground of the base of the aurora was fifty kilometres (thirty-four miles) at Cape Thorsden, Spitsbergen; at this height the pressure of the nitrogen of the atmosphere would be only about one-tenth of a millimetre, and Moissan and Deslandres have found that in atmospheric air at pressures less than one millimetre the rays of nitrogen and oxygen fade and are replaced by those of argon and by five new rays which Stassano identifies with rays of the more volatile gases measured by us. Also Collie and Ramsay's observations on the distance, to which electrical discharges of equal potential traverse different gases throw much light on the question. They find that, while for helium and neon this distance is from 250 to 300 mm., for argon it is $45\frac{1}{2}$ mm., for hydrogen it is 39 mm., and for air and oxygen still less.

This indicates that a good deal depends on the very constitution of the gases themselves, and certainly helps us to understand why neon and argon, which exist in the atmosphere in larger proportions than helium, krypton, or xenon, should make their appearance in the spectrum of auroras almost to the exclusion of nitrogen and oxygen.

How much depends not only on the constitution, and it may be temperature, of the gases, but also on the character of the electric discharge, is evident from the difference between the spectra at the cathode and anode in different gases, notably in nitrogen and argon, and not less remarkably in the more volatile compounds of the atmosphere.

Without stopping to discuss that question, it is certain that changes in the character of the electric discharge produce definite changes in the spectra excited by them. It has long been known that in many spectra the rays which are inconspicuous with an un-condensed electric discharge become very pronounced when a Leyden jar is in the circuit. This used to be ascribed to a higher temperature in this condensed spark, though measurements of that temperature have not borne out the explanation. Schuster and Hemsalech have shown that these changes of spectra are in part due to the oscillatory character of the condenser discharge, which may be enhanced by self-induction, and the corresponding change of spectrum thereby made more pronounced.

If we turn to the question what is the cause of the electric discharges which are generally believed to occasion auroras, but of which little more has hitherto been known than that they are connected with sun-spots and solar eruptions, recent studies of electric discharges in high vacua, with which the names of Crookes, Röntgen, Lenard, and J. J. Thomson will always be associated, have opened the way for Arrhenius to suggest a definite and rational answer. He points out that the frequent disturbances which we know to occur in the sun must cause electric discharges in the sun's atmosphere far exceeding any that occur in that of the earth. These will be attended with an ionisation of the gases, and the negative ions will stream away through the outer atmosphere of the sun into the interplanetary space, becoming, as Wilson has shown, nuclei of aggregation of condensable vapours and cosmic dust. The liquid and solid particles thus formed will be of various sizes ; the larger will gravitate back to the sun, while those with diameters less than one and a half thousandths of a millimetre, but nevertheless greater than a wave-length of light, will, in accordance with Clerk-Maxwell's electromagnetic theory, be driven away from the sun by the incidence of the solar rays upon them, with velocities which may become enormous, until they meet other celestial bodies, or increase their dimensions by picking up more cosmic dust, or diminish them by evaporation. The earth will catch its share of such particles on the side which is turned towards the sun, and its upper atmosphere will thereby become negatively electrified until the potential of the charge reaches such a point that a discharge occurs, which will be repeated as more charged particles reach the earth.

[J. D.]

LONDON : PRINTED BY WILLIAM CLOWES AND SONS, LIMITED
GREAT WINDMILL STREET, W., AND DUKE STREET, STAMFORD STREET, S.E.

WEEKLY EVENING MEETING,

Friday, January 23, 1903.

Sir WILLIAM CROOKES, F.R.S., Honorary Secretary and
Vice-President, in the Chair.

TEMPEST ANDERSON, M.D. B.Sc. F.G.S. *M.R.I.*

Recent Volcanic Eruptions.

THERE is a remarkable similarity between the islands of St. Vincent and Martinique. Both are roughly oval in form, with the long axis almost north and south. The north-west portion of each is occupied by a volcano, the Soufrière and Mont Pelée, which have many points in common. Both volcanoes show a single or practically single vent, and a remarkable absence of parasitic cones and a scarcity of dykes. In both a transverse valley exists to the south of the volcanoes, and the main discharge of ejecta during the recent eruptions, which have often been nearly synchronous, has been into this depression, and especially into its westerly portion. In both islands, the recent eruptions have been characterised by paroxysmal discharges of incandescent ashes, with comparatively few larger fragments and a complete absence of lava.

There are, however, a few points of difference. The eruptions of St. Vincent have been altogether on a much larger scale than those in Martinique. The area devastated was considerably larger, the amount of ashes ejected probably ten times as great, and if the loss of life was not so large, this is accounted for by the absence of a populous city at the foot of the mountain. While both volcanoes show practically a single vent, this is much more markedly the case of St. Vincent, where, excepting the new crater, which is practically part of the old or main one, there is not a single parasitic cone. We saw no fumaroles, no hot springs, nor any trace of radial cracks or fissures.

On Mont Pelée, it is true, the main activity is confined to a restricted area about the summit of the mountain, and the top of the great fissure which extends or extended from this down in the direction of the Rivière Blanche; and there are no parasitic cones comparable, for instance, to those which are so numerous on Etna; but there are many fumaroles which Professor Lacroix and his colleagues speak of as emitting gases hot enough to melt lead though not copper wire. A telegraph cable has been three times broken at about the same place, and the broken ends on one occasion, at any rate, showed marks of fusion. There are also several hot springs. Judging from these and other indications, it is most probable that radial cracks entered deeply through the substance of the mountain, and penetrated even the submarine portion of its cone.

The local distribution of erupted material in Martinique is accounted for by the great fissure at the top of the valley of the Rivière Blanche, which communicated with the main pipe of the volcano, and out of which the eruptions took place. This fissure, which was mentioned as existing in the eruption of 1851, pointed almost directly towards St. Pierre, and as the erupted material flowed out almost like a fluid, it was directed straight down on the doomed city. The lowest portion of the lip of the crater of the Soufrière was much broader and more even, so the incandescent avalanche which descended from it was spread much more widely.

The latest accounts from Professor Lacroix indicate that the recent small eruption of Mont Pelée has filled up the highest parts of the fissure and formed a cone, the foot of which covers up the former crater ring. In any further eruption, therefore, the avalanche of incandescent sand will not be confined to the district of the Rivière Blanche, but may descend on any side of the mountain.

The accompanying photograph of Mont Pelée in eruption was obtained from a ten-ton sloop in a sea way, and is therefore not quite sharp. Attention was directed to the eruption by a peculiar black cloud which appeared over the volcano and then rolled down the side of the mountain to the sea. The cloud was formed of surging, rolling, expanding masses, in shape much like those of the previous cauliflowers, but quite black, and full of lightning-flashes and scintillations, while small flashes constantly struck from its lower surface on to the sea. The upper slopes of the mountain cleared somewhat, and some big red-hot stones were thrown out; then the triangular crack became red, and out of it poured a surging mass of incandescent material, reminding us of nothing so much as a big snow-avalanche in the Alps, but at a vastly different temperature. It was perfectly well defined, did not at all tend to rise like the previous cauliflowers, but flowed rapidly down the valley in the side of the mountain which had clearly been the track of previous eruptions, until in certainly less than two minutes it reached the sea, and was there lost to view behind the remains of the first black cloud, with which it appeared to coalesce. There and on the slopes of the mountain were doubtless deposited the greater part of the incandescent ash, while the steam and gases, with a certain portion of still entangled stones and ash, came forward in our direction as a black cloud, but with much greater rapidity than before. The cloud got nearer and nearer; it was well defined, black and opaque, formed of surging masses of the cauliflower type, each lobe rolling forward, but not all with one uniform rotation; bright scintillations appeared, some in the cloud itself and some like little flashes of light vertically between the cloud and the sea on which it rested. These were clearly the phenomena described by the survivors of the St. Vincent eruption as "fire on the sea," occurring in the black cloud which overwhelmed the windward side of that island. We examined them carefully, and are quite clear that they were electric discharges. The scintillations in the body of the cloud became less numerous and more defined, and gradually took the form of vivid flashes of forked lightning darting from one part of the cloud to another. When the cloud had got within perhaps half a mile or a mile of us—for it is difficult to estimate distances at sea and in a bad light—we could see small material falling out of it in sheets and festoons into the sea, while the onward motion seemed to be chiefly confined to the upper part, which then came over our heads and spread out in advance and around us, but left a layer of clear air in our immediate neighbourhood. It was ablaze all the time with electric discharges.

As soon as it got overhead, stones began to fall on deck, some as big as a walnut, and we were relieved to find that they had parted with their heat and were quite cold. Then came small ashes and some little rain. The cloud was also noticed at Fort de France. It was described as like those in the previous eruptions, but was the only one in which electric scintillations had been noticed. Two unbiassed observers, who had seen it and that of May, declared this was the larger of the two.

As to the mechanism of the hot blast and the source of the power which propelled it, both Dr. Flett and I are convinced of the inadequacy of previous explanations, such as electricity, vortices, or explosions in passages pointing laterally and downwards, or explosions confined and directed down by the weight of the air above. Such passages into the mountain, which, to be effective, would require to be closed above, do not exist in the case of the Soufrière, and we are not aware that they have been observed in Mont Pelée, and as to the weight of the air, this did not prevent the explosions in the pipe of the Soufrière from projecting sand and ashes right through the whole thickness of the trade-winds until they were caught by the anti-trade current above and carried to Barbados. Moreover, the black cloud, as we saw it emerge from Mont Pelée, seemed to balance itself at the top of the mountain, start slowly to descend and gather speed in its course, and the second incandescent discharge followed the same rule. We believe that the motive power for the descent was gravity, as in the case of any ordinary avalanche.

The accepted mechanism of a volcanic eruption is that a molten magma rises in the volcano chimney. It consists of fusible silicates and other more or less refractory minerals, sometimes already partly crystallised, and the whole highly charged with water and gases, which are kept in a liquid state by the immense pressure to which they are subjected. When the mass rises near the surface and the pressure is diminished, the water and gases expand into vapour and blow a certain portion of the heavier and less fusible materials to powder, or, short of this, form pumice stone, which is really solidified froth, and they are violently discharged from the crater.

When the greater part of the steam and gases have been discharged, the lava, still rising, gets vent either over the lip of the crater or often through a lateral fissure, and flows quietly down the side of the mountain.

It is quite recognised that these phenomena may occur in various relative proportions. We believe that in these Pelean eruptions, the lava which rises in the chimney is charged with steam and gases, which explode as usual, but some of the explosions happen to have only just sufficient force to blow the mass to atoms and lift the greater part of it over the lip of the crater without distributing the whole widely in the air. The mixture of solid particles and incandescent gas behaves like a heavy liquid, and before the solid particles have time to subside, the whole rolls down the side of the mountain under the influence of gravity, and consequently gathers speed and momentum as it goes. The heavy solid particles are gradually deposited, and the remaining steam and gases, thus relieved of their burden, are free to ascend.

The effect of avalanches in compressing the air before them and setting up a powerful blast, the results of which extend beyond the area covered by the fallen material, has long been recognised. A group of large trees was overthrown by the blast of the great avalanche from the Attels on the Gemmi pass in 1895; all lay prostrate in directions radiating away from the place where the avalanche came down.

[T. A.]

WEEKLY EVENING MEETING,

Friday, January 30, 1903.

The Right Hon. Sir James Stirling, M.A. LL.D. F.R.S.,
Vice-President, in the Chair.

Professor W. E. Dalby, M.A. B.Sc. M. Inst. C.E.

Vibration Problems in Engineering Science.

Vibration Problems in Engineering Science arise from the fact that
the different parts forming engines and machinery are always changing their motion relatively to the frame, the change being in general
continuous and periodic.

To change the motion of a body requires the action of a force, and
it is the equal and opposite aspects of the accelerating forces acting
on the moving parts of an engine or machine which re-acting on the
frame tend to set it in motion and with it the foundations to which
it is attached. In this way vibration in the surroundings may be
set up.

My first object is to make this principle clear to you, partly by
the aid of diagrams, partly by simple experiments.

From Newton's Laws of Motion we know that :—

The natural mode of motion of a body is in a straight line with
uniform speed.

To change the speed in the line of motion requires the action of
a force.

To turn the body out of its straight path and to compel it to
move in a curved path requires the action of a force, although the
speed in the curved path may be uniform.

The word force as usually understood does not convey the idea
that a force acting one way must necessarily be accompanied by an
equal force acting in the opposite way. Yet such is the case. A push
must always be accompanied by an equal and opposite push, a pull
by an equal and opposite pull, a twist by an equal and opposite twist.

A very cursory examination of a machine or engine will show
that none of the parts are moving in the natural mode. None of
them are passing over equal intervals in equal times in a straight
line. The forces which must act to produce these changes are necessarily accompanied by equal and opposite forces acting on the frame.
These forces try to pull the frame first in this direction then in that,
they try to twist it first in one way then in another, and the consequent unrest of the frame is communicated to its supports and
foundations.

This may be illustrated by the following simple experiments:

Experiment 1.—Stretching a tyre under the tension induced by rapid rotation.

Each part of the tyre is compelled to move in a circle. Therefore a force must act towards the centre. The equal and opposite aspect of this force can only appear as a tension in the tyre. You notice as the speed is increased how the tension lengthens the tyre until it is hanging quite loose. Reducing the speed it gradually contracts and finally grips the wheel tightly.

Experiment 2.—A mass of 1 lb. attached to a truly turned wheel and rapidly revolved caused great vibration.

The question now arises, How are these forces to be dealt with in order to stop the vibration? The answer is, Apply another mass to the wheel in such a way that the force its rotation produces on the frame is exactly equal and opposite to the force from the first mass. Applying such a mass you see the wheel runs quietly at a much higher speed than I dared to run it before.

These experiments sufficiently indicate the nature of the problems engineers have to deal with.

A model of a locomotive was freely supported on rollers and driven, the parts being entirely unbalanced. The consequent vibration indicated the unbalanced forces.

Balance weights were then applied and the model stood still on the rollers.

Since revolving balance weights are added the part of the reciprocating masses which they balance horizontally appears as an unbalanced vertical force, but the weight of the model masks this force entirely. Relieving the weight the force soon makes itself evident by the vertical oscillation it produces. This vertical force appears as a variation of pressure on the rail in an actual case, the weight on the wheel in ordinary practice being more than sufficient to mask the force. The extent to which the pressure is varied is shown by the slide in the screen. The curves are drawn for an actual case of an express locomotive running at 60 miles per hour.

To obtain balance in the longitudinal direction and without the " hammer blow," as this variation of vertical force is called, Mr. Webb has arranged the cranks of his recent compound locomotives in the way shown in the model of the engine on the table. The model has kindly been sent from Crewe by Mr. Webb for this occasion. There is still a tendency to twist about a vertical axis, but this cannot be eliminated without setting the crank at different angles, and the arrangement shown is preferred on account of the mechanical simplicity which results.

This four-crank model of a marine engine shows the arrangement which must be adopted if both unbalanced force and unbalanced twist or torque is to be eliminated.

The conditions which must be fulfilled are :—

1. The forces acting on the frame shall have no resultant.

2. The moments of these forces about any point in the axis of
rotation shall have no resultant.

The model is running now in a state in which neither condition
is satisfied, and the consequent vibration is manifest.

Adding this mass I know, by previous calculation, that the two
conditions are separately satisfied. Running it now it seems to stand
quite still, although it is supported on these flexible springs.

There are few people who have not had some experience of steamship vibration. In recent years the vibration problem has forced
itself into notice because of the increase in engine speeds and the
relative lightness of the hulls. The generally-held belief, not so
many years ago, was that the vibration in ships was produced entirely
by the propeller. To disprove this current notion Mr. Yarrow carried
out a series of beautiful and costly experiments on a first-class tor-
pedo-boat. The propeller was removed and the boat moored in still
water, and the engines were run under different conditions of
balancing. The consequent vibration disclosed itself by the ripples
surrounding the hull. Unbalanced engines caused violent disturbance; balanced, the ripples were smoothed out in the way you can
see from the photographs of the two cases on the screen, the slide
for which has been kindly lent by Mr. Yarrow. These experiments
proved conclusively that one cause of steamship vibration may be
removed by balancing the engines.

Hitherto I have tacitly assumed that the pistons move with simple
harmonic motion. Really the forces required for the acceleration of
the pistons are different from those in simple harmonic motion because
of the obliquity of the connecting rod. Whatever law the force may
follow, however, it has one property, namely, that it is continuous
and periodic, and may thus be represented by a Fourier series. It
may be shown that the force on any one piston is given by an expression of the form—

$$F = Mw^2r \{ \cos \theta + A \cos 2\theta + B \cos 4\theta + C \cos 6\theta \}$$

The values of the co-efficients have been worked out by Mr.
Macalpine, and for the case where the rod is $3\frac{1}{2}$ times the length
of the crank, the series becomes—

$$F = Mw^2r \{ \cos \theta + 0 \cdot 29 \cos 2\theta - 0 \cdot 006 \cos 4\theta + 0 \cdot 0002 \cos 6\theta \}$$

Looking at these terms, their value decreases very rapidly after the
second. By the methods I have indicated the first term of the series
is eliminated from all the series belonging to the respective pistons.
This is called balancing for *primary effects*, and is conditioned by
four simultaneous equations.

To eliminate the second term four more equations have to be
satisfied, making eight in all. These equations are now on the
screen, and the next slide shows the solution.

This solution, as you will readily perceive, is not one from which
a practicable four-crank engine can be made. If the last two of the

equations are omitted a solution of the remaining six can be found, and it is such a solution which is the basis of what is known as the Yarrow-Schlick-Tweedy method of balancing an engine. The secondary moments are left unbalanced, otherwise the first two terms in the respective series are eliminated, and the engine is said to be balanced for Primary Forces and couples and for Secondary Forces. A large number of engines have been built on this plan.

The steam turbine is of course a balanced machine, and its running causes no unbalanced forces to act on the framework.

A good many marine engines have been built with cranks at right angles, and with their reciprocating masses approximately equal.

This model is arranged in that way, and its running discloses at once that there is an unbalanced torque acting on the frame. In fact four cranks at right angles is just the one sequence of angles for which it is impossible to balance the reciprocating parts of a four-crank engine amongst themselves, even for primary effects.

[W. E. D.]

GENERAL MONTHLY MEETING,
Monday, February 2, 1903.

SIR JAMES CRICHTON-BROWNE, M.D. LL.D. F.R.S., Treasurer
and Vice-President, in the Chair.

Dr. Louisa Garrett Anderson, M.D. B.Sc.
Dr. Henry T. Böttinger,
Walter Child, Esq.
Mrs. Alexander Ionides,
Miss Edna Lea-Smith,
Sir Frederick Treves, Bart. K.C.V.O. C.B. F.R.C.S.
Walter Whitehead, Esq. F.R.C.S.
were elected Members of the Royal Institution.

The following Resolution, passed by the Managers, was read and
adopted :—

Resolved, That the Managers of the Royal Institution of Great Britain desire
to record their sense of the loss sustained by the Institution in the decease of
Mr. James Wimshurst, F.R.S.

He became a Member of the Royal Institution of Great Britain in 1884, and
was elected a Visitor in 1887 and a Manager in 1902. He delivered an interesting Friday Evening Discourse on "Electrical Influence Machines" in 1888, and
presented a large machine of his own invention to the Institution.

The Managers desire to offer to the family the expression of their sincere
sympathy with them in their bereavement.

The PRESENTS received since the last Meeting were laid on the
table, and the thanks of the Members returned for the same, viz. :—

FROM
The Secretary of State for India—
  Geological Survey of India—
    The G.T. Survey of India, Vol. XXIX. 4to. 1902.
*Accademia dei Lincei, Reale, Roma—*Classe di Scienze Fisiche, Matematiche e
    Naturali. Atti, Serie Quinta: Rendiconti. Vol. XI. 2° Semestre,
    Fasc. 10–12. 1902. Vol. XII. 1° Semestre, Fasc. 1. 1903. 8vo.
  Classe di Scienze Morali, Storiche, etc. Seria Quinta, Vol. XI. Fasc. 9–10.
    8vo. 1902.
*Ador, Dr. E. Hon. Mem. R.I. (the Editor)—*Œuvres de J. C. Galissard de Marig-
    nac, Tome II. 1860–87. 4to. 1903.
*American Academy of Arts and Sciences—*Proceedings, Vol. XXXVIII.,
    Nos. 1–3. 8vo. 1902.
*Amsterdam, Royal Academy of Sciences—*Verslag, Deel X. 8vo. 1901–2.
  Verhandlungen, Erste Sect. Deel VIII. Nos. 1, 2; Tweede Sect. Deel VIII.;
    Deel IX. Nos. 1–3. 8vo. 1901–2.
  Jahrboek, 1901. 8vo.
  Proceedings of Section of Sciences, Vol. IV. 8vo. 1902.
  Catalog von Stermen, von Dr. N. M. Kan. 4to. 1901.
*Asiatic Society, Royal—*Journal, Jan. 1903. 8vo.
*Astronomical Society, Royal—*Monthly Notices, Vol. LXIII. Nos. 1–2. 8vo. 1902.
*Automobile Club—*Journal for Jan. 4to. 1903.
*Bankers, Institute of—*Journal, Vol. XXIII. Part 9. 8vo. 1902. Vol. XXIV.
    No. 1. 8vo. 1903.

Belgium, Royal Academy of Sciences—Bulletin, 1901; 1902, 9-11. 8vo.
　Mém. Cour. et des savants étrang. Tome LX. Fasc. 1; Tome LXII. 4to. 1902.
　Mém. Cour. et autres Mém. Tome LVI. and LXII. Fasc. 2-3. 8vo. 1902.
Berlin, Academy of Sciences—Sitzungsberichte, 1902, Nos. 41-53. 8vo.
Bombay, Government of—Progress Report of the Archæological Survey of
　Western India, 1902. fol.
Boston Public Library—Monthly Bulletin for Dec. 1902. 8vo.
Boston Society of Medical Sciences—Journal of Medical Research for Nov. 1902.
　8vo.
Bright, Charles, Esq.—Life of Sir Charles Tilston Bright. 2 vols. 8vo. 1902.
British Architects, Royal Institute of—Journal, Third Series, Vol. X. Nos. 3-6.
　4to. 1902.
British Astronomical Association—Journal, Vol. XIII. No. 2, 3. 8vo. 1902.
British Museum Trustees—Southern Cross Collections. 8vo. 1902.
　Guide to the Coral Gallery. 8vo. 1902.
　Handbook of Instructions for Collectors. 8vo. 1902.
Buenos Ayres, City—Monthly Bulletin of Municipal Statistics, Oct. 1902. 4to.
Cambridge Philosophical Society—Proceedings, Vol. XI. Part 7. 8vo. 1902.
　Vol. XII. Part 1. 1903. 8vo.
Canada, Royal Society of—Proceedings, Series II. Vol. VII. 8vo. 1901.
Canadian Institute—Proceedings, New Series, Vol. II. Part 5. 8vo. 1902.
　Transactions, Vol. VII. Part 2. 8vo 1902.
Chemical Industry, Society of—Journal, Vol. XXI. Nos. 22-24; Vol. XXII.
　Nos. 1, 2. 8vo. 1902.
Chemical Society—Journal for Jan.-Feb. 1903. 8vo.
　Proceedings, Vol. XVIII. Nos. 257-259. 8vo. 1902.
Clowes, Prof. F. M.R.I.—Bacterial Treatment of Crude Sewage. Fourth Report
　to London County Council on. 4to. 1902.
Cornwall Polytechnic Society, Royal—Sixty-ninth Annual Report. 8vo. 1901.
·*Cracovie, l'Académie des Sciences*—Bulletins, Classe des Sciences Mathématiques
　et Naturelles, 1902, Nos. 8-10. 8vo.
　Classe de Philologie, 1902, Nos. 8-10. 8vo.
Devonshire Association—Reports and Transactions, Vol. XXXIV. 8vo. 1902.
　Calendar of Devonshire Wills, Part IV. 8vo. 1902.
Dewar, Prof. M.A. F.R.S. M.R I.—Œuvres de J. C. Galissard de Marignac,
　Tome II. 1860-87. 4to. 1903.
Editors—Aeronautical Journal for Dec. 1902 and Jan. 1903. 8vo.
　American Journal of Science for Jan. 1903. 8vo.
　Astrophysical Journal for Nov.-Dec. 1902.
　Athenæum for Dec. 1902 and Jan. 1903. 4to.
　Author for Dec. 1902 and Jan. 1903. 8vo.
　Board of Trade Journal for Dec. 1902 and Jan. 1903. 8vo.
　Brewers' Journal for Dec. 1902 and Jan 1903. 8vo.
　Chemical News for Dec. 1902 and Jan. 1903. 4to.
　Chemist and Druggist for Dec. 1902 and Jan. 1903. 8vo.
　Electrical Engineer for Dec. 1902 and Jan. 1903. fol.
　Electrical Review for Dec. 1902 and Jan. 1903. 4to.
　Electrical Times for Dec. 1902 and Jan. 1903. 4to.
　Electricity for Dec. 1902 and Jan. 1903. 8vo.
　Engineer for Dec. 1902 and Jan. 1903. fol.
　Engineering for Dec. 1902 and Jan. 1903. fol.
　Feilden's Magazine for Dec. 1902. 8vo.
　Homœopathic Review for Dec. 1902. 8vo.
　Horological Journal for Jan.-Feb. 1903. 8vo.
　Invention for Dec. 1902 and Jan. 1903. fol.
　Ironmonger for Jan. 1903. 8vo.
　Journal of the British Dental Association for Dec. 1902. 8vo.
　Journal of State Medicine for Dec. 1902 and Jan. 1903. 8vo.
　Law Journal for Dec. 1902 and Jan. 1903. 8vo.

Editors—continued.
London Technical Education Gazette for Nov.-Dec. 1902. fol.
Machinery Market for Dec. 1902 and Jan. 1903. 8vo.
Model Engineer for Dec. 1902 and Jan. 1903. 8vo.
Mois Scientifique for Dec. 1902. 8vo.
Motor Car Journal for Dec. 1902 and Jan. 1903. 8vo.
Musical Times for Jan. 1903. 8vo.
Nature for Dec. 1902 and Jan. 1903. 4to.
New Church Magazine for Dec. 1902 and Jan. 1903. 8vo.
Nuovo Cimento for Nov.-Dec. 1902. 8vo.
Page's Magazine for Dec. 1902. 8vo.
Pharmaceutical Journal for Dec. 1902 and Jan 1903. 8vo.
Photographic News for Dec. 1902 and Jan. 1903. 8vo.
Physical Review for Dec. 1902 and Jan. 1903. 8vo.
Popular Astronomy for Jan. 1903. 8vo.
Public Health Engineer for Dec. 1902 and Jan. 1903. 8vo.
Science Abstracts for Dec. 1902. 8vo.
Travel for Dec. 1902. 8vo.
Zoophilist for Dec. 1902. 4to.
Franklin Institute—Journal, Vol. CLIV. No. 6; Vol CLV. No. 1. 8vo. 1902-3.
Geographical Society, Royal—Geographical Journal for Dec. 1902 and Jan. 1903.
    8vo.
Geological Society—Quarterly Journal, Vol. LVIII. Part 4. 8vo. 1902.
    Abstract of Proceedings, Session 1902-3, No. 770.
Hipkins, A. J. Esq. (the Author)—Dorian and Phrygian. 8vo. 1902.
Hugues, Lachiche, Esq. (the Author)—Un Seul Champignon sur le Globe. 8vo.
    1902.
Imperial Institute—Imperial Institute Journal for Dec. 1902.
Iron and Steel Institute—Journal, Vol. LXII. No. 11. 8vo. 1902.
Johns Hopkins University—Circulars, No. 160. 4to. 1902.
    American Journal of Philology, Vol. XXIII. No. 3. 8vo. 1902.
Life-Boat Institution, Royal National—Journal for Feb. 1903. 8vo.
Linnean Society—Journal, Botany, Vol. XXXVI. No. 249; Zoology, Vol. XXVIII.
    No. 185. 8vo. 1903.
Liverpool Literary and Philosophical Society—Proceedings, No. 56. 8vo. 1902.
Manchester Literary and Philosophical Society—Proceedings, Vol. XLVII. Parts
    1, 2. 8vo. 1902.
Mechanical Engineers, Institution of—Proceedings, 1902, No. 3. 8vo.
Mensbrugghe, Prof. G. van der, M.R.I. (the Author)—Une Triple Alliance Natu-
    relle, 2e Partie. 8vo. 1902.
Meteorological Society, Royal—Meteorological Record, Vol. XXI. No. 86. 8vo.
    1902.
Metropolitan Asylums Board—Annual Report, 1901. 2 vols. 8vo. 1902.
Microscopical Society, Royal—Journal for Dec. 1902. 8vo.
Musical Association—Proceedings, Twenty-eighth Session. 8vo. 1902.
Navy League—Navy League Journal for Dec. 1902 and Jan. 1903. 8vo.
New South Wales, Agent-General for—Year Book of New South Wales, 1903. 8vo.
    The Wealth and Progress of New South Wales. By T. A. Coghlan. 8vo.
    1900-1.
New Zealand, Agent-General for—Papers and Reports relating to Minerals and
    Mining, 1901-2. 4to.
    Report of the Department of Lands for the Years 1900-1 and 1901-2. 4to.
    Statistics of the Colony for 1901. 2 vols. 4to. 1902.
    Official Year Book for 1902. 8vo.
    Settlers' Handbook. 8vo. 1902.
North of England Institute of Mining and Mechanical Engineers—Transactions,
    Vol. LII. No. 1. 8vo. 1902.
Odontological Society—Transactions, Vol. XXXV. No. 2. 8vo. 1902.
Onnes, Dr. H. K.—Communications, No. 80. 8vo. 1902.

Pennsylvania, University of—Contributions from the Zoological Laboratory, 1900 and 1901. 8vo.
Publications, Astronomical Series, Vol. I. Parts 2, 3. 4to. 1901.
Pharmaceutical Society of Great Britain—The Calendar, 1903. 8vo.
Photographic Society, Royal—Photographic Journal for Nov. 1902. 8vo.
Rayleigh, Lord, M.A. F.R.S. M.R.I. (the Author)—Scientific Papers, Vol. IV. 4to. 1903.
Rio de Janeiro Observatory—Monthly Bulletin for April-June, 1902. 8vo.
Rome, Ministry of Public Works—Giornale del Genio Civile for Aug. Sept. 1902. 8vo.
Royal College of Surgeons Museum—Catalogue of Physiological Series, Vol. II. 8vo. 1902.
Royal Irish Academy—Transactions, Vol. XXXII. Section A, Parts 3-5; Section B, Part 1. 4to. 1902.
    Proceedings, Vol. XXIV. Section A, Part 1; Section B, Parts 1, 2; Section C, Part 1. 8vo. 1902.
Royal Society of Edinburgh—Proceedings, Vol. XXIV. No. 3. 8vo. 1902.
    Transactions, Vol XL. Part 2; Vol. XLII. 4to. 1902.
Royal Society of London—Philosophical Transactions, A, Nos. 323-327. 4to. 1902.
    Proceedings, Nos. 468-470. 8vo. 1902.
    International Catalogue of Scientific Literature, C. Physics, Part 1. 8vo. 1902.
Salford, County Borough of—Fifty-fourth Report of the Museum, Libraries, and Parks Committee. 8vo. 1902.
Saxon Society of Sciences, Royal—
    Mathematisch-Physische Classe—
    Berichte, 1902, Nos. 3-5. 8vo.
    Abhandlungen, Band XXVII. Nos. 7-9. 4to. 1902.
    Philologisch-Historiche Classe—
    Berichte, 1902, Nos. 1, 2. 8vo.
Scottish Microscopical Society—Proceedings, Vol. III. No. 3. 8vo. 1902.
Selborne Society—Nature Notes for Dec. 1902. 8vo.
Shand, Mason, & Co. Messrs. (the Publishers)—Works Fire Brigades, their organisation and equipment. 8vo. 1902
Smith, B. Leigh, Esq. M R.I.—The Scottish Geographical Magazine, Vol. XVIII. No. 12. 8vo. 1902.
Smithsonian Institution—Annual Report, 1901. 8vo. 1902.
Society of Arts—Journal for Dec. 1902 and Jan. 1903. 8vo.
Statistical Society—Journal, Vol. LXV. Part 4. 8vo. 1902.
St. Pétersbourg, L'Académie Impérial des Sciences—Comptes Rendus des Séances de la Commission Sesmique Permanente, Livraison I. 4to. 1902.
Tacchini, Prof. P. Hon. Mem. R.I. (the Author)—Memorie della Società degli Spettroscopisti Italiani, Vol. XXXI. Disp. 11. 4to. 1902.
United Service Institution, Royal—Journal for Dec. 1902 and Jan. 1903. 8vo.
United States Department of Agriculture—Experiment Station Record, Vol. XIV Nos. 2-4. 8vo. 1902.
    Monthly Weather Review, Aug.-Oct. 1902. 4to.
United States Geological Survey—Twenty-first Annual Report, 1901, Vols. I-VII. 4to.
    Monographs, Vol. XLI. 4to. 1902.
United States Patent Office—Official Gazette, Vols. XCV.-CI.; Vol. CII. Nos. 1-3. 4to. 1900-2.
Verein zur Beförderung des Gewerbfleisses in Preussen—Verhandlungen, 1902, Heft 10. 8vo.
Vienna, Imperial Geological Institute—Verhandlungen, 1902, Nos. 11-13. 8vo.
    Abhandlungen VI. Band 1, Sup. Heft. 4to. 1902.
Washington Academy of Sciences—Proceedings, Vol. V. pp. 1-37. 8vo. 1903.
Western Society of Engineers—Journal, Vol. VII. No. 6. 8vo. 1902.
Wheeley, A. T. Esq. (the Author)—An Inquiry into the meaning of the Word Ing. 8vo. 1903.
Yorkshire Archæological Society—Journal, Part 66. 8vo. 1903.

WEEKLY EVENING MEETING,

Friday, February 6, 1903.

Sir James Crichton-Browne, M.D. LL.D. F.R.S., Treasurer
and Vice-President, in the Chair.

The Right Hon. Sir Herbert Maxwell, Bart., M.P.
LL.D. F.R.S. F.S.A.

George Rómney and his Works.*

I have undertaken to speak to you to-night about one who rose
from exceedingly humble circumstances to the first rank of British
painters, and to show how he encountered and overcame difficulties
which almost invariably bar the road against persons much more
propitiously situated.

In the middle of the eighteenth century there lived in the little
farm of Beckside, close to Dalton-in-Furness, a carpenter, cabinet-
maker, and cultivator in a small way, named John Romney. There is
still in that village a carpenter of the name of Romney, grandson or
great-grandson of a cousin of the original family at Beckside.
Family names always have a meaning, if we can read it; and at first
it seemed to me that, like many others, this family must have taken
their name from some place. But, although there is Romney in
Kent and Rhymney in South Wales, there is no place of that name
in the north of England; and I am informed by one who bears the
surname that he believes it to be neither less nor more than Romany,
a gipsy. You may remember that it was in the Border country that
the gipsies, or Egyptians as they were termed in the old penal laws
of both England and Scotland, mustered most strongly of yore,
finding it expedient to be able to slip across the marshes into one
dominion when the warders and magistrates of the other showed
diligence in enforcing the statutes against vagrants. Both in Scotland
and England many gipsies abandoned their wandering habits and
settled down as industrious members of the community. If that be
the origin of the family of Romney or Romany, then we may be
inclined to trace some of the unrest of George Romney's life to the
wandering instinct in his ancestors.

The carpenter of Dalton was known among his neighbours as
"Honest John," and was a man of some capacity in his craft, being

---

* The discourse was fully illustrated by lantern slides. For fuller information see 'George Romney,' in the Makers of British Art Series, published by
Walter Scott and Co.

credited with having turned out the first cart with spoked wheels in Furness. Before his day, all the carts had solid or "clog" wheels. But he was a bad book-keeper, negligent in collecting his debts, and remained a poor man all his days. His wife was Anne Simpson, daughter of a Cumberland statesman owning Sladebank. She bore him eleven children, whereof the second was born 15th (26th) December, 1734, and was named George. This lad was sent to school very young at Dendron, where the Rev. Mr. Pell taught him the humanities for the modest fee of 5s. a quarter, and one, Gardner, boarded him at the rate of 4l. 10s. a year. George was not a promising scholar, and at ten years old his father took him from school and employed him in the carpenter's shop.

Now "Honest John," besides being a cabinet-maker and farmer, was also a bit of an architect. Therefore drawing materials were always to be found in his shop, and George soon showed a natural turn for using them.

One of the workman in the shop, Sam Knight by name, used to take in an illustrated magazine, which he lent to George, who diligently copied the engravings in the same.

But this stupid schoolboy possessed another latent gift, which one of his neighbours, a watchmaker named Williams, being an enthusiast in music, helped him to develop. For several years it seemed doubtful of which muse young George Romney was most inclined to become the disciple—the muse of painting or the muse of harmony. Only one thing seemed certain, that he would never be anything but a bad carpenter. This became quite clear when, about the age of eighteen, he took employment under another cabinet-maker, Wright of Lancaster. Wright gave the lad a fair trial, and then informed his father that he could make nothing of him; that the only thing he did well was sketch portraits of his fellow workmen. He recommended John Romney to apprentice his son to an artist.

Here then was George, on the threshold of manhood, having failed as a scholar and failed in the calling he had adopted—what chance was there that he should ever be heard of again? Not much, you will admit. Nevertheless, when opportunity presented itself, George did not miss it.

Opportunity came to him in the guise of an eccentric character named Christopher Steele, an itinerant portrait painter, who happened to be working in Kendal when George Romney was twenty years old. Steele had been taught painting in Paris by Carlo Vanloo, and having acquired at the same time some of the manners of a third-rate French fop, was known in the neighbourhood as the Count. To this worthy George was entered as apprentice in March 1755.

Now Count Steele was a flighty, improvident fellow, moving from town to town to avoid his creditors. Nevertheless, he imparted some sound principles of the preparation and use of colour, which stood Romney in good stead throughout his life. Paints in those days were not sold in patent collapsible tubes as we have them now; artists employed their apprentices to grind and mix them for their daily use.

The Count also found less legitimate employment for his apprentice. A young lady, who happened to be taking drawing lessons from him, possessed a small fortune, which exercised a powerful charm upon the impecunious Count, who paid his addresses secretly to her, and employed George as a go-between. Finally, he persuaded the heiress to elope with him to Gretna Green; and George, thus left in the lurch, presently fell into a fever. His landlady was a widow of very humble means, one Mrs. Abbot, whose daughter Mary nursed the young fellow through his illness, who out of gratitude gave her his heart. During his convalescence they became betrothed —the first step in the long tragedy of their lives.

Next came a summons from the Count to his apprentice bidding him join him at York, where he had set up his studio. George would not leave without fulfilling his plighted troth, and on 14th October, 1756, being of the mature age of one-and-twenty, he married Mary Abbot and hurried off to resume his duties with Steele, leaving his bride in Kendal. Of course he had, as yet, no means of keeping a wife; on the contrary, while working in York, Mary used to send him from time to time half-a-guinea hidden in the seal of a letter. Still he worked both ardently and steadily, so that when the Count decamped hurriedly from York, in 1757, and the apprenticeship was prematurely broken, Romney on returning to Kendal was able to earn something on his own account. Kind and helpful encouragement came from country gentlemen in the neighbourhood, and portraits painted by Romney at this period may still be seen in their houses and in the Town Hall of Kendal. Of course, you will not expect to behold in these juvenile works much of that ease and grace which are apparently distinguished in his later ones. They are more like the handiwork of Hogarth than the painter we know as George Romney.

In Kendal, George Romney remained working steadily for five years. Steadily, but not hopefully. He was eight-and-twenty by this time, and, although conscious of the scope of his natural powers, the very insight which he had obtained into the requirements of his art showed him that those powers could not be brought to perfection without wider acquaintance with the work of other painters.

To raise funds for his journey to London, in 1762, the young painter disposed of the pictures he had by him, about twenty in all, by lottery. This brought in about 100*l*., half of which he left with his wife, the other half he took with him, riding on horseback to London. Arriving there on the eighth day after his departure, he was kindly treated by his old schoolfellow, Thomas Greene, who was in good practice as an attorney of Gray's Inn, and by another old Westmorland friend, Daniel Braithwaite of the Post Office. Mr. Pennington of Muncaster also was of service to him, and Romney soon had a few sitters, for there were no photographers in those days to gratify the pardonable curiosity which everybody feels about his own appearance. Thus Romney was able to keep the wolf from his own door, to send home, we may suppose, money to his wife, and also, as we know, to help three of his brothers who were in difficulties.

Romney spent six happy and fruitful weeks in Paris during 1764. In 1767 he paid a flying visit to his wife; returning to London, found plenty of work to do, and continued to exhibit in the Free Society of Artists. The Royal Academy was founded about this time, but injudicious friends persuaded Romney that Joshua Reynolds, its first President, was prejudiced against him. Romney's suspicions were ever on the alert: he refused all invitations to send pictures to Somerset House, and so it came to pass, that no work of his was exhibited on the walls of the Royal Academy until 1871, sixty-nine years after his death.

Romney's most productive period was from 1776–87. His studio was besieged by distinguished and fashionable sitters, and even from the moderate fees he charged he never earned less than at the rate of 3000*l.* a year.

The period of his decline dates from 1791. About that time he began to weary of what he called the drudgery of portrait painting, and gave up his days to working on historical and Shakespearian subjects. His irritability and suspicious habits increased to an alarming extent; he nursed mighty projects of a painting gallery, which he carried out at an imprudent expense in a new building at Hampstead in 1796.

Romney died at Kendal on 15th November, 1802.

I cannot close this brief sketch of your distinguished countryman without allusion to one feature in his life-work which I believe to be without parallel. It has often happened that pictures which could not find a purchaser during the painter's life, have risen to almost fabulous value when he was no more. To mention only one example. The French peasant-painter, Jean Francois Millet, died in poverty in 1875, after failing to obtain 80*l.* for his *Angelus*, which had been painted sixteen years before. Fourteen years after his death that picture was bought by the French nation for 60,000*l.*

Again, there are plenty of cases in which a painter has been popular with the public during his life, but has fallen out of all esteem after his death. What is remarkable—unique—in Romney's case is this, that his painting enjoyed an enormous vogue during his active life, fell into utter neglect after his death, and of late years has found feverish favour again with collectors.

[H. M.]

WEEKLY EVENING MEETING,

Friday, February 13, 1903.

DONALD W. C. HOOD, C.V.O. M.D. F.R.C.P., Vice-President,
in the Chair.

PROFESSOR SHERIDAN DELÉPINE, M.B. B.Sc.

Civilisation and Health Dangers in Food.

"Wherefore it appears to me necessary to every physician to be skilled in nature, and to strive to know, if he would wish to perform his duties, what man is in relation to the articles of food and drink."—*Hippocrates*, 460 B.C.

Civilisation as a Disturbing Element of Health Factors and of Natural Evolution.

WHAT was said of the duties of the medical man, over two thousand years ago, may now be considered to be applicable to those of the statesman.

One of the effects of civilisation has been the gradual abandonment of rural life by a large number of individuals and their aggregation in cities.

With this shifting of population have been associated new conditions of life. These new conditions do not seem to have been altogether favourable to the evolution of our race; at any rate statistics of mortality from disease seem to point clearly to the fact that mortality is greater in populous centres than in thinly populated rural districts placed under similar climatic and hygienic conditions.

As we create for ourselves new conditions of life, we must at the same time consider carefully not only the benefits but also the dangers associated with these changes. Modern life is gradually increasing the distance between us and nature.

It is obviously necessary that we should keep a careful watch upon the artificial modification of natural factors which we introduce in our existence, so as to eliminate those which are detrimental before the race as a whole has suffered deterioration.

Quality of Food a Matter of National Importance.

Food has often to travel over great distances, or to be stored up for a considerable time, before it is consumed. Various methods of preservation, some distinctly harmful, have therefore been devised. Many natural products are considerably altered by artificial processes, which deprive them of some of their normal qualities; some articles of food are entirely manufactured by chemical processes.

VOL. XVII.   (No. 97.)   S

All these things are often carried out by persons who have no clear conception of the dangers associated with the use of certain substances, and who look upon the preparation of articles of food simply as a commercial matter.

Artificial Production of Food Stuffs. Arsenical Contamination of Beer as the result of the use of a Manufactured Sugar.

It is to this kind of ignorance that the severe outbreak of arsenical poisoning, which afflicted the northern counties mostly, at the end of the year 1900 was due.

A certain manufacturer of brewing sugars did not take sufficient care to ascertain the quality of the ingredients used in the manufacture of his sugars ; in his ignorance he used large quantities of very impure oil of vitriol, perfectly unfit for the preparation of an article of food. The manufacturer of sulphuric acid also committed the indiscretion of sending, without sufficient warning, to the glucose manufacturer a very impure sulphuric acid instead of the purer article that he had usually supplied. Now this impure sulphuric acid contained a large amount of arsenic, a fact which he knew, and that the glucose manufacturer should have known, and yet neither the one nor the other suspected that their carelessness was endangering thousands of lives. The brewers who used the sugars should have known that glucose was liable to contain arsenic owing to the use of sulphuric acid in its preparation, and yet it did not occur to them that such a possibility invited careful supervision on their part.

As a result of this ignorance many hundred people were rendered very ill, and not a few died.

Here we have a good example of the need of more knowledge of the effects which may result from the use of artificial methods in the preparation of food stuffs.

It seems to me that, under a state of things which involves such risks, the State may well exact a certain amount of knowledge on the part of persons who undertake the manufacture of products which may become such a source of danger.

Arsenic in Fuel. Contamination of Malt, and of the Air of Towns.

The same inquiry led to other observations, which also show one of the gradual changes which are taking place in our environment. The search for arsenic in beer revealed the fact that much of the malt prepared in the northern counties of England, and also elsewhere, contained a fairly large amount of arsenic, and that this arsenic was mostly derived from the fumes of the impure coke and coal burnt in kilns where malt is heated and dried. The large amount of arsenic present in certain samples of malt, led me to estimate the amount of arsenic which accumulated in the flues of certain chimneys where Yorkshire coal was burnt, and of stoves where gas coke was consumed. I found in a sample of coal soot over 5 grains, and in a sample of coke soot about 28 grains, of arsenic per pound. It was natural to expect that such an amount of arsenic in soot would lead to a distinct contamination of the atmosphere of a town such as Manchester, where soot abounds. This, early in 1901, I found to be actually the case in Manchester, for I discovered that dust deposited from the atmosphere in inhabited rooms, in uninhabited lofts, upon the leaves of trees and shrubs, contained very material amounts of arsenic.

During the month of March 1902 there was a heavy fall of snow in Manchester, this was immediately followed by a heavy fog lasting one day. The snow, which was perfectly white before, was grey after the occurrence of the fog. I collected the superficial snow over a surface measuring exactly one meter square, and, from the quantity found in a sample of the snow water, calculated that during a single day about 0·0000132 grammes of arsenic had been deposited on the small patch of snow under investigation. What may be the effect of the continual breathing of air containing frequently a distinct trace of arsenic? I cannot say, but the fact remains that the existence of large cities is associated with the contamination of the air with many products of which I have indicated only one. To satisfy myself that this inference was correct I collected, under conditions similar to those I had observed in Manchester, samples of dust in one of the London suburbs, in the small town of Montreux in Switzerland, and in the open country near Ambleside; and found that the London suburban dust contained a material amount of arsenic, though less than the dust collected in the centre of Manchester. On the other hand, the dust collected in the Lake district and in Switzerland contained either no arsenic or an inappreciable trace of the poison.

I do not wish to labour the argument, nor to suggest that the presence of a small amount of arsenic in the air of a town is certainly a source of danger. I wish rather to indicate the insidiousness than the potency of dangers associated with the gradual modification of our surroundings.

It might appear that air is not a food and should not be introduced into this discourse, but I contend that air and water are as much foods as beef and mutton, for without oxygen and water metabolism is impossible.

Disposal of Sewage. Pollution of Water and of Shell-Fish.

Man in his nomadic state was not troubled with the question of sewage disposal, for then he had a movable home. Now that large populations have become fixed over limited areas, the disposal of sewage has become a problem of such difficulty and magnitude, that it taxes to the utmost the ability of engineers, of municipal authorities, and even of state departments, without speaking of Royal Commissions.

Certain cities are so situated that their sewage has to be thrown into the sea. Wherever a large number of individuals are congregated, certain infectious diseases such as typhoid fever are liable to occur, and this leads to a more or less extensive infection of the sewage, and of the tidal waters into which the sewage is discharged.

The tragic events which have lately followed the Mayoral banquets at Winchester and Portsmouth have roughly shaken the public indolence, and the moral of facts which scientific people had pointed out for many years has at last been understood.

Dr. H. Timbrell Bulstrode, in his able Report on Oyster Culture in Relation to Disease, which appeared in 1896 * had clearly stated " that there are cases where the risk of sewage pollution to oysters is so great and so considerable, that nothing short of complete diversion of the sewers or drains, or withdrawal of existing fattening beds or ponds from use, can be regarded as satisfactory in the public interest."

The Emsworth storage ponds belonged obviously to the category of dangerous ponds, as shown by Dr. Bulstrode's report and by the map accompanying it. The evidence of pollution was so great that the late Sir R. Thorne Thorne mentioned Emsworth specifically as an example of much polluted pits. But for six years nothing seems to have been done to remove the danger, and it was necessary that several valuable lives should be lost to cause an action to be taken, an action for which the Fishmongers' Company is to be praised, provided that it does not stop short of a general exclusion from our markets of all the polluted oysters.

The general condition of things is well indicated by the following statement, appearing also in the report already alluded to: "A superficial glance at the maps with which Dr. Bulstrode's report is illustrated, might lead to the hasty conclusion that sewage is deemed to be of value—if indeed it is not actually sought for—in connection with the process of oyster-fattening and storage.†

Manufacture of Prepared Foods on a Large Scale.

I must now direct your attention to a danger of another kind, but still attributable to modern modes of living. The supply of fresh food to large cities is a matter of difficulty, which has given rise to the increased use of prepared articles of food which can be preserved for various lengths of time. Preservation by cooking or other means allows of a more economical utilisation of meat than the simple division of a carcase into joints.

Derby Outbreak of Pork Pie Poisoning.

During the first week in September 1902, Dr. Howarth, the Medical Officer of Health for the Borough of Derby, was informed

---

* Supplement to the 24th Annual Report of the Local Government Board, 1894-95, p. 82.
† R. Thorne Thorne, pp. xiv. and xv.

that several persons who had partaken of certain pork pies were seriously ill, being affected with a kind of diarrhœal disease usually associated with food poisoning. In the course of the enquiries that followed, it was ascertained that at least 221 persons who had partaken of pies purchased at one shop in the course of a few days, had been similarly affected, and that four of them had died. Of the 221 persons, 131 were living in Derby, and 90 were taken ill in other towns. The four fatal cases belonged to the latter group, and it was apparent that pies which had been rapidly consumed in Derby had been less fatal than those which had been taken, or sent, to a distance.

Careful investigation revealed the fact that the great majority of cases, if not all, could be connected with one batch of pies, baked on Tuesday, September 2.

On inspecting the premises where the pies had been prepared Dr. Howarth ascertained that the owner conducted his business with more than ordinary care, and that there was no evidence of any contravention of any of the borough bye-laws. There was no evidence to show that the flesh of any diseased pig had been used.

The results of this first part of the investigation were: (1) that the outbreak of illness was clearly connected with the consumption of a certain batch of pork pies; and (2) that these pies contained a noxious substance of some unknown origin. Outbreaks of food poisoning have, during the last few decades, been investigated both in this country and abroad; those connected with the ingestion of animal food have generally been attributed to the use of the flesh of diseased animals, or to putrefactive changes which had given rise to the formation of poisonous alkaloids, known generally under the name of ptomaines. In this case I found no evidence of the presence of ptomaines in the pies, and I think that careful investigation of outbreaks of food poisoning will show that true ptomaine poisoning is a very rare occurrence indeed.

On microscopical examination of the pies I found that the meat showed no evidence of having been derived from diseased swine. Many of the small pieces of meat which I examined were, however, partly covered with a layer of bacteria. These bacteria had invaded the surrounding jelly, which I found crowded with colonies of a bacillus. After a careful study of the character of the bacillus, I found that it resembled closely one described by Gaertner in 1888 under the name of *Bacillus enteritidis,* and I named mine *Bacillus enteritidis Derbiensis.* The distribution of this microbe led me to the conclusion that the meat had been exposed to fæcal pollution after being chopped up, and before being baked. To test the correctness of my theory it was necessary to find out whether the meat had actually been exposed to such pollution, and whether bacilli of the kind discovered could have survived the baking process.

A visit to the premises where the pies had been prepared allowed me to realise more clearly than I had done before how many were the chances of infection, and how desirable it was that in the interest of manufacturers of comestibles, and more especially of the public generally, the sanitary authorities should be in position to regulate the preparation of food exposed to contaminations.

On inspection of the premises where the fatal pies had been prepared I found that opportunities for pollution were numerous. I may enumerate them as follows :—

1. The meat, hung in an entrance passage, was exposed to occasional contact with persons and pigs going into the yard.

2. It was also exposed in the same place to dust from the street and from the yard.

3. *The dust from the yard, which was used as a passage by men and animals (and where there were accumulations of refuse), was liable to be blown into the chopping-house* where the meat and jelly were prepared, also through ground level gratings into the cellar where the jelly was often left for several days.

4. *Live pigs were kept in the slaughter-house for at least one day*, and sometimes for a longer time, before being slaughtered ; the excreta of these pigs were allowed to gravitate towards a gutter in the centre of the slaughter-house, so that a pig affected with infectious enteritis would have soiled the floor of the slaughter-house. After being scalded, the carcase of the pig had to be hung over the same gutter, and as it was transferred from the tub to the hook some of its parts frequently touched the ground. The nature of the cleansing which followed permitted of the carcase and its parts being smeared over with a thin layer of filth, possibly not noticeable to the eye.

5. *The men employed in the abattoir passed from there to the chopping-house*, where the meat was chopped or minced and where the jelly was prepared. They soiled the floor with their boots, which had come in contact with excreta in the slaughter-house. It was said that they changed their boots and washed thoroughly, *but I found no evidence of their realising the great importance of special care* in that matter ; and I saw that the floor of the chopping-house, which had recently been washed, was being rapidly covered with soil at the time of my visit.

6. *The tub where the bowels and other parts were soaked, was in close proximity to the mincing-machine.*—It was the usual practice to clean bowels in the chopping-house.*

7. The jelly-tins or bowls, and the scoops used for taking jelly out of the copper, were either *regularly, or at least frequently, laid on the floor of the chopping-house.* The vessels in which jelly and meat were placed *remained frequently on that floor for hours.*

---

* Some of the water from a bucket containing the bowels was obtained by Dr. Howarth. This water had a dirty appearance. In order to find out whether this water might be fairly considered a constant possible source of pollution, I tested it for the presence of the bacillus coli, and found that this organism was, as expected, present in large numbers.

8. No special precaution was taken to prevent hands and un-sterilised vessels from coming in contact with the luke-warm jelly poured into the finished pies.

All these things, and many others which I need not mention in more detail, indicate numerous sources of pollution which should not be tolerated. Meat or jelly are quite as susceptible to pollution as milk, and a place where such things are exposed should be kept as scrupulously clean as a model dairy, or as an operating theatre. No surgeon would think of passing directly from a post-mortem room, where he had performed an autopsy, to the operating room, or he would expose the person operated upon to dangerous infection. Dead flesh and jelly are far more prone to be affected by bacterial infection than living flesh.

Beneficial Effects of Complete Cooking.

The only thing which has saved the consumer of pork pies, and other dainties prepared in the pork shop, from even more frequent disasters than have occurred is the sterilisation to which these articles are submitted during the process of cooking. What I say of pork applies also to many other eatables. *It is obvious that it is possible by preventive measures to guard against such serious sources of pollution,* and thus to prevent the recurrence of such outbreaks as the one in question. A large amount of illness, the source of which is not always so evident, is undoubtedly due to similar pollution of other articles of food, and I have for many years attributed much of the epidemic diarrhœa, so fatal to young children, to contamination by bacilli of the milk identical with or allied to those found in the pies.

Temperature reached by the various parts of a Pork Pie during the process of Baking.

Baking is not so great a safeguard as might be supposed. The pies are, it is true, placed in a very hot oven, the temperature of which is more than sufficient to kill non-sporing bacilli, such as the bacillus enteritidis, in a few seconds. But a pie is so constructed that its central parts are reached but slowly by heat. The pie is surrounded in the oven by hot, comparatively dry air, the rapid evaporation taking place from the surface of the moist crust keeps for a time the temperature of the rest of the pie comparatively low. The meat does not constitute a homogeneous mass, it is separated from the crust by a layer of air, there is also more or less air between the small pieces of meat occupying the centre of the pie. All these things prevent the rapid penetration of heat. To satisfy myself that these views were not purely theoretical, I have made, with Dr. Howarth and Mr. Cope, observations upon the temperature reached by various parts of pies baked in Mr. Cope's own oven. For the purpose of this experiment Mr. Cope had a certain number of pies prepared in the usual way, and placed in the oven, some at 4.30 p.m. and some at 5.20 p.m., on September 22. These pies were removed respectively at 6.17 p.m. and 6.11 p.m. before Dr. Howarth and myself, and we immediately took the temperature of the central parts of the pies, and also of the meat close to the bottom and top crust.

I could not devise any more reliable way of doing this than by plunging the bulb of the thermometer rapidly into the pie through the vent hole. A continuous record of the changes of temperature taking place in the pies during baking could not have been obtained on the premises, and would not have offered great guarantees of accuracy. The distance between the centre of the pie and the top crust having been measured on the stem of the thermometer, the instrument was plunged into the meat so that its bulb should reach at once the centre of the pie. This undoubtedly allowed a little of the boiling fluid which was on the surface of the meat to follow the bulb, so that the *temperature observed must have, in every case, been higher than the actual temperature of the centre of the pie*; but the error so produced was in the right direction, for it did not tend to make one under-estimate the temperature. After allowing the mercurial column to rise to the utmost, and seeing that it remained stationary, the temperature was read, the bulb of the thermometer was then pushed as far as the bottom crust, the temperature being again taken, the bulb was now withdrawn so as to come almost in contact with the top crust. As each of the ovens is heated chiefly by means of radiators situated above the pies, it was to be expected that the top crust would be hotter than the bottom crust.

In a *first set of observations* I found that the highest temperature reached in the central part of the pies was 47·2° C. These pies were said to have been under-baked (fifty-one minutes instead of the usual ninety minutes), but they had the appearance of well-baked pies.

In a *second set of observations* I found that the highest temperature in the centre of the pies was 86·6° C. These pies were said to be over-baked (107 minutes instead of 90 minutes), and they looked distinctly too much baked, their colour being much darker than is usual.

Several facts were brought out by these experiments.

1. The temperature of the centre of a *pie said to be under-baked, but having all the external appearances of being well-baked*, may not exceed 47·2° C. A batch of *pies prepared in a hurry* might therefore be so cooked that *bacteria might continue to grow* in their centre during the greater part of their stay in the oven, and the bacteria *would certainly not be killed*.*

2. The temperature of the centre of a pie obviously over-baked, and acknowledged to be so, had not gone beyond 86·6°C., which

---

* The appearance of the portions of pies which I received from Dr. Howarth at the beginning of this enquiry, gave me the impression that these pies which caused the Derby epidemic had been under-baked.

means that, even after excessive baking, the *temperature reached in the centre of a pie does not in any case exceed by many degrees that at which bacteria of low resistance are killed.*

3. There was a difference of several degrees between the temperature of various pies.

The importance of these results will be better understood in the light of some experiments which I have conducted, with the assistance of Dr. A. Sellers, to ascertain the resistance of the *Bacillus Derbiensis* to heat. In these experiments care was taken to ascertain the exact duration of exposures to certain temperatures. We found that the bacillus isolated from the pies, and cultivated in broth, was not killed when exposed in that fluid for twenty-four hours to a temperature of 50°C. (121°F.) in four experiments out of six.

It was only when a temperature of 60° C. (150° F.) was reached, that death of the bacillus was usually obtained in less than five minutes.

It is therefore obvious that the bacillus could easily resist the temperature to which the central parts of several of the pies which I examined had been raised, and that pollution of the meat in the chopping-house was quite sufficient to explain the Derby outbreak.

Another property of the bacillus explains how large masses of food may rapidly be infected. Its rate of multiplication is extraordinarily high. Thus I found that, with a small particle of a pure culture of the *Bacillus enteritidis*, I was able to infect throughout several ounces of broth, meat jelly or milk, in less than two hours; 2 milligrammes of culture are capable of infecting in two hours 150 grammes of these materials, when they are kept at temperatures ranging between summer temperature and blood heat. In other words, 1 part of infective material may easily infect in two hours 70,000 parts of one of the foods mentioned. A single drop of polluted water, or particle of excreted matter, would therefore be capable in summer to infect a gallon of milk, broth or jelly in a few hours.

Carriage of Food from a Distance.

If time permitted, I would be able to show how frequently cows' milk is infected at the farm or through dirty milk-cans, and how infectious bacilli multiply in the milk sent from the country to towns in hot railway vans. In this way a quantity of infectious matter, originally too small to cause a definite danger, is capable of increasing to such an extent as to render milk distinctly noxious.

During the past few years a large number of samples of milk sent to Manchester have been tested in my laboratory, by means of inoculation experiments. A small quantity of wholesome fresh milk injected under the skin of a guinea-pig causes no inconvenience to the animal, but infectious milk produces various forms of illness, some of which are rapidly fatal. Many of the cases of fatal illness are due to bacilli resembling those which I have mentioned in connection with the pork pies. It seems, therefore, desirable that infants should not be fed on such milk.

The effects of refrigeration, during transit, upon the properties of milk received in Manchester from various counties, offer a proof of the dangers connected with this multiplication of bacteria in milk, and also the that the danger is not without remedy.

About 50 per 1000 guinea-pigs, inoculated with non-refrigerated milk coming from a distance, died within ten days after inoculation. Not more than 3 per 1000 died when inoculated with milk which had been kept cold during transit. If the milk had been refrigerated immediately after milking, no death at all would have occurred, unless the milk had been obtained from much-diseased cows.

Preservation by means of Chemical Preservatives.

Another source of danger, which I must mention before concluding, is that which results from the addition of preservatives to perishable food stuffs, such as meat, milk, cream, butter, etc. Dealers in those articles are exposed to serious losses owing to putrefactive changes, which occur more or less rapidly in such articles. In most cases these putrefactive changes can easily be prevented by refrigeration ; but in many cases that method is not economical or convenient, and therefore chemical substances, such as salicylic acid, boracic acid, formalin, which arrest putrefaction, are extensively used by various trades. Some of these substances, when taken in sufficient quantities, have been shown to be more or less detrimental to health ; and as a matter of fairness to the consumer, it is obviously desirable that the addition of such substances should be made known to the purchaser who is not anxious to try upon himself experiments regarding the action of drugs taken in small doses over an unlimited period of time. It is not, however, to this aspect of the question that I wish to attract attention specially. There is another aspect which is of more importance.

Some of the preservatives in common use, although they are able to check the growth of putrefactive organisms, are unable to arrest that of some of the disease-producing bacteria which may be present in food ; and this is a serious source of danger.

During my investigation of the Derby outbreak, I was struck with the absence of putrefactive changes in the jelly of the pies, and previous experiments led me to infer that a preservative had been added to it in fairly large proportion. On analysis I found that the preservative used had been boracic acid.

If such a preservative had not been used the pies would rapidly have become stale, and few of the stale pies would have been eaten. Instead of this, the pies retained the appearance of freshness whilst the deadly bacillus was multiplying in their midst. It will have been noticed, that the pies which had been kept for some time had proved more noxious than those which had been consumed at an earlier date.

To obtain more accurate data on that point, I added to some broth and to some jelly four times more boracic acid than is considered sufficient to preserve food, and then inoculated those media with the *Bacillus enteritidis.* The bacillus grew abundantly in this comparatively strong solution of boracic acid. Moreover, I found that the bacillus remained alive for three months in broth to each 500 parts of which 1 part of boracic acid had been added (140 grains per gallon).

Apart, therefore, from the action which some preservatives may have upon the human frame, it seems evident that their use is attended with a definite danger; for, although they may be able to check putrefaction, they do not prevent the multiplication of certain infectious germs. The consumer, being deprived of the useful index of staleness which putrefaction offers, is therefore exposed to consume dangerous articles which he would have rejected otherwise.

Conclusion.

I must now conclude these remarks. I have dealt with a few instances only of the dangers lurking in food, such as it reaches us under the complicated conditions by which civilisation has surrounded us. The nature of these dangers indicates clearly, I think, that they can be met only by thorough legislation and administration. When one considers that the *question of the purity of food is only one of the many with which our Public Health service has to deal,* and how important such questions are to a nation, one is tempted to ask why the importance of this work is not more fully recognised. It seems that the magnitude and technical difficulties of the task should be enough to occupy fully a State Department, headed by a Cabinet Minister of great ability.

[S. D.]

WEEKLY EVENING MEETING.

Friday, February 20, 1903.

SIR FREDERICK BRAMWELL, Bart., D.C.L. LL.D. F.R.S. M.Inst.C.E. Vice-President, in the Chair.

PRINCIPAL E. H. GRIFFITHS, Sc.D. F.R.S.

The Measurement of Energy.

[No Abstract.]

WEEKLY EVENING MEETING,

Friday, February 27, 1903.

His Grace The Duke of Northumberland, K.G. D.C.L. F.R.S.,
President, in the Chair.

Adolf Liebmann, Esq., M.A. Ph.D. *M.R.I.*

Perfumes, Natural and Artificial.

History teaches that perfumes have been known and used ever since historical records have been kept. Even the Ayur Vedas (the book of life), which contains the earliest traditions of the Sanskrit literature, mentions attar of roses, oil of antropogon and of calmus, and the way of producing them.

In the same way, the mythical documents of the ancient Persians prove that perfumes were known to them, but foremost among the ancient nations who were acquainted with perfumes, and their mode of production, were the Egyptians, who probably obtained them in a purified form.

The Egyptian culture spreading to Europe, imparted the knowledge of the East and the love for scents to their western neighbours, and Greece, as well as Rome, used, especially during the prosperous times, the extracts of flowers for perfumery as well as for medicinal purposes.

Later on, the Moors contributed largely to the evolution of knowledge on this subject, and Messue described in his book, 'Antidotarium,' in a special chapter 'De Oleis,' the preparation of essential oils.

In the middle ages, distillation, rediscovered in the fifteenth century, introduced the possibility of obtaining purer and more fragrant products, and as a consequence the production of perfumes proper was started as an industry in the south-west of France.

Scientific exploration of the chemistry of essential oils was, however, only possible after the great discoveries of Cavendish, Priestley and Lavoisier. From that time chemists began to devote their attention to perfumes, and their researches, although not explaining the structure and nature of these compounds, nevertheless contributed considerably to our knowledge of the subject. It was left to Wallach's and Bayer's classical researches to demonstrate the nature of terpenes, and their derivatives, which form part of nearly all essential oils.

Terpenes are cyclic compounds. Semler proved, on examining a number of alcohols and aldehydes, geraniol, linalool, citral and citronellal, frequently found in essential oils, that they belong to the aliphatic series; and Thiemann, through his studies of vanilline, oil of iris and of irone, arrived at the synthesis of vanilline and ionone, or artificial violet. Many more might be mentioned as having contributed to our knowledge of this subject.

The chief sources of natural perfumes are the flowers, plants, aromatic roots, the sweet spices of the East, etc., from which the active principle, the essential oil, is extracted in different ways. The oldest method which is still in use for substances readily decomposable, and to which distillation with steam cannot be applied without destruction, is to treat the blossoms with vaseline or with molten fat, or press them between two sheets saturated with fat until the perfume has been completely dissolved in the fat (*enfleurage*).

The perfume can be separated from its fatty solvent by cold alcohol, in which the fats are insoluble. The essential oils of jasmine and tuberoses are, for instance, prepared in this manner. The mode is practically identical with the one described by Plinius in the second half of the first century.

The second mode of obtaining esssential oils, and the one which is most generally used, is to distil the blossoms with steam. Attar of roses, oil of neroli, oil of lavender and many others are produced by this process. There are three distinct operations:

(1) Preparation of the raw material.
(2) Distillation.
(3) Purification of the crude oil.

The distillation can be carried out

(*a*) With high-pressure steam;
(*b*) With superheated steam;
(*c*) With water;

all of which may be used, but the selection must always depend on the properties of the oil.

The purification of the distillate is very important and must depend on the special impurities which may contaminate the oil.

A third mode of obtaining the perfume is to extract it from the raw material with low boiling liquids such as ether, petroleum ether or acetone, and to remove the solvent by evaporation, which is usually effected in vacuo.

An effort has been made to explain the fragrant properties of scents by the presence of aromatophoric groups in a similar way as the colour of substances is due to chromophoric groups, but not sufficient data have been collected to allow of any fixed conclusions.

A variety of classes of compounds have been isolated from essential oils such as hydrocarbons, alcohols, ethers, aldehydes, ketones, acids and their esters.

The most important hydrocarbons belong to the class of terpenes, and their complex structure has been explained by a series of classical researches by Wallach; they are cyclic compounds, and have the general formula $C_{10}H_{16}$.

Pinene, camphene, limonene, hipentine, phellandrene, sylvestrene, terpinene, are the terpenes more frequently occurring in essential oils.

Other hydrocarbons found in oils are sesquiterpenes, $C_{15}H_{24}$, and cadinene and caryaphillene are its chief members. Alcohols occur frequently either combined as acid esters, phenolic esters, or in the free state.

Of the greatest interest and importance are the diolefinic alcohols geraniol and linalool $C_{20}H_{18}O$, and the olefinic alcohol citronellol. Geraniol is the chief ingredient of attar of roses, and linalool is also present, but in smaller quantity.

Several modes of preparing these alcohols in pure state have been devised, and they can now be obtained in the market as chemical individuals.

Whilst these alcohols are open chain compounds, essential oils contain a number of cyclic alcohols which are true hydroxy derivatives of terpenes proper. Amongst the more important, terpineol, borneol and menthol may be mentioned.

Open chain aldehydes are substances of strong odour, not always very pleasant. Octyl and nonyl aldehydes are constituents of the German attar of roses and of oil of lemon. Citral, however, is more important, and occurs in nearly all essential oils, which are distinguished by the characteristic lemon odour of this substance.

It is contained, to the extent of from 70 to 80 per cent., in lemon grass oil, and its chief importance lies in the fact that it forms the raw material for the preparation of ionone or artificial violet.

Citronellal frequently occurs along with citral, and has similar properties.

A number of aromatic aldehydes, benzaldehyde, salicylic aldehyde, cumaric aldehyde, anis aldehyde, vanilline, heliotropin and cinnamic aldehyde belonging to the benzene series, occur in nature, but most of them are now prepared synthetically.

Of aliphatic ketones only methyl heptenone, distinguished by the fruity odour of amylacetate, and methyl nonyl ketone are of any importance. Cyclic ketones are, however, important constituents of some essential oils, for example, fenchone, tujone, pulegone, all distinguished by a characteristic smell. But the most valuable in the industry of perfumes is irone, the active principle obtained from the roots of violets, and ionone or artificial violet.

Acids as such do not assist the perfuming qualities of essential oils. They are either odourless or have an objectionable smell; but a number of esters have been found in natural perfumes and possess highly valuable properties; amongst them, methyl salicylate, methyl anthranilate, etc., and a number of esters of fatty acids.

Lactones are also represented amongst perfumes; sedanolide with a strong odour of celery is present in this plant, and cumarine with a refreshing odour of new hay, occurs in many plants, such as woodruff, tonca beans, etc., and imparts to them their fragrance.

Phenols and phenolic ethers are very important members of the substances contributing to produce perfuming qualities.

Especially important are those which contain an olefinic side chain, such as anethol, eugenol, safrol, etc.

The examination of the natural perfumes was useful, not only as an extension of our knowledge of organic chemistry, but also for commercial reasons, that is, for the introduction of the synthetical essential oils into the market, reconstructed from the individuals found in the natural product. We are now able to buy a number of synthetical products, attar of roses, oil of jasmine, oil of neroli, etc.

But not only were the complex products of nature reconstructed synthetically, but the synthetical preparation of a number of chemical individuals was also successfully achieved, and vanilline was the pioneer in this direction.

Thiemann obtained it first by oxidation of coniferin or coniferil alcohol, and later on by the oxidation of acetisoeugenol. A number of other modes are now known, but Thiemann's latter process is technically still the most important one. The price of this aldehyde has fallen from 150*l.* in 1876 to 1*l.* 5*s.* to-day.

Heliotropin, protecatechu-aldehyde-methylene-ester, is likewise prepared synthetically by the oxidation of piperonic acid or by the oxidation of isosafrol; it forms, like vanilline, an example of the changes which are usually the result of competition, technical improvements and increasing consumption. Its price, which was once 75*l.*, has gone down to 15*s.* per pound.

Anisaldehyde and cinnamic aldehyde are now prepared in the laboratory. All these products, although artificial, are yet in another sense products of nature; that means, they are prepared synthetically, but they occur all of them in natural perfumes, and they have only been artificially produced for economic reasons.

But there are two scents of very great importance which are artificial in every meaning of the word: artificial musk and ionone, artificial violet.

Artificial musk, discovered by Baur, is trinitro-isobutyl-toluene or xylene, and is obtained by nitration of isobutyltoluene, isobutylxylene, and isobutylhydrindene.

There are a variety of these penetrating scents in which one of the nitro-groups is replaced by other groups, such as the cyanogen, halogen, ketonic and other groups, but the result from a perfuming point of view is identical.

Ionone, a ketone of the formula $C_{13}H_{20}O$, was discovered by Thiemann and forms the subject of a series of the most remarkable researches which this great scientist published.

It is obtained by condensation of citral with acetone, and by converting the new open chain ketone, which Thiemann calls pseudo-donone, into its isomeric ketone by treatment with acids.

Ionone is one of the finest perfumes in existence: it has the characteristic smell of violets and consists of two isomers, alpha and beta ionone, which in their perfuming qualities are practically identical.

There are some perfumes which are produced by animal life, such as musk, ambergris, but they have not yet been explored from a chemist's point of view.

GENERAL MONTHLY MEETING,

Monday, March 2, 1903.

Sir James Crichton-Browne, M.D. LL.D. F.R.S., Treasurer and Vice-President, in the Chair.

Rev. Charles H. Bowden,
Lord Robert Cecil, K.C.
William Chattaway, Esq. F.I.C. F.C.S.
Cumberland Clark, Esq.
Henry R. Goring, Esq.
James B. Hilditch, Esq.
Charles E. Kayler, Esq.
Mrs. C. E. Kayler,
James R. MacDonald, Esq. L.C.C.
Henry Forster Morley, Esq. M.A. D.Sc.
Miss H. Nunes,
Lady Russell Reynolds,
Julius Rheinberg, Esq. F.R.M.S.
T. Kirke Rose, Esq. D.Sc. F.C.S.
Horatio H. Shephard, Esq. M.A. LL.D.
James Knox Spence, Esq. C.S.I.
Frederick T. Trouton, M.A. D.Sc. F.R.S.
Mrs. Vlasto,
John Augustus Voelcker, Esq. Ph.D. F.C.S.
William Weddell, Esq.
John William Western, Esq.
Miss E. Willmott,

were elected Members of the Royal Institution.

The Special Thanks of the Members were returned to The Worshipful Company of Clothworkers for their donation of £100 to the Fund for the Promotion of Experimental Research at Low Temperatures.

The PRESENTS received since the last Meeting were laid on the table, and the thanks of the Members returned for the same, viz. :—

FROM

The Secretary of State for India—
  Geological Survey of India—
    Memoirs, Vol. XXXII. Part 3; Vol. XXXIV. Part 2; Vol. XXXV. Part 1.
    8vo. 1902.
Accademia dei Lincei, Reale, Roma—Classe di Scienze Fisiche, Matematiche e
    Naturali. Atti, Serie Quinta: Rendiconti. 1º Semestre, Vol. XII. Fasc. 1,
    Nos. 2. 3. 8vo. 1903.
Agricultural Society of England, Royal—Journal, Vol. LXIII. 8vo. 1902.
American Academy of Arts and Sciences—Proceedings, Vol. XXXVIII. No. 4.
    8vo. 1902.
American Geographical Society—Bulletin, Vol. XXXIV. No. 5. 8vo. 1902.
Astronomical Society, Royal—Monthly Notices, Vol. LXIII. No. 3. 1903. 8vo.
Automobile Club—Journal for Feb. 1903. 4to.
Bankers, Institute of—Journal, Vol. XXIV. Part 2. 8vo. 1903.
Batavia, Meteorological Observatory—Regenwaarnemingen in den Nederlandsch-
    Indischen Archipel, Deel 23, 1901. 8vo.
Belgium, Royal Academy of Sciences—Bulletin, 1902, No. 12. 8vo.
Boston Public Library—Monthly Bulletin for Feb. 1903. 8vo.
    Annual List of Books, 1901-1902. 8vo. 1903.
Botanic Society of London, Royal—Quarterly Record, Vol. VIII. No. 92. 8vo. 1902.
British Architects, Royal Institute of—Journal, Third Series, Vol. X. Nos. 7, 8.
    4to. 1902.
British Astronomical Association—Journal, Vol. XIII. No. 4. 8vo. 1903.
    Memoirs, Vol. XI. Part 2. 8vo. 1903.
British South Africa Company—Report on the Present Condition of Rhodesia.
    4to. 1903.
Buenos Ayres, City—Monthly Bulletin of Municipal Statistics, Dec. 1902. 4to.
Canada, Geological Survey of—Contributions to Canadian Palæontology, Vol. III.
    4to. 1902.
Cassell & Co. Messrs. (the Publishers)—Living London, Part 33. 8vo. 1903.
Chemical Industry, Society of—Journal, Vol. XXII. No. 3. 8vo. 1903.
Chemical Society—Proceedings, Vol. XIX. No. 260. 8vo. 1903.
    Catalogue of the Library. 8vo. 1903.
Civil Engineers, Institution of—Minutes of Proceedings, Vol. CL. 8vo. 1902.
Dax: Société de Borda—Bulletin, 1902, 2e Trimestre. 8vo.
Editors—American Journal of Science for Feb. 1903. 8vo.
    Astrophysical Journal for Jan. 1903. 8vo.
    Athenæum for Feb. 1903. 4to.
    Author for Feb. 1903. 8vo.
    Board of Trade Journal for Feb. 1903. 8vo.
    Brewers' Journal for Feb. 1903. 8vo.
    Chemical News for Feb. 1903. 4to.
    Chemist and Druggist for Feb. 1903. 8vo.
    Dioptric Review for Feb. 1903. 8vo.
    Electrical Engineer for Feb. 1903. fol.
    Electrical Review for Feb. 1903. 4to.
    Electrical Review (New York) for Jan. 1903. 4to.
    Electrical Times for Feb. 1903. 4to.
    Electricity for Feb. 1903 8vo.
    Engineer for Feb. 1903. fol.
    Engineering for Feb. 1903. fol.
    Feilden's Magazine for Feb. 1903. 8vo.
    Homœopathic Review for Feb. 1903. 8vo.
    Horological Journal for March, 1903. 8vo.
    Journal of the British Dental Association for Feb. 1903. 8vo.

Editors—continued.
Journal of State Medicine for Feb. 1903. 8vo.
Law Journal for Feb. 1903. 8vo.
London Technical Education Gazette for Jan. 1903.
Machinery Market for Feb. 1903. 8vo.
Model Engineer for Feb. 1903. 8vo.
Mois Scientifique for Feb. 1903. 8vo.
Motor Car Journal for Feb. 1903. 8vo.
Nature for Feb. 1903. 4to.
New Church Magazine for Feb.-March, 1903. 8vo.
Page Magazine for Feb 1903. 8vo.
Photographic News for Feb. 1903. 8vo.
Physical Review for Jan. and Feb. 1903. 8vo.
Public Health Engineer for Feb. 1903. 8vo.
Tea for Feb. 1903. 4to.
Zoophilist for Feb. 1903. 4to.
Electrical Engineers, Institution of—Journal, Vol. XXXII. Part 1. 8vo. 1903
Franklin Institute—Journal, Vol. CLV. No. 2. 8vo. 1903.
Geneva, Société de Physique—Comptes Rendus des Séances, XIX. 1902. 8vo.
Geographical Society, Royal—Geographical Journal for Feb. 1903. 8vo.
Geological Society—Quarterly Journal, Vol. LIX. Part 1. 8vo. 1903.
     Abstracts of Proceedings, No. 771. 1903.
Harlem, Société Hollandaise des Sciences—Archives Néerlandaises, Tome VIII.
     1e Liv. 8vo. 1903.
Horticultural Society, Royal—Journal, Vol. XXVII. Parts 2, 3. 8vo. 1902.
Johns Hopkins University—American Journal of Philology, Vol. XXIII. No. 4.
     8vo. 1903.
Manchester Geological Society—Transactions, Vol. XXVIII. Parts 1-3. 8vo. 1903.
Meteorological Society, Royal—Quarterly Journal, Vol. XXIX. No. 125. 8vo. 1903.
Microscopical Society, Royal—Journal for Feb. 1903. 8vo.
Mitchell, Messrs. C. & Co. (the Publishers)—Newspaper Press Directory, 1903. 4to.
Moscow University—Le Physiologiste Russe, Vols. I. II. 1898-1902. 8vo.
Navy League—Navy League Journal for Feb. 1903. 8vo.
North of England Institute of Mining and Mechanical Engineers—Transactions,
     Vol. LII. No. 2. 8vo. 1903.
Odontological Society—Transactions, Vol. XXXV. No. 3. 8vo. 1903.
Onnes, Professor H. K.—Communications, No. 82. 8vo. 1902.
Pharmaceutical Society of Great Britain—Journal for Feb. 1903. 8vo.
Philadelphia, Academy of Natural Sciences—Proceedings, Vol. LIV. Part 2. 8vo.
     1902.
Photographic Society, Royal—Photographic Journal for Jan. 1903. 8vo.
Rochechouart, La Société les Amis des Sciences et Arts—Bulletin, Tome XI. No. 6;
     Tome XII. Nos. 1, 2. 8vo. 1902.
Royal Irish Academy—Proceedings, Vol. XXIV. Section C, Part 2. 8vo. 1902.
Royal Society of London—Philosophical Transactions, A. No. 328; B. Nos. 211-
     213. 4to. 1903.
     Proceedings. No. 471. 8vo. 1903.
     Year Book, 1903. 8vo.
St. Bartholomew's Hospital—Reports, Vol. XXXVIII. 1902. 8vo. 1903.
Sanitary Institute—Journal, Vol. XXIII. Part 4. 8vo. 1903.
     Supplement, Vol. XXIII. Part 4. 8vo. 1903.
Selborne Society—Nature Notes for Jan. Feb. 1903. 8vo.
Smith, B. Leigh, Esq M.R I.—The Scottish Geographical Magazine, Vol. XIX.
     No. 2. 8vo. 1903.
Society of Arts—Journal for Feb. 1903. 8vo.
Sutcliffe, J. H. Esq. (The Editor)—The Dioptric Review, Vol. VI. 1902. 8vo.
Tacchini, Prof. P. Hon. Mem. R.I. (the Author)—Memorie della Società degli
     Spettroscopisti Italiani, Vol. XXXI. Disp. 12. 4to. 1902.
United Service Institution, Royal—Journal for Feb. 1903. 8vo.

United States Army, Surgeon-General's Office—Index Catalogue to Library Vol. **VII.** 4to. 1903.
United States Coast and Geodetic Survey—Magnetic Declination Tables for 1902. 8vo.
United States Department of Agriculture—Monthly Weather Review, Nov. 1902. 4to.
United States Patent Office—Official Gazette, Vol. CII. Nos. 4–6. 1903.
    Alphabetical List of Patents. Vol. 100. 1903.
Verein zur Beförderung des Gewerbfleisses in Preussen—Verhandlungen, 1903. Heft 1, 2. 8vo.
Vienna, Imperial Geological Institute—Jahrbuch, 1901, Band LI. Heft 3, 4. 8vo. 1902.
Yale University—Transactions of the Astronomical Observatory, Vol. I. Part 6. 4to. 1902.

WEEKLY EVENING MEETING.

Friday, March 6, 1903.

GEORGE MATTHEY, Esq., F.R.S. Vice-President, in the Chair.

PROFESSOR JOHN GRAY McKENDRICK, M.D. LL.D. F.R.S. F.R.C.P.

Studies in Experimental Phonetics.

[No Abstract.]

WEEKLY EVENING MEETING,

Friday, March 13, 1903.

THE RIGHT HON. SIR JAMES STIRLING, M.A. LL.D. F.R.S.,
Vice-President, in the Chair.

Professor KARL PEARSON, F.R.S.

Character Reading from External Signs.

[ABSTRACT.]

REFERRING to the delineation of character by physiognomy, phren-
ology, palmistry, handwriting, etc., he said that the only way of
determining whether there was any truth in these methods was the
dull way of statistics. To this end he had for six years been
obtaining observations on school children, and had now some 6,000
records, about half being of boys and half of girls; the observations
on the boys were, however, the only ones that so far had been reduced.
In connection with the problem he thought that folk-belief also should
be considered. For instance, so far as he knew, the colour of Judas
Iscariot's hair was not recorded; yet by the old masters Judas was
painted with red hair, and in directions for passion plays his
representative was ordered a red wig. In proverbs a general prejudice against red hair was apparent, and the curly-haired person
also seemed to be objectionable. Again, why was roundness of face
associated with foolishness (e.g. *Antony and Cleopatra*, Act III.,
sc. 3)? What about "fat-headed" people? and why were shrewd
people called " long-headed "?

The rest of the lecture consisted in the exhibition of statistical
tables showing the degrees of correlation between certain mental
characters and certain physical features, as indicated by the records
mentioned above, which were the results of observations and measurements of a number of schoolmasters. The comparison of hair-colour
and eye-colour with temper, health, conscientiousness, intelligence,
popularity, etc., yielded the general conclusion that, on the whole,
the red boys were more conscientious, more quick-tempered, and
more delicate, the black ones being less conscientious, more sullen,
and less delicate. There seemed to be a good deal in the attempt
to read character by handwriting, and the investigation of 2000
specimens of handwriting (classified into seven classes according to
general goodness) indicated that bad writing was a warning note.
With regard to the connection of size of head with intelligence, the
data taken from 1000 Cambridge undergraduates showed that on the
whole the head was longest in those who took first-class honours
and shortest in poll-men, but there was not enough correlation to

render prediction possible: all that could be said being that, if we considered two in 100 to be men of exceptional ability, their average head-length would be greater than 54 per cent. of the community and smaller than 44 per cent.

Again, there was a high correlation between intelligence and aptitude for athletics, and the tendency was for the athletic boy to be healthy, intelligent, quick-tempered, not sullen, self-conscious, noisy above the average, and distinctly popular.

In conclusion, Professor Pearson thought a real science of judging character was to a certain extent possible, though we should never be able to predict of the individual or to say more than that there was a measurable probability of his falling into a certain class. A known probability, not definite knowledge, was the only basis, however, upon which we were able to act in nine cases out of ten in practical life.

WEEKLY EVENING MEETING,

Friday, March 20, 1903.

Sir James Crichton-Browne, M.D. LL.D. F.R.S., Treasurer and Vice-President, in the Chair.

Professor E. A. Schäfer, LL.D. F.R.S.

The Paths of Volition.

On the 31st of March, 1596—just three hundred and seven years ago—there was born of a noble family, at La Haye in Touraine, René Descartes (whose portrait I here place before you), one of the greatest thinkers of that or any other age. This was, be it remembered, the age of Shakespeare, of Rembrandt and of Galileo; a period of extraordinary progress in literature, art and science. The education of Descartes he himself testifies to have been excellent of its kind. He was brought up from the age of eight at the College of La Flèche, "one of the most celebrated schools of Europe," conducted by the Jesuits, who have always been famous for the thoroughness of their teaching. But since the kind of education which they imparted was similar to that which still survives in the public school system of this country, it was little calculated to satisfy the inquiring spirit of the future philosopher, who left the college at the age of sixteen "loaded with laurels, and still more," says one of his biographers, "with philosophic doubts." The story of his disgust—at a much later period of his life—at hearing that Queen Christina was spending several hours a day in the study of Greek, and his remark that it was no evidence of learning to have an acquaintance with Latin no better than that which was possessed by the Roman populace, have often been quoted.

After spending three years in Paris he determined to educate himself by seeing the world, rightly considering the proper study of man to be mankind. Solely, it would appear, with this object in view he enrolled himself as a volunteer in the army of Prince Maurice of Nassau, assisting at the siege of Breda. Two or three years later he transferred his services to the Duke of Bavaria, then commanding the Catholic forces in the Thirty Years War, and was present at the battle of Prague. Under all these strange conditions Descartes continued the mathematical investigations which he had already commenced in Paris, and whilst still occupied with soldiering he found time for the pursuit of the studies to which he had resolved to devote his energies. It was at this time that he excogitated his famous Method of pursuing the study of science, and the earliest result of that excogitation was the discovery of the application of algebra to the solution of geometrical problems—a discovery which was made at the age of 23. Quitting the army in 1621 he travelled in various parts of Europe for a time, and presently found his way again to Paris, where he made acquaintance with many of the prominent literary and scientific men of the day. France, however, was not a safe place to reside in for anyone who was inclined to be speculative in the regions of science and metaphysics. He accordingly made up his mind to settle in Holland, and in 1625 he proceeded thither and resided there for nearly 20 years. It was here that he published eventually his famous 'Dissertatio de Methodo,' as well as a series of essays embracing metaphysics and nearly all the natural sciences. Having acquired some knowledge of anatomy he began to formulate original ideas on the subject of human physiology, which were published (with other essays on metaphysics and physics) in an essay which is contained in the ancient tome upon the table before me, the title-page of which I show you in the lantern. It was in this essay that he set forth his conception of the nervous system. The brain had long been recognised as the seat of intelligence; Descartes proceeded to further localisation, considering one particular spot in the centre of the brain, the conarium or pineal gland, to be more especially the seat of the soul. He was acquainted with the fact that through the spinal marrow, and the nerves issuing from it, the brain was brought into connection with sensory surfaces and with muscles, and he formulated a scheme of the manner in which a sensory impression may be transformed into movement. Looking upon the nerves as tubular cords, along which might flow that entity to which the older writers applied the term "animal spirits"—and which we in these modern times, without really knowing anything more about it than they did, now speak of as "nervous impulses"—Descartes supposed that they (the nerves) might possess some kind of valvular arrangement by which the animal spirits are directed into this or that channel, with the view of producing this or that movement; in a modified form this supposition has been revived in modern physiology. As is plain from the diagrams which I here show you, he clearly conceived that when a voluntary movement is performed it is the result of impulses which are carried to the brain along sensory nerves, and which, being transferred within the brain along motor paths, cause such movement. He explains, in the page here produced, how contact with the fire, A, produces an effect which is propagated along the nervous cord, C (which he compares to the cord of a bell), so that instantaneously at the other end of the cord an effect is produced, at the seat of intelligence, of such a nature that the "animal spirits," or "nervous impulses," are made to pass by appropriate nervous channels so as to cause such a movement as to fend off the fire. Descartes did not clearly distinguish—this was not indeed done until nearly two and a half centuries later—between what is commonly spoken of as a reflex action (not necessarily accompanied by consciousness) and volitional action which we are accustomed to associate with conscious sensation. It is, in fact, one of the most difficult problems in physiological psychology to differentiate in all cases between these forms of movement; although in many instances the distinction is evident enough. Many actions lie on the borderland, and, to include these, psychologists have invented the term "psychical reflex," by which they mean a kind of action which in a measure partakes of the nature of a purely reflex action, but which is nevertheless set going as the result of consciousness. But since it must be admitted that every voluntary action, however spontaneous it may appear, is the result of antecedent impressions which have reached the brain by sensory paths, the distinction between a psychical reflex and a volitional act is one not easily maintained.   There is no doubt that volitional impulses originate in and emanate from the cortex of the brain. But when we use the word originate, we must be careful to remember that no such impulse would ever emanate from the cerebral cortex unless sensory impulses of one kind or another had at some time previously been communicated to that cortex. It is only in this sense that we can speak of a voluntary impulse as originating in the brain. In many cases it is quite obvious that this is not their real origin. If I, for example, receive a slap in the face, my immediate impulse is to return the blow with interest, and this would probably be termed a "psychical reflex," but if the circumstances are such that it is impossible at the moment to perform that action, which has obviously been started by the sensory impression which I have received, this impression is stored for a while within the grey matter of the brain, to be brought out at a convenient opportunity and converted into the "volitional impulses" which will enable me to effect my purpose. In the one case, as in the other, the movement is called into play by the same sensory impulse as originally reached the brain.   What we call volition is, as a matter of certainty, the resultant of sensory impressions which have been previous received, and which may or may not have been stored for a long time within the cerebral grey matter.   The part of that grey matter from which they emanate is at the present day fairly well known.   The observations on cortical localisation which were begun by Fritsch and Hitzig in 1870, and brilliantly continued by Ferrier some three years later, have been of late years extended still further, both by clinical observations upon the human subject and by the work of many experimentalists. As a result of the most recent investigations, those of Professor Sherrington and Dr. Grünbaum upon the brain of the chimpanzee, it has been made clear that it is from the part of the cerebral cortex which lies immediately in front of the fissure of Rolando that impulses for the volitional movement of all parts of the body emanate. The problem that I propose to put before you this evening is, the consideration of the paths which are followed within the nerve centres by these volitional impulses, in the passage from the grey cortical layer of this part of the cerebrum to the grey matter of the spinal cord and other lower nerve centres, from which the motor nerve fibres which carry those impulses to the muscles directly emerge.

But before actually proceeding to consider this question, I must say a word as to the methods which are employed to solve it. On the evening of Friday, the 31st of May, 1861, a discourse was delivered in this theatre, which contained an account of a discovery of first-rate importance in the physiology of the nervous system. This discourse was prepared by Augustus Waller,* the distinguished father of a distinguished son, himself well known to members of the Royal Institution as having occupied for a time the position of Fullerian Lecturer in Physiology. Waller had noticed (some ten years previously) that after any nerve had been cut, one part of the cut nerve, that namely which was furthest from the nerve centre, showed after a few days a remarkable degeneration of its fibres, which has ever since been known as the " Wallerian degeneration " (Diagram 7). The drawing which I here exhibit to you (7) is one made by Mrs. Waller from one of her husband's preparations. I have had it, as well as the portrait which you have just seen, copied from ' Some Apostles of Physiology,' which I have placed upon the table, and which we owe to the erudition and literary research of Professor Stirling and the liberality of Dr. Whitehead of Manchester ; a book which is not only notable on account of its artistic and literary merit, but also by reason of the fact that only 100 copies of it have been printed (and none of these placed on the market), so that it will no doubt be, in time to come, an object of particular interest to the bibliomaniac.

Waller, as I have said, observed this granular degeneration in the peripheral part of the cut nerve, whereas in the central part there was no such appearance to be seen. Proceeding further, he noticed that when a nerve upon which there is a ganglion (a collection of nerve cells, such, for example, as that upon a posterior root of one of the spinal nerves) is cut so as to sever the part of the nerve which is connected with the spinal cord from the ganglion, it is now the more central part of the nerve which undergoes degeneration, whereas the part which is still in connection with the ganglion remains unaltered. The Wallerian doctrine, stated in present-day language, postulates that a nerve fibre undergoes degeneration when it is cut off from the cell with which it is connected, and from which it has originally grown. From it and in connection with it has developed a corollary which postulates that, at any rate within the nerve centres, the nerve fibre only conducts nervous impulses away from the cell with which it is naturally connected. This is, I may point out in passing, the revival of the valvular idea relating to nerve paths which had been suggested by Descartes, and to which I have already alluded. If the doctrine of nerve degeneration is true, it follows that if we cut away

---

* In point of fact the discourse, although prepared by Waller, was read for him, as he was prevented by illness from being himself present.

a nerve cell or group of nerve cells, all the fibres which emanate from them will undergo the degeneration in question. And this is, in fact, what is found to occur. For a long time it was a matter of difficulty to trace with exactitude the course of single nerve fibres, which, in consequence of removal of their nerve cells, or of their being cut off from these nerve cells, had undergone Wallerian degeneration. But a method of staining the granules of the degenerated fibres was devised some years ago by an Italian histologist, Dr. Marchi, by means of which the granules become stained intensely black, while normal nerve fibres remain unstained. By this method, it becomes possible to trace degenerated fibres in all parts of the nervous centres, and not only in battalions but even as single spies. It is hardly likely that Waller himself appreciated the full significance of his discovery. He could not have foreseen what a masterkey he had produced, nor how many doors of the secret passages of the nervous system were to be opened by it. Although it is the Wallerian method which has mainly contributed towards the recent progress of our knowledge regarding the paths of conduction in the central nervous system, it has been assisted by another method, that of Flechsig, which depends upon the fact that different tracts of fibres in the central nervous system undergo development at different periods of growth.

It will of course be understood, from what I have said, that in order to employ the Wallerian method it is necessary to effect in animals section of the nerve fibres, the course of which it is proposed to follow, or to remove altogether the cells with which those nerve fibres are connected; that is to say, the Wallerian method is an experimental method applied to the elucidation both of the structure and functions of the nervous system. Frequently Nature herself makes such experiments for us, when for example, as the result of disease, lesions become established in the brain or spinal cord, which sever nerve tracts or destroy groups of nerve cells. But such experiments are rarely as definite in their character as those which we ourselves plan out and make to order, although they are of value in establishing for the human subject the truth of observations which have been already made in animals Rarely have such natural experiments by themselves led to the actual discovery of an unquestioned fact. An exception must be made in connection with the subject with which I am dealing to-night, namely, "The Path of Volitional Impulses from the Brain to the Spinal Cord;" for it was first noticed (by Türck) in the human subject that lesions of the cerebral hemisphere occasioned by disease are followed by the occurrence of Wallerian degeneration along a particular tract within the brain and spinal cord, which it has become customary to speak of as the pyramidal tract. The fibres of this tract originate in the motor cortex of the brain. Hence we can trace them through the subjacent white matter; through the broad band of nerve fibres which is known as the internal capsule; through the isthmus of brain substance, which is known as the *brain stem,* in which they lie at the surface in a ventral situation, forming the greater part of the mass of nerve fibres, which has been termed the foot of the stem (pes pedunculi) or crusta; through the pons Varolii, where they are covered and concealed by the fibres which emerge from the hemispheres of the cerebellum; and down the ventral part of the medulla oblongata, where they form two bulging projections on either side of the middle line. These projections were termed by the older anatomists the anterior pyramids, and this name has given rise to the designation *pyramidal tracts* for the fibres which we are tracing and which are contained within these pyramids. Finally, from the pyramids of the medulla oblongata we can trace these same fibres crossing the middle line in large masses; and passing into the upper part of the spinal cord, where they are found to have deserted the ventral or anterior surface to which in the medulla oblongata they were confined, and to lie deeply in the nervous substance, occupying a large triangular area in the opposite lateral column of the cord   The general structure of the parts which have just been enumerated is shown in the accompanying photographs (8 to 17), and the manner in which the Wallerian degeneration can be followed by the Marchi method is illustrated in the slides which I now show you, which are made from preparations by Dr. Sutherland Simpson, and illustrate the degenerations which occur in the pyramidal tract of the cat as a result of injury or removal of the motor cortex of the brain. Not all the fibres in the pyramidal tract, however, go into the opposite lateral column of the cord. A few pass into the lateral column of the same side, and a variable number, in man and the anthropoid apes (but not in any other animals), remain for a while in the neighbourhood of the ventral or anterior fissure, and only gradually disappear from it as we trace them down from the cervical region. Ever since this great pyramidal tract of fibres has been recognised, it has been held that along it must pass those volitional impulses which originate in the grey matter of the brain, and which produce the movements of the muscles which are innervated from the spinal cord. And indeed this appears to be established by incontestible evidence, for it is certain that a lesion in any part of the pyramidal tract produces a corresponding paralysis in the lower regions of the body. If the lesion be above the place where the pyramidal fibres cross to the opposite side (at the junction of the medulla oblongata and spinal cord), the paralysis will be a paralysis of the opposite half of the body. If the lesion affect the fibres of the pyramidal tract below the decussation, the paralysis will be upon the same side.

We have seen that these pyramidal tract fibres originate in cells of the cortex of the brain, and that they terminate within the grey matter of the spinal cord, and within corresponding parts of the medulla oblongata, pons Varolii and middle brain. We have also seen that the impressions which they carry emerge from the motor nerve cells in the grey matter of the cord and pass out along the motor nerves to the muscles. The pyramidal tract fibres do not, however, go directly to the anterior horns of the cord, where the large motor cells are situated, but terminate in what is commonly regarded as the sensory part of the grey matter, viz. the base of the posterior horn. Here they end in fine ramifications, and the nervous impulses which they convey must be carried to the cells of the anterior horn by processes of other nerve cells, which are found here. In this way the pyramidal tract of fibres connects the cortex of the brain with the motor nerve cells of the spinal cord. The supposition that this is the path which volitional impulses take, in order to pass from the brain to produce movements of the muscles, thus rests upon a broad basis of observation.

We are met, however, in further considering this question, by the remarkable fact that the pyramidal tracts, well marked as they are in mammals, are entirely absent in all vertebrates below mammals. And yet no one will deny to the fish, the frog, the lizard, and least of all to the bird, the power of exercising volition over its muscles. There must, therefore, exist in these animals some path other than the pyramidal tracts, which, as we have seen, are not represented at all in them, by which the volitional impulses may pass from the cortex of the brain to produce movements of their voluntary muscles.

The path which connects the brain cortex of the bird with its spinal cord has been the subject of study by various observers—in this country especially by Professor Boyce and Dr. Warrington— who have made out a distinct and important system of fibres by which, on the one hand, the cerebral cortex is connected with the middle brain, and, on the other hand, the middle brain is connected with the lower centres in the spinal cord and medulla oblongata. The situation of these fibres, which do not form an uninterrupted path as in the case of the pyramidal fibres, but are, as we have seen, interrupted in the middle brain, is not the same as that taken by the pyramidal fibres of mammals. Now since it is certain that mammals have been developed from a lower type of vertebrates, it is, to say the least, not improbable that the same tracts as exist in the lower vertebrates would be retained and serve to some extent the same purpose in the mammal. In other words, there may be an alternative path in the mammal by which volitional impulses can be carried from the brain to the spinal cord.

This supposition is rendered the more probable by an observation, which has been made by more than one experimentalist, to the effect that although a section of the pyramidal tracts in the medulla oblongata or cord does in fact produce voluntary paralysis of the parts of the body below the section, yet, if the animal be kept alive, in the course of a few days or weeks there is considerable recovery from the resulting paralysis, and ultimately volitional movements may be performed in much the same manner as before the lesion. It follows from this observation that there must exist in mammals an alternative path for volitional impulses, and it is one of the problems which I have lately set myself to endeavour to determine within what part of the medulla oblongata and spinal cord that alternative path runs. The method which I have used has been the production of an artificial lesion or section in a definite portion of the medulla oblongata or spinal cord, and the observation of the effects, whether structural or functional, produced by such lesions.

But before giving you the results of these experiments, I wish to bring before your notice an important tract of nerve fibres which, in conformity with the law that nervous impulses follow degeneration, may, like the pyramidal tract, be regarded as one which carries descending impulses from higher to lower parts. This tract of fibres, which was first described by Löwenthal, occupies a position in the anterior part of the cord extending somewhat into the lateral region; it is known as the antero-lateral descending tract, or tract of Löwenthal. The termination of its fibres is in the anterior horn, where their ramifications come into close relationship with the large motor cells which are so characteristic of that part of the grey matter. The origin, on the other hand, of the fibres of the tract in question appears to be twofold. For, if we trace them upwards in sections of the medulla oblongata and pons Varolii, we find that, while a large number of them emanate from a group of cells which are accumulated in grey matter at the side of the medulla oblongata and pons Varolii, and which is known as the nucleus of Deiters, others come from a yet higher part of the brain stem, probably from the part known as the anterior corpora quadrigemina, although their origin has not been quite satisfactorily determined. Each cell of the nucleus of Deiters sends its nerve fibre towards the middle line, and the fibre there bifurcates, one branch passing upwards towards the mid-brain, where it ends by ramifying amongst the motor cells which give origin to the nerves to the eye muscles; the other branch turning downwards and passing into the spinal cord, where we see it as one of the fibres of the antero-lateral descending tract, and, as has already been mentioned, terminating amongst the motor cells of the anterior horn. The bundle which these fibres form lies on either side of the middle line; and from the fact that in the upper portion of its course it is placed in the dorsal or posterior part of the brain-stem, it has come to be known as the *posterior longitudinal bundle.* But as this posterior longitudinal bundle passes downwards into the spinal cord, it comes to occupy a ventral rather than a dorsal situation, and the name posterior longitudinal bundle now becomes a misnomer, since the fibres lie in the anterior or ventral column.

Those fibres of the antero-lateral descending tract which are traceable higher up in the brain stem, form a bundle lying ventral to the posterior longitudinal bundle, and, as we have seen, originating apparently in the mid-brain, where they are found crossing over from the opposite side. Their exact origin in the mid-brain is, as already stated, not known, but it is believed that they arise in the grey matter of the corpora quadrigemina, which correspond with the optic lobes of birds. Now it is precisely the optic lobes in birds which furnish an important intermediate station for the main tract of fibres connecting the cerebral hemisphere with the lower nerve centres, and on this ground alone it is extremely probable that the bundle of fibres which we are now discussing, and which is known as the *ventral longitudinal bundle*, is a part of the alternative path by which volitional impulses may pass from the cerebral cortex to the spinal cord. Moreover, in mammals also, as well as in birds, there exists a system of fibres which is known to connect the cerebral cortex with the middle brain, and which has quite recently been the subject of a special investigation by Dr. Beevor and Sir Victor Horsley.

I now come to the results of my own experiments. I have, in the first place, been able to confirm the statement that section of one or both of the pyramids of the medulla oblongata, involving as it does section of all the fibres of the pyramidal tract of one or both sides, produces paralysis below the lesion, which, although complete at first, is not lasting. This point settled, the next object was to determine, what, under these circumstances, is the alternative path along which volitional impulses from the brain to the anterior horn now pass; and in order to do this I have, both as a primary lesion and also as a lesion secondary to the injury of the pyramidal tract, cut the antero-lateral descending tract upon one or both sides of the cord. For, from the anatomical relations of these tracts, and especially the close connection which they have with the motor cells of the cord, it seemed most probable that the alternative volitional path would pass along them. This supposition appears to be confirmed as the result of the experiment, for, in all cases in which the antero-lateral descending tracts are cut, there is very pronounced voluntary paralysis in the parts of the cord below the section.

It might be thought that this experiment definitely settles the point, and that we may conclude as a result of it that there are two paths for volitional impulses, the one following the pyramidal tract and the other following the antero-lateral descending tracts. The matter, however, is not so simple as it seems; and in order to illustrate this I have only to bring before your notice a remarkable experiment which was performed by Dr. Mott and Professor Sherrington, which is itself a modification of a much older experiment of Sir Charles Bell. Sir Charles Bell found after he had cut the sensory nerve of the face in an ass, that the result of the section was not only to abolish sensation on that side of the face, but also to abolish voluntary action of the facial muscles. Mott and Sherrington extended this observation by cutting in a monkey the posterior roots of all the spinal nerves which supply fibres to one arm. These posterior roots contain, as is well known, only sensory fibres, and yet the result of their section was the production of just as complete a paralysis for voluntary motion as if the anterior roots, the fibres of which directly conduct impulses for volition, were themselves severed. Mott and Sherrington were inclined to regard their experiment as indicating that volitional impulses can only be originated in the brain in consequence of a sensation being conducted to the brain from the part which is to be moved, a sensation probably emanating from the muscles of the part themselves; but it appears to me that the true explanation of the experiment is that which has been given by Dr. Charlton Bastian, who has pointed out that the severance of the posterior roots, by cutting off all afferent impressions from the limb to the grey matter of the cord, will abolish that condition of the muscles of the limb which is known as *muscular tonus,* by virtue of which the muscles are always kept in a condition of readiness to contract, and without which far stronger impulses from the brain are required to act upon the motor nerve cells of the cord in order to induce movements of the muscles.   Mott and Sherrington indeed noticed that when an animal was executing movements in an energetic fashion, the paralysed arm participated in the movements, and they also found that the apparently paralysed muscles can be called into activity by electrical excitation of the corresponding part of the cerebral cortex.

If we apply the teaching of this experiment to the experiment which I have just placed before you—viz., that section of the antero-lateral columns of the cord produces paralysis of voluntary motion below the lesion—we see that it is possible to explain the paralysis not necessarily by the supposition that we have severed a volitional path along the antero-lateral tract, but by supposing that the severance of these fibres of the antero-lateral tract has cut off impressions which were proceeding to the anterior horn cells; impressions which were helping to keep them in such a condition of tonus as would render them prepared to send out nervous impulses with great facility along the corresponding motor nerves. That the section of the antero-lateral tracts may produce paralysis as a consequence of such loss of tone, is rendered even more probable by an observation made by Professor Ewald, who found that if that part of the auditory nerve—the vestibular branch—which comes from the semicircular canals is severed, or if the semicircular canals themselves are destroyed, there is produced, besides other symptoms, diminution or loss of tone in the muscles of the body generally.   Now many of the fibres of the vestibular nerve pass to the nucleus of Deiters, from which, as we have seen, the fibres of the posterior longitudinal bundle, which go to the cord, originate ; and it is therefore very possible that these fibres may carry the impressions derived from the vestibular nerve which help to maintain the tone of the motor cells of the cord. Of course the cutting of the antero-lateral descending tracts would cut off such impressions, and the tone of the motor-cells would be thereby diminished. Further, it must be borne in mind, that the antero-lateral descending tracts of the cord also include the fibres of the ventral longitudinal bundle coming from the mid-brain, and descending in all probability from cells with which the fibres of the optic nerve are in close connection. And it is probable that impulses arriving along this sensory path also assist in maintaining the tone of the motor-cells of the cord and lower level centres.   One must, therefore, admit the possibility that the paralysis which results from section of the antero-lateral descending fibres, may be at least in some measure connected with this diminution of tone of the motor apparatus of the cord ; and one of the questions which we have to consider is whether this by itself is a sufficient explanation of the resulting paralysis. The question is not an easy one to answer, but it may be stated that the paralysis which is produced by severance of the antero-lateral descending fibres, is at first so complete that nervous impulses which are produced by direct electrical excitation of the cerebral cortex are unable to affect the motor apparatus of the cord, whereas in Mott and Sherrington's experiment excitation of the cortex produced the usual contractions of the voluntary muscles.   It seems, therefore, probable that the fibres of the antero-lateral descending tract do serve to convey 'impulses from the volitional centres of the brain to the grey matter of the cord (even although we admit that by way of the vestibular nerve and the nucleus of Deiters, and by way of the optic nerve and the *corpora quadrigemina*, impulses may also pass along those fibres to assist in maintaining the tone of the motor centres), and thus furnish a path for volitional impulses other than that furnished by the pyramidal tracts ; a path which in all vertebrates below mammals is sufficient for the entire conduction of volitional impulses from the cortex of the brain to the motor apparatus of the cord.

[E. A. S.]

WEEKLY EVENING MEETING,

Friday, March 27, 1903.

GEORGE MATTHEY, ESQ., F.R.S., Vice-President, in the Chair.

PROFESSOR W. A. HERDMAN, D.Sc. F.R.S.

The Pearl Fisheries of Ceylon.

THE celebrated pearl "oysters" of Ceylon are found mainly in certain parts of the wide shallow plateau which occupies the upper end of the Gulf of Manaar, off the north-west coast of the island and south of Adam's Bridge.

The animal (*Margaritifera vulgaris*, Schum. = *Avicula fucata*, Gould) is not a true oyster, but belongs to the family Aviculidæ, and is therefore more nearly related to the mussels (*Mytilus*) than to the oysters (*Ostræa*) of our seas.

The fisheries are of very great antiquity. They are referred to by various Classical authors, and Pliny speaks of the pearls from Taprobane (Ceylon) as "by far the best in the world." Cleopatra is said to have obtained pearls from Aripu, a small village on the Gulf of Manaar, which is still the centre of the pearl industry. Coming to more recent times, but still some centuries back, we have records of fisheries under the Singhalese kings of Kandy, and subsequently under the successive European rulers—the Portuguese being in possession from about 1505 to about 1655, the Dutch from that time to about 1795, and the English from the end of the eighteenth century onwards. A notable feature of these fisheries under all administrations has been their uncertainty.

The Dutch records show that there were no fisheries between 1732 and 1746, and again between 1768 and 1796. During our own time the supply failed in 1820 to 1828, in 1837 to 1854, in 1864 and several succeeding years, and finally after five successful fisheries in 1887, 1888, 1889, 1890 and 1891 there has been no return for the last decade. Many reasons, some fanciful, others with more or less basis of truth, have been given from time to time for these recurring failures of the fishery; and several investigations, such as that of Dr. Kelaart (who unfortunately died before his work was completed) in 1857 to 1859, and that of Mr. Holdsworth in 1865 to 1869, have been undertaken without much practical result so far.

In September 1901, Mr. Chamberlain asked me to examine the records and report to him on the matter, and in the following spring I was invited by the Government to go to Ceylon with a scientific assistant, and undertake any investigation into the condition of the banks that m ght be considered necessary. I arrived at Colombo in January

1902, and as soon as a steamer could be obtained proceeded to the pearl banks. In April it was necessary to return to my university duties in Liverpool, but I was fortunate in having taken out with me as my assistant, Mr. James Hornell, who was to remain in Ceylon for at least a year longer, in order to carry out the observations and experiments we had arranged, and complete our work. This programme has been carried out, and Mr. Hornell has kept me supplied with weekly reports and with specimens requiring detailed examination.

The ss. *Lady Havelock* was placed by the Ceylon Government at my disposal for the work of examining into the biological conditions surrounding the pearl oyster banks ; and this enabled me on two successive cruises of three or four weeks each to examine all the principal banks, and run lines of dredging and trawling and other observations across, around and between them, in order to ascertain the conditions that determine an oyster bed. Towards the end of my stay I took part in the annual inspection of the pearl banks, by means of divers, along with the retiring Inspector, Captain J. Donnan, C.M.G., and his successor Captain Legge. During that period we lived and worked on the native barque *Rangasamee-porawee*, and had daily opportunity of studying the methods of the native divers and the results they obtained.

It is evident that there are two distinct questions that may be raised,—the first as to the abundance of the adult " oysters," and the second as to the number of pearls in the oysters, and it was the first of these rather than the frequency of the pearls that seemed to call for investigation, since the complaint has not been as to the number of pearls per adult oyster, but as to the complete disappearance of the shell-fish. I was indebted to Captain Donnan for much kind help during the inspection, when he took pains to let me see as thoroughly and satisfactorily as possible the various banks, the different kinds and ages of oysters, and the conditions under which these and their enemies exist. I wish also to record my entire satisfaction with the work done by Mr. Hornell, both while I was with him and also since. It would have been quite impossible for me to have got through the work I did in the very limited time had it not been for Mr. Hornell's skilled assistance.

Most of the pearl oyster banks or " Paars " (meaning rock or any form of hard bottom, in distinction to " Manul," which indicates loose or soft sand) are in depths of from 5 to 10 fathoms and occupy the wide shallow area of nearly 50 miles in length, and extending opposite Aripu to 20 miles in breadth, which lies to the south of Adam's Bridge. On the western edge of this area there is a steep declivity, the sea deepening within a few miles from under 10 to over 100 fathoms; while out in the centre of the southern part of the Gulf of Manaar, to the west of the Chilaw Pearl Banks, depths of between one and two thousand fathoms are reached. On our two cruises in the *Lady Havelock* we made ρ careful examination of the ground in several places outside the banks to the westward, on the chance of finding beds of adult oysters from which possibly the spat deposited on the inshore banks might be derived. No such beds, outside the known "Paars," were found; nor are they likely to exist. The bottom deposits in the ocean abysses to the west of Ceylon are "globigerina ooze," and "green mud," which are entirely different in nature and origin from the coarse terrigenous sand, often cemented into masses, and the various calcareous neritic deposits, such as corals and nullipores, found in the shallow water on the banks. The steepest part of the slope from 10 or 20 fathoms down to about 100 fathoms or more, all along the western coast seems in most places to have a hard bottom covered with Alcyonaria, sponges, deep-sea corals and other large encrusting and dendritic organisms. Neither on this slope nor in the deep water beyond the cliff did we find any ground suitable for the pearl oyster to live upon.

Close to the top of the steep slope, about 20 miles from land, and in depths of from 8 to 10 fathoms, is situated the largest of the "Paars," the celebrated Periya Paar, which has frequently figured in the inspectors' reports, has often given rise to hopes of great fisheries, and has as often caused deep disappointment to successive Government officials. The Periya Paar runs for about 11 nautical miles north and south, and varies from one to two miles in breadth, and this—for a paar—large extent of ground becomes periodically covered with young oysters, which, however, almost invariably disappear before the next inspection. This paar has been called by the natives the "mother-paar" under the impression that the young oysters that come and go in fabulous numbers migrate or are carried inwards and supply the inshore paars with their populations. During a careful investigation of the Periya Paar and its surroundings we satisfied ourselves that there is no basis of fact for this belief; and it became clear to us that the successive broods of young oysters on the Periya paar, amounting probably within the last quarter century alone to many millions of millions of oysters, which if they had been saved would have constituted enormous fisheries, have all been overwhelmed by natural causes, due mainly to the configuration of the ground and its exposure to the south-west monsoon.

The following Table shows, in brief, the history of the Periya Paar for the last twenty-four years:—

Feb. 1880. Abundance of young oysters.
Mar. 1882. No oysters on the bank.
Mar. 1883. Abundance of young oysters, 6 to 9 months old.
Mar. 1884. Oysters still on bank, mixed with others of 3 months old.
Mar. 1885. Older oysters gone, and very few of the younger remaining.
Mar. 1886. No oysters on bank.
Nov. 1887. Abundance of young oysters, 2 to 3 months.
Nov. 1888. Oysters of last year gone and new lot come, 3 to 6 months.
Nov. 1889. Oysters of last year gone; a few patches 3 months old present.
Mar. 1892. No oysters on the bank.

Mar. 1893.   Abundance of oysters of 6 months old.

Mar. 1894.   No oysters on the bank.

Mar. 1895.   Ditto.

Mar. 1896.   Abundance of young oysters, 3 to 6 months.

Mar. 1897.   No oysters present.

Mar. 1898.   Ditto.

Mar. 1899.   Abundance of oysters, 3 to 6 months old.

Mar. 1900.   Abundance of oysters 3 to 6 months old; none of last year's remaining.

Mar. 1901.   Oysters present of 12 to 18 months of age, but not so numerous as in preceding year.

Mar. 1902.   Young oysters abundant, 2 to 3 months. Only a few small patches of older oysters (2 to 2½ years) remaining.

Nov. 1902.   All the oysters gone.

It is shown by the above that since 1880 the bank has been naturally re-stocked with young oysters at least eleven times without yielding a fishery.

The 10-fathom line skirts the western edge of the paar, and the 100-fathom line is not far outside it. An examination of the great slope outside is sufficient to show that the south-west monsoon running up towards the Bay of Bengal for six months in the year, must batter with full force on the exposed seaward edge of the bank and cause great disturbance of the bottom. We made a careful survey of the Periya Paar in March 1902, and found it covered with young oysters a few months old. In my preliminary report to the Government written in July, I estimated these young oysters at not less than a hundred thousand millions, and stated my belief that these were doomed to destruction, and ought to be removed at the earliest opportunity to a safer locality further inshore. Mr. Hornell was authorised by the Governor of Ceylon to carry out this recommendation, and went to the Periya Paar early in November with boats and appliances suitable for the work; but found he had arrived too late. The south-west monsoon had intervened, the bed had apparently been swept clean, and the enormous population of young oysters, which we had seen in March, and which might have been used to stock many of the smaller inshore paars, was now in all probability either buried in sand or carried down the steep declivity into the deep water outside. This experience, taken along with what we know of the past history of the bank as revealed by the inspectors' reports, shows that whenever young oysters are found on the Periya Paar, they ought, without delay, to be dredged up in bulk and transplanted to suitable ground in the Cheval district—the region where the most reliable paars are placed.

From this example of the Periya Paar it is clear that in considering the vicissitudes of the pearl oyster banks, we have to deal with great natural causes which cannot be removed, but which may to some extent be avoided, and that consequently, it is necessary to introduce large measures of cultivation and regulation in order to increase the adult population on the grounds, give greater constancy to the supply, and remove the disappointing fluctuations in the fishery.

There are in addition, however, various minor causes of failure of the fisheries, some of which we were able to investigate. The pearl oyster has many enemies, such as star-fishes, boring sponges which destroy the shell, boring Molluscs which suck out the animal, internal Protozoan and Vermean parasites and carnivorous fishes, all of which cause some destruction and which may conspire on occasions to ruin a bed and change the prospects of a fishery. But in connection with such zoological enemies, it is necessary to bear in mind that from the fisheries point of view their influence is not wholly evil, as some of them are closely associated with pearl production in the oyster. One enemy (a Plectognathid fish) which doubtless devours many of the oysters, at the same time receives and passes on the parasite which leads to the production of pearls in others. The loss of some individuals is in that case a toll that we very willingly pay, and no one would advocate the extermination of that particular enemy.

In fact the oyster can probably cope well enough with its animate environment if not too recklessly decimated at the fisheries, and if man will only compensate to some extent for the damage he does by giving some attention to the breeding stock and " spat," and by transplanting when required the growing young from unsuitable ground to known and reliable " paars."

Those were the main considerations that impressed me during our work on the banks, and therefore, the leading points in the conclusions given in my preliminary report (July 1902) to the Governor of Ceylon ran as follows :—

1. The oysters we met with seemed on the whole to be very healthy.

2. There is no evidence of any epidemic or of much disease of any kind.

3. A considerable number of parasites, both external and internal, both Protozoan and Vermean, were met with, but that is not unusual in Molluscs, and we do not regard it as affecting seriously the oyster population.

4. Many of the larger oysters were reproducing actively.

5. We found large quantities of minute " spat " in several places.

6. We also found enormous quantities of young oysters a few months old on many of the Paars. On the Periya Paar the number of these probably amounted to over a hundred thousand million.

7. A very large number of these young oysters never arrive at maturity. There are several causes for this :—

8. They have many natural enemies, some of which we have determined.

9. Some are smothered in sand.

10. Some grounds are much more suitable than others for feeding the young oysters, and so conducing to life and growth.

11. Probably the majority are killed by overcrowding.

12. They should therefore be thinned out and transplanted.

13. This can be easily and speedily done, on a large scale, by dredging from a steamer, at the proper time of year, when the young oysters are at the best age for transplanting.

14. Finally there is no reason for any despondency in regard to the future of the pearl oyster fisheries, if they are treated scientifically. The adult oysters are plentiful on some of the Paars and seem for the most part healthy and vigorous; while young oysters in their first year, and masses of minute spat just deposited, are very abundant in many places.

To the biologist two dangers are however evident, and, paradoxical as it may seem, these are *overcrowding* and *overfishing*. But the superabundance, and the risk of depletion are at the opposite ends of the life cycle, and therefore both are possible at once on the same ground—and either is sufficient to cause locally and temporarily a failure of the pearl oyster fishery. What is required to obviate these two dangers ahead, and ensure more constancy in the fisheries, is careful supervision of the banks by some one who has had sufficient biological training to understand the life-problems of the animal, and who will therefore know when to carry out simple measures of farming, such as thinning and transplanting, and when to advise as to the regulation of the fisheries.

In connection with cultivation and transplantation, there are various points in structure, reproduction, life-history, growth and habits of the oyster which we had to deal with, and some of which we were able to determine on the banks, while others have been the subject of Mr. Hornell's work since, in the little marine laboratory we established at Galle.

Although Galle is at the opposite end of the island from the pearl banks of Manaar, it is clearly the best locality in Ceylon for a marine laboratory—both for general zoology and also for working at pearl oyster problems. Little can be done on the sandy exposed shores of Manaar island or the Bight of Condatchy—the coasts opposite the pearl banks. The fisheries take place far out at sea, from 10 to 20 miles off shore; and it is clear that any natural history work on the pearl banks must be done not from the shore, but, as we did, at sea from a ship during the inspections, and cannot be done at all during the monsoons because of the heavy sea and useless exposed shore. At such times the necessary laboratory work supplementing the previous observations at sea can be carried out much more satisfactorily at Galle than anywhere in the Gulf of Manaar.

Turning now from the health of the oyster population on the "paars," to the subject of pearl formation, which is evidently an unhealthy and abnormal process, we find that in the Ceylon oyster there are several distinct causes that lead to the production of pearls. Some pearls or pearly excrescences on the interior of the shell are due to the irritation caused by boring sponges and burrowing worms.

Minute grains of sand and other foreign bodies gaining access to the body inside the shell, which are popularly supposed to form the nuclei of pearls, only do so, in our experience, under exceptional circumstances. Out of the many pearls I have decalcified, only one contained in its centre what was undoubtedly a grain of sand; and from Mr. Hornell's notes taken since I left Ceylon, I quote the following passage, showing that he has had a similar experience :—

"February 16, 1903—*Ear-pearls.* Of two decalcified, one from the anterior ear (No. 148), proved to have a minute quartz grain (micro. preparation 25) as nucleus."

It seems probable that it is only when the shell is injured, as, for example, by the breaking off or crushing of the projecting "ears," thereby enabling some fine sand to gain access to the interior, that such inorganic particles supply the irritation which gives rise to pearl formation.

The majority of the pearls found free in the tissues of the body of the Ceylon oyster contain, in our experience, the more or less easily recognisable remains of Platyelmian parasites; so that the stimulation which causes eventually the formation of an " orient " pearl is, as has been suggested by various writers in the past, due to infection by a minute lowly worm, which becomes encased and dies, thus justifying, in a sense, Dubois' statement that—" La plus belle perle n'est donc, en définitive, que le brillant sarcophage d'un ver." *

To Dr. Kelaart (1859) belongs the honour of having first connected the formation of pearls in the Ceylon oyster with the presence of Vermean parasites. It is true that Filippi seven years before (in 1852), showed that the Trematode *Distomum duplicatum* was the cause of pearl formation in the fresh-water mussel *Anodonta*, and Küchenmeister (1856), Moebius (1857), and others extended the discovery to some of the larger pearl oysters, and to other parasites; but it is probable that Kelaart knew nothing of these papers and that he made his discovery in regard to the Ceylon oyster quite independently. He (and the Swiss zoologist Humbert, who was with him at a pearl fishery) found " in addition to the filaria and cercaria, three other parasitical worms infesting the viscera and other parts of the pearl oyster. We both agree that these worms play an important part in the formation of pearls; and it may yet be found possible to infect oysters in other beds with these worms, and thus increase the quantity of these gems."

Thurston, in 1894, confirmed Kelaart's observation, finding in the tissues, and also in the alimentary canal, of the Ceylon oyster, "larvæ of some Platyhelminthian (flat-worm)."

Garner (1871) associated the production of pearls both in the pearl oysters and also in our common English mussel (*Mytilus edulis*) with the presence of Distomid parasites; Giard (1897) and other

---

* Comptes Rendus, 14th Oct., 1901.

French writers have made similar observations in the case of *Donax* and other Lamellibranchs; and Dubois (1901) has more recently ascribed the production of pearls in mussels on the French coast, to the presence of the larva of *Distomum margaritarum*. Jameson (1902) then followed with a more detailed account of the relations between the pearls in *Mytilus* and the Distomid larvæ, which he identifies as *Distomum (Brachycœlium) somateriæ* (Levinson). Jameson's observations were made on mussels obtained partly at Billiers (Morbihan), a locality at which Dubois had also worked, and partly at the Lancashire Sea-Fisheries marine laboratory at Piel in the Barrow Channel. Finally, Dubois has just published a further note * in which, referring to the causation of pearls in *Mytilus*, he says (p. 178): " En somme ce que ce dernier [Garner] avait vu en Angleterre en 1871, je l'ai retrouvé en Bretagne en 1901. Quelques jours après mon départ de Billiers, M. Lyster Jameson, de Londres, est venu dans la même localité et a confirmé le fait observé par Garner et par moi." But Jameson has done rather more than that. He has shown that it is probable (his own words are " there is hardly any doubt ") that the parasite causing the pearl-formation in our common mussel (not in the Ceylon " pearl oyster ") is the larva of *Distomum somateriæ*, from the eider duck and the scoter. He also believes that the larva inhabits *Tapes* or the cockle as a first host before getting into the mussel.

We have found, as Kelaart did, that in the Ceylon pearl oyster there are several different kinds of worms commonly occurring as parasites, and we shall I think be able to show in our final report that Cestodes, Trematodes and Nematodes are all concerned in pearl formation. Unlike the case of the European mussels, however, we find so far, that in Ceylon the most important cause is a larval Cestode of the Tetrarhynchus form. Mr. Hornell has traced a considerable part of the life history of this parasite, from an early free-swimming stage to a late larval condition in the file fish (*Balistes mitis*) which frequents the pearl banks and preys upon the oysters. We have not yet succeeded in finding the adult, but it will probably prove to infest the sharks or other large Elasmobranchs which devour *Balistes*.

It is only due to my excellent assistant, Mr. James Hornell, to state that our observations on pearl formation are mainly due to him. During the comparatively limited time (under three months) that I had on the banks I was mainly occupied with what seemed the more important question of the life-conditions of the oyster, in view of the frequent depletion of particular grounds.

It is important to note that these interesting pearl-formation parasites are not only widely distributed over the Manaar banks, but also on other parts of the coast of Ceylon. Mr. Hornell has

---

* Comptes Rendus Acad. d. Sci., 19th. Jan., 1903.

found *Balistes* with its Cestode parasite both at Trincomalie and at Galle, and the sharks also occur all round the island, so that there can be no question as to the probable infection of oysters grown at these or any other suitable localities.

There is still, however, much to find out in regard to all these points, and other details affecting the life of the oyster and the prosperity of the pearl fisheries. Mr. Hornell and I are still in the middle of our investigations, and this must be regarded as only a preliminary statement of results which may have to be corrected, and I hope will be considerably extended in our final report.

It is interesting to note that the 'Ceylon Government Gazette' of December 22 last, announced a pearl fishery, to commence on February 22, during which the following banks would be fished :—

The South-East Cheval Paar, estimated to have 49 million oysters.

The East Cheval Paar, with 11 millions.

The North-East Cheval Paar, with 13 millions.

The Periya Paar Kerrai, with 8 million—making in all over 80 million oysters.

That fishery is now in progress, Mr. Hornell is attending it, and we hope that it may result not merely in a large revenue from pearls but also in considerable additions to our scientific knowledge of the oysters.

As an incident of our work in Ceylon, it was found necessary to fit up the scientific man's workshop—a small laboratory on the edge of the sea, with experimental tanks, a circulation of sea-water and facilities for microscopic and other work. For several reasons, as was mentioned above, we chose Galle at the southern end of Ceylon, and we have every reason to be satisfied with the choice. With its large bay, its rich fauna and the sheltered collecting ground of the lagoon within the coral reef, it is probably one of the best possible spots for the naturalist's work in Eastern tropical seas.

In the interests of science it is to be hoped, then, that the Marine Laboratory at Galle will soon be established on a permanent basis with a suitable equipment. It ought, moreover, to be of sufficient size to accommodate two or three additional zoologists, such as members of the Staff of the Museum and of the Medical College at Colombo, or scientific visitors from Europe. The work of such men would help in the investigation of the marine fauna and in the elucidation of practical problems, and the laboratory would soon become a credit and an attraction to the Colony. Such an institution at Galle would be known throughout the scientific world, and would be visited by many students of science, and it might reasonably be hoped that in time it would perform for the marine biology and the fishing industries of Ceylon very much the same important functions as those fulfilled by the celebrated Gardens and Laboratory at Peradeniya for the botany and associated economic problems of the land.

[W. A. H.]

WEEKLY EVENING MEETING,

Friday, April 3, 1903.

His Grace the Duke of Northumberland, K.G. D.C.L. LL.D.
F.R.S., President, in the Chair.

The Right Hon. Lord Rayleigh, O.M.
M.A. D.C.L. LL.D. Sc.D. F.R.S.
PROFESSOR OF NATURAL PHILOSOPHY, R.I.

Drops and Surface Tension.

Lord Rayleigh introduced his subject by showing the well-known experiment of water rising inside a capillary tube to a higher level than that at which it stood outside, and explained the phenomenon as a compromise between the tendency of the water to come in contact with the glass, and thus creep upwards, and the tendency of gravity to pull it downwards. Water was a liquid which tended to wet glass, but if one, such as mercury, which did not have that tendency, were employed, the opposite effect was to be seen, and the liquid did not rise inside the tube so high as it did outside.

Lord Rayleigh then illustrated the effect of capillary attraction, or surface tension, in determining the formation of drops, and mentioned the part it played in soap-films. He next discussed some interesting phenomena depending on the contact of materials with water not perfectly pure. For example, fragments of camphor dropped on perfectly clean water immediately were set in rapid rotation. But if the surface were at all greasy, even to the extent that could be produced by dipping the finger in the water for a few seconds, the rotation stopped, to begin again if the greasiness were removed. He had calculated that a thickness of oil amounting to two-millionths of a millimetre was sufficient to stop the rotation from taking place.

Extremely small amounts of grease had no effect on the surface tension; the first degrees of contamination produced no alteration at all, and it was only after a certain quantity of grease had been added that the alteration was noticeable, though it then increased very rapidly. About half the amount of oil necessary to stop the camphor rotating was required to affect the surface tension. But one-millionth of a millimetre we might suppose to be about the diameter of an oil-molecule; hence, short of the point where the surface tension altered, there was only a single layer of oil-molecules on the water.

Why the surface tension was altered by a greater number might be indicated by an analogy. If a few marbles were floating sparsely on mercury, they did not offer any particular resistance if one pushed them together to one side of the vessel; but there was a resistance if they were so numerous as completely to cover the mercury surface, and if on being pushed together they had to mount one on top of the other.

Finally, Lord Rayleigh showed his audience the effect of dirt or grease in the liberation of gas from champagne or soda-water. The adherence of the bubbles, he was sorry to tell them, was a sign of the dirtiness of the glass, as he showed by putting in soda-water two iron rods, one treated so as to be free from grease and the other not, when the bubbles were seen adhering to the latter, but not to the former.

GENERAL MONTHLY MEETING,

Monday, April 6, 1903.

His Grace The DUKE OF NORTHUMBERLAND, K.G. D.C.L. F.R.S.,
President, in the Chair.

Miss Margaret Gordon Anderson,
Harold Brown, Esq.
Thomas Willes Chitty, Esq.
Herbert Thomas Crosby, Esq. M.A. M.B. B.C.
Robert Abbott Hadfield, Esq. M.Inst.C.E.
William Waller Pope, Esq.
Miss Sophy Felicité de Rodes,
Mrs. George Orr Wilson,

were elected Members of the Royal Institution.

The PRESENTS received since the last Meeting were laid on the
table, and the thanks of the Members returned for the same, viz. :—

FROM

The Lords of the Admiralty—Nautical Almanac for 1906. 8vo.
Accademia dei Lincei, Reale, Roma—Classe di Scienze Fisiche, Matematiche e
 Naturali. Atti, Serie Quinta: Rendiconti. Vol. XII. 1º Semestre, Fasc.
 4–6. 1903. 8vo.
 Classe di Scienze Morali, Storiche, etc. Seria Quinta, Vol. XI. Fasc. 11, 12.
 8vo. 1902.
American Academy of Arts and Sciences—Proceedings, Vol. XXXVIII. Nos. 5–9.
 8vo. 1902.
American Philosophical Society—Proceedings, Vol. XLI. No. 171. 8vo. 1902.
Antiquaries, Society of—Archæologia, Vol. LVIII. Part 1. 4to. 1902.
 Proceedings, Vol. XIX. No. 1. 8vo. 1903.
Asiatic Society, Royal (Bombay Branch)—Journal, Vol. XXI. No. 58. 8vo. 1902.
Astronomical Society, Royal—Monthly Notices, Vol. LXIII. No. 4. 1903.
Automobile Club—Journal for March. 4to. 1903.
Bankers, Institute of—Journal, Vol. XXIV. No. 3. 8vo. 1903.
Boston Public Library—Monthly Bulletin for March, 1903. 8vo.
British Architects, Royal Institute of—Journal, Third Series, Vol. X. Nos. 9–11.
 4to. 1903.
British Astronomical Association—Journal, Vol. XIII. No. 5. 8vo. 1903.
Brooklyn Institute of Arts and Sciences—Science Bulletin, Vol. I. No. 3. 8vo.
 1902.
Buenos Ayres, City—Monthly Bulletin of Municipal Statistics, Jan. 1903. 4to.
Cassell & Co. (the Publishers)—Familiar Wild Birds. By W. Swaysland. Part I.
 8vo. 1903.
Chemical Industry, Society of—Journal, Vol. XXII. Nos. 4–6. 8vo. 1903.
 List of Members, 1903. 8vo.
Chemical Society—Journal for March–April, 1903. 8vo.
 Proceedings, Vol. XIX. Nos. 262, 263. 8vo. 1903.

Chicago, Field Columbian Museum—Publications: Anthropological Series, Vol. III. No. 3; Botanical Series, Vol. I. No. 7; Zoological Series, Vol. III. No. 7. 8vo. 1902.

Congress, Library of, Washington—Report of the Librarian, 1902. 8vo.

Cracovie, l'Académie des Sciences—Bulletins, Classe des Sciences Mathématiques et Naturelles, 1903, No. 1. 8vo.

Dublin Society, Royal—Scientific Transactions, Vol. VII. Nos. 14–16; Vol. VIII. No. 1. 4to. 1902.

    Scientific Proceedings, Vol. IX. Part 5. 8vo. 1903.

    Economic Proceedings, Vol. I. Part 3. 8vo. 1903.

Edinburgh, Royal College of Physicians—Laboratory Reports, Vol. VII. 8vo. 1903.

Editors—American Journal of Science for March, 1903. 8vo.

    Astrophysical Journal for March, 1903.

    Athenæum for March, 1903. 4to.

    Author for March–April, 1903. 8vo.

    Board of Trade Journal for March, 1903. 8vo.

    Brewers' Journal for March, 1903. 8vo.

    Chemical News for March, 1903. 4to.

    Chemist and Druggist for March, 1903. 8vo.

    Dioptric Review for March, 1903. 8vo.

    Electrical Engineer for March, 1903. fol.

    Electrical Review for March, 1903. 4to.

    Electrical Times for March, 1903. 4to.

    Electricity for March, 1903. 8vo.

    Engineer for March, 1903 fol.

    Engineering for March, 1903. fol.

    Feilden's Magazine for March, 1903. 8vo.

    Homœopathic Review for March-April, 1903. 8vo

    Horological Journal for April, 1903. 8vo.

    Journal of the British Dental Association for March, 1903. 8vo.

    Journal of Physical Chemistry for Feb.–March, 1903. 8vo.

    Journal of State Medicine for March, 1903. 8vo.

    Law Journal for March, 1903. 8vo.

    London Technical Education Gazette for March, 1903. fol

    Machinery Market for March, 1903 8vo.

    Model Engineer for March, 1903. 8vo.

    Mois Scientifique for March, 1903. 8vo.

    Motor Car Journal for March, 1903. 8vo.

    Musical Times for Feb.–March, 1903. 8vo.

    Nature for March, 1903. 4to.

    New Church Magazine for April, 1903. 8vo.

    Nuovo Cimento for Jan. 1903. 8vo.

    Page's Magazine for March, 1903. 8vo.

    Photographic News for March, 1903. 8vo.

    Physical Review for March, 1903. 8vo.

    Popular Astronomy for Feb.–March, 1903. 8vo.

    Public Health Engineer for March, 1903. 8vo.

    Science Abstracts for Jan. Feb. and March, 1903. 8vo.

    Terrestrial Magnetism for Dec. 1902. 8vo.

    Travel for Feb. 1903. 8vo.

    World's Fair Bulletin for March, 1903. 4to.

    Zoophilist for March, 1903. 4to.

Florence, Biblioteca Nazionale—Bulletin, March, 1903. 8vo.

Florence, Reale Accademia dei Georgofili—Atti, Vol. XXV. Disp 3, 4. 8vo. 1903.

Franklin Institute—Journal, Vol. CLV. No. 3. 8vo. 1903.

Geographical Society, Royal—Geographical Journal for March–April, 1903. 8vo.

Geological Society—Abstracts of Proceedings, Session 1902–3, Nos. 772–775.

Göttingen, Royal Academy of Sciences—Nachrichten : Mathematisch-physikalische Klasse, 1902, Hefts 1-6, 1903, Heft 1 ; Geschäftliche Mitteilungen, 1902, Heft 1, 2. 8vo.

Hutton, Captain F.˙W. (the Author)—Presidential Address to the Australian Association for the Advancement of Science. 8vo. 1902.

Johns Hopkins University—Circulars, No. 161. 4to. 1903.

Kansas University—Science Bulletin, Vol. I. Nos. 5-9. 8vo. 1902.
    University Quarterly, Vol. X. No. 4. 8vo. 1901.

Leighton, John, Esq. F.S.A. M.R.I.—Journal of the Ex-Libris Society, Feb. 1903. 8vo.

Madras Government Museum—Bulletin, Vol. IV. No. 3. 8vo. 1903.

Massachusetts Institute of Technology—Technology Quarterly, Vol. XV. No. 4. 8vo. 1902.

Mullick, Promatha Nath. Esq. (the Author)—History of the Vaisyas of Bengal. (2 copies). 12mo. 1902.

Munich, Royal Bavarian Academy of Sciences—Sitzungsberichte, 1902, Heft III. 8vo. 1903.

Navy League—Navy League Journal for March, 1903. 8vo.

North of England Institute of Mining and Mechanical Engineers—Transactions, Vol. LII. No. 3. 8vo. 1903.

Odontological Society—Transactions, Vol. XXXV. No. 3. 8vo. 1903.

Peru, Cuerpo de Ingenieros de Minas—Boletin, No. 1. 8vo. 1902.

Pharmaceutical Society of Great Britain—Journal for March, 1903. 8vo.

Photographic Society, Royal—Photographic Journal for Feb. 1903. 8vo.

Rio de Janeiro Observatory—Monthly Bulletin for July–Sept. 1902. 8vo.

Rome, Ministry of Public Works—Giornale del Genio Civile for Oct.-Nov. 1902. 8vo.

Royal Medical and Chirurgical Society of London—Medico-Chirurgical Transactions, Vol. LXXXV. 8vo. 1902.

Royal Society of London—Philosophical Transactions, A, Nos. 329-335. 4to. 1903. Proceedings, No. 472. 8vo. 1903.
    The Sub-Mechanics of the Universe. By Osborne Reynolds. 8vo. 1903.

Selborne Society—Nature Notes for March, 1903. 8vo.

Smith, B. Leigh, Esq. M.R.I.—The Scottish Geographical Magazine, Vol. XIX. Nos. 3, 4. 8vo. 1903.

Society of Arts—Journal for March, 1903. 8vo.

Statistical Society—Journal, Vol. LXVI. Part 3. 8vo. 1903.

Swedish Academy—Handlingar, Band 35. 4to. 1901-2.
    Bihang, Band 27, Parts 1-4. 8vo. 1901-2.
    Ofversigt, Vol. LVIII. 8vo. 1901.

Tacchini, Prof. P. Hon.Mem.R.I. (the Author)—Memorie della Società degli Spettroscopisti Italiani, Vol. XXXII. Disp. 1. 4to. 1903.

United Service Institution, Royal—Journal for March, 1903. 8vo.

United States Department of Agriculture—Report of the Chief of the Weather Bureau, 1901-2. 4to. 1903.
    Monthly Weather Review, Dec. 1902. 4to.

United States Patent Office—Official Gazette, Vol. CII. Nos. 6-8 ; Vol. CIII. Nos. 1-4. 4to. 1903.

Verein zur Beforderung des Gewerbfleisses in Preussen—Verhandlungen 1903, Heft 3. 4to.

Vienna, Imperial Geological Institute—Verhandlungen, 1903, Nos. 14-18. 8vo. Abhandlungen VI. Band 1, Sup. Heft. 4to. 1902.

Washington Academy of Sciences—Proceedings, Vol. V. pp. 39-98. 8vo 1903.

Western Society of Engineers—Journal, Vol. VIII. No. 1. 8vo. 1903.

WEEKLY EVENING MEETING,

Friday, April 24, 1903.

HIS GRACE THE DUKE OF NORTHUMBERLAND, K.G. D.C.L. F.R.S.,
President, in the Chair.

THE HON. R. J. STRUTT, M.A., Fellow of Trinity College,
Cambridge.

Some Recent Investigations on Electrical Conduction.

WE have here a gold-leaf electroscope, which you can see projected on the screen. I charge it with electricity, and the leaves remain divergent. If, however, I touch the knob of the electroscope with my finger, or with any other conductor, the leaves at once collapse.

Now the knob is at all times in contact with the air of the room. We may infer therefore that if the air of the room conducts at all under the condition of this experiment, it can be only to a very slight extent.

If, however, air be submitted to very much greater electrical stress, quite a different state of things sets in. I have here a tube, containing rarefied air, which we will expose to a powerful electric stress, by connecting its terminals to those of an induction coil. You see at once that its insulation breaks down, and that it conveys the electric current, which produces brilliant and complex luminous effects in it.

These phenomena are of great interest and importance, and some light has been thrown on their cause and nature by recent investigations. But I do not propose to enter into such difficult questions to-night. We shall confine our attention to the behaviour of gases under small electromotive forces, such as are insufficient to produce luminous discharge.

I have explained that under these conditions the conductivity is very slight, or the leaves of the electroscope could not remain divergent. If, however, we expose the air to Röntgen rays, an immediate and most striking change in its electrical behaviour takes place.

You will see, in the gallery of the theatre, a bulb capable of emitting Röntgen rays. I charge the electroscope, and as soon as the bulb, many yards away, is set in action, the leaves collapse, showing that they have lost their charge. The air of the room, traversed by Röntgen rays, has lost its power of insulation, and the charge of the electroscope quickly leaks away through it. Almost immediately after the rays are turned off, the air recovers its insulating power, and as you see, the electroscope is again able to retain its charge. Although the recovery of insulation is very rapid, it is not absolutely instantaneous. I have here an experiment bearing on this point. This metal box has a window of aluminium, through which the Röntgen rays can pass, so that the air in the box is exposed to them. This air is blown out through a considerable length of tubing on to the electroscope, and you see that it is able to discharge it. So that it is evident that some of the conducting power is retained during the time that the air takes to pass through about 2 feet of tubing, a considerable fraction of a second.

There is another way in which air can be made to lose its insulating power, and that is by exposing it to the action of the mysterious rays given out by radio-active bodies, notably by radium salts.

I charge the electroscope again, and you see that when I bring near it this sample of radium salt, the leaves fall together, as under the influence of Röntgen rays.

We may now consider more in detail the behaviour of gases made conducting by these methods.

When the electric current passes through a metal or an electrolyte, the relation between the current and the electromotive force applied is the simplest possible—the current is proportional to the electromotive force. The conductor is said to obey Ohm's law. But with a gas under the influence of Röntgen or Becquerel rays it is far otherwise. The current increases at first in proportion to the E.M.F., but when the E.M.F. is increased beyond a certain point, the current no longer increases correspondingly. Finally, when the E.M.F. is very large, a maximum value for the current is reached, and further increase in the E.M.F. is without influence upon it. The value of this limiting current, and the E.M.F. necessary to produce it, will of course depend on the strength of the rays. It is evidently of interest to compare the maximum or saturation current with different gases. Such comparisons have been carried out, and I will give you in a table some of the results.

Gas.	Relative Density.	Relative Saturation Current.	
		Röntgen Rays.	Becquerel Rays.
Hydrogen	·07	·10	·16
Air	1·0	1·00	1·00
Sulphur dioxide . . .	2·19	8·73	2·32
Methyl iodide . . .	5·05	70·3	5·18

You will observe that, under the action of Becquerel rays, the saturation current is nearly proportional to the density, while, under the action of Röntgen rays, a gas containing an element of high molecular weight, such as iodine, gives a current out of all proportion greater.

I wish now to return to a question which was too lightly passed over at the beginning of this lecture.

I said that the leakage of electricity through the air in its normal condition was very slight. But the existence of such a leakage has been incontestably proved.

The difficulty of establishing the conclusion is this. It is necessary to make use of some solid insulating support for the gold leaves of the electroscope. Now, if the leaves are observed to slowly collapse, it is difficult, with ordinary arrangements, to determine whether this leakage is really through the air or whether it takes place through the insulating support. This ambiguity of the experiment has been ingeniously overcome by Mr. C. T. R. Wilson, of Cambridge. His method was to carry the further end of the insulating support on a piece of metal which was at a higher potential than the gold leaves. Any failure in the insulation would then cause the leaves to diverge more than at [first. It was found, however, that in fact the leaves collapsed in the course of a few hours. There could, therefore, be no doubt whatever that the charge was escaping through the air.

I have here an experiment which shows the same thing, though perhaps not quite so satis-

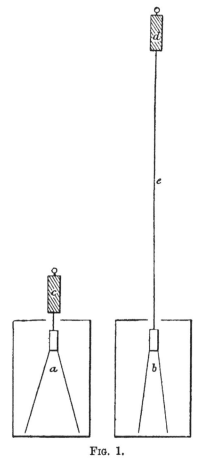

Fig. 1.

factorily. There are two pairs of gold leaves, *a, b* (Fig. 1), in all respects similar. These are supported by clean ebonite insulators, *c, d*, exactly alike for each. The left-hand pair, however, hangs immediately from the ebonite support, while the right-hand one hangs from the support by a long wire *e*. The right-hand charged system, therefore, has much better access to the air than the left-hand one. And in the course of half-an-hour you will see that its leaves have collapsed much further, in spite of its greater electrical capacity.

Mr. Wilson made a series of measurements of this electrical leakage through various gases, and he came to the interesting conclusion that the rates of leak were in the same ratio to one another as those which I had found for the same gases under the action of Becquerel rays. As a rule, the leakages were proportional to the densities of the gases, but as in the case of Becquerel rays, hydrogen was an exception, giving about twice as great a leakage as it ought to, if this law were exactly obeyed.

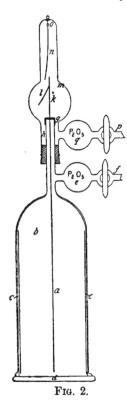

FIG. 2.

This curious agreement naturally suggested that the leakage ordinarily occurring was due to the same cause as the leakage under Becquerel rays. In other words, that the walls of the vessel containing the electroscope were giving off rays of this kind, although of course only to a very slight extent. In order to test whether this was really the case, I carried out a series of experiments on the rate of leak from a charged wire, when surrounded by cylinders of a uniform size, but of different materials. It soon became apparent that the rate of discharge depended on the nature of the surrounding wall.

I will now show you a diagram of the apparatus used (Fig. 2). *a* is a charged wire in the axis of the cylindrical vessel *b*. The walls of *b* could be lined with any desired material by inserting a cylinder *c c* composed of it. This could be done by removing the glass plate *d* at the end, which was cemented on. The vessel could be exhausted through the stopcock *f* if desired. *e* was a drying bulb, containing phosphoric anhydride. The wire *a* passed air-tight through the brass cap *g*, cemented on to the neck *h* of *b*. *h* was made of lead glass, on account of the superior insulating qualities of this kind of glass. The cap *g* carried a brass strip *k*, to which the gold leaf *l* was attached. The whole was surrounded by a vessel *m*, as shown. *n* was an iron wire, attached to a platinum wire *o* sealed through the glass. The iron wire could be brought into contact with *k* by means of an external magnet, in order to charge the system. *m* could be exhausted through the stopcock *p*, and dried by means of the phosphoric anhydride contained in *q*. The position of the gold leaf was read by a microscope with micrometer eye-piece, focussed upon it.

Before making an experiment, the insulation, which is all-important, was tested. *m* was permanently exhausted, and the stopcock *p* closed. *b* was also exhausted for the time, and a charge given to the brass strip *k*, so that the leaf *l* diverged. It was found that the leaf moved over less than ⅙ of a scale division in the course of three hours.

As soon as any air was admitted into *b*, a leakage of electricity from the wire *a* was observed.

The following table gives the results, with various materials:—

Material of Cylinder.	Current, in scale divisions per hour.
Zinc	1·2
Phosphoric acid on glass	1·3
Aluminium	1·4
Silver, chemically deposited	1·6
Copper oxide	1·7
Copper	2·3
Tinfoil, 1st sample	2·3
„ 2nd sample	3·3
Platinum, 1st sample	2·0
„ 2nd sample	2·9
„ 3rd sample	3·9
Uranium nitrate	12,000
Strongest radium preparations	1,200,000,000

The number for uranium nitrate was obtained by cementing a small piece of the substance on to the inner wall of the cylinder, and determining the rate of leak. The number thus obtained was corrected to the value which would correspond to covering the whole cylinder with uranium nitrate. The value for radium is merely computed from the comparisons which have been made between radium and uranium.

We conclude, then, that the leakage always found in the air is not an essential property of the air itself, but is due to feeble radioactivity in the surrounding solid objects.

In the experiment I showed you, with the pair of gold leaves hung from a long electrified wire, the leakage was due to Becquerel rays emitted by the walls and floor of the room, and possibly even by the persons of the audience.

So far we have been speaking of gases of the ordinary kind, that is, the vapours of volatile inorganic substances.

There are, however, special reasons for thinking that a metallic vapour should behave differently in its electrical relations.

We will take mercury vapour as an example.

Let us suppose some mercury placed in a hermetically closed vessel, and let the vessel be heated. The mercury will partially evaporate, and, at any given temperature, there will be a definite density of vapour which is in equilibrium with the liquid, so that no further evaporation takes place. When the temperature is raised, the equilibrium vapour density increases, while the density of the liquid diminishes, so that, if the temperature be increased sufficiently, the liquid and vapour will have the same density, and will be indistinguishable from one another. The temperature at which this happens is called the critical temperature. Now we know that the liquid mercury is an excellent conductor of electricity. Mercury vapour, however, as obtained by boiling mercury at the ordinary pressure of the air, does not conduct at all, or at least it only possesses the very feeble conducting power conferred on it by the Becquerel rays from the vessel walls, as I explained to you before. And yet, at the critical temperature, it must conduct as much as the liquid at that temperature, for indeed the two, liquid and vapour, are indistinguishable.

It is evident, therefore, that at these high temperatures and pressures, some very profound change must occur in the electrical properties of mercury. Either the liquid must be very much less conducting, or else the vapour immeasurably more_ so, than it would be under ordinary conditions. It might seem an easy matter to put this question to the test of experiment. But the practical difficulties are very great. It is necessary to confine the mercury in a closed vessel. This vessel must be capable of standing an enormous pressure; it must be able to stand a very high temperature without melting, and it must be made of electrically insulating material. These qualities cannot be found in sufficient degree in any known material. But they are most closely approached by vitreous silica, obtained by fusing rock-crystal in the oxy-hydrogen blow-pipe, and working it into tubes by the ingenious methods which were not long ago explained in this Institution by Mr. W. A. Shenstone.

I have here a tube of quartz, with thick walls. Some mercury has been hermetically sealed up in it. I place it over the flame of the blowpipe, and you see that the mercury, instead of boiling at a moderate temperature, as it would in an open vessel, is heated to full redness. In experimenting privately, I have been able to raise the temperature to a yellow heat. At that point the mercury vapour begins to show a steely blue absorption tint. Soon after this appears, the strongest tubes burst with the pressure of the vapour.

I will now show you a diagram of the tube used for measuring the electrical conductivity of mercury, and its vapour at a red heat (Fig. 3). The quartz tube took the form of an inverted Y, *a b b*. It was constricted to a very small diameter at the parts *d d* for a length of about 1 cm. on either side of the joint. The lower part of the limbs *b b* were of much larger diameter. The tube was filled with mercury up to the level *c*, the current being led in and out by iron wires *e e*, which projected some distance up, inside the arms *b b*.

The iron wires terminated in brass cups *f f*, carrying appropriate binding screws. These cups were filled with sealing-wax, which cemented them to the quartz tube. This sealing-wax had been sucked up the limbs while hot for a considerable distance, nearly up to the points *g g*, so as to fill the space between the iron wires *e e* and the lower parts of the quartz tube. The tops of the iron wires projected out through the sealing-wax, making contact with the mercury. The electrical resistance between the electrodes *f f* lay mainly in the narrow portion $d\,d$, and this alone, with the branch a, was kept hot.

It was found that at a full red heat, the resistance of the liquid mercury was about doubled. The resistance of the saturated vapour was taken with the same apparatus, the narrow part $d\,d$ being in this case filled with the vapour instead of the liquid; it was still ten

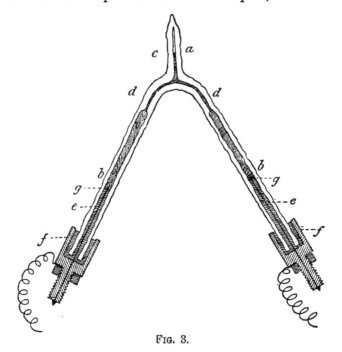

Fig. 3.

million times as great as that of the liquid. But the vapour did conduct very appreciably; and a current easily measurable with the galvanometer could be sent through it with a single battery cell. I think that in all probability, if we could trace the charge up to the critical temperature, we should find that the saturated vapour would approach in its electrical behaviour to the liquid metal.

ANNUAL MEETING,

Friday, May 1, 1903.

SIR JAMES CRICHTON-BROWNE, M.D. LL.D. F.R.S., Treasurer and
Vice-President, in the Chair.

The Annual Report of the Committee of Visitors for the year
1902, testifying to the continued prosperity and efficient management
of the Institution, was read and adopted, and the Report on the Davy
Faraday Research Laboratory of the Royal Institution, which accompanied it, was also read.

Sixty-two new Members were elected in 1902.

Sixty-three Lectures and Twenty Evening Discourses were
delivered in 1902.

The Books and Pamphlets presented in 1902 amounted to about
211 volumes, making, with 683 volumes (including Periodicals bound)
purchased by the Managers, a total of 894 volumes added to the
Library in the year.

Thanks were voted to the President, Treasurer, and the Honorary
Secretary, to the Committees of Managers and Visitors, and to the
Professors, for their valuable services to the Institution during the
past year.

. The following Gentlemen were unanimously elected as Officers
for the ensuing year:

PRESIDENT—The Duke of Northumberland, K.G. D.C.L. F.R.S.
TREASURER—Sir James Crichton-Browne, M.D. LL.D. F.R.S.
SECRETARY—Sir William Crookes, F.R.S.

MANAGERS.

Henry E. Armstrong, Esq. Ph.D. LL.D. F.R.S.
Sir Benjamin Baker, K.C.B. K.C.M.G. LL.D.
F.R.S. M.Inst.C.E.
Shelford Bidwell, Esq. M.A. Sc.D. LL.B. F.R.S.
Sir Alexander Binnie, M.Inst.C.E.
Sir Frederick Bramwell, Bart. D.C.L. LL.D.
F.R.S. M.Inst.C.E.
The Hon. Sir Henry Burton Buckley, M.A.
The Right Hon. The Earl of Halsbury, M.A.
D.C.L. F.R.S.
Donald William Charles Hood, C.V.O. M.D.
F.R.C.P.
The Right Hon. Lord Lister, O.M. M.D.
D.C.L. LL.D. F.R.S.
George Matthey, Esq. F.R.S.
Edward Pollock, Esq. F.R.C.S.
Sir Owen Roberts, M.A. D.C.L. F.S.A.
Sir Thomas Henry Sanderson, G.C.B.
K.C.M.G.
Sir Felix Semon, C.V.O. M.D. F.R.C.P.
Sir John Isaac Thornycroft, LL.D. F.R.S.
M.Inst.C.E.

VISITORS.

John B. Broün-Morison, Esq. J.P. D.L. F.S.A.
(Scot.)
Horace T. Brown, Esq. LL.D. F.R.S.
Frank Clowes, Esq. D.Sc. F.C.S.
James Mackenzie Davidson, Esq. M.B. C.M.
Frederick William Fison, Esq. M.P. M.A.
F.C.S.
Francis Fox, Esq. M.Inst.C.E.
Francis Gaskell, Esq. M.A. F.G.S.
James Dundas Grant, M.D. F.R.C.S.
Lord Greenock, D.L. J.P.
James E. Horne, Esq. M.A.
Maures Horner, Esq. J.P. F.R.A.S.
Wilson Noble, Esq. M.A.
Arthur Rigg, Esq.
George Johnstone Stoney, Esq. M.A. D.Sc.
F.R.S.
George Philip Willoughby, Esq. J.P.

WEEKLY EVENING MEETING,

Friday May 1, 1903.

SIR WILLIAM CROOKES, F.R.S., Honorary Secretary and
Vice-President, in the Chair.

PROFESSOR WILLIAM J. POPE, F.R.S.

Recent Advances in Stereochemistry.

IN the year 1803, just a century ago, John Dalton delivered in the
Royal Institution a series of scientific lectures during the course of
which he doubtless laid before his audience a theory which he had
just devised for the purpose of connecting together the vast number
of isolated chemical facts known at the beginning of the nineteenth
century. This theory, of which the centenary is being celebrated
during the present month by the Manchester Literary and Philosophical Society, is known as the Atomic Theory, and was destined
to form the foundation upon which the whole superstructure of
modern chemistry has been built up. For our present purpose,
Dalton's theory may be briefly stated in the form of the following
two principles :—(1) Every element is made up of homogeneous atoms
of which the mass is constant; (2) Chemical compounds are formed
by the union of atoms of the various elements in simple numerical
proportions. In accordance with Dalton's hypothesis, chemical substances may be mentally pictured by imagining the atoms as small
spheres which have the power of aggregating themselves together
under suitable conditions to form complexes or 'molecules'; thus,
taking two similar spheres representing hydrogen atoms, in conjunction with a sphere of a different kind representative of an atom of
oxygen, a chemical representation can be given of the compound
water, the molecule of which is composed of two atoms of hydrogen
and one of oxygen.

The original atomic theory offers no explanation of the observed
fact that the atoms combine together in different proportions; this
deficiency was remedied by the doctrine of 'Valency' enunciated by
the late Sir Edward Frankland in 1852. Frankland supposed that the
atoms of certain elements, such as hydrogen and chlorine, are unable to combine with more than one atom of any other element; these
elements are termed monovalent. Other atoms, such as those of
barium and zinc, can become directly attached to at most two other
atoms; these are the divalent elements. Tri-, tetra-, penta-, hexa-,
hepta-, and octa-valent elements can be similarly distinguished, the
valency of hydrogen being taken as unity, in order to measure and

define the saturation-capacity or the atom-fixing power of the atoms of the other elements. It will be clear that for rough diagrammatic purposes we may provide the spheres representing the atoms with as many wooden pegs as the element itself exhibits units of valency; compound molecules can then be represented by fitting the atoms together by means of the pegs representing the number of valency units possessed by the various constituent atoms. By so doing a great advance is made upon the atomic theory of Dalton's time and a mental picture is obtained of the way in which the atoms are connected together within the molecule itself.

During the early part of the nineteenth century it became evident, principally from the work of Liebig and Wöhler in Germany and of Faraday at the Royal Institution, that substances exist which possess totally different properties, but nevertheless have the same molecular composition; as this became slowly realised, the atomic theory was naturally called upon to furnish some adequate explanation. In view of the proven identity of molecular composition the required explanation could only be sought for in differences in the atomic arrangement within the molecules of the several substances. That such differences can be successfully illustrated by the aid of the atomic models will be seen on considering some specific case. Ordinary ethyl alcohol and methyl ether differ greatly from each other—the first is a liquid whilst the second is a gas at ordinary temperatures—but possess the same molecular composition, the molecule in each case consisting of two atoms of carbon, six of hydrogen and one of oxygen. These two substances have to be represented on the assumption that hydrogen is monovalent, carbon tetravalent and oxygen divalent. By joining wooden spheres together in the order shown in the figures—in which the valencies of the component atoms are carefully respected—diagrammatic representations are obtained which illustrate to the chemist the differences existing between ethyl alcohol and methyl ether. Substances related to each other in this

<pre>
    H H   H   H
    | | | |
H—C—C—O—H   H—C –O—C—H
    | | | |
    H H   H   H
  Ethyl alcohol   Methyl ether
</pre>

way are said to be isomeric; they have the same molecular composition but different molecular *constitutions*. The step in advance which is involved in thus writing molecular constitutions or in constructing molecular models was taken by Kekulé in 1858.

Two great stages in the development of chemical theory have now been indicated. First, that contributed by Dalton, who regarded constancy of molecular composition as characteristic of a chemical substance; secondly, that further stage, attained as a result of the labours of Liebig, Wöhler, Faraday, Frankland and Kekulé, which involved the introduction of the idea that the chemical individuality of a substance is dependent upon its molecular constitution as well as upon its molecular composition. A third great development in the atomic theory had yet to take place.

Whilst the theoretical views which culminated in Kekulé's constitutional formulæ were at first found sufficient to explain numerous observed cases of isomerism, instances soon began to accumulate of substances which exist in so many isomeric forms that the Kekulé method of representation is incapable of accounting for them all. At an early date, Pasteur showed clearly that substances exist which have the same molecular composition and the same molecular constitution, but which nevertheless differ in important respects. A crisis was ultimately reached when, in 1870, Wislicenus demonstrated the existence of three isomeric lactic acids, all having the molecular composition, $C_3H_6O_3$, and the molecular constitution—

$$\begin{array}{c} \text{H} \quad \text{O-H} \quad \text{O} \\ | \quad\quad | \quad\quad \| \\ \text{H—C—C—C} \\ | \quad\quad | \quad\quad \diagdown \text{O—H} \\ \text{H} \quad\quad \text{H} \end{array}$$

and contended that he had amply proved the insufficiency of Kekulé's method of writing constitutional formulæ.

The step needed to rid the atomic theory of these apparent anomalies was indicated by van't Hoff and Le Bel in 1874; they pointed out that the weakness of the Kekulé method lies in the tacit assumption that the molecule is spread out upon a plane surface: that by throwing this assumption aside and taking a rational view of the way in which the molecule is extended in space, all difficulties immediately vanish. The considerations put forward by van't Hoff and Le Bel form the basis of the subject now known as *Stereochemistry*, the branch of science which deals with the manner in which the atoms are distributed within the molecule in three-dimensional space; they deal, in the first place, with the arrangement of the constituent atoms in the simple organic compound, methane, the molecule of which has the composition, CH_4, or consists of one carbon atom and four hydrogen atoms. The Kekulé constitutional formula pictures the component atoms of the methane molecule as if joined together in one plane (Fig. 1), whilst according to the new view, the four hydrogen atoms are imagined situated at the four apices of a regular tetrahedron of which the carbon atom occupies the centre (Fig. 2). This is conveniently illustrated with the aid of a few cardboard models.

Consider now the result of replacing three of the four hydrogen atoms present in the methane molecule by three different groups of atoms—the three groups, CH_3, OH, and CO_2H, for example. One of the most striking results which has accrued from the chemical investigation of the past century has been the demonstration of the remarkable rigidity with which the atoms are held together in the molecule; it might therefore be anticipated that by actually making all the isomerides having the constitution indicated above, some means would be afforded of judging whether the van't Hoff-Le Bel

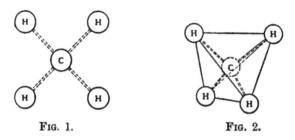

Fig. 1.   Fig. 2.

or the Kekulé view forms the closest approximation to the truth. Kekulé's constitutional formulæ indicate the existence of two isomeric compounds of the following types—

$$CH_3-\overset{\overset{\textstyle OH}{|}}{\underset{\underset{\textstyle H}{|}}{C}}-CO_2H \quad \text{and} \quad HO-\overset{\overset{\textstyle CH_3}{|}}{\underset{\underset{\textstyle H}{|}}{C}}-CO_2H$$

whilst, on the van't Hoff-Le Bel view, two isomerides should exist in which the four groups, H, CH_3, OH, and CO_2H, are arranged about the central carbon atom in the manner indicated in Figs. 3 and 4.

Although in each case two isomerides would be obtained, the examination of the two kinds of figure reveals very essential differences. The solid figure isomerides differ only in that the one is the image in a mirror of the other—they are related in the same kind of

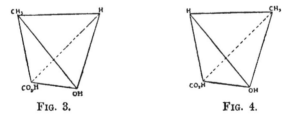

Fig. 3.   Fig. 4.

way as a right- and a left-hand glove. The differences observable between two molecules thus related should consequently not be differences of an ordinary chemical nature, but differences involving merely a kind of chemical, physical and mechanical right- and left-handedness. The Kekulé constitutional formulæ, on the other hand, would indicate—if they indicate anything—that the substances to which they refer differ in the more gross way in which ordinary chemical isomerides differ in chemical, physical and mechanical respects. That carbon atom which was present in the original methane molecule is, in these new compounds, now attached to four different atomic groups, and such a carbon atom is termed an asymmetric carbon atom. It is in the case of substances containing an asymmetric carbon atom that a lack of agreement is observed between the facts and the kind of isomerism indicated by the Kekulé formulæ, and in these cases also the species of isomerism indicated by the solid models exhibited is found to correspond closely with the facts.

To illustrate this, we may refer to a somewhat complicated substance termed tetrahydroquinaldine, which has the appended constitution and the molecule of which contains an asymmetric carbon

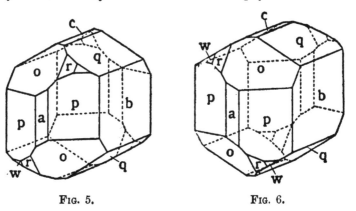

Tetrahydroquinaldine.

atom, that, namely, which is printed in heavy type. Three different isomeric forms of this substance exist and are quite indistinguishable by any of the ordinary methods of chemical or physical identification;

Fig. 5.   Fig. 6.

one of these is a loose kind of compound of the other two, and may therefore be disregarded for the moment. The remaining two have the same melting point, the same boiling point and correspond exactly in all ordinary properties; they yield, however, series of derivatives

as dextro- and lævo-rotatory. Van't Hoff and Le Bel declared that the molecules of all naturally occurring substances which exhibit optical activity when in the fluid state contain asymmetric carbon atoms. All substances, the molecules of which contain an asymmetric carbon atom must possess enantiomorphous molecular configurations—similar to those assigned to the two lactic acids—because they exhibit properties of an enantiomorphous character.

A very beautiful experiment which the late Sir G. Gabriel Stokes devised and doubtless exhibited in this room may be so modified as to serve for the demonstration of optical activity. Stokes' experiment consists in passing a plane polarised beam of light through a tall cylinder containing water which has been rendered very slightly turbid by the addition of a little alcoholic solution of resin; a spectrum is then seen spread out in the column of liquid, and spread out in a way which is not enantiomorphous—the water possessing no optical activity. The modification of Stokes' experiment consists in replacing the non-enantiomorphous water by some enantiomorphous liquid—conveniently by a 70 per cent. aqueous solution of the dextro-rotatory cane-sugar or by a 50 per cent. solution of the lævo-rotatory fruit-sugar; on making this change, it is seen that instead of the spectrum lying in the cylinder * vertically and therefore non-enantiomorphously, it winds spirally or corkscrew-wise round the column of the enantiomorphous liquid. These spirals or helices are clearly enantiomorphous and the two liquids of opposite optical activity give rise in this experiment to oppositely wound spirals—to spirals which are related to each other like the right- and left-handed corkscrews shown in the lantern slide. The opposite sign of the rotatory power exhibited by the cane-sugar and the fruit-sugar solutions is even more clearly shown by rotating the polarising prism in its mount, when the two spirals wind in opposite directions.

Although cases of optical activity are very frequently met with among chemical substances of animal or vegetable origin, it must be noted that no purely laboratory product or substance prepared without the use of enantiomorphous operations or materials is, in the ordinary way, optically active. The reason of this needs but little seeking if the solid models are once more consulted. Starting with a non-enantiomorphous substance is equivalent to starting with a methane derivative of the constitution—

and replacing one of the two X groups by the group Q, so as to obtain a compound containing an asymmetric carbon atom. Obviously,

---

* In performing this experiment it is convenient to use a glass cylinder about 4 feet in length and 2½ inches in diameter, and to throw the polarised beam vertically through the column of liquid by means of a 45° prism.

unless some power of selection of an enantiomorphous nature is exercised in replacing X by ꞯ, the doctrine of chance will ensure the one X group being replaced the same number of times as the other in an enormous number of tiny molecules. Thus there will result just the same amount of the right-handed optically active substance as of its left-handed isomeride. When an optically active substance is prepared in the laboratory it is therefore obtained as a mixture of two enantiomorphously related isomerides; such a mixture is said to be compensated, because the right-handedness of the one component is just counterbalanced by the left-handedness of the isomeric constituent. These compensated substances are represented by the third tetrahydroquinaldine previous referred to but not further discussed.

Since one of the great problems with which chemistry is grappling involves the synthetic preparation of naturally occurring optically active substances, it is of the utmost importance that the chemist should be in possession of working methods for resolving these compensated mixtures into their optically active components. All kinds of methods applicable to such resolutions necessarily involve the introduction of enantiomorphism—either of method or of material. Three types of method were introduced by Pasteur, namely, (1) spontaneous resolution by crystallisation, (2) resolution by combination with optically active substances, and (3) resolution by the action of living organisms.

The first kind of method depends upon the fact that on crystallising a compensated substance it sometimes deposits crystals of the dextro- and the lævo-isomeride side by side, and of such size that they can be mechanically sorted. The enantiomorphous factor determining the separation in this kind of method is obviously the enantiomorphous intelligence which has the power of discriminating between right and left-handedness. This sort of method is, unfortunately, but rarely applicable, owing to the fact that two enantiomorphously related substances usually crystallise together in the form of a loose chemical compound.

The second kind of Pasteur method is applicable to the resolution of compensated acids and bases, and depends upon the following considerations. On combining a compensated basic substance, viz. a mixture of d-B and l-B,* with an optically active acid, say d-A, a mixture of two salts, namely, d-B, d-A and l-B, d-A, will be obtained. These salts, however, are not enantiomorphously related, as will be realised on substituting, for illustrative purposes, a hand for the base and a glove for the acid. The combination d-B, d-A will then be represented by a right hand in a right-handed glove, whilst the combination l-B, d-A will correspond to a left hand in a right-handed glove. The struggles of the left hand with the right-handed glove will not be a factor in determining the behaviour of the appropriately sorted right hand and right-handed glove. So also, the properties of the

* The prefixes dextro- and lævo- are conveniently abbreviated to the initials d- and l-.

substance d-B, d-A—its solubility, melting point, etc.—will be conditioned by an enantiomorphous relationship of quite a different order from that determining the corresponding properties of the salt l-B, d-A ; the solubilities, being determined by different factors, will naturally also differ, and the two salts will therefore be separable by crystallisation. The first resolution of a compensated base was effected by Ladenburg in 1885 and consisted in resolving the synthetic alkaloid coniine into its optically active components—one of which proved to be identical with the alkaloid contained in the juice of the hemlock—by crystallising it with d-tartaric acid. Since this date the methods of resolving compensated bases have been materially improved by the application of optically active acids derived from camphor for use in place of the dextro-tartaric acid, and an experiment in illustration can now be shown on the lecture table. On adding a solution of ammonium dextro-bromocamphorsulphonate to a solution of compensated tetrahydro-β-naphthylamine hydrochloride, a white crystalline precipitate of d-tetrahydro-β-naphthylamine d-bromocamphorsulphonate,—the salt d-B, d-A,—is thrown down, whilst the lævo-tetrahydronaphthylamine remains in the solution as its hydrochloride. The resolution in this, and in many other cases, can thus be very rapidly effected, and by still further applying the optically active camphorsulphonic acids a considerable extension of the original van't Hoff-Le Bel theory has become possible. These workers traced all cases of optical activity to the presence of an asymmetric carbon atom and deduced from their work the conclusion that the environment of the carbon atom in methane is a tetrahedral one. It is true that all the optically active substances which have yet been obtained from natural sources owe their optical activity to the presence of an asymmetric carbon atom, but it is important to note that by applying the second Pasteur method to the investigation of synthetic materials, compounds owing optical activity to the presence of asymmetric atoms other than those of carbon can be prepared.

Thus, ammonium iodide has the molecular composition NH_4I and, like methane, contains in its molecule four hydrogen atoms, which are replaceable by other atoms or groups of atoms ; on replacing these hydrogen atoms by the four groups of atoms or radicles, methyl, allyl, phenyl, and benzyl, a substance is obtained which is conveniently named methylallylbenzylphenylammonium iodide and has the following constitution—

$$CH_3 \diagdown \quad \diagup C_6H_5$$
$$N$$
$$C_3H_5 \diagup \quad | \quad \diagdown C_7H_7$$
$$I$$

On replacing the iodine atom in this molecule by an optically active group of atoms, viz. by the dextrobromocamphorsulphonic residue, two salts are obtained, each of which contains an optically active basic part and an optically active acidic part; these are salts of the kinds, d-B, d-A, and l-B, d-A, and can be separated by crystallisation from a convenient solvent, and after separation has been effected, each salt may be reconverted into the iodide. These regenerated iodides are found to be optically active in solution, and the conclusion is consequently drawn that optical activity is an attribute of the asymmetric pentavalent nitrogen atom as well as of the asymmetric tetravalent carbon atom. The optical activity of this substituted ammonium compound indicates that its molecule has an enantiomorphous configuration and is extended in three-dimensional space; the exact nature of this configuration is not yet known, inasmuch as a space arrangement of five groups is concerned, but the environment of the nitrogen atom in ammonium salts is clearly not a simple tetrahedral one. Just as enantiomorphism has been proved to be an attribute of the asymmetric nitrogen atom, we have also demonstrated that asymmetric tetravalent atoms of sulphur, selenium and tin give rise to optical activity; optically active substances having the constitutions shown below have been prepared, and we are thus well on the way towards obtaining a complete stereochemical scheme embracing all the elements.

It has been mentioned that optically active substances occur as such, rather than in the compensated form, in many animal and vegetable products, and also that when a substance containing an asymmetric carbon atom is prepared synthetically in the laboratory it is of necessity obtained in the compensated form or as a mixture in equal proportion of the dextro- and the lævo-isomerides. Taken together, these two facts have a very interesting bearing upon our speculations as to the origin of animal and vegetable life. Optically active substances have been isolated as products of the vital activity of all forms of animal or vegetable life which have been properly examined, but in spite of this, they are never obtained directly as laboratory products; some enantiomorphous influence has always to be exerted in their synthetic preparation, just as Pasteur applied enantiomorphism, either of method or of material, to the resolution of compensated substances. It was very strenuously argued by Professor Japp in his Presidential Address to the Chemical Section of the British Association in 1898, that no matter how successful we may be in reducing the problems relating to vital processes to mere questions of physics and chemistry, a residuum will always evade explanation by such means; this residuum will involve the discussion of the way in which the first enantiomorphous substance was resolved into its optically active components. This question involves the introduction of an enantiomorphous agency at some period during the evolutionary development of living matter. In attributing difficulty to the solution of this residuary problem, Dr. Japp implies that the enantiomorphous agency, the co-operation of which is essential, must be an intelligent agency. Let us ask ourselves whether the enantiomorphous agency premised is necessarily other than one acting fortuitously. The assumption of a fortuitously enantiomorphous agency is certainly all that need be made to explain the building up of many enantiomorphous systems. The dead universe itself, as we know it, is enantiomorphous, but this fact has never been regarded as a valid argument against the current hypothesis as to the cosmic origin of our planet. Some degree of obscurity is, however, introduced into the discussion of the primitive origin of the optically active substances now produced by animals and plants by the probability that ages of evolution have transformed the primeval optically active substance into multitudes of other and more complex products, have, in fact, accentuated the enantiomorphism to such an extent that physiological chemistry is now almost entirely the chemistry of enantiomorphous substances. If in any particular case, however, we can show that an optically active substance can be locally accumulated, by the aid of some enantiomorphous agency acting purely fortuitously, it will be clear that the formation of the first optically active substance was not necessarily the work of an intelligent enantiomorphous agency. Such a species of separation of an optically active substance from a compensated one can be readily brought about in the laboratory. Pasteur showed that on crystallising the sodium ammonium salt of compensated tartaric acid (racemic acid) at ordinary temperature, large crystals separate; each of these consists of the salt of one or other of the d- and l- tartaric acids, the separation being brought about by the first of the Pasteur methods. If one of these crystals be selected casually, without the exercise of any selective intelligence, and used as a nucleus for inducing the crystallisation of further large quantities of the original solution, it will cause the separation of salt of its own kind; and ultimately a large quantity of salt of one of the optically active tartaric acids can be accumulated as the result of the introduction of an enantiomorphous agency such as might act fortuitously in a non-living universe. The probability of such a fortuitous agency arising would naturally be far greater in a living universe.

Again, suppose that, at its origin, life was carried on non-enantiomorphously, and that it involved the consumption and the production only of non-enantiomorphous substances and of compensated mixtures; it may well be foreseen that a stage in development might arise when each individual, in view of the increasing complexity of his vital processes, would have to decide to use only the one enantiomorphous component of his compensated food and so evade an otherwise necessary duplication of his digestive apparatus.

Acting unintelligently or fortuitously, one half of the individuals would become dextro-beings whilst the other half would become lævo-individuals; the succeeding generations would thus be of two enantiomorphously related configurations. It is, however, very difficult to believe that the natural selective operations which have been instrumental in conducting living organisms to their present stage of development, would allow the perpetuation of this state of affairs for any considerable period; some fortuitous enantiomorphous occurrence would temporarily give the one configuration the advantage over the other, an advantage which would be quickly accentuated and would involve the permanent disappearance of the weaker configuration.

The kind of difficulties involved in the existence, side by side, of individuals of dextro- and of lævo-tendencies, may be shown by a simple illustration. There is no reason connected with human enantiomorphism why vehicular traffic should be forced to keep to one side of the road rather than to the other; as, however, the conditions of civilised life have gradually become more complex, economic reasons have arisen forcing us to make an enantiomorphous selection and in this country we arbitrarily force the traffic to keep to the left; other countries also make an arbitrary, and sometimes a different, selection. Even if, when legislation on this matter first became necessary, the population had been equally and obstinately divided upon the question of the rule of the road, we cannot doubt that by this time the difficulty would have been satisfactorily and finally settled by the extermination of one or other of the enantiomorphously inclined parties without the co-operation of any intelligent enantiomorphous agency.

I mentioned that Pasteur gave a third method for the resolution of compensated substances, a method depending upon the selection exercised by living organisms upon the enantiomorphously related components of the mixture. He found, for instance, on allowing the mould *Penicillium glaucum* to grow in a solution containing compensated tartaric acid, that the mould used the d-tartaric acid as a food-stuff and rejected the lævo-isomeride, which latter could ultimately be separated from the solution. The kind of method thus indicated has been applied with success in a great number of cases, and is, in the end, merely a special application of Pasteur's second method. During recent years a considerable change has taken place in our views concerning the action of the lower organisms upon their food-stuffs. It was formerly supposed, for example, that the fermentation of sugar by an ordinary beer yeast is a part of the vital process of the organism itself, that the sugar taken in as food by the organism is finally thrown out in the form of carbon dioxide and alcohol; it is now clear, however, that the formation of these two products is in no way a vital process. On triturating yeast with powdered quartz so as to shatter the cell walls, and expressing the pulp thus produced, Buchner succeeded in obtaining a solution which, when mixed with sugar solution converts the sugar into carbon dioxide and alcohol. This fermentation is therefore not a vital phenomenon but is a chemical action induced by some non-living substance contained in the expressed juice of the yeast cells. This substance—zymase—has been isolated in the solid state and belongs to the class of substances known as unorganised ferments or enzymes. Although many enzymes are known, each active in inducing the occurrence of some particular chemical change or changes, nothing is as yet known as to their molecular constitutions; ages of evolution have given such complexity to these substances that a century or less of chemical investigation has contributed practically nothing towards elucidating their nature.

During the investigation of great numbers of cases of animal and vegetable vital activity, instances of the action of enzymes have been found, the function of the enzyme being to bring about the molecular degradation and, in certain cases, the molecular complication, of more or less complex materials used or produced in the organism. As an example of molecular degradation due primarily to enzymic action, the action of zymase on grape sugar—d-glucose—may be quoted. In aqueous solution, one molecule of grape sugar becomes directly converted into two molecules of alcohol and two molecules of carbon dioxide, in accordance with the equation—

$$C_6H_{12}O_6 = 2C_2H_6O + 2CO_2,$$

by the enzyme zymase. The enzyme itself suffers no permanent change as a result of exercising the power of causing this chemical reaction to take place, so that a comparatively minute quantity of the enzyme, acting for a more or less prolonged period, is able to convert an unlimited quantity of grape sugar into alcohol and carbon dioxide. The power which the enzyme possesses of inducing the occurrence of some chemical reaction which otherwise does not take place is not peculiar to enzymes; many substances, which are all classed together as the so-called catalytic agents, are known to exercise the same sort of influence in assisting a chemical reaction to occur. Thus the action of finely divided platinum in causing certain inflammable gases to ignite in air at the ordinary temperature is a catalytic action. The particular function exercised by enzymes in animal or vegetable life consists in bringing about chemical change quietly and continuously without necessitating the application of any violent chemical effects such as we are in the habit of using in the laboratory. Although they proceed so quietly, the chemical changes thus effected are, in many cases, changes which we have not yet succeeded in carrying out without the assistance of an enzyme; in the conversion of sugar into alcohol and carbon dioxide, zymase is performing a reaction which has never yet been brought about by the use of the ordinary laboratory methods.

Without quoting more specific instances, it may be generally stated that most of the cases of enzymic action hitherto investigated are cases in which a large molecular complex is degraded or broken down into substances of lower molecular weight. But it is important to note that the organism is also the seat of processes which result in the building up of very complex molecules from simpler ones, such for instance, as the formation of starch from carbon dioxide and water. A specific case in which enzymic action leads to the production of a complex substance from simpler ones has been recently worked out by Fischer and Armstrong, who show that the enzyme, lactase, converts the sugar galactose, $C_6H_{12}O_6$, into a new sugar, iso-lactose, $C_{12}H_{22}O_{11}$, of nearly twice the molecular weight of the former.

All the enzymes with which we are acquainted appear to be enantiomorphous bodies; they are perhaps substances to which no definite molecular composition can ever be assigned, inasmuch as they may be systems consisting of a number of different true chemical compounds, the system being one which becomes endowed with extraordinary chemical activity when placed in a suitable environment. The enantiomorphism of the enzymes has been repeatedly demonstrated during the course of Emil Fischer's remarkable synthetic work on the sugars. Fischer succeeded in preparing fruit sugar by purely synthetical methods as a mixture of the dextro- and the lævo-isomerides; in order to isolate the previously unknown l-fructose he applied the third Pasteur method in that he cultivated a yeast in the solution of the compensated fructose. The yeast enzyme—presumably zymase—has arrived at its present stage of development by passing through countless generations, all of which have been fed upon sugars of the dextro-configuration, these being the only ones occurring in nature. In Fischer's experiment, the enzyme therefore readily devoured the d-fructose but refused to touch the l-fructose, which had never before been presented to it. The l-fructose was of course subsequently isolated from the solution. The need for compatibility between the enzyme and the material upon which it has to act is very elegantly illustrated by considering the effect of yeast upon a number of optically active and isomeric sugars. In the table are given the constitutions of a number of sugars of the composition $C_6H_{12}O_6$, the configurations of the three or four asymmetric carbon atoms present in the molecule being indicated by writing the hydrogen atoms on the right or the left of the figure, as the case may be.

COH	CH₂OH	COH	COH	COH
H·C·OH	C:O	HO·C·H	H·C·OH	HO·C·H
HO·C·H	HO·C·H	HO·C·H	HO·C·H	HO·C·H
H·C·OH	H·C·OH	H·C·OH	HO·C·H	HO·C·H
H·C·OH	H·C·OH	H·C·OH	H·C·OH	H·C·OH
CH₂OH	CH₂OH	CH₂OH	CH₂OH	CH₂OH
d-Glucose.	d-Fructose.	d-Mannose.	d-Galactose.	d-Talose.

The beer yeast ferments d-glucose, d-mannose and d-fructose, each of which contains a similar set of three asymmetric carbon atoms in the molecule, with about equal facility; d-galactose is, however, only fermented with difficulty—of its four asymmetric carbon atoms, one differs in configuration from the corresponding one in the easily fermentable d-glucose; d-talose, in which two of the asymmetric carbon atoms differ in configuration from the corresponding atoms in d-glucose, is quite unaffected by the yeast. It is just as if the enzyme were provided with hands which enable it to grip the sugar molecule before tearing it to pieces; with these hands it grips the corresponding hands of, and so obtains a firm hold upon, the molecules of the first three sugars. The enzyme can only, however, grip the d-galactose molecule by two hands and so obtains a less firm hold. Owing to the greater incompatibility between the zymase and the d-talose the former obtains too feeble a hold on the latter to enable it to make a successful assault, and the sugar therefore remains unfermented.

The fact that the chemical reactions of animal and vegetable physiology consist in the main, of the production or destruction of optically active substances through the agency of enantiomorphous enzymes, is one of enormous importance. The complex substances concerned, such as starches, albumins and food-stuffs generally, occur in nature in but one of the enantiomorphously related configurations; all the albumins are lævo-rotatory, all the starches and sugars are derived from dextro-glucose. Since Fischer's work teaches us that none of the sugars derived from lævo-glucose are fermentable by yeast, it would seem to follow as a legitimate conclusion that, whilst d-glucose is a valuable food-stuff, we should be incapable of digesting its enantiomorphously related isomeride, l-glucose. Humanity is therefore composed of dextro-men and dextro-women. And just as we ourselves would probably starve if provided with nothing but food enantiomorphously related to that to which we are accustomed, so, if our enantiomorphously related isomerides, the lævo-men, were to come among us now, at a time when we have not yet succeeded in preparing synthetically the more important food-stuffs, we should be unable to provide them with the food necessary to keep them alive.

[W. J. P.]

GENERAL MONTHLY MEETING,

Monday, May 4, 1903.

Sir James Crichton-Browne, M.D. LL.D. F.R.S., Treasurer and
Vice-President, in the Chair.

The Chairman announced that His Grace the President had
nominated the following Vice-Presidents for the ensuing year:—

Sir Benjamin Baker, K.C.B. K.C.M.G. LL D. D.Sc. F.R.S.
Sir Frederick Bramwell, Bart. D.C.L. LL.D. F.R.S.
The Rt. Hon. The Earl of Halsbury, M.A. D.C.L. F.R.S.
Donald William Charles Hood, C.V.O. M.D.
The Rt. Hon. Lord Lister, O.M. M.D. D.C.L. LL.D. D.Sc. F.R.S.
George Matthey, Esq., F.R.S.
Sir James Crichton-Browne, M.D. LL.D. F.R.S. *Treasurer.*
Sir William Crookes, F.R.S. *Honorary Secretary.*

William Macdonald Bird, Esq.
Sir Edward Hamer Carbutt, Bart., J.P. M.Inst.C.E.
Constantine Craies, Esq.
Harold Malcolm Duncan, Esq.
Alfred Pearce Gould, Esq., F.R.C.S.
John C. Hawkshaw, Esq., M.A. M. Inst. C.E.
J. G. Hossack, Esq.
Captain Henry Arthur Johnstone,
Lieut.-Col. W. R. Murphy, D.S.O.
Major George J. W. Noble,
Mrs. George J. W. Noble,
Heseltine Augustus Owen, Esq.
Charles John Radermacher, Esq.
J. B. L. Stilwell, Esq.
Robert Græme Watt, Esq.
Arthur Charles Wombwell, Esq.

were elected Members of the Royal Institution.

The Special Thanks of the Members were returned to Mrs. Frank
Lawson for her Donation of £50 to the Fund for the Promotion of
Experimental Research at Low Temperatures.

The PRESENTS received since the last Meeting were laid on the table, and the thanks of the Members returned for the same, viz.:—

FROM

The Secretary of State for India—
 Geological Survey of India—
  Memoirs, Vol. XXXIII. Part 3. 8vo. 1902.
  Archæological Survey of Western India, Vol. IX. Northern Gujarat. 4to. 1903.
Abderhalden, E. Esq. (*the Author*)—Hydrolyse des Krystallisirten Oxyhämo-globins aus Pferdeblut. 8vo. 1903.
  Hydrolyse des Edestins. 8vo. 1903.
  Hydrolyse des Krystallisirten Serumalbumins aus Pferdeblut. 8vo 1902.
  Oxydation von Eiweiss mit Permanganat. 8vo. 1903.
Accademia dei Lincei, Reale, Roma—Classe di Scienze Fisiche, Matematiche e Naturali. Atti, Serie Quinta: Rendiconti. 1o Semestre, Vol. XII. Fasc. 1. No. 6. 8vo. 1903.
American Academy of Arts and Sciences—Proceedings, Vol. XXXVIII. Nos. 10–18. 8vo. 1902.
American Geographical Society—Bulletin, Vol. XXXV. No. 1. 8vo. 1903.
Asiatic Society, Royal—Journal for April, 1903. 8vo.
Astronomical Society, Royal—Monthly Notices, Vol. LXIII. No. 5. 1903. 8vo.
Automobile Club—Journal for April, 1903.
Bankers, Institute of—Journal, Vol. XXIV. Part 4. 8vo. 1903.
Basel, Naturforschende Gesellschaft—Verhandlungen, Band XV. Heft 1; Band XVI. 8vo. 1903.
Belgium, Royal Academy of Sciences—Bulletin, 1903, Nos. 1, 2. 8vo.
  Mém. Cour. et autres Mém. Tome LXII. Fasc. 4; Tome LXIII. Fasc. 1–3. 8vo. 1903.
  Mém. Cour. et des savants étrangers, Tome LIX. Fasc. 4; Tome LXII. Fasc. 2. 4to. 1903.
Boston Public Library—Monthly Bulletin for April, 1903. 8vo.
British Architects, Royal Institute of—Journal, Third Series, Vol. X. No. 12. 4to. 1903.
British Astronomical Association—Journal, Vol. XIII. No. 6. 8vo. 1903.
  Memoirs, Vol. XII. Part 1. 8vo. 1903.
British South Africa Company—Reports on the Administration of Rhodesia, 1900–1902, with Appendix. 4to. 1903.
Buenos Ayres, City—Monthly Bulletin of Municipal Statistics, Feb. 1903. 4to.
Cambridge Philosophical Society—Proceedings, Vol. XII. Part 2. 8vo. 1903.
Chemical Industry, Society of—Journal, Vol. XXII. Nos. 7, 8. 8vo. 1903.
Chemical Society—Proceedings, Vol. XIX. Nos. 264, 265. 8vo. 1903.
  Journal for May, 1903. 8vo.
Columbia, Republic of (*Ministerio de Relaciones Exteriores*)—Anales Diplomaticos y Consulares, Tome II. 8vo. 1901.
Coste, Eugene, Esq. (*the Author*)—The Volcanic Origin of Natural Gas and Petroleum. 8vo. 1903.
Cracovie Academy of Sciences—Bulletin, Classe des Sciences Mathématiques, 1903, No. 2; Classe de Philologie, 1903, Nos. 1, 2. 8vo.
Editors—Aeronautical Journal for April, 1903. 8vo.
  American Journal of Science for April, 1903. 8vo.
  Analyst for April, 1903. 8vo.
  Astrophysical Journal for April, 1903. 8vo.
  Athenæum for April, 1903. 4to.
  Author for May, 1903. 8vo.
  Board of Trade Journal for April, 1903. 8vo.
  Brewers' Journal for April, 1903. 8vo.
  Chemical News for April, 1903. 4to.

Editors—continued.
Chemist and Druggist for April, 1903. 8vo.
Electrical Engineer for April, 1903. fol.
Electrical Review for April, 1903. 4to.
Electrical Times for April, 1903. 4to.
Electricity for April, 1903. 8vo.
Engineer for April, 1903. fol.
Engineering for April, 1903. fol.
Feilden's Magazine for April, 1903. 8vo.
Homœopathic Review for April, 1903. 8vo.
Horological Journal for May, 1903. 8vo.
Journal of the British Dental Association for April, 1903. 8vo.
Journal of State Medicine for April, 1902. 8vo.
Law Journal for April, 1903. 8vo.
London Technical Education Gazette for April, 1903.
London University Gazette for April, 1903. 4to.
Machinery Market for April, 1903. 8vo.
Model Engineer for April, 1903. 8vo.
Mois Scientifique for April, 1903. 8vo.
Motor Car Journal for April, 1903. 8vo.
Motor Car World for April, 1903. 4to.
Musical Times for April, 1903. 8vo.
Nature for April, 1903. 4to.
New Church Magazine for May, 1903. 8vo.
Nuovo Cimento for Feb. 1903. 8vo.
Page's Magazine for April, 1903. 8vo.
Photographic News for April, 1903. 8vo.
Physical Review for April, 1903. 8vo.
Popular Astronomy for April, 1903. 8vo.
Public Health Engineer for April, 1903. 8vo.
Travel for April, 1903. 8vo.
World's Fair Bulletin for April, 1903. 4to.
Zoophilist for April, 1903. 4to.
Electrical Engineers, Institution of—Journal, Vol. XXXII. Part 2. 8vo. 1903.
Florence, Biblioteca Nazionale—Bulletin for April, 1903. 8vo.
Franklin Institute—Journal, Vol. CLV. No. 4. 8vo. 1903.
Geographical Society, Royal—Geographical Journal for May, 1903. 8vo.
Geological Society—Quarterly Journal, Vol. LIX. Part 1. 8vo. 1903.
  Abstracts of Proceedings, No. 776. 1903.
Johns Hopkins University—American Journal of Philology, Vol. XXIII. No. 4 8vo. 1903.
  University Circulars, Vol. XXII. No. 162. 4to. 1903.
  University Studies, Series XX. Nos. 2-12, and Extra No. 8vo. 1902.
Leighton, John, Esq. M.R.I.—Ex-Libris Journal for April, 1903. 8vo.
Linnean Society—Journal, Botany, Vol. XXXVI. April, 1903. 8vo.
Manchester Literary and Philosophical Society—Proceedings, Vol. XLVII. Part 3. 8vo. 1903.
Manchester Steam Users' Association—Boiler Explosions Acts, 1882 and 1890, Reports, No. 1320-1387. fol. 1902.
Mechanical Engineers, Institution of—Proceedings, 1902, No. 4. 8vo. 1903.
  List of Members, 1903. 8vo.
Meteorological Society, Royal—Meteorological Record, Vol. XXII. No. 87. 8vo. 1903.
Mexico, Imperial Geological Institute—Bulletin, No. 16. 4to. 1902.
Mexico, Sociedad Cientifica "Antonio Alzate"—Memorias, Tome XVII. Nos. 4-6; Tome XVIII. Nos. 1, 2; Tome XIX. No. 1. 8vo. 1902.
Microscopical Society, Royal—Journal for April, 1903. 8vo.
Navy League—Navy League Journal for April, 1903. 8vo.
New Jersey Geological Survey—Annual Report for 1902. 8vo. 1902.

New South Wales, Agent-General for—The Seven Colonies of Australia. By T. A. Coghlan. 8vo. 1902.
New Zealand, Registrar-General—Report on the Census of New Zealand, 1901. 4to. 1902.
Odontological Society—Transactions, Vol. XXXV. No. 4. 8vo. 1903.
Onnes, Professor H. K.—Communications, No. 81. Supplement, Nos. 4, 5. 8vo. 1903.
Pennsylvania University—
    Publications: Astronomy, Vol. II. Part 1. 4to. 1903.
        University Bulletin, Third Series, No. 2. Parts 1, 2. 8vo. 1903.
Peru, Cuerpo de Ingenieros de Minas—Boletin, 1902, No. 2. 8vo.
Pharmaceutical Society of Great Britain—Journal for April, 1903. 8vo.
Photographic Society, Royal—Photographic Journal for March 1903. 8vo.
Pickering, E. C. Esq. (the Author)—A Plan for the Endowment of Astronomical Research. 8vo. 1903.
Quekett Microscopical Club—Journal for April, 1903. 8vo.
Righi, Professor A. (the Author)—Sulla Ionizzazione dell'Aria. 4to. 1903
    Il Moto dei Ioni. 8vo. 1903.
Royal Irish Academy—Transactions, Vol. XXXII. Section B, Part 2. 4to. 1903.
Royal Society of Edinburgh—Proceedings, Vol. XXIV. No. 4. 8vo. 1903.
Royal Society of London—Philosophical Transactions, B, No. 214. 4to. 1903.
    Proceedings, No. 473. 8vo. 1903.
    Year Book, 1903. 8vo.
Sanitary Institute—Journal, Vol. XXIV. Part 1. And Supplement, 8vo. 1903.
Scottish Society of Arts, Royal—Transactions, Vol. XV. Part 4. 8vo. 1903.
Selborne Society—Nature Notes for April, 1903. 8vo.
    Annual Report, 1902-3. 8vo. 1903.
Society of Arts—Journal for April, 1903. 8vo.
Stonyhurst College Observatory—Results and Observations, 1902. 8vo. 1903
Sweden Royal Academy of Sciences—Öfversigt, Vol LIX. 8vo. 1902-3.
Tacchini, Prof. P. Hon. Mem. R.I. (the Author)—Memorie della Società degli Spettroscopisti Italiani, Vol. XXXII. Disp. 3. 4to. 1903.
Toronto, University of—
    Studies: Biological Series, No. 3. 8vo. 1902.
        Geological Series, No. 2. 8vo. 1902.
        Psychological Series, Vol. II. No. 1. 8vo. 1902.
United Service Institution, Royal—Journal for April, 1903. 8vo.
United States Department of Agriculture—Annual Summary, 1902. 4to. 1903.
    Monthly Weather Review, Jan. 1903. 4to.
United States Patent Office—Official Gazette, Vol. CIII. Nos. 5-8. 8vo. 1903.
Verein zur Beförderung des Gewerbfleisses in Preussen—Verhandlungen, 1903. Heft 4. 8vo.
Vienna, Imperial Geological Institute—Verhandlungen, 1903, Nos. 2-4. 8vo.

WEEKLY EVENING MEETING,

Friday, May 8, 1903.

Sir James Crichton Browne, M.D. LL.D. F.R.S., Treasurer and
Vice-President, in the Chair.

H. Rider Haggard, Esq.

Rural England.

The subject upon which I have the honour to address you to-night is
so vast that I confess I hardly know how it should be approached.
Recently I have found two years of incessant labour very inadequate
to the collection and consideration of the body of facts which are
treated of in my work "Rural England," and some twelve hundred
pages of close print scarcely sufficient to their record.

How then am I to deal with them intelligently under the same
great title within the space of a single hour?

I must confine myself to certain aspects of the subject—that is
clear—and even so, ask you to forgive me if I omit much.

In those aspects are involved the difficult and debated question of
small-holdings; their relation to national health and well-being;
the methods which would best promote their establishment and
multiplication, and their working in practice illustrated by certain
examples with which I am acquainted. Also, if time allows, I should
like to say a few words on the present position of the agricultural
labourer, the very important matter of rural housing, and the pallia-
tives that I propose for some existing evils.

Speaking broadly, although of course its physical conditions are
the same, the England of the past must have been a very different
country from that with which we are acquainted to-day. Then, the
land was the main strength of Britain, now commerce is its strength.
Then, except for occasional foreign wars, our energies were concentrated at home, now they are diffused over every quarter of the globe.
Then, Empire abroad was but a dream—in the beginning not so much
as that—now, although some argue that it brings no adequate material
advantage to our toiling millions, we think and talk of little else, and
in a sense are lost—as well we may be—in the contemplation of the
glory of a world-wide rule. Then, we grew the food we ate, and at
times even exported some, now we import three parts of it, and, in the
event of a few defeats at sea, must begin to starve within a dozen
weeks. Then, we were land-dwellers and the towns were small; now
the great bulk of us are city-dwellers, most of whom know nothing
of the land, and, as it does not concern their purses or their plenty,
are absolutely indifferent to its welfare. Then, the country side was

occupied by small gentry, yeomen and peasant-holders, all of whom in the majority of districts have disappeared, to be replaced by large owners, tenant farmers, and an ever-lessening number of agricultural labourers working for a weekly wage with but little prospect of raising themselves in the world.

Quite recently—by which I mean to within the last forty years and later—thousands of families of each of which the head for the time being was called "Squire," lived from generation to generation upon their own properties and, even where the estates were small— say from one to two thousand acres—in most cases without extraneous resources maintained a certain position with sufficiency, and fulfilled its duties on the whole to their own credit and to the advantage of the country. Now members of this class are rare. Rents have fallen, charges have increased ; more is expected of the minor gentry, who can no longer live in the old simple fashion; mortgages, jointures and portions, often based upon a too liberal estimate of the productive powers of the estate, have accumulated. The result is that these little squires have vanished, or, where they still remain, support themselves upon outside means. I can recall none of moderate estate who still exist upon the revenues of that estate alone, except perhaps a few in Yorkshire and in the richest parts of southern and western England, whereas even those who own large properties and nothing else, are frequently much crippled.

The yeomen, too, who once formed the very backbone of the nation, have gone. In the prosperous days many of them were bought out at tempting prices by their richer neighbours, in order that their parcels of land might be absorbed into the great estates. In the evil years that followed, which began about 1879, the majority of those who remained were starved out, with the result that the yeoman farmer living on and out of his own acres is now hard to find. In all my wanderings through England I met but very few of them.

The class of agricultural labourers is in much the same case. Thus, as is pointed out in the final report of the Royal Commission on Agriculture published in 1897, the population of Great Britain between the years 1871 and 1891 increased by nearly seven millions, whereas the agricultural labourers of Great Britain during the same period decreased by about 242,000. I have not at hand the figures for the decade between 1891 and 1901, and am not certain whether they are as yet fully available, but if I may trust to the results of my own knowledge and observation, I believe that decrease to have been progressive. I read in fact, that in England and Wales their number has further dwindled by about 160,000. Also I think that I am right in saying that this last Census shows a shrinkage of the population in no less than 401 of the rural districts of England and Wales. When it is remembered that in 1891 about 77 per cent. of their inhabitants were already living in cities or their suburbs, as against 23 per cent. living in the country, it will be admitted that this fresh depletion of the villages is a very significant fact. A lessening of the number of agricultural labourers and shepherds from 1,253,786 in 1851 to 620,000 in 1901 (which I understand to be the figures) can scarcely be ignored.

Of course it may be answered that even in districts that are purely rural in character and include no towns of any importance, more folk dwell to-day than were to be found there in the past. But to this the rejoinder is that such matters should surely be judged proportionately. Also I incline to the belief that the rural population of England—say in the middle ages and after them—was more numerous than is commonly supposed.

Any observant traveller in Northamptonshire and other midland counties will have noticed that many tens of thousands of acres which are now under grass, have at some time in our past history been beneath the plough, as is shown by the endless high-ridged "lands" or "stetches" of an equal width. Now even to-day with our improved implements, to cultivate so much arable would require a far larger population than is to be found in the tiny villages of the grass counties. What, then, must it have required when ploughs were cumbersome wooden instruments, dragged, after the South African fashion, by a team of six or more oxen which would probably need a leader as well as a ploughman? Of course we have no sure means of knowing at what period this great extent of country ceased to produce corn and began to produce grass. That, at any rate in certain instances, this was a long time ago I have, however, proved to my own satisfaction, thus :—

It occurred to me that as a plough cannot be dragged through the trunk of a timber, if I could find trees growing upon the exact crests of continuous and undoubted plough-lands now under grass, it would show that those particular "lands" were already under grass when such trees seeded themselves, or were planted. Acting upon this thought I searched diligently. In Oxfordshire I found trees of about 200 years standing in the required position, and showing therefore that these fields where they grew, were grass two centuries ago. In Northamptonshire I found trees of at least four hundred years similarly situated, showing that four centuries ago that land was grass. Finally on a farm about ten miles from York, standing in a pasture upon the very crest of a distinct and undoubted "land," I discovered an ancient pollard oak that in my judgment—and I have given some attention to the age of timber—must have sprung from the acorn at least seven centuries ago, showing that about the year 1200 this particular field was already under grass. Of that oak a photograph, taken at the time, may be found in the second volume of my work, ' Rural England.'

On the whole I incline to the belief that it was after the Black Death that the bulk of these districts went down naturally to grass, and that before this the population upon them was comparatively dense. It is, however, possible, since the shape of heaped-up earth varies little, even in the course of many centuries, that all this cultivation may be attributed to a far earlier period, perhaps even to Roman, or pre-Roman days.

To come to more modern instances of a population that has vanished, I will quote two examples which I was fortunate enough to be able to discover. The first of these was demonstrated by a map that I saw of the parish of Feckenham in Worcestershire, which was executed by the order of Queen Elizabeth at the date of the Disforestation in 1591—or to be accurate, I saw a copy of this map made by one John Dobarte in 1744. According to this chart, in the days of Elizabeth 3000 out of about 7000 acres which compose the parish, were held by no fewer than sixty-three different owners. To-day nearly the entire parish—namely 6000 acres of it—is held by six owners. It is a curious case of the passing away of the land from the people into the hands of a single class.

In the parish of Western Colvile in Cambridgeshire I came across an even more remarkable example. Here I was shown a large detailed map executed in 1612 which had been found by the owner of the land hidden away in a cottage. This map tells us in unmistakeable fashion that in 1612 about 2000 acres of the area of Western Colvile were held by some three hundred different owners—say an average of six acres to an owner. Now the said 2000 acres, over which I drove, is, I think, owned by one person and occupied by three, and the tiny strips into which it was divided are merged into vast, lonely fields.

How, even in the time of James I., when wants were few, these three hundred owners earned a living, with the aid of but little manure, from soil so light that in places it is almost a blowing sand, is a mystery that I do not pretend to solve. I suppose, however, that they forced the land to bear sufficient corn and meat for their sustenance and that of their families, and sufficient wool for their clothing. It must not be supposed that I suggest that these small holders of the past always, or indeed often, lived in plenty or comfort. My belief, on the contrary, is that their existence must always have been hard, and in bad seasons their lot one of great misery. Still they *did* live, and in considerable numbers, in places where the population is sometimes scant enough to-day.

Now, owing to many causes (among which may be numbered our system of primogeniture, the working of the Enclosures Acts, the introduction of Free Trade, which has made agriculture in England when practised on a small, and indeed on a large scale, a somewhat doubtful and unprofitable business), they have for the most part vanished. Nor has the land and the industry of its cultivation prospered of late years. According to an estimate arrived at by the Royal Commission on Agriculture, the value of that land in the United Kingdom decreased between the years 1875 and 1895 by some 834,000,000*l.*, or 50 per cent., although since then it may again have risen a little. Further, I think that it produces to-day something less than half the wheat that it produced in 1850, whilst, though the grass area has largely increased, the head of stock kept upon it seems to he diminishing, since arable will carry more cattle and sheep than does pasture. A good deal of the poorer soil also is now so badly tilled by an impoverished race of farmers, that it might almost as well be out of cultivation. Lastly, the character of the husbandry throughout England, or at least in most of the twenty-seven counties in which I pursued my investigations, has, I believe, in the main deteriorated. Of course I speak with many exceptions, but tenant farmers are not what they were thirty years ago. Their capital is not so large, their energy and enterprise are often less ; notwithstanding the great fall in rents, their prosperity on the whole has dwindled. Farms still let, at a price—indeed, the market for them is rather better than it has been, for to the man with a little capital, inherited, acquired by marriage, or even borrowed, the independent life of a farmer offers many advantages. But few of them make more than a living, which, after all, is as much as most people can expect now-a-days; and some—often, it is true, through their own fault—do less than that. Still the tenant of land has not been so hard hit as the owner of the land, who is oppressed by tithe, taxes and death duties, all of which he must meet out of an enormously lessened income. In some counties his acres are now frequently let for not more than a fair interest on the capital value of the improvements effected on them by himself or his predecessors, that is to say, the land itself is thrown in for nothing—which I take it, is something smaller than that prairie-value whereof we hear so much. And all this, be it remembered, has come about in a country full of the finest markets in the world, that clamours for produce and imports it from abroad to the value of tens of millions of pounds annually; a country moreover, where irrigation is not needed, and of which the soil, if fairly treated, remains of unimpaired fertility. Could any development of events be more bewildering, more inexplicable ? Yet such are the facts as I have found them, and if I do not know nearly as much about these matters as any other single man in England, it is my own fault.

What is the reason of this strange state of affairs to which there seems to be no obvious and sufficient answer ? Is it national senility? Is it that agricultural prosperity is a prerogative of young peoples ? China, Egypt, and to some extent, India, seem to contradict such a theory, but the East has always been different from the West. Of course, ninety-nine British farmers out of a hundred would shout " Free Trade " with a single voice, and doubtless this is a product of high civilisation which is always at war with a primitive art such as agriculture. But I think that we must look further afield, since even from other European countries where protection is still in force agricultural prosperity seems to be slipping away.

For instance, has not our system of large estates something to do with the matter, our rooted and almost ineradicable notion that the holding of land is mainly a right of the rich, to be used by them for the purposes of pleasure and not of profit, for the shooting of pheasants, the hunting of foxes, and as sites for " the stately homes of England," and so forth ?

Would not matters be improved if more individuals held less land and worked it themselves to earn a living out of it, thus getting rid of one of the three profits that it has to bear ? If, in short, instead of supporting their acres by extraneous means, owners were forced, as in the past history of England, to make those acres support them ?

This is not at all the feudal idea that still has so firm a hold upon our national life, which finds perfect expression in such a document as the will of that great man, the late Mr. Cecil Rhodes, who therein defined " country landlords " as a class " who devote their efforts to the maintenance of those on their own property," I suppose out of means acquired otherwise than from their own property —in English land.

Ought they not rather to devote their efforts to enabling those on their estates to support themselves without the aid of any extraneous or artificial assistance ? In fact, is not all this lingering feudalism— built, as it must be in our conditions, upon a false basis, and often maintained, where it still exists, with money drawn from trade, or the Stock Exchange, or American heiresses, or African mines—a grave mistake ?

The question is large, and, I admit, difficult to answer, yet it may —at any rate to some extent—be dealt with by contrast. When I was engaged upon my researches, I visited Jersey and Guernsey, places with some agricultural advantages, though these are by no means so pronounced as is perhaps generally imagined.

Now in the Channel Islands the land system is totally different from our own.

Thus: Real property, on the death of the head of a family, must be divided among all the children in certain proportions, the eldest son by virtue of his prior birth having only an extra right to the principal house and some twenty perches of land. Consequently there are a multitude of small freeholders, the average holding in Guernsey amounting to about one acre only, and in Jersey to fifteen or twenty, many there being under ten and few over fifty. If we are to believe half of what is preached to us as to the disadvantages of such holdings, this state of affairs should mean misery to all concerned.

And yet what are the facts, unless they have changed since 1901 ? In Guernsey, land to be used for agricultural purposes fetches as much as 500*l.* an acre, and lets at from 4*l.* to 9*l.* the acre. In Jersey I saw a little farm of about twenty-three acres which had been sold not long before, also for agricultural purposes, for 5760*l.* Compare this state of affairs with that which prevails—or prevailed at that date—in Wiltshire, where land sells often enough for 6*l.* the acre, and lets at from 5*s.* to 15*s.* the acre. Yet, in the Channel Islands, the people are as a whole, extraordinarily prosperous, and often accumulate fortunes.

Can this be said to be the case in Wiltshire ?

Of course it will be objected that the circumstances of these islands are exceptional, inasmuch as Guernsey chiefly depends upon glass, and Jersey upon early potatoes. But Guernsey was prosperous before there was a greenhouse in the island, and I believe that the same may be said of Jersey before it grew a potato. Also, to say nothing of France, early potatoes, which to a great extent, compete with those of Jersey, can be grown in Cornwall, in Cheshire and elsewhere : the island has no alsolute monopoly in that business.

As for glass, the whole of the south of England might be covered with it, in competition with Guernsey, over which it has, indeed, enormous advantages in the shape of cheaper coal for the greenhouse boilers, and freedom from sea-carriage. Yet it is in Guernsey that most money is made from these houses ; in Guernsey that folk can begin as labourers and end in the possession of fortunes—no infrequent occurrence there, as I was assured.

We do not often hear of such things among the agricultural labourers of England. Moreover, if this good fortune is exceptional, the general average of prosperity is extraordinarily high in the Channel Islands. The people who, according to theory, should be plunged in wretchedness because of their system of land-division, live in great comfort, for the most part in houses of their own which are a pleasure to look at. Of course, they work hard—very hard, as men do who work for themselves and not for a master. But then they reap the reward of their work in ownership, increased comfort and resources.

Now does not all this suggest that there is perhaps something to be said for the tenure of the land by the people instead of by a very limited class, and that it is better to cultivate one acre thoroughly, making it produce as much as it possibly can, than to treat ten or twenty in the fashion that is common enough in England?

Within the last few months, among many of the leaders of thought in this country, there has been much searching of spirit on the matter of our food supply in time of war, which is now seen to constitute a national peril of the gravest sort. Indeed, I myself am a member of a Committee or Association which has persuaded the Government to inquire into the whole subject. Various remedies have been proposed, of which two of the chief are the raising of the Navy at whatever cost to such a degree of strength that it could keep the seas open to our corn ships in any conceivable war ; and, as an alternative scheme, the establishment of national granaries in which vast quantities of grain could be stored, as Joseph stored it in Egypt, and the knightly orders stored it in the pits that may still be seen, and I think are still used, in Malta—a very ancient and primitive expedient.

Upon the question of these remedies I do not propose to enter further than to say, I believe that no navy we could build would effect this object, for the simple reason that ships of war cannot control the schemes and combinations of foreign, and I may add of home speculators in food-stuffs. The corn might be got into the country, but at what price would our millions be able to purchase it ?

As regards the storage question, I am convinced that no Government would face its expense in time of peace, while in time of war it would be too late for them to do so. Pharaoh and Joseph did this on the strength of a dream, but if they are ever visited by such visions, British ministries pay no attention to them. Rather do they wait until the facts have declared themselves in a fashion unmistakable by the humblest man in the street, and then, after the horse has been stolen, with much public pomp and solemnity proceed to lock the stable door. Therefore, unless there are other easy and practicable solutions with which I am unacquainted, it would seem that in this matter of the lack of food reserves, we must remain where we are and take our chance.

My reason for touching on the subject at all is to point out, however, that the easiest and most practicable solution of them all is never even alluded to, or if anybody has thought of it, is promptly dismissed as unworthy of serious consideration. It is—that we should grow the food, or most of it, ourselves.

I think it was that shrewd observer, Prince Kropotkin, who demonstrated that if the soil of this country were only cultivated as it is on an average cultivated in Belgium, we could produce as much as our people need to eat, and have some surplus over. Well, I believe —and perhaps you will allow the opinion some weight as that of a person who has long and earnestly investigated these questions connected with the land, and, what is more important here, the land itself —that Prince Kropotkin is right, at any rate to a large degree, and that if our acres were cultivated as they might be cultivated by the aid of sufficient energy, capital, intelligence and labour, the produce from them of all foodstuffs could be well-nigh doubled.

Allow me to give you a single example, which after all is worth more than many words of argument. I am myself a farmer, and I dare say my neighbours would tell you that I farm as well as my bailiff will allow me to do, which means that to my own fancy I might farm a great deal better. Now at Ditchingham I let off to a small tenant at my gate, a builder by trade, a four-acre arable field of no better quality than the average of my land, and with it perhaps two acres of old pasture. When I took the farm in hand a dozen or more years ago, this field, like the rest, was left in very poor condition by the outgoing tenant. Indeed, I remember that my friend the builder showed me a gigantic pile of dock roots many yards in length by three or four feet in thickness and as much in height, which he had extracted from it at a cost, he said, of about 14*l.*, and was salting down to use as manure.

A couple of years ago this man came to consult me. It appeared that he had sold his wheat to a person in the neighbouring town who had promptly gone bankrupt and left him mourning for his money. I could only recommend him to console himself with appropriate reflections upon the uncertainty of all human affairs, especially where the recovery of debts is concerned, and then asked him how much

wheat he had thrashed out and handed over to his customer from the
the one acre that he had under that crop.

To my astonishment he replied—and afterwards I satisfied myself
of the truth of the statement—nine and a half quarters! Now I
consider five quarters of wheat to the acre a good crop, and six a
large one (indeed I do not think that I have ever grown so much);
nine and a half being of course phenomenal. Pursuing my investigations, I discovered that from a second acre of his field he had
secured eight quarters of oats—a splendid return; from a third acre
under clover, three tons of hay—nearly double what I average; and
from the fourth acre a very heavy crop of roots—of the weight of
which he had kept no exact record. If only I could show similar
returns from the four hundred acres or so that I farm, I at least
should have no cause to complain of agricultural depression, even in
these days of low prices.

How, then, is the mystery to be explained? Thus. The builder
farms his four acres of arable much better than I do my four hundred.
He goes in for no out-of-the-way crops. He uses no artificial manure,
only that produced by his few horses, pigs, fowls and young stock,
with the addition of soot, which I suppose he can purchase from the
sweep at 6d. the bushel; but of these dressings he does use plenty,
more in proportion than I do. Further, when building is slack his
two sons and his trade horses are continually at work upon the land,
in which not a weed is to be seen. Indeed I will go so far as to say
that if all Great Britain were cultivated as that small man cultivates
his four acres of plough and his two of pasture, the question of food
supply in time of war would no longer sit like a nightmare on the
national breast. For there it—or a great deal of it—would be in our
granaries and stockyards, or growing in our fields and pastures.

But how is it cultivated in fact? Let any one who understands
the question take a drive through certain counties that I could name
to you and judge for himself. There he may see thousands and tens
of thousands of acres which used to produce four or five quarters of
corn to the acre, gone down to a miserable apology for pasture.
Indeed this laying away of land to grass, which does not produce so
much meat as land under crop even if the pasture be good, is one of
the great features of our modern husbandry, and, although I am
obliged to practise it myself, I may add, one of the worst.

The result is that our output of foodstuffs, or so the Agricultural
Returns seem to show, is actually diminishing. It has been calculated that to produce the amount of food that we import, would
require the cultivation of another 23,000,000 acres yielding our
present average returns. These of course are not available. But
supposing that the production of the 47,800,000 acres now under
cultivation in the United Kingdom could be increased by nearly one-
half, it would seem that the same result would be attained—although
of course a larger area than at present would have to be devoted to
the growth of wheat.

Taking these figures as approximately correct, I ask whether such a result is beyond the bounds of possibility? In my opinion the answer is, No.

We are confronted, then, as I hope I have shown to your content, by two great national needs. The need of increasing the numbers of our rural population, and the need of increasing the productive powers of our land—which needs, in my opinion, are more or less dependent for their satisfaction upon each other.

How can the rural population be increased? How even can the exodus from the countryside be stayed?

To deal with this question, we must begin by considering the conditions—vastly improved of late years, it is true—under which the great majority of actual earth-tillers still exist in England. I will state them as briefly as I can.

Firstly, they live upon a weekly wage which must be earned by very hard and continuous manual labour, involving a great deal of skill which can only be acquired by years of experience. Nothing is more common than the idea that the agricultural labourer is a foolish, thick-headed person (generically known as "Hodge") whose work could be done by anybody. Let anybody try it. I have used my hands in my time, and know what it is to toil at some heavy monotonous labour for eight or ten hours at a stretch, but I should be sorry to attempt the daily tasks that the farm-hand carries out with so much skill, such as ploughing, hedge-laying, draining, stacking, or even manure-turning—I mention but a few of them.

In reward for these, our labourer, when he is not out of place, or sick which fortunately for him is seldom, receives a wage that, including his harvest and haysel, varies from about 16s. a week in the eastern counties, to 1l. or a little more in Yorkshire, the south of England and the districts near London. Out of this, unless he is a stock or horseman, who have their houses free, he must pay 1s. 6d. or 2s. a week for a cottage, the contribution to his sick-club, and support and clothe a family, often large and thriving. Obviously, therefore, he has few holidays, since holidays mean the loss of the daily half-crown, without which his accounts will not balance. In short, his life from childhood to the grave, is one of incessant, unvarying exertion, carried out in every sort of weather, with little change or recreation, and involving rising soon after the light in summer, and long before it in winter; coarsely prepared food, clothing that is not too ample; and perhaps an old age spent in the poor-house; to which must be added all the other anxieties and cares that are the common lot of humanity. Unless he is a very exceptional man, he has in most districts small prospect of rising in the world. Where he began, there he must end, after a lapse of half a century or so, only perchance a little lower down. Nobody took any interest in his birth, and nobody will take any interest in his funeral, nor outside of the parish register will any record remain of his existence upon this troubled sphere, except of course in the ditches that he has dug and the trees that he has planted—for the benefit of somebody else.

For some years I employed upon my farm a curious old fellow of the not uncommon name of Smith.   The other day I met a funeral and inquired whose it was, to be informed that Smith was inside the coffin.   I had seen him working away a few days before, but nobody had taken the trouble to inform me that since then he had fallen sick and died.   Apparently it was not thought a matter worthy of notice, and already somebody else occupied his place in the fields that he had trod for sixty years.   Thus exit Smith and all his class.

Their lot sounds somewhat depressing, although of course it has alleviations, especially that of the health with which an out-door life endows them.   Still it is not to be wondered at if, incited thereto by a board-school education imparted to them by town-bred teachers, they strive to escape from it to a town.   There wages are higher for the young and strong.   There are music-halls, processions and other glorious things, including water-taps, and pavements upon which the children may walk dryshod to school, and there, as they think, they may rise.   Of course, in the majority of instances, they are disappointed ; they do not rise, they sink.

Let those who doubt it consult Mr. Rowntree's study upon the urban population of York, which is not a very large city.   Even in the case of the fortunate amongst these immigrants, the extra money that they win is more than absorbed in the extra expenses of their new life, while the cottage that they inhabited in their native village, possibly tumbled-down enough—for the lack of good cottages is one of the great causes of the rural exodus—proves to be a paradise when compared to the one or two rooms in a London court for which they pay three times the rent.

These facts are generally recognised and need not be dwelt on— or at least I have no space to dwell on them.   So much is this so, that before I read you this paper I had promised to consult with the Committee of the Charity Organisation Society, as to what steps can be taken to transport immigrants from the land—or their children—back to the land.   But such folk look at the beginning and not at the end of things, as indeed all of us do in our degree, leaving the future to Fate, an opportunity of which Fate generally avails itself—to their disadvantage.

So for various reasons they go, and of those reasons, the chief is, in my opinion, a lack of prospects if they elect to stay at home.

What is the prospect that they require ?

I take it that of being able to lift themselves up in life, of being able to farm land on their own account, upon however small a scale, and of ending their days " living upright," as we say in Norfolk, that is, living on their own means.

Now is this an impossibility for the agricultural labourer ?

At present, in most cases—yes ; but I hold that it should not be so, and that it need not be so.   The man who is hard-working, provident and ambitious, ought to be able to raise himself in the world. You may ask how it is possible upon his scanty wages? I answer that where opportunities exist it is done, meaning by opportunities the chance of acquiring small holdings, either as owners or as tenants.

I will give you some examples. Epworth, in the Isle of Axholme, has an area of 5741 acres held by 294 occupiers, of whom 235 hold under 20 acres apiece, 115 holding under 10, and 80 under 2 acres. When I visited the place in 1901 nine of the largest farmers in it, working from 200 down to 40 acres respectively, were stated to have risen from the position of farm labourers to that which they held at that date, and the same facts obtain, as I was informed, amongst very many of the smaller owners and occupiers, both in Epworth itself and in the neighbouring villages.

Most of these men did not confine themselves to potato growing, for which the isle is famous, but followed a course of general farming, and according to my informants raised more grain per acre than the large farmers, quite an equal weight of beef, more pork and vegetables, but less mutton, small holdings being unadapted to sheep. It was also said that their little properties are not by any means so heavily mortgaged as is commonly supposed.

Another nest of small holders is to be found at Downham in Cambridgeshire. These for the most part hold their own land under a copyhold tenure, of which the fines payable on death or sale press heavily upon them and must generally be met by raising money upon mortgage. The result is that they have to work very hard to earn a living, harder indeed than the ordinary farm servant. Yet the results of my personal inquiries were that none of them would have been willing to exchange their lot for that of a labourer; that they did live, and reared large families; and that very few of the young men migrated to the towns, as most of them hoped in due course to hold land of their own.

To take a third instance, which is particularly interesting, since here the land is not remarkable for richness, being chiefly a thin black earth upon the chalk, and there existed no hereditary leaning towards small holdings. In 1888 Sir Robert Edgcumbe purchased the Rew Farm of 343 acres in Martinstown, Dorset. This he sold on an instalment system extending over 9 years, in lots of varying sizes to 27 purchasers, of whom 24 have in his own words "made a thorough success of their holdings," which average about 11 acres each. Within 6 years all the purchase money had been paid off, with the exception of about 500*l.* Also 14 of the purchasers had erected dwellings on their lots. In 1888, when the farm was bought, the total population on it amounted to 21 souls. In 1902 there were nearly 100, and it was still increasing. In 1888 the rateable value was 215*l.* In 1902 it was 346*l.*, that is to say, 60 per cent. more. During the same period the rateable value of the entire parish in which the farm lies fell 26 per cent.

I had hoped to dwell upon the results of Major Poore's experiments at Winterslow on the Wiltshire border, which are even more remarkable, and also on other instances, among them to those districts in which fruit is grown, but find that I have no time. Still those that I have given above may suffice for my argument.

It amounts to this—that where small-holdings exist the exodus from the land is comparatively little; that labour is more plentiful, since the small-holders are generally glad to work for others in their spare time; that there are more children; that the production from the land is larger; and that an altogether better and more hopeful tone prevails than is the case elsewhere.

If these things are true—and I believe them to be true—it follows that it would be well for England if small-holdings were largely multiplied, and that it ought to be the object of all sound statesmanship to multiply them accordingly.

I wish to make it clear, however, that I do not advocate the cutting up of the whole country into such holdings. England is very large, and in it there is room for every kind and class of estate. But I do advocate the employment of much land, which at present produces but very little and is useless even for sporting purposes, in this fashion. It seems to me further that many thousands of our population would be better employed in cultivating that land and adding to the national wealth by the production of foodstuffs, than in walking the streets of London and other great cities shaking money boxes in the faces of the passers-by.

The problem is how to get them there. Or, if that be not possible, how to prevent new thousands from joining their melancholy multitude—a problem, I may add, that no Government seems to have thought worthy even of consideration. I do not urge the artificial creation of small-holdings, which in my opinion would be bound to end in failure. But there are things that might be done if only our rulers could be persuaded to do them.

Thus, Co-operative Credit Banks on the Raffeisen system which have worked such wonders on the Continent, might be established under Government inspection and control without the risk of loss to the nation, from which banks intending small-holders could borrow upon the well-tested and approved principles of mutual guarantee. Loans might be advanced to landowners or corporations at a moderate rate of interest, which would include a sinking fund by which they would be automatically repaid in an agreed term of years, to enable such individuals or bodies to erect the cottages and buildings necessary to small-holdings. The Housing of the Working Classes Act of 1890 is, I may remark, practically a dead letter in the country districts, chiefly owing to the high minimum rate of interest charged by the Treasury.

Co-operative associations might be fostered, or even inaugurated by the Board of Agriculture to facilitate the collection and disposal of the produce of small-holders at a reasonable but remunerative rate of charge. Let those who doubt the advantages of co-operation as applied to agricultural products, study the recent story of the movement in Denmark, Ireland and elsewhere.

Last, but by no means least, the principle of the Parcels Post might be extended to a limit say of 100 lbs. weight, so as to make it easy for the small-holder to put his produce daily upon the great markets where it would be dealt with by the co-operative associations aforesaid.

If these humble remedies, to deal with which adequately would require a separate lecture, were adopted—as they could be, I believe, without the nation losing a penny-piece thereby—they would, I am convinced, result during the next generation in the establishment of thousands and even tens of thousands of happy and industrious workers upon the land who now come to compete with those of the over-crowded cities, and in keeping millions of money at home which at present are paid to foreigners for produce which we could grow as well as they do. And remember—they involve no form of protection!

I venture to urge them upon your attention and upon that of the public. The land and its interests receive little care from Government because those interests are divided and do not speak with any certain political voice. Yet I believe that now, more than ever in our history, are they vital to the welfare and safety of the nation.

No question of the day is more neglected and thrust aside than that of the flocking of our population from the fields to the cities, and yet none is more important. It means that the character of the race is changing; it means that their physique is deteriorating; it means that fewer men of the best class are available for purposes of national defence; it means the crowding together of vast hordes of people, who, at the first breath of panic or of serious trade depression will cease to be able to earn their livelihood and will have to be supported and controlled. In short, it means a new race of city-bred Englishmen, the victims to various diseases which, it is alleged, kill them out within three generations. We know what country-bred Englishmen have done in the past and the height to which they have lifted their race and nation. Can as much be expected from their descendants if they are to be herded together in the most dismal and airless quarters of our great cities ?

I have treated of a very large subject most inadequately—the time at my disposal would enable me to do no more. Still I trust that I may have said enough to convince you of its importance to our country, and to persuade some of you, according to your opportunities, to help me in my struggle to call attention to these evils, and to secure measures of relief that may tend to palliate if not to obliterate them. I do not wish to complain, but at times I find this somewhat lonesome task too heavy for me, and grow discouraged. It is, therefore, the more needful that those who have greater authority and ability should take it up, for if it is not taken up and carried to a successful issue, I think that our country must suffer in the near future.

The land should be in more hands than it is at present. It should produce more. It should support many more homes and rear many more sturdy, even-minded, children to carry the traditions of our flag and race down to a distant future—all of which things it well can do. At present we look to trade to maintain our wealth and greatness, perhaps wisely. But to rely upon this alone is not wise, for our trade, or part of it, may leave us. By all means cultivate your trade, but do not neglect to cultivate your land.

The land is the mother of men and women of that sort without whom no nation can remain great, and when all is said and done, such men and women are more than any money, which is but an appanage.

Whosoever helps this work forward and forces our Governments to recognise its paramount necessity, will, I think, play the part of a good citizen, and however humble, however unacknowledged may be his efforts, deserve well of this great and ancient country, whose history is his heritage, and whose glory it is his duty to help to pass on undimmed to the hands of future generations.

[H. R. H.]

WEEKLY EVENING MEETING,

Friday, May 15, 1903. :

SIR FREDERICK BRAMWELL, Bart., D.C.L. LL.D. F.R.S.,
Vice-President, in the Chair.

D. H. SCOTT, Esq., M.A. Ph.D. F.R.S.

The Origin of Seed-bearing Plants.

WHEN Linneus, in 1735, brought out his famous sexual system of
classification, which for so long dominated systematic botany, 23
out of his 24 classes were occupied by Flowering plants, and one only
was left for the Flowerless plants or Cryptogamia. As the name
" Cryptogamia " indicated, a thick veil of mystery still hung over the
reproductive processes of these flowerless plants. When this
obscurity became gradually dissipated, with the aid of improved
microscopes, by the brilliant researches of Hedwig, Mirbel, Nägeli,
Pringsheim, Cohn, Thuret, and above all, Hofmeister, and the
" Cryptogamia," to quote a phrase of Professor Sachs', became the
true " Phanerogamia," their relative importance received better
recognition. In a recent classification—that of Professor Warming
—out of 23 classes, no less than 18 are assigned to Cryptogams.

In spite of our vastly increased knowledge of the Cryptogamia,
the Flowering plants are still in the majority as regards species.
According to a recent census, out of about 175,000 known species of
plants, about 100,000 or four-sevenths are Phanerogamic. For our
present purpose we may speak of the Flowering plants as the seed-
bearing plants or Spermophyta, for, at least in recent vegetation, the two
characters, the grouping of the reproductive leaves in a flower and
the formation of a seed, go together and the latter is the more definite
and constant feature. The Cryptogams, such as Ferns, Mosses,
Seaweeds and Fungi, may in contradistinction be spoken of as the
spore-bearing plants or Sporophyta. In the vegetation then of the
present day, the seed-bearers are enormously predominant, not so
much in number of species as in importance, including with few
exceptions all plants of utility to man and almost all those of conspicuous stature, and occupying vastly the greater part of the earth's
land surface.

To what do the now dominant seed-plants owe their success?

This is a difficult question, for all organisms are well adapted or
they could not exist, and nothing requires more careful discrimination
than the attempt to determine the exact factors which constitute the

relative superiority of one group over another in the struggle for life. Everything depends on the conditions of the contest.

In the simpler of the higher Cryptogams, such as ordinary Ferns, the spores are all of one kind and on germination give rise to an independent plantlet, the prothallus, on which the sexual organs are borne. Fertilisation requires the presence of water for the actively moving male cells, the spermatozoids, to swim in. This condition may be something of a handicap to the plant, but if water is present, reproduction is fairly well ensured. In the more advanced spore-plants, such as the Selaginellas so commonly grown in our green-houses, the differentiation of the sexes begins earlier, for the spores themselves are of two kinds. There are numerous male spores of very small size (microspores) and comparatively few female spores of relatively large size (megaspores). In the group of the Water-ferns (Hydropterideæ) only one of these large spores is produced in each spore-sac, which then, if provided with a special envelope, as in *Azolla*, may closely simulate a seed.

In the microspores, the prothallus is scarcely developed ; the spore has practically nothing else to do but to produce the spermatozoids. On the female side, provision has to be made for the nutrition of the embryo, and here there is a comparatively bulky prothallus, though as compared with that of the Ferns, it tends to lose the character of an independent plant and to become a mere storehouse of food-materials. There are certain obvious advantages in this heterosporous condition. The male spores are kept small for easy dispersal and can be produced in correspondingly large numbers. The prothallial tissue is economised and only formed where it is wanted, i.e. in connection with the egg-cells from which the embryos arise.

The differentiation of microspores and megaspores is in fact comparable to that earlier differentiation of minute mobile spermatozoids and large stationary ovum, which took place far back in the history of both animals and plants, and laid the foundation of sex.

At the same time, the heterosporous arrangement, as we find it in Cryptogams, puts a new obstacle in the way of the successful accomplishment of the act of fertilisation. In order that this may happen, it is necessary that the two kinds of spores should germinate together as well as in the presence of an adequate water supply. The necessary association of the large and small spores is, as a rule, left to chance, the small spores being produced in enormous numbers, so that the chance may be a good one.

In the case of the great cryptogamic trees of the Palæozoic period, the difficulty must have been a serious one. We know that their large and small spores often differed in mass in the proportion of at least 100,000 to 1, and when bodies of such diverse weights were scattered by the wind from the tops of lofty trees, the chances must have been enormously against their coming to rest at the same spot. It was perhaps to this difficulty that the series of adaptations leading up to seed-formation owed their first inception.

If the microspores could be brought to the megaspores while the latter were still attached to the parent plant, much greater certainty of their union would be gained, for adaptations would now become possible for catching the small spores and retaining them in position. Some of the Cryptogams now living have got as far as this; the work of an American lady, Miss Lyon, has shown that in some species of *Selaginella* the microspores and megaspores meet and the spermatozoids are discharged within the sporangium; fertilisation is effected, and even an embryo may develop before the megaspore is shed. In this last respect these *Selaginellas* go beyond the Seed-plants of the Palæozoic period, as we shall presently see.

The first advantage, then, to be secured, was the occurrence of fertilisation, or rather the bringing together of the two kinds of spore on the parent plant. This is one of the constant characteristics of the seed-bearing plants; the process is spoken of as *pollination*, for what we call the pollen-grains are nothing but the microspores of the Spermophyta.

We will now see how the process actually goes on in some of the simpler Seed-plants of the present day.

The Seed-plants, as is well known, are divided into two great classes: the Angiosperms, in which the seeds are inclosed in a seed-vessel; and the Gymnosperms, in which they are exposed. In the former, fertilisation is effected by the growth of the pollen-tube through the tissues of the young seed-vessel; in the Gymnosperms the pollen falls directly on to the young seed or ovule, and the pollen-tube has only a short way to grow before reaching the egg-cell.

The Angiosperms (Monocotyledons and Dicotyledons) include practically all our familiar Flowering plants, but with them we are not concerned this evening. The question of the origin of Angiosperms is one of the great unsolved problems of Botany, but it does not immediately touch our present subject. It is to the simpler Seed-plants—the Gymnosperms—that we must turn for light on the origin of the Seed-plants as a whole. The Gymnosperms are enormously the more ancient of the two classes, extending back through the whole of the Carboniferous period into the Devonian, while the Angiosperms, so far as we know, only appeared quite late in the Mesozoic period.

The most familiar of the Gymnosperms—the Coniferæ or cone-bearing trees—are themselves too advanced on the seed-bearing line for our purpose; we will concentrate our attention on a family, which of all living Flowering plants stands nearest to the Cryptogams, namely the Cycads. This group, not very well known to the non-botanist, but of which a splendid collection will be found in the Palm-house at Kew, is now a small one, including 9 genera and about 70 species distributed over the tropical and sub-tropical regions of both the old and new worlds. In habit these plants, which may rise to the stature of small trees, bear some superficial resemblance to Palms; the agreement with Ferns is however much more striking.

In the genus *Stangeria* from tropical Africa, the leaves bear so close a resemblance to those of some Ferns in form and veining, that the plant, before its fructification was known, was described by competent botanists as a species of the Fern genus *Lomaria*.

In all Cycads the male fructifications are in the form of cones ; the pollen-sacs are borne in great numbers on the under surface of the scales of the cone. In all the genera but one, the female fructifications are also cones, each scale bearing two large ovules. In the type genus *Cycas*, however, there is no specialised female cone at all, the fertile leaves are borne in rosettes on the main stem, alternating with zones of the ordinary vegetative leaves.

The fertile leaves are of large size and compound form, and usually each of them bears several ovules, which, whether fertilised or not, grow to a great size, sometimes as big as an egg-plum. They are in some species of a bright red colour, and, contrasting with the yellow woolly leaves on which they are borne, are conspicuous and beautiful objects.

In thus bearing its seeds on leaves so little modified and springing like the ordinary leaves from the main stem, *Cycas* is the most fern-like genus of Flowering plants.

The ovule, at the time when pollination takes place, is about the size of a small hazel nut. It consists of an outer envelope and a central body, the two being closely joined together, except towards the top, where the envelope leaves a narrow passage open, leading down to the central body. The apex of the latter becomes excavated into a hollow pit, the pollen-chamber, a feature almost peculiar to Cycads amongst *living* plants, discovered by our countryman Griffith as long ago as 1854,[*] though the credit is often wrongly given to later French or German investigators.

The pollen, blown by the wind or possibly conveyed by insects, is received in the opening of the envelope by a drop of gummy substance, and as this evaporates, the pollen-grains are drawn down through the narrow passage into the pollen-chamber below. There each grain anchors itself by sending out a tube into the neighbouring tissue of the ovule ; thus pollination is accomplished. Fertilisation, i.e. the actual union of the male and female cells, takes place some months later, when the ovule, now to all external appearance a seed, has reached its full size. In the meantime, the single megaspore or embryo-sac, imbedded in the tissue of the central body of the seed, has grown to enormous dimensions, filled itself with prothallus and developed the egg-cells at its upper end, which are so large as to be easily seen with the naked eye.

The pollen-grain behaves like a cryptogamic microspore and develops two large spermatozoids each with a spiral band bearing numerous cilia—the organs of motion. The pollen-tube becomes

---

[*] Icones Plant. Asiat., Part iv., Plates 377 and 378; Notulæ ad Plant. Asiat., pp. 6–8.

distended with water, bursts and sets free the sluggishly moving spermatozoids, which by aid of the water discharged from the pollen-tubes, are able to swim to the egg-cells and effect fertilisation.

This remarkable process, first discovered in 1896 by two Japanese botanists, Ikeno and Hirase, and independently in 1897 by the American Webber, occurs not only in the Cycads but also in that strange plant the Maidenhair-tree, *Ginkgo*, a form now completely isolated, certainly rare in a wild state, and said to have been only saved from extinction by cultivation around Buddhist temples in China and Japan, but which has a long geological history.

The Cycadean method of fertilisation holds exactly the middle place between the purely cryptogamic process, where the active male cells accomplish the whole journey to the egg by their own exertions, and the method typical of Seed-plants, where these cells are little more than mere passengers carried along by the growth of the pollen-tube.

The adaptations, which in the Cycads allow of pollination and fertilisation on the plant, are chiefly three.

1. The envelope of the seed with its narrow opening down which the pollen-grains are guided.

2. The pollen-chamber below, in which they are received.

3. The pollen-tube, which however plays a somewhat less important part here than in the higher Flowering plants, and which in the Palæozoic allies of the Cycads may perhaps have been dispensed with altogether.

There are, however, other points in which the ovule of a Cycad differs from the spore-sac of a Cryptogam. Not only is the megaspore solitary—that is a condition already reached among the Water-ferns—but it is firmly imbedded in the surrounding tissue. It is no longer a mere spore destined to be shed, but remains throughout an integral part of the ovule, while the ovule ripens into a seed and ultimately germinates. Thus the whole development of the prothallus takes place within the seed, and this requires special methods of food-supply, involving a complexity of structure far beyond that of any cryptogamic spore-sac. When the time for dispersal comes the seed is shed as a whole.

There is, however, another character commonly regarded as essential to the definition of a seed : a seed should contain an embryo ; this implies, that after the egg-cell has been fertilised, the young plant develops to a certain extent while still within the seed and before it is shed. In the ripe seed the embryo passes into a resting stage and only resumes its development when the seed begins to germinate and the embryo becomes a seedling. Usually too, the ripening of the seed itself is dependent on the development of the embryo ; if there is no fertilisation, there is no true seed, only an abortive ovule.

In the Cycads this is not the case ; the ovule ripens into a full-sized and apparently normal seed, even if fertilisation has failed. In our hot-houses Cycads are seldom fertilised; yet the conspicuous scarlet seeds of *Cycas revoluta* or the crimson seeds of *Encephalartos* are familiar objects to many Kew visitors. Further, the degree of development of the embryo at the time the seed is shed is very inconstant; sometimes, although fertilisation has taken place, the embryo is scarcely to be detected.

The definite resting stage of the young plant in the dry seed, so characteristic of the higher Phanerogams, is unknown to these primitive seed-bearers, the Cycads and the Maidenhair-tree, and the same appears to hold good for the seeds found in the Palæozoic rocks. Such seeds are common in certain localities, as in the Coal Measures of Central France, and to a less degree in our own coal-beds. In petrified specimens the structure is often beautifully preserved, yet in no single case has a Palæozoic seed been found to contain an embryo. It is not merely a matter of preservation, for that is not unfrequently so good that the delicate egg-cells can still be recognised. Thus there is no " seed " of Palæozoic age, which, according to current definitions, strictly deserves the name. Technically, the term " ovule " would be more appropriate, but the obvious maturity of the integument makes the word " seed " seem more natural. So far the case is parallel to that of our recent Cycads or Maidenhair-tree.

It is of course possible that any day we may light on some Palæozoic seed with an embryo; it may be that the specimens hitherto found were all unfertilised, though the frequent presence of pollen-grains in the pollen-chamber makes this explanation unlikely. It seems not improbable that the development of an embryo in the ripening seed was a later device; that in the older seed-plants the period of rest came immediately after fertilisation, and that the growth of the embryo, when once started, went on rapidly and continuously to germination. In that case a seed with a recognisable embryo would rarely be preserved.

We are now in a position to see what are the chief advantages gained by a plant in adopting the seed habit, they are:—

1. Pollination on the parent plant, and consequently greater certainty in bringing together the two kinds of spore.

2. Fertilisation either on the plant, or at least within the spore-sac, giving greater certainty of success, and protection at a critical moment.

3. Protection of the young prothallus from external dangers.

4. A secure water-supply during its growth.

5. Similar protective and nutritive advantages for the young plant developed from the egg-cell. This last end, however, was very probably not yet fully attained in the earlier seed-bearing plants.

We may now go on to consider our main subject for this evening, —the historical question, From what group of spore-bearing plants were the Seed-plants derived ?

One thing is plain : the stage of heterospory was the immediate precursor of seed-formation, and it was from some group of Cryptogams producing spores of two kinds, that the Seed-plants sprang. Such heterosporous groups are however known in three of the main phyla of the higher Cryptogams.

In the Lycopod series, we have, among their living representatives, pronounced heterospory in *Selaginella* and *Isoetes* ; among the Palæozoic Lycopods it was commoner still. Within the class of the Ferns we have the heterosporous Water-ferns ; in the third series, that of the Horsetails, we have, it is true, only homosporous forms now living, but in Palæozoic times a well-marked differentiation of micro- and megaspores was attained, though less extreme than in the other two lines.

So far, therefore, there is no reason why the early Seed-plants might not have had family relations with any of the great pteridophytic phyla, and as a matter of fact, all three lines have been championed by one botanist or another as the probable ancestors of the Seed-plants.

The Horsetail stock, though it attained an extraordinary development, shows no further sign of transition towards the higher plants.

The case for the Lycopods is stronger, and indeed they were long the " favourites " and were commonly regarded as lying nearest the true line of spermophytic descent. This idea was specially based on the mode of development of the spore-sacs, which has much in common with that of the pollen-sacs and ovules of Phanerogams, and this, combined with the occurrence of well-marked heterospory in some genera, appeared to point to a relationship. But the former character (the development of the spore-sac from a group of cells instead of from a single one) is now known to be common to certain Ferns, and to just those Ferns (the Marattiaceæ, etc.) which prove to be the most ancient, so that this argument has lost its weight. It has lately been found, indeed, that some of the Carboniferous Lycopods produced seed-like organs, presenting the most striking analogies with true seeds, but the plants which bore them were in all other respects Lycopods pure and simple, and the case appears to have been one of homoplastic modification. There is no indication, as yet, of any forms really transitional between the Lycopods and the Spermophyta.

The one line, which, so far, has yielded truly intermediate types, is that of the Ferns.

Among recent plants the Cycads, as we have seen, offer some points of agreement with Ferns, sufficient to have led certain distinguished botanists, for example Sachs and Warming, to strongly maintain their Fern-ancestry. The chief points of agreement are :—

1. The Fern-like foliage in some Cycads, and in many the mode of folding of the leaflets in the bud.

2. The arrangement of the pollen-sacs in groups on the under-side of the cone-scales, like that of the spore-sacs of Ferns on the under-side of the leaves.

3. The carpels or fertile leaves of *Cycas*, which, though bearing true seeds, are more like fertile Fern-fronds than any other reproductive leaves.

By themselves these characters, though suggestive, would be inconclusive; the anatomy is not directly comparable with that of any living Ferns.

What, then, do we know of the history of this family in past times? The Cycads are now a small isolated group; in the Mesozoic period, from the Trias to the Lower Cretaceous, they were one of the dominant types of vegetation, and spread all over the world. Of the fossil species recorded from the Oolite of the Yorkshire coast, and from the Wealden of the South of England, one-third are referred to Cycads, and they were equally abundant in the Mesozoic Floras of North America, India and other countries. If they existed in the same proportion now as then, they would have about 35,000 species instead of 70! The Cycads of the Mesozoic, however, were not, as now, a single family, but a great class (the Cycadophyta of Nathorst) embracing very diverse types, often with organs of reproduction widely different from those of their surviving relatives, and showing a certain parallelism with angiospermous fructifications. But with all this there was on the whole a remarkable uniformity in habit, just as we find a general similarity in outward characters among so many Dicotyledonous trees of the present day though belonging to the most diverse families. In the Mesozoic rocks we also find a certain number of plants (known only from their foliage) as to which it remains doubtful whether they belonged to Cycads or Ferns, or to some intermediate group.

Besides the Cycadophyta, seed-plants were represented in Mesozoic days by a great number of Coniferæ, more or less allied to those still living, and by various forms akin to the Maidenhair-tree, perhaps the most ancient type surviving in the recent Flora.

When we go further back, to the Palæozoic rocks, it is only in their uppermost strata that we find forms clearly referable to Cycads or Conifers.

The best known seed-bearing plants of the older rocks are those of the family Cordaiteæ, which stretches back to the Devonian. They were tall, branched trees, bearing great simple leaves, sometimes a yard long. The anatomy of stem and root resembled that of an Araucarian Conifer, but the leaves had just the structure of the leaflets of a Cycad. Male and female flowers were borne in little spikes or catkins, and may best be compared with those of the Maidenhair-tree. The seeds, of which the structure is known, closely resemble those of that plant, or of recent Cycads.

The Cordaiteæ, however, ancient as they are, were already pronounced gymnospermous seed-plants; by themselves they give no direct clue to the origin of Spermophyta; we must look elsewhere for the key to our main problem.

The vast number and variety of Fern-like remains throughout the Palæozoic strata, wherever land-plants are known, is familiar to all. Almost every form of recent Fern-frond can be matched from the impressions in the Carboniferous and Devonian rocks. A considerable number of these fossil Fern-fronds are known to have really belonged to Ferns, for typical Fern-fructifications are found upon them.

An experienced collector of Coal-Measure plants, Mr. Hemingway, once told me that he reckoned on finding about 20 per cent. of the specimens of a true fossil Fern in the fertile state. When, therefore, a common fossil Fern-frond (so-called) is *never* found fertile, a strong suspicion is awakened that the plant must have had some kind of fructification other than that of an ordinary Fern. This is the case with a surprisingly large proportion of the Palæozoic plants commonly described as Ferns, and holds good of certain entire "genera": the important genera *Alethopteris, Neuropteris, Mariopteris, Callipteris, Tæniopteris,* and others, have never yet been found, in any of their species, with fertile fronds, if we except one or two specimens so questionable and obscure that no conclusion can be drawn from them. It is probably under the mark to say that one-third of the so-called Ferns of Palæozoic age afford no evidence from fructification that they were really Ferns, as we now define them.

The absence of recognisable fertile fronds may, it is true, be partly accounted for by dimorphism. Many Ferns, both recent and fossil, bear their reproductive organs on modified portions of the frond, or even on special fronds, very different from the vegetative foliage. Fossil remains are usually fragmentary, and when the sterile and fertile fronds are found isolated there may be nothing to show that the one belonged to the other. But, allowing for this, there are very many "Fern-fronds" which offer no evidence, even from association, of any Fern-like fructification, while the fructifications actually associated with them are often anything but Fern-like. There are in fact a number of unassigned seeds from the Coal Measures, some of which are commonly associated with certain of the quasi-Ferns of which we are speaking.

On the whole, however, we have, up to this point, had before us merely negative evidence, indicating that many of the leaves, so familiar to palæobotanists, classed on account of their form and veining as Fern-fronds, may really have belonged to some group different from the true Ferns. Negative evidence is notoriously weak; at most it only justifies us in taking up a position of philosophic doubt, though in this case, it was enough to induce the distinguished Austrian palæobotanist Stur to suspect that the genera *Alethopteris, Neuropteris* and others were no Ferns, but Cycads.

During the last thirty years, however, positive evidence has been accumulating, proving that certain of the Fern-like Palæozoic plants were at any rate something distinct from true Ferns, as we now know them.   This evidence is derived from a study of the anatomical structure, which in Cycads and Ferns as they now exist, is sufficiently different to prevent any possible confusion between the two groups. A single section from the leaf-stalk of the Fern-like Cycad *Stangeria* would be enough to show that it is a true Cycad and no Fern, and conversely, a single section from the frond of *Lomaria*, with which *Stangeria* was once confused, would show it to be a true Fern and not a Cycad.

A common Coal-Measure plant, named *Lyginodendron Oldhamium*, was one of the first of the Palæozoic quasi-Ferns to be examined anatomically   We owe this work, like so many other great advances in fossil botany, to the late Professor Williamson, who thus led the way to the solution of the problem which occupies us this evening.

Externally, the plant is wholly Fern-like; its characteristic highly compound foliage is that of a *Sphenopteris* (*S. Höninghausi*) with a *Davallia*-like habit.   The large fronds were borne, at intervals, on a somewhat slender stem, which rooted freely.   The slender proportions and the presence of spines everywhere, on leaf and stem, suggest that the plant may have been a scrambling climber, like *Davallia aculeata*, for example, among recent Ferns.

The structure of all the vegetative parts of the plant-stem, leaf and root is now known as perfectly, perhaps, as in any plant now living.   The leaves turn out to be true "Fern-fronds" in structure as well as in external aspect.   The vascular bundle traversing the petiole, for example, is of the "concentric" type characteristic of Ferns, and any differences there may be are in details only.

A section of the stem, however, bears at first sight no resemblance to that of a Fern; outside the pith, we find a broad zone of wood and bast with its cells arranged regularly in radial series, like that of an ordinary "exogenous" tree, and in detail approaching especially the Cycadean structure.   At the border of the pith there are distinct strands of wood, and this region, which was laid down before the radially arranged zone, recalls the structure of an *Osmunda*. The bundles in the cortex of the stem, on their way out to the leaves, have in this part of their course exactly the structure of the strands in the leaf-stalk of a Cycad—a structure found, in this form, in no other living plants.

The roots, when young, resemble those of certain Ferns (Marattiaceæ), but as they grew older they also formed secondary wood and bast like the roots of Gymnosperms.

On the ground of this remarkable combination of structural characters, it was inferred that *Lyginodendron* could not have been a true Fern, but must have occupied a position intermediate between the Ferns and the Cycadean Gymnosperms.

A similar association of diverse anatomical characters has now been proved to exist in various other quasi-Ferns of Palæozoic age.

In *Heterangium*, for example, a genus also investigated by William-son, leaves and roots resemble those of the previous genus, but the stem is more obviously Fern-like, agreeing in its earlier stages with that of a *Gleichenia*, but acquiring, with advancing age, a zone of secondary wood and bast of the Cycadean type. This plant like-wise bore foliage of the *Sphenopteris* form (*S. elegans*).

In *Medullosa*, on the other hand, to which the *Alethopteris* and *Neuropteris* foliage belonged, the primary ground-plan of the tissues in the stem is like that of a complex Fern, but the structure of leaves and roots, and the secondary structure of the stem itself, is almost purely Cycadean. And we might continue the list much further. Wherever one of these quasi-Ferns has been examined anatomically, a similar combination of characters has been found. It may be pointed out in passing that while many of these inter-mediate forms lead on towards the Cycadophyta themselves, others approach more nearly to the extinct family Cordaiteæ, and indicate that they also, though so different from Ferns in habit, may yet have sprung from the same stock.

But so far the positive evidence has been wholly anatomical, and botanists are not yet altogether in agreement as to the value of anatomical characters. The anatomist very naturally thinks that there is nothing like anatomy, but the pure systematist will not be satisfied without the characters on which he has been accustomed to rely, and his faith in which has been so amply justified, those namely drawn from the reproductive organs. Darwin, however, who neglected nothing, was fully alive to the importance of anatomical evidence; he expresses his interest in an anatomical character in an amusing way in one of his lately published letters (1861), saying, " The destiny of the whole human race is as nothing compared to the course of vessels in Orchids ! "

Until the present year, we had no satisfactory knowledge of the fructification in any one of the Cycadofilices, as we now call them, of the Palæozoic period. There is, it is true, some reason to believe that a form of fructification with long tufted spore-sacs belonged to *Lyginodendron*, but we know nothing as yet as to the details—it may prove to represent the male reproductive organs of the plant. Among the unidentified seeds of the Coal Measures, there are some—the great seeds known as *Trigonocarpon*—which are not only associated with *Medullosa*, but which show a certain structural resemblance to some of its tissues. But still the indications were slight—so slight that Professor Zeiller of Paris, than whom there is no higher authority, has recently expressed a doubt whether these Cycadofilices were, after all, anything more than a peculiar group of Ferns.

Within the last few months, however, an altogether new light has fallen on our subject. Among the seeds discovered by Williamson in the English Coal Measures were three species which he placed in his genus *Lagenostoma*. These, as we shall see, are characteristic seeds of complex structure. One of them, named *L. Lomaxi* by

Williamson, though not described by him, has lately been re-investigated, in the first instance by my friend Professor F. W. Oliver. The great peculiarity about it is that the seed itself was borne in a little calyx-like cup, fitting loosely round it, just as a hazel-nut is borne in its husk. The cup, or cupule, which is deeply lobed, bears very peculiar glandular bodies, usually with a short thick stalk and a round head, which is empty, as if the secretory tissue had broken down. These glands on the cupule of the seed have been found to agree exactly in dimensions, form and structure with the glands borne on the leaves and stems of the particular form of *Lyginodendron Oldhamium* with which the seeds are associated.

Suppose that in some tropical forest where the trees were too lofty for their leaves and fruits to be reached, seeds and leaves and twigs were found scattered together on the ground, and that they all proved to bear exactly similar glandular outgrowths, of a kind unknown elsewhere. Suppose further that the structure of the envelope of the seed turned out to agree in other respects with that of the vegetative fragments, should we hesitate to conclude that the seeds belonged to the same plants as the leaves and twigs, though we had never seen them actually in connection? Such is the argument with regard to the relation of the seed *Lagenostoma* to the plant *Lyginodendron*. Short of finding the vegetative and repro-ductive organs in continuity the proof is as strong as it can be, and I think we need not hesitate to conclude that the one belonged to the other.*

But, if this be so, the question as to the nature of the Palæozoic Cycadofilices is settled, at least as regards one member of the group. *Lyginodendron* was already a seed-bearing plant. The seeds are highly organised and, broadly speaking, of the Cycadean type. The integument and central body of the seed are closely joined to near the tip, and along the line of junction run the strands which conveyed the water-supply. The upper part of the integument has a curious chambered structure—the central body terminates in a large pollen-chamber of peculiar bell-shaped form, in which the pollen-grains are sometimes found. The neck of the pollen-chamber fits into the opening of the integument and reaches the surface. The middle of the seed is occupied by the large megaspore or embryo-sac, in which remains of prothallial tissue can sometimes be detected. The seed, in fact, is as highly differentiated as any seed of its period, lacking only an embryo, as do all its contemporaries.

But if *Lyginodendron*, with all its Fern-like characters, was thus a true seed-plant, we cannot doubt that other quasi-Ferns of that period, exhibiting a similar combination of characters, had also entered the ranks of the Spermophyta, and we may confidently

---

* Oliver and Scott, on *Lagenostoma Lomaxi*, the Seed of *Lyginodendron*, Proc. Roy. Soc., vol. lxxi. (1903) p. 477.

expect that one by one, many of the as yet unowned Palæozoic seeds will be traced to their Fern-like possessors.

Further positive indications of this are already presenting themselves. For example, there is a specimen in the British Museum collection, showing a cast of a branched rachis accompanied by a multitude of ribbed seeds, many of which are in clear connection with the rachis itself. At one place we see a leaflet of *Sphenopteris obtusiloba*, a well-known Coal-Measure " Fern," and everything indicates that we have here the fertile, seed-bearing rachis of that species. There are other specimens which point in the same direction, and now that the eyes of collectors are opened to the possibility of their so-called " Fern-fronds " bearing seeds—an idea which before seemed too improbable to be entertained—more of such specimens will doubtless find their way into our museums.

The present position, then, of our question is this. Some, probably many, of the Fern-like plants of Palæozoic age bore seeds, of the same general structure as those of the Cycads among living Gymnosperms. The plants in question were not merely Fern-*like*; their anatomical structure proves them to have had so much in common with true Ferns that there can be no doubt of their affinity with them. In fact, apart from the newly discovered seeds, these plants for the most part show a balance of characters on the Fern side.

The evidence thus points unmistakeably to the conclusion that the Cycadophyta—the most primitive of the seed-plants—sprang from the Fern stock. Thus the origin of the great mass of Cycadean forms which overspread the world during Mesozoic epoch is accounted for: they were doubtless derived from the more primitive Cycad-ferns of the preceding Palæozoic age, and through them from some early Filicinean ancestry. The first divergence from this original cryptogamic stock must have occurred very far back; the seeds of *Lyginodendron* and other Carboniferous seeds referable to the Cycadofilices, are, as we have seen, already highly organised, and the stages of their evolution from the cryptogamic sporangium are still to be discovered.

The origin of Seed-plants from the Fern phylum will probably prove to hold good for other groups besides the Cycadophyta. The great Palæozoic family Cordaiteæ combines the characters of Cycads and Coniferæ, and at the same time shares certain of those anatomical features which first betrayed the true nature of the Cycadofilices. There is thus a strong presumption that the Cycadophyta, the Cordaiteæ, and the Coniferæ themselves had a common origin, or at least that they all sprang, directly or indirectly, from the great plexus of modified Ferns which played so large a part in Palæozoic vegetation.

Hence, so far as the gymnospermous Seed-plants are concerned, we are led to the conclusion that they were derived, at a very early period, from the Fern stock. The following up of the clue, which, as I believe, we have now grasped, will afford a pursuit of the utmost interest and promise.

But the other great problem—the origin of the angiospermous Seed-plants, which are now supreme in the vegetable world, is as yet untouched. And so, though real progress has been made, it will be long before we can hope for a complete answer to the question which we have had before us this evening.

WEEKLY EVENING MEETING,

Friday, May 22, 1903.

SIR JAMES CRICHTON-BROWNE, M.D. LL.D. F.R.S.,
Treasurer and Vice-President, in the Chair.

J. A. H. MURRAY, Esq., M.A. LL.D. D.C.L. D.Litt. Ph.D.

Dictionaries.

THE subject on which I have been asked to discourse this evening is
that of "Dictionaries." I have met people who knew nothing about
Dictionaries; they knew only that there was a book called "The
Dictionary," just as there is a book called the Bible, another called
the Prayer-book, another the Koran; when they saw or heard a word
that was new to them, they wondered if it was "in the Dictionary,"
just as they might wonder whether a particular text was in the Bible,
or a particular person's name in *Who's Who*. Like these other books,
the dictionary might be in different types or in different bindings, but
it was still an individualised work—The Dictionary. That there are
dictionaries *and* dictionaries, or that, sad to say, dictionaries differ,
had not yet dawned upon their apprehension. But of late years the
merits and excellencies of rival dictionaries have been thundered
upon us by daily papers from *The Times* downward, so loudly and so
long, that the number of these ingenuous people must be greatly
diminished. At any rate, your Institution, in asking me to discourse
upon "Dictionaries," has recognised at once their plurality and their
variety.

The singular of "Dictionaries" is a *Dictionary*; what is a dictionary? The term is formed from the Latin *dictio*, a word originally
meaning *saying* or *speaking*; but already by the later Latin Gram-
marians used in the sense of *verbum* or *vocabulum*, a word, whence we
have the originally synonymous term *vocabulary*. A *Dictionary* then
is a repertory of *dictiones* or *words*. The Latin form, *Dictionarium*
or *Dictionarius*, does not occur in ancient Latin; so far as I know
it was formed, or at least first used, in 1225, by an Englishman,
Johannes de Garlanda, or John of Garland, as the title of a collection
of Latin vocables, arranged according to their subjects, in sentences,
for the use of learners. About a century later, Peter Berchorius,
who died at Paris in 1362, made a *Dictionarium morale utriusque
Testamenti*, or Moral Dictionary of the Old and New Testament,
consisting of Moralisations on the principal words of the Vulgate,
for the use of students in theology. The first to use "Dictionary" as
an English word in its modern sense, was apparently Sir Thomas
Elyot in the title of his Latin-English Dictionary of 1538. The

early dictionaries indulged in divers titles : the *Promptorium* or store-room, the *Catholicon* or general manual, the *Medulla* or marrow of the language, the *Hortus Vocabulorum* or garden of words, the *Alvearie* or beehive, the *Abecedarium*, or alphabetic (A B C) book, the *Table Alphabetical*, the *English Expositor*, the *World of Words*, the *Glossographia*, the *Gazophylacium* ; and it would have been im-possible to predict in the sixteenth century that the name *Dictionary* would supplant all the others, and even rise superior to the better-born word *Vocabulary*. Dictionaries and vocabularies were not all, at first, in alphabetical order; the words were more often arranged under subject-headings, so as to bring together all the names of things of the same kind, e.g. things in heaven, things in the air, names of times and seasons, names of beasts, of kindred and relation-ship, of trades, of clothing, of things in the garden, of church furniture, &c. But it was gradually found that for purposes of reference, there was nothing so simple as the alphabetical order, although it brings together strange bedfellows, and as the old lady is said to have remarked of the quotations in Johnson's dictionary, though interesting, it is rather disconnected. Since the end of the sixteenth century, therefore, all ordinary dictionaries of European languages have been in A B C order ; and this has so taken hold of the popular mind as to be considered an essential feature of a dictionary, with the curious result of extending the name *Dictionary* to any book which gives information of any kind in A B C order, although it does not treat of *dictiones* or *words*, but of *things*, persons, places, history, or branches of science. Thus we have in modern use dictionaries of chronology, of geography, of music, of commerce, of manufactures, of London, of the Thames, of Christian antiquities, of national biography. In the etymological and historical sense of the word, these are not dictionaries at all, but *Abecedariums*, text-books or reference-books on their special subject, with their articles arranged in alphabetical order ; modern usage, however, has sanctioned this application of *dictionary*, and in questions of nomenclature usage has to be accepted. But I assume that in asking me to discourse on "Dictionaries," your Institution expected me to take the word in its strict sense of "books explaining the meaning of words."

Dictionaries explaining words are, broadly, of two kinds ; those which give the meaning of the words of one tongue in the language of another, as an English Dictionary of Latin, Greek, or French; and those which deal with the words of their own language, setting forth their current spelling, pronunciation, meaning and use, and sometimes telling more or less of their origin and history.

Of these two kinds, the former was naturally the earlier ; we have been told that even among the brick tablets of ancient Assyria, there exist remains of dictionaries or vocabularies explaining the earlier Sumerian or Accadian in the later language of the country ; we know that vocabularies of Greek words explained in Latin appeared not many centuries after the Christian Era. But on these ancient essays at dictionary-making I have no time to dwell, and must confine myself to the history of English dictionaries and their immediate antecedents, in itself a long story, for " The English dictionary, like the English constitution is the creation of no one man and no one age; it is a growth that has slowly developed itself adown the ages. Its beginnings lie far back some 1200 years ago, in times almost prehistoric. And these beginnings themselves were, strange to say, neither English nor yet dictionaries."

All my present audience have been at school; probably all of them, when there, made some attempts at learning some foreign tongue. Most of them when they found a word in their translation book difficult to remember, have written the meaning over it in an unobtrusive form. The learned name for such an interlinear explanation of the text is a *gloss*. The early monks were addicted to writing glosses over difficult words in their Latin books. These glosses were sometimes in easier Latin; sometimes in their own vernacular. Sometimes books, especially scriptural and service books, were glossed so completely as to have a sort of interlinear translation; at other times only individual hard words at long distances were glossed. Afterwards, collections of these glossed words, with their glosses, were made; these were *glossaries*. The earliest glossaries contained the words just as they were found and copied down; later ones were re-arranged more or less alphabetically. One of the oldest extant is the *Épinal Glossary* of the seventh century; it has in the page three columns of glossed words followed by their glosses, mostly Latin, but a small proportion English. In later glossaries, the English explanations more and more predominated, till by the tenth century these formed one of the important foundations of Latin-English lexicography. The other source was the formation, even in Anglo-Saxon times, of short vocabularies of Latin words with their English meanings, arranged in subject-classes, for the use of young scholars. These, in course of time, were added to or incorporated with the glossaries, so that combined they formed, by the eleventh century, more or less extensive Latin-English Dictionaries.

Then the Norman Conquest overthrew English learning and literary culture, and for more than 300 years English lexicography stood still. But from the end of the fourteenth century English was again in the ascendant, and many Latin-English Vocabularies were compiled, some alphabetical, but mostly arranged under subject-headings. The fifteenth century made a great step forward by the production about 1440 of the earliest English-Latin Vocabulary, the *Promptorium Parvulorum* of Brother Geoffrey of King's Lynn, followed by the *Catholicon Anglicum* c. 1483. With the Renascence of Ancient Learning such work rapidly increased. Sir Thomas Elyot published his Latin-English Dictionary in 1538. Enlarged and re-edited by Thomas Cooper, afterwards Bishop of Winchester, this became at length Cooper's *Thesaurus*. J. Withall published in 1554 *A short dictionary for young beginners*, arranged in subject-classes. The most interesting of these works was Baret's *Alvearie* of 1573, the preface of which relates in such naive language the circumstances in which it originated, and the difficulties its compiler had to encounter.

Of all these works Latin was an essential part; but Latin was now becoming insufficient to express the wants of the modern world, and in 1530 Mayster John Palsgrave, a Londoner, who had studied in Paris, composed his *Esclaircissement de la Langue Françoyse*—Grammar and Dictionary combined—for the Princess Mary Tudor, when she was married to Louis XII. of France. Then followed a Dictionary of English and Welsh in 1547, the famous Italian-English *World of Words* of Florio in 1598, Percival's Spanish-English Dictionary in 1599, and the no less renowned French-English Dictionary of Randall Cotgrave in 1611.

Thus by the close of the sixteenth century English was supplied with dictionaries of the important neighbouring languages, but it had not yet been found necessary to have a dictionary to teach men their own English tongue. And when the want was felt, it was natural that only the difficult words, the "ink-horn terms" were contemplated. All the dictionaries of the seventeenth century were only in purpose "Dictionaries of Hard Words." The first was the *Table Alphabetical* of Robert Cawdrey 1604; the second the *English Expositor* of John Bullokar 1616; then came the first book calling itself *The English Dictionarie* — by Henry Cockeram of Exeter, in 1623. This was divided into three parts: 1, the hard words explained by easier words; 2, the easy words, rendered by hard words, so as to enable the writer to acquire a learned style; 3, a classified list of the names of History, Mythology, Natural History, etc., to which allusions occurred in learned writings; here appears the celebrated account of the *Crocodile*, explaining the meaning of the phrase "crocodile's tears," i.e. those wept by the crocodile over the head of the man whose body it has devoured, before it proceeds to eat the head itself. Other seventeenth-century dictionaries were Blount's *Glossographia*, Phillips's *New World of Words*, originally plagiarised from Blount, and the Dictionary of Cocker the famous writing master and arithmetician of Southwark; also the three editions of Kersey, whence Chatterton learned his Elizabethan words.

The next great step forward was made by Nathanael Bailey, who first endeavoured to make the dictionary contain *all* words, not indeed for the sake of the explanation or even the spelling of the common words, but in order to indicate their etymology. Bailey's was the greatest name in lexicography during the first half of the eighteenth century, and his fame was not for a long time eclipsed even by that of Johnson. He had many rivals and imitators, most of whom tried like him to make "complete" dictionaries, and the competition in the number of words now began. But before many years elapsed, the feeling began to prevail that it was time there was prepared a Standard Dictionary, in order to "fix the language" in the state of perfection which it was considered to have now attained.

The same notion had led to the formation of the French Academy, and of the *Accademia della Crusca* at Florence, both of which bodies had as a main object the formation of a standard dictionary to "fix the language" and prevent its deterioration or change—a fond idea borrowed from the artificial preservation of an unchanged Ciceronian Latin in the schools. As England had no Academy, it was thought that the work must be done here by some eminent man of letters, and it was hoped that it might be done by Pope, who is said to have entertained the idea, and even to have drawn up a list of the authors who were to be accepted as standards. But he died in 1744, without anything further being done. The matter was then pressed upon the attention of Samuel Johnson, and in 1747 a syndicate of London publishers contracted with him to produce the desired Standard Dictionary. This was to be done in three years, but it occupied nearly nine; the work appeared at last in 1755. There is no time within the limits of this lecture to describe the varied interest of Johnson's great work, and it is only necessary to point out that its characteristic feature was the support and illustration of every word and sense, as far as possible, by quotations drawn from accredited writers. These were all selected by the lexicographer himself from books read and marked by him, whence they were copied out by his clerkly assistants, or, in many cases, from the stores of his own memory— the latter plan being facilitated by the fact that he gave no reference beyond the author's name, which in its turn left a considerable room for inaccurate quotation. Books belonging to Johnson's library are in existence which show his method of reading and marking books; one of these is among the treasures of my scriptorium.

Johnson's work raised English lexicography to altogether a higher level. In his hands it became a department of literature. The value of his Dictionary was recognised from the first by men of letters; a second edition was called for the same year. But it hardly became a popular work, or even a work of popular fame, before the beginning of the nineteenth century. For forty years after its appearance, new editions of Bailey continued to be issued in rapid succession; and new dictionaries of the size of Bailey, often largely indebted to Johnson's definitions, appeared. An important new feature was moreover added in 1773, when Dr. William Kenrick gave in his *New Dictionary* the Orthoepy or Pronunciation. But of these and many other dictionaries I can say little to-night, except to mention the names of Noah Webster and Charles Richardson. Webster was a great original genius and an independent worker, great as a definer, but weak and often puerile in his derivations, while Richardson's splendid work exalted the function and use of literary quotations far beyond the notion of Dr. Johnson. And I cannot dwell upon the *New English Dictionary on historical principles,* founded upon materials collected by the Philological Society, and prepared at the expense and under the fostering care of the University of Oxford, except to say that it marks another great stage in the development of lexicography, in which the treatment becomes strictly historical, and the dictionary comes to be truly a *history* of every English word, telling of its age, its source, the circumstances in which it arose in English or came into English, its original form and sense, and all the changes of form and development of sense that it has passed through. As this means an enormous collection of facts—for history consists wholly of facts and inferences from facts—the work was, of course, far beyond what could be overtaken by any one man. Accordingly, at the suggestion of the late Archbishop Trench, then Dean of Westminster, the whole English-speaking world was invited to co-operate in this great task, and more than two thousand men and women have actually lent their aid in systematically reading and excerpting some 100,000 books, and thus furnishing the six millions of quotations which supply the facts required; in arranging these in alphabetical order, in classifying them chronologically and according to their senses, in sub-editing sections of the alphabet, and preparing them for final editorial treatment. It would have been of interest, if time had allowed me, to detail the methods by which this has been done, to tell of the multitude of readers, of the number of scholars and men of technical knowledge who have put their special knowledge at our disposal, of the many curious things that have happened to us in the course of twenty years; but it is really not possible, in a discourse upon dictionaries generally, to give a full account of the ideal dictionary, and as my time is exhausted I will only put on the screen a few of the slides which give a glimpse of certain phases of the work.[*]

These show portraits of Dr. Trench, Mr. Herbert Coleridge, and Dr. F. J. Furnivall, great names in the history of the Philological Society's preparatory work; views of the interior of the original Scriptorium at Mill Hill, of a corner of that at Oxford showing the pigeon-holes full of slips, and the editorial staff at work; copies of the slips themselves of various kinds, ordinary and extraordinary; autograph letters on dictionary matters from George Eliot and Lord Tennyson; pages of the dictionary in proof, in revise, and in final.

The second half of the nineteenth century has seen the undertaking of four great dictionaries of the modern languages, which are naturally compared with the New English Dictionary. The famous German dictionary, *Deutsches Wörterbuch*, of Jacob and Wilhelm

---

[*] The discourse up to this point had been illustrated by photographic lantern slides of examples of Old English glosses and beautiful glossed manuscripts, of the Old and Middle English manuscript glossaries and vocabularies referred to, quaint title-pages and interesting typical pages or openings of the early dictionaries, Latin-English, English-Latin, English-Foreign, and of all the Seventeenth and Eighteenth century dictionaries mentioned, with portraits of dictionary-makers, ancient and modern. Portraits of the Brothers Grimm and of M. Littré were also shown later.

Grimm, after several years of preparation began to be printed in 1852. It had advanced to the word *Frucht* when Jacob Grimm died in 1863. After forty years more of work, two generations of his successors have brought it to the end of S, and its completion may be expected, perhaps, in twenty years. The great Dutch dictionary, *Woordenboek der Nederlandsche Taal*, was commenced by Matthias de Vries in 1852 ; its first volume, containing half of A, was published in 1882, and it is now about half finished. Of the new edition of the great Italian *Vocabolario Della Crusca*, which is to a certain extent on historial principles, vol. i., containing A, was published in 1863, and vol. viii., completing I, in 1899, at which rate of progress it may be finished in thirty years. The famous French dictionary of M. Littré, which is, however, a much less extensive, though perhaps better arranged work than any of the preceding, occupied the author twenty years, and its last volume was published in 1873, with a supplement in 1881. None of these great works embraces so long a period of the language, or is so strictly historical in plan as the New English Dictionary ; they are in truth all dictionaries of the modern tongues since the sixteenth century, with a helping glance at the earlier stages of the language. But the immense pre-eminence of the historical method of our dictionary, with its exact dates and absolute chronological order, have been recognised by scholars in every land, who look forward to the time when it will be possible to follow its model in dealing with their own languages ; and already a great dictionary of the Swedish language has been commenced on the pattern of the English, one of the editors of which spent several weeks in Oxford in order to gain a complete insight into our methods. It may be useful to append a table giving the chief landmarks in the progress of English dictionary-making.

A.D.

600. Hard Latin words glossed or explained in easier Latin.

800. Latin English glossaries compiled.

1000. Extensive vocabularies explaining Latin words in English.— *Ælfric.*

1430. Vocabularies rendering English words into Latin begin.— *Promptorium, Catholicon, Elyot, Huloet, Baret, Cooper.*

1530. English Dictionaries of French, Spanish, Italian, Welsh.— *Palsgrave, Percival, Florio, Salesbury.*

1604–1720. Vocabularies of hard English words explained in easier English.—*Cawdrey, Cockeram, Blount, Phillips, Cocker, Kersey.*

1721. Dictionaries of all English words. Derivations given.— *Bailey.*

1755. Illustrative quotations added to show literary usage.— *Johnson.*

1773. Orthoepy or Pronunciation added.—*Kenrick, Perry, Sheridan, Walker.*

A.D.

1806–28. New and independent work at definition.—*Webster.*

1818–36. Attempt to make words explain themselves by abundant Quotations.—*Richardson.*

1857. Philological Society's Dictionary projected. — *Trench, Coleridge, Furnivall.*

1884. First part of New English Dictionary on Historical principles published.

[J. A. H. M.]

WEEKLY EVENING MEETING,

Friday, May 29, 1903.

SIR WM. CROOKES, F.R.S., Honorary Secretary and Vice-President,
in the Chair.

J. Y. BUCHANAN, ESQ., F.R.S. *M.R.I.*

(In the absence of H.S.H. the Prince of Monaco.)

Historical Remarks on Some Problems and Methods of
Oceanic Research.

[Sir William Crookes read a telegram from H.S.H. The Prince of Monaco,
expressing his regret that owing to his accident he was unable to deliver the
discourse, but that he was in a fair way to recovery.]

THE telegram which Sir William Crookes has just read must re-
assure you, as it has reassured me, in regard to the possible couse-
quences of the accident which has befallen H.S.H. the Prince of
Monaco.

It happened only four days before the great Paris-Madrid race
in which so many people were killed and injured, and it must be a
matter of congratulation to all of us that the accident to the Prince,
though serious enough in itself, was of small account compared with
these, and that it is unlikely that it will interfere in any way with
his oceanographical work of this season, which is to deal with an
economic question of the greatest importance to France, namely, the
migration of the sardine. The disappearance of the sardine from
many of the usual fishing grounds has caused much distress amongst
the French fishing folk. When the matter was brought under His
Highness' notice by the French Government he immediately offered
the services of his ship and her appliances, and of himself and his
staff to be devoted exclusively for the whole season to the elucidation
of this question.

This fact alone will show you how great the Prince is as a man
and how world-wide are his sympathies.

You will easily understand that I must claim the greatest
indulgence from you in my endeavour to fill the place left vacant
owing to his unfortunate accident.

It would manifestly be impertinent if I were to try to imagine
what he would have said and to give it to you at second hand. I
may, however, be permitted, before passing on to matters more closely
connected with my own experience, to say a few words about the
history of the Prince's connection with the sea, and how he has come
to be the scientific centre as well as the powerful patron of everything connected with oceanic research. He began his sea life as a

midshipman in the Spanish navy, in which he rose to the rank of lieutenant, after which, as a future reigning sovereign, he could not well continue in the service of a foreign country. He then took to the sea on his own account and, in various very small vessels, he made adventurous cruises in the North Atlantic. It is remarkable that in the course of one of these early cruises he came to be lying in his yacht in the Tagus when the *Challenger* anchored at Lisbon in January 1873, being the first port of call in her long voyage.

The Prince's first yacht of any size was the *Hirondelle*, and she was only a schooner of 100 tons. In her he made numerous trips across the Atlantic, prosecuting his researches always by the way. In these cruises he showed what can be accomplished by patience and perseverance; and at the present day it is almost incredible that he carried out successfully both sounding and dredging by hand in depths of as much as 2000 fathoms. It is only those who know what it is to do work of this kind with all the assistance of steam power, who can truly appreciate the qualities of the man who can get it done by sailors labouring round a capstan perhaps for 24 hours or longer. Last summer some of these early soundings were repeated in his present perfectly fitted yacht, and they were found to be quite exact.

It is important to remember that, like the *Challenger* expedition, the Prince's life-long expedition was begun before there was a science of oceanography, indeed before the word was invented, consequently the development which oceanic research received at his hand was mainly determined by his personal originality. He has always been devoted to the chase in all its forms, whether with gun and rifle, or with fishing rod, net and spear, or with the various instruments and methods known to the experienced trapper, and it was mainly the instinct of the trapper backed by the love of natural history that led him to extend the territory of the chase from the worn-out land to the still virgin sea. On board the *Princesse Alice* and in the new palatial *Musée d'Océanographie* at Monaco all departments of oceanographical research are cultivated, but the Prince's especial line and the one in which he stands ahead of everybody else is the pursuit and capture of the animals inhabiting the sea. With the instinct of the true trapper his most deadly weapon is his knowledge of the habits of the animal that he is pursuing. Every advance which he has made in the art of trapping has enriched science by the discovery of new classes of animals and of new regions, previously believed to be desert, which teem with life. For instance, the extension of the use of the lobster-pot or trap to the greatest oceanic depths has revealed a fauna which eluded all attempts to take it with the dredge or trawl, and the habit of examining the contents of the stomach of larger animals has led to the discovery that the intermediate depths of the ocean are inhabited by a gigantic race of octopus more wonderful than any of the fabled animals of antiquity.

I hope that at no distant date His Highness will be able to lay before us in this place the marvellous results of his unceasing activity in this field of research. Meantime in endeavouring to fill the vacant place I will ask you to let me take you back a matter of 30 years, when the *Challenger* expedition was about to start from this country.

The *Challenger* sailed from Portsmouth on December 21, 1872, and this may not unfairly be taken as the date of the birth of the science of oceanography. Since this date a new generation has sprung up, and it may be interesting to recall some of the circumstances under which the expedition came into being.

I pass over the early work done by different nations (most of it suggested by the obvious importance of bridging the ocean with telegraphic cables) because it would take too much time to do it justice. The real movement for the development of deep-sea research began with the cruise of the *Lightning* in 1868, followed by those of the *Porcupine* in 1869 and 1870. The scientific work of these expeditions was directed by one or other of three men, Dr. Carpenter, Mr. Gwynn Jeffreys and Professor Wyville Thomson. The work done in these expeditions is brought together in 'The Depths of the Sea,' by Professor Wyville Thomson, published on the eve of the sailing of the *Challenger*. The *Porcupine* established a record for the time by dredging in the Bay of Biscay in a depth of 2345 fathoms, where abundance of life was found. The results of these two expeditions aroused public interest in deep-sea research, and it was powerfully supported by the increasing importance of submarine telegraphy. It was in these favouring circumstances that Dr. Carpenter threw himself with heart and soul into the work of promoting an expedition on a great scale, for the exploration on broad lines of the whole of the ocean-covered surface of the globe. It was to his well-directed importunity that the government of the day yielded, and that the pioneer scientific exploration of the ocean fell to the lot of Great Britain. It was probably not unfortunate that the Chancellor of the Exchequer of the day was Robert Low, whose long colonial experience gave him a greater breadth of view than is commonly found at home. In his first letter to Professor Stokes, then Secretary of the Royal Society, Dr. Carpenter puts forward the organisation of an expedition on a large scale as a national duty. The first sentence of his letter of June 15, 1871, runs:—" The information we have lately received as to the activity with which other nations are now entering upon the physical and biological exploration of the deep sea, makes it appear to my colleagues and myself that the time is now come for bringing before our own Government the importance of initiating a more complete and sytematic course of research than we have yet had the means of prosecuting."

The result of Dr. Carpenter's labours was that the expedition was resolved on before the end of 1871, and preparations for its equipment could be commenced at once. Although the formal appointments were not made till much later, most of the members of the scientific staff had been selected and had begun their preparatory

work before the beginning of 1872.   The late Professor Sir Wyville
Thomson was made director of the scientific staff, and I was fortunate
enough to be selected as the chemist and physicist of the expedition,
for which I owe him a debt of gratitude that I can never repay.

On the naval side the expedition was particularly lucky.   Admiral
Richards was the hydrographer and was probably the most experienced surveying officer of the time.   He entered into the proposed
expedition with the greatest enthusiasm, and in Captain Nares he
appointed the best officer of his department to the command of the
ship.   Captain Nares and the officers who had served with him in
various ships and who followed him to the *Challenger* had already
had abundance of experience of sounding and other observations in
great depths, so that when the *Challenger* sailed it was with a naval
staff which thoroughly knew its business.   Owing to the early
selection of the civilian scientific staff it also was able to start
thoroughly knowing its business in so far as that could be known at
the time, and the principal thing that it had to acquire was its sea
legs.   The tempestuous weather of the first ten days of the cruise
enabled it to acquire these without delay.

Almost all the men to whose influence the expedition was due
are dead.   Those who at present are most active in the furtherance
of the science of oceanography were then children, and notwithstanding the voluminous reports of the voyage, it is difficult for
the student of to-day to realise what were the views and expectations
thirty years ago which determined the procedure of the *Challenger* in
breaking ground in the vast dominion of the sea.   I propose to-night
to illustrate this with one or two examples.   The subject is more
fully dealt with in a paper entitled ' A Retrospect of Oceanography
during the past Twenty Years' published in the Report of the sixth
International Congress at London in 1895.

Continuity of the Chalk.—The effect of the work of the *Bulldog*
and of Dr. Wallich's report on it was to produce the belief which was
generally expressed by saying, that at the present time *chalk* is
being laid down all over the deep ocean, that therefore, geologically
speaking, the bottom of the ocean is a contemporary cretaceous
formation.   Thus Wyville Thomson,* after pointing out that whereas the chalk of our downs consists of almost pure carbonate of
calcium with no silica, the chalk mud of the Atlantic contains as
much as twenty or thirty per cent. of silica, and that English chalk
is the very purest of its kind, he goes on to say, " There can be no
doubt whatever that we have forming at the bottom of the present
ocean a vast sheet of rock which very closely resembles chalk; and
there can be as little doubt that the old chalk, the cretaceous formation which, in some parts of England, has been subjected to enormous
denudation, and which is overlaid by the beds of the Tertiary series,
was produced in the same manner and under closely similar circum-

---

* ' Depths of the Sea,' p. 470.

stances; and not the chalk only, but most probably all the great limestone formations."

Further on (p. 495) he says, " I have said at the beginning of this chapter that I believe the doctrine of the continuity of the chalk, as understood by those who first suggested it, now meets with very general acceptance; and in evidence of this I will quote two passages in two consecutive anniversary addresses by presidents of the Geological Society, and we may have every confidence that the statements of men of so great weight, made under such circumstances, indicate the tendency of sound and judicious thought.   Professor Huxley, in the address for the year 1870, says, " Many years ago (*Saturday Review*, 1858) I ventured to speak of the Atlantic mud as 'modern chalk,' and I know of no fact inconsistent with the view which Professor Wyville Thomson has advocated, that the modern chalk is not only the lineal descendant of the ancient chalk, but that it remains, so to speak, in possession of the ancestral estate ; and that from the cretaceous period (if not much earlier) to the present day, the deep sea has covered a large part of what is now the area of the Atlantic.   But if Globigerina and Terebratula, *Caput serpentis*, and Beryx, not to mention other forms of animals and of plants, thus bridge over the interval between the present and the Mesozoic periods, is it possible that the majority of other living things underwent a sea change into something new and strange all at once ? "

The other quotation is from Mr. Prestwich, in 1871, but that from the late Professor Huxley will suffice to show the tendency of opinion at the time.

This was undoubtedly the prevalent view when the *Challenger* sailed from England.   In connection with this it must be noted that the only really oceanic bottom samples that had been examined were those of the *Bulldog*, and were obtained in the moderate depths which are now known to usually cover deposits of *globigerina ooze*. That this is by no means the universal deposit on the bottom of the ocean was proved at the very outset of the cruise.

Having left Portsmouth on December 21, 1872, the *Challenger* called at Lisbon, Gibraltar, Madeira and arrived at Santa Cruz de Tenerife in the Canaries on February 7.   Shortly after the ship dropped anchor a salute was fired on shore.   As the *Challenger* had no guns for saluting purposes an officer was sent ashore to ask the reason of the salute, and he was informed that the mail from Europe which had just arrived had brought news of the birth of a Prince. It is remarkable that while the salute was being fired King Amadeo had already resigned the throne and had left Spain.   The Prince whose birth was thus celebrated was the Duke of the Abruzzi who has made so distinguished a name for himself as an Arctic explorer. Oddly enough, about two years later it was the *Challenger* which brought the news to Manila of the restoration of King Alfonso. Neither of these groups of islands were then, or till long afterwards, on the submarine telegraph system.   On the way to the Canaries the work done by the ship was preparatory, getting the gear and the hands into working order. Regular work commenced on February 13, 1872, when she left the Canary Islands to run her first line of sounding and research stations across the Atlantic to the Island of Sombrero in the West Indies, and fundamental discoveries were made in the very first week. On the second day out, February 15, a sounding in lat. 25° 41′ N., long. 20° 14′ W. gave a depth of 1525 fathoms, but the sounding tube brought back no sample of the bottom. It is remarkable as indicating the views of the time, that the absence of a sample of mud was not supposed to afford evidence that the bottom was not soft, but only that the tube had not acted. It was considered inconceivable that any part of the bottom of the ocean could avoid being covered with mud.

Oceanic ideas, if I may say so, had not yet been born : thought ran only on coasting lines. Notwithstanding the absence of evidence of the nature of the bottom, the dredge was put over and it afforded perhaps the most remarkable haul of the whole cruise. The dredge came up full of masses of jet black branching coral attached to black banded fragments of mineral matter resembling brown coal. In every branch of coral siliceous sponges were sticking like huge birds' nests. I have seen many dredgings and trawlings since, but none that was so striking as this. I see it now before me as clearly as I did thirty years ago on the deck of the *Challenger*. On that day we learned two new things; namely, that " hard ground " occurs in the open ocean as well as in the tidal waters, and that peroxide of manganese and allied ochres are amongst the important oceanic formations. The experience of the *Challenger* was that manganese nodules are found all over the ocean, but principally in the great depths where calcareous deposits are rare or absent. On September 23, 1878, while prosecuting oceanographical researches in the Firth of Clyde * on the steam yacht *Mallard*, which I had built in that year expressly for such work, I found in Loch Fyne in water of about 100 fathoms a rich deposit of mud which contained over 20 per cent. of its bulk of manganese nodules, which, in outward appearance and characteristics as well as in chemical composition, were not to be distinguished from oceanic nodules. This was a very important discovery. Some years later these nodules were found in other parts of the Firth of Clyde. The submarine manganese nodules are a distinct geological formation. Their essential constituent is an ochre, that is, a higher hydrated oxide of one or more of the metals of the iron group. The hydrates of the peroxides of manganese and of iron are present in preponderating quantity, and they are always accompanied by the homologous oxides of nickel and cobalt. The association of these four metals and the constancy of character observed in the nodules, suggested to me as a first idea that they were perhaps simply the products of the oxidation of meteorites. Further

---

* ' Manganese Nodules in Loch Fyne,' ' Nature ' (1878) xviii., p. 628.

acquaintance with them rendered this explanation very improbable. A characteristic feature of the nodules is that when heated in the closed tube they emit a strongly empyreumatic odour and give off steam which condenses to an alkaline liquid. As my attention was thus early directed to the formation of ochres, I carefully studied every occurrence of them. The organic matter revealed by heating in the closed tube was as invariably present as the ochres, and in the many instances, principally in the Pacific, where large fragments of pumice were brought up from great depths, these masses were perforated by annelids and the holes produced were almost always clothed with a black ochreous lining of the same composition as these manganese nodules, and the pumice in the neighbourhood of the holes was stained of blackish brown colour from the same cause.

This frequent occurrence of the ochreous formation in connection with the deep sea annelids and the invariable occurrence of organic matter in freshly collected nodules, suggested the connection of the formation of the ochreous deposits with the organic life on the bottom. Ochres, especially hydrated ferric oxide, are essential constituents of the oceanic " red clay." When the sounding tube brings up a sample of bottom from one of these regions, it is quite usual to find that the upper layers of the samples are of a red colour, while the mud immediately below is of a bluish-black colour. As the dredge furnished evidences of the abundance of life in the mud, as the difference of colour of the upper and lower layers of the mud was evidently due to a different state of oxidation of the iron in it, and as the water in contact with the surface of the mud always showed a deficiency of oxygen, there was little difficulty in concluding that the existence of animal life in the mud had some effect in modifying its chemical composition.

When a freshly collected sample of submarine mud is carefully washed with a jet of water, until the finer flocculent particles are removed, the mud which remains is in the form of elongated casts of ellipsoidal form. Pressure with the finger breaks them up into flocculent particles, which can be washed away with the jet of water, leaving still some ellipsoids. By continuing this treatment, finally all the flocculent matter can be washed away, but the ochreous deposits thus freshly collected and carefully examined are always found to be made up of these ellipsoids which are nothing more nor less than coprolites. The animals, which live in abundance in the mud, live by passing it through their bodies and extracting from it what nutriment they can. The trituration of the mud in the interior of the animals and in contact with living organic matter reduces the sulphates of the sea-water to sulphides. These, in contact with the ferric oxide of the mud, reduce it to ferrous sulphide with separation of sulphur. Hence the mud not immediately in contact with the water has a bluish-black appearance. When it comes in contact with the water which contains free oxygen, the ferrous sulphide is oxidised and the surface layer becomes red. If this is the true explanation we ought to be able to find traces of free sulphur in the mud, although the finely divided sulphur which is produced in this class of reaction is easily oxidised. Acting on this idea, and connecting it with Oscar Peschel's brilliant application of the *Relicten fauna* of lakes and rivers in the diagnosis of morphological terrestrial changes, I treated a series of oceanic muds and manganese nodules with chloroform for the extraction and determination of any sulphur that they might contain. The experiment was successful in every case, and the results are given in a paper * on the occurrence of sulphur in marine muds, read before the Royal Society of Edinburgh. When surveying the Gulf of Guinea in 1886 in the *Buccaneer*, I found this coprolitic character of the mud near the mouth of the Congo so highly developed that in the reports of the soundings I had to introduce a new designation for this class of mud, namely *coprolitic mud*. Returning, however, to the first cruise of the voyage, namely that from Tenerife to Sombrero, it was firmly established that the nature of the deposits in the open ocean varies in a definite way according to the depth of the water. Between 1000 and 1500 fathoms we have the pteropod shell. At greater depths they disappear and the calcareous portion of the mud consists of shells of foraminifera to a depth of about 2500 fathoms. Beyond this depth the foraminifera rapidly disappeared, and at a depth of over 3000 fathoms the mud consisted almost entirely of red ochreous and argillaceous matter. During the second year of the voyage the diatomaceous ooze of the Antarctic ocean was discovered, and later still in the Pacific the radiolarian ooze. Starting therefore with the expectation of finding a more or less universal chalk formation at the bottom of the ocean, the result of the *Challenger's* work in the first two years was to open up a new geological world and to show its dependence on the physical condition of the oceans.

Bathybius.—When the *Challenger* started on her voyage, it was not only expected that the bottom of the sea would be found everywhere covered by a calcareous deposit, but it was believed that it had been shown that the mud at the bottom of the ocean was everywhere associated with an all-pervading organism to which Huxley,† its discoverer, had given the name of *Bathybius*.

The following extract from Wyville Thomson's 'Depths of the Sea,' p. 410, gives a description of a mud in which this mysterious being was believed to be present.

" In this dredging, as in most others in the bed of the Atlantic, there was evidence of a considerable quantity of soft gelatinous organic matter, enough to give a slight viscosity to the mud of the

---

* 'On the Occurrence of Sulphur in Marine Muds and Nodules, and its bearing on their Mode of Formation.' By J. Y. Buchanan, F.R S., Proceedings of Royal Society, Edinburgh (1890), vol. xviii., pp. 17–39.

† Journal of Microscopical Science (1868), vol. viii., p. 1.

surface layer. If the mud be shaken with weak spirit of wine, fine flakes separate like coagulated mucus ; and if a little of the mud in which this viscid condition is most marked be placed in a drop of sea-water under the microscope, we can usually see, after a time, an irregular network of matter resembling white of egg, distinguishable by its maintaining its outline and not mixing with the water. This network may be seen gradually altering in form, and entangled granules and foreign bodies change their relative positions. The gelatinous matter is therefore capable of a certain amount of move-ment, and there can be no doubt that it manifests the phenomena of a very simple form of life.

"To this organism, if a being can be so called which shows no trace of differentiation of organs, consisting apparently of an amor-phous sheet of a protein compound, irritable to a low degree and capable of assimilating food, Professor Huxley has given the name of *Bathybius haeckelii.* If this has a claim to be recognised as a distinct living entity, exhibiting its mature and final form, it must be referred to the simplest division of the shell-less rhizopoda, or if we adopt the class proposed by Professor Haeckel, to the monera. The circumstance which gives its special interest to *Bathybius* is its enormous extent; whether it be continuous in one vast sheet, or broken up into circumscribed individual particles, it appears to extend over a large part of the bed of the ocean ; and as no living thing, however slowly it may live, is ever perfectly at rest, but is continually acting and reacting with its surroundings, the bottom of the sea becomes like the surface of the sea and of the land,—a theatre of change, performing its part in maintaining the "balance of organic nature."

Although *Bathybius* was discovered by Huxley it was Haeckel who popularised it. His paper on 'Bathybius und das freie Proto-plasma der Meerestiefen,'* is one of the most fascinating memoirs that has ever been written.

In reviewing Huxley's article, he says that the most important fact brought out by Huxley's investigations is that the bottom of the open ocean, even in the greatest depths, is covered with enormous masses of free-living protoplasm which exists there in the simplest and most original form, that is, it has no definite shape and is hardly individualised. The fact that these enormous masses of living pro-toplasm cover the great depths of the ocean in preponderating quantity and under quite peculiar conditions, suggests so many re-flections that a book could be written on them. Haeckel asks, "What is this *Bathybius* for an organism? How did it come into being? What becomes of it? What place are we to accord to it in the economy of nature in these abysses?"

Haeckel recognised clearly the far-reaching importance of the discovery. He concludes with the inquiry, "Have we not here the

---

* Haeckel, 'Zeitschrift für biologische studien.'

case of protoplasm coming continuously into being by creation?
We stand here face to face with a series of dark enigmas, the answer
to which we must hope to receive from future investigations."

It must be remembered that the material for the study of *Bathybius* was rare and valuable. Specimens of the mud from the bottom
of the open ocean were then very scarce and were jealously guarded.
It was quite legitimate for Haeckel to look forward for more light
when material would be more plentiful.

In the early part of the cruise all the naturalists sought for
Bathybius, but they found nothing answering to it which showed
motion. Apart, however, from the motion, the white gelatinous
matter like coagulated albumen seemed to be present.

It was obvious that, if an organic body like albumen were present
all over the bottom of the sea, the water taken from the bottom must
necessarily contain enough of it to show clear evidence of organic
matter when evaporated to dryness. Experiments which I made
repeatedly in this direction gave a negative result.

As chemist of the expedition I looked at the matter from a different point of view from that of the naturalists. The nature of the
experiments which I made and their result are best given by quoting
from my report to Professor Wyville Thomson, which was published
in the Proceedings of the Royal Society.*

"If the jelly-like organism which had been seen by some eminent
naturalists in specimens of ocean bottoms and called *Bathybius* really
formed, as was believed, an all-pervading organic covering of the
sea-bottom, it could hardly fail to show itself when the bottom water
was evaporated to dryness and the residue heated. In the numerous
samples of bottom water which I have so examined, there never was
sufficient organic matter to give more than a just perceptible greyish
tinge to the residue, without any other signs of carbonisation or burning.
Meantime, my colleague, Mr. Murray, who had been working according to the directions given by the discoverers of *Bathybius*, had
actually observed a substance like 'coagulated mucus,' which
answered in every particular except the want of motion to the description of the organism, and he found it in such quantity that, if it were
really of the supposed organic nature, it must necessarily render the
bottom water so rich in organic matter that its presence would be
abundantly evident when the water was treated as above described.

"There remained then but one conclusion, namely that the body
which Mr. Murray had observed was not an organic body at all; and
on examining it and its mode of preparation, I determined it to be
sulphate of lime, which had been eliminated from the sea-water,
always present in the mud, as an amorphous precipitate on the

---

* 'Report on chemical work done on board H.M.S. *Challenger*,' by J. Y.
Buchanan, Proceedings of Royal Society (1876), vol. xxiv., p. 605.

addition of spirits of wine. The substance when analysed consisted of sulphuric acid and lime; and when dissolved in water and the solution allowed to evaporate, it crystallised in the well known form of gypsum, the crystals being all alike, and there being no amorphous matter amongst them."

Haeckel relied chiefly on its faculty of being stained by carmine as evidence that the body which he was examining was organic. Sulphate of lime as prepared by the precipitation of an aqueous solution of a calcium salt by alcohol is a perfectly amorphous flocculent precipitate which is coloured intensely by carmine, and the colour is fast as against treatment with spirit. The naturalists on board had great difficulty at first in believing that this reaction was not, as Haeckel thought it was, absolutely decisive of the organic character of the body.

To remove this view, however, it was only necessary to point out that the production of pigments by the staining of amorphous mineral precipitates with organic colouring matters was a very old chemical industry. The pigments so produced are called by the generic name of *Lakes*; and the mineral precipitate most commonly used is hydrate of alumina; but many other substances can be used for the purpose, and it appeared that sulphate of lime when freshly precipitated by alcohol was to be added to the list.

I have dwelt at considerable length on these two doctrines relating to the conditions at the bottom of the ocean, namely that of the continuity in time of the chalk and that of the continuity in space of organic plasma, not only because they characterise the views held by leading naturalists between the years 1868 and 1873, but also because the proving of these doctrines was the immediate motive of much of the early work done on board the *Challenger*. That the result showed that it was impossible to uphold either doctrine, diminishes in nothing the usefulness of their having been put forward as hypotheses, nor does it afford any reason for their being allowed to pass into oblivion.

The Ship and her Equipment.—Before concluding I should like to say a word or two about the *Challenger* as a ship and her equipment. She was a spar-decked corvette, and, when serving an ordinary commission, she carried twenty-one guns. These had been removed and the large ports enabled the ship to enjoy the most perfect ventilation. She was ship-rigged and her engines were able to drive her 11 knots at full speed. Her displacement was about 2300 tons. Like all the men-of-war of her time she was built of wood, with very solid timbers. Her screw propeller could be hoisted up out of the water. This was a great convenience because all the *passage* was made under sail. The whole amount of coal which she could carry was very little more than that required for manœuvring the ship at the sounding and dredging stations. The work at a station generally took the whole day from sunrise to sunset, and every one familiar with steamers knows how expensive in coal is the operation of keeping station.

The material collected at each station had to be examined, preserved and stored, before the ship arrived at the next one. The stations were generally about 200 miles apart, so that in the passage from one port to another a station was made about every second day. This was easily accomplished under sail and it added enormously to the comfort and the interest of the voyage. All the advantages of having a wooden sailing ship were not fully realised at the time. It was not until I had taken part in one or two expeditions in well found iron or steel ships in tropical waters that I found out the discomfort which we escaped by being on board of an " old wooden ship." The temperature of the air in the ship was, of course, never lower than that of the air outside ; but, on the other hand it was never higher. Nothing astonished me more than the perfect uniformity of temperature of the air of the main deck of the ship in the tropics. I was able to make experiments on the effects of pressure on the deep sea thermometers in a hydraulic apparatus on the main deck, which I could not have made anywhere else. The temperature of the air did not vary by one-tenth of a degree (C.) during the whole of the day.

Iron or steel ships, even the magnificent yacht of the Prince of Monaco, get heated through by the sun in the course of the day, and at first they do not cool as much during the night. They are like a black bulb thermometer, they do not lose as much heat as they gain until their temperature has risen a good many degrees above the mean temperature of the air, and that can be pretty high. The voyage of the *Challenger* lasted three years and five months. Of this time three years were spent between the parallels of 40° S. and 40° N., and the greater part of that time, between the tropics. I have no hesitation in saying that the work could not have been carried on continuously in these tropical seas for such a length of time in any other kind of ship. The principal points of advantage were, the thick wooden walls, which completely prevented over heating and over cooling, the splendid ventilation which was provided by the twenty gun ports on the main deck, and the practice of making the passage under sail.

A word with regard to the equipment. Throughout the voyage hemp sounding line and hemp dredging rope were used, and much of the success of the expedition is due to this fact. There was really no temptation to use anything else, because wire sounding had not passed the experimental stage, and all that was known for certain was that the same wire could not be expected to be used in many soundings without breaking. As our sounding line was on every occasion to carry a load of valuable instruments, this risk could not be run. Captain Nares knew that he could do all that was wanted with sounding line, and he was brilliantly justified by the result. There was no question of using wire rope for dredging. It was first used in America by Agassiz in 1877, and on this side of the Atlantic by myself in 1878.

Wire is suitable only for sounding, pure and simple; and the detailed investigation of the form of the bottom of the ocean cannot be carried out with anything else. As there is nothing at the end of the wire but a sounding tube or a sinker, it can be hove up as quickly as the engine will run, and the loss owing to a breakage of the wire is insignificant. For deep-sea research it is entirely unsuitable, because it has to carry valuable instruments. Their value increases largely with the number of times that they are employed, and when they are lost, for instance, in the Pacific, it takes the best part of a year to replace them, and then they are replaced by new instruments which have no history.

The record of the *Challenger* in freedom from accidents is a very brilliant one. She was supplied with two qualities of sounding line designated Nos. 1 and 2, of which No. 2 was inferior in size and quality. It was used during the first two months of the cruise and was found so untrustworthy that it was condemned, and from February 1873 line No. 1 was used exclusively. Before reaching Lisbon three No. 2 lines were lost, with three thermometers. After leaving Lisbon the No. 2 line was used for the last time on February 19, 1873, when it parted, and two thermometers were lost. From this date until the end of the voyage in May 1876, No. 1 line was used exclusively, and it parted on five occasions, namely, April 28, 1873, losing one thermometer; on May 1, 1873, losing one thermometer; on August 16, 1873, losing two thermometers; and on June 14 and June 16, 1874, losing on each of these days two thermometers. After June 16, 1874, no sounding line was lost. In all, therefore, nine sounding lines were lost, and along with them thirteen thermometers. During the whole voyage only two temperature lines were lost with eight thermometers. This immunity from accident was due not only to the excellence of the line, but in a far greater degree to the constant care which was taken of it by the officers and men who had charge of it.

The immunity from breakage of the sounding line enabled the temperature at the bottom of the ocean to be determined time after time with the same thermometer, and even with the same pair of thermometers. The following is a list of the thermometers which were used more than ten times for determining the bottom temperature.

Thermometer . . No.	93	69	68	87	86	83	92	66	89
Times used	78	74	31	29	23	15	13	20	11

In the following table will be found the number of times that certain pairs of thermometers were used together for the determination of the bottom temperature at the same locality.

Thermometers . . . . Nos.	69 68	69 83	69 66	93 86	93 87	93 92
Times used together . . .	28	13	12	23	22	13

The advantage of being able to use the same thermometers for determining the bottom temperature at so many different localities, does not require to be pointed out to any scientific man.

It was part of the regular routine of a station to determine the temperature of the intermediate water from the surface down to 1500 fathoms at every hundredth fathom. In addition, that of the water between the surface and a depth of 100 fathoms was generally determined at every tenth fathom. Sometimes it was determined at every twenty-fifth fathom from the surface to a depth of 300 fathoms. It was usual to use from six to eight thermometers on the line at once, so that the temperature at every hundredth fathom down to 1500 fathoms was effected in two operations. The temperatures at closer intervals, in the water near the surface, were determined also at the rate of eight per operation. Therefore in obtaining the inter-mediate temperatures, first 1500 fathoms had to be run out and hauled in; then 700 fathoms; then one or perhaps two shorter lengths, according to the number of temperatures near the surface which were desired. This service alone entailed the handling of something like 2500 fathoms of line, and, as we have seen, during the whole voyage only two accidents occurred to the temperature line. The following table shows the work which was done by the temperature line.

Interval of Depth.	Number of Separate Temperatures Observed.	Mean of the Depths Corresponding to these Temperatures.
Fathoms.		Fathoms.
From 10 to 100 incl.	1278	59·4
„ 110 „ 200 „	992	160
„ 225 „ 300 „	522	278
„ 400 „ 1000 „	1396	700
„ 1100 „ 1500 „	711	1300
From 10 to 1500 incl.	4899	465

The number of stations which furnished these 4899 temperature observations was 262, which gives an average of 18·7 observations per station. A full station in deep water included either three or four operations and furnished from 20 to 30 temperatures. At many stations the depth was less than 1500 fathoms, and it was not neces-sary always to take temperatures at such close intervals. The main result of the above table is to show that the enormous number of nearly five thousand deep-sea temperatures was obtained with an expenditure of only eight thermometers and one line. It was at the rate of 612 observations for every thermometer lost.

Although specimens of the earliest pattern of Negretti and Zambra's reversing thermometer were received on board the *Challenger* before the end of the voyage, all the temperatures above referred to were taken with the *Millar-Casella maximum and minimum* protected thermometer. It is owing to the use of this instrument that the temperature work of the *Challenger* during little over three years is comparable in amount with all the temperature work which has been done by other ships in the thirty years since her date. The reversing thermometer is an indispensable instrument for observations in isolated depths, and for series of temperatures in the very restricted localities where the great law of the decrease of temperature with increase of depth does not hold. The total extent of these localities is less than one-tenth of that of the whole ocean. They cover the two polar areas and the neighbouring waters which are affected by the presence of ice. In lower latitudes they include only the so-called *enclosed basins*, the largest of which is the Mediterranean, and in these the law holds rigorously down to a definite depth. The whole of the open ocean lying between the parallels of 50° N. and 50° S. can be thoroughly investigated with the protected maximum and minimum thermometer, and if hemp line be used, seven or eight of them, as experience showed, can be safely risked in each operation. The actual pattern to be used is the one with which I supplied the late Mr. Casella on the return of the *Challenger*; and with it I have made all my later temperature investigations, notably, the thermal survey of Loch Lomond* and other Scottish Lakes, as well as that of the Gulf of Guinea which I carried out on board the *Buccaneer* † in the early part of the year 1886. It differs from the original *Challenger* pattern in being longer and having two scales. The one scale carries either Celsius' or Fahrenheit's degrees on enamel slips fixed to the vulcanite backing of the thermometer and close alongside the stem. The other is a scale of millimetres, etched on the stem itself. This is the real scale of the instrument, and the value of its divisions is determined by careful comparison with a standard thermometer. At every observation both scales are read and the readings recorded, and the one always corrects the other in the case of a misreading. In my instruments the length of one Fahrenheit's degree was from 2·5 to 3 millimetres, which enables the temperature of very deep water to be determined with great exactness. An exploring ship should always carry some thermometers reversing by messenger to test cases where from the indications of the maximum and minimum thermometer the law of decrease of temperature with increase of depth seems to be departed from.

Samples of intermediate water were collected at depths of 800, 400, 300, 200, 100, 50 and 25 fathoms, and a separate operation was required for each depth. At a full station this necessitated the handling of 1875 fathoms of line. This service was performed *without the loss of any material whatever.*

---

* Proceedings of Royal Society, Edinburgh, 1885, vol. xiii., p. 403.
† 'The Scottish Geographical Magazine,' April and May, 1888.

The mean depth corresponding to 251 soundings with temperatures, tabulated in the Report on 'Deep-sea Temperatures,' is 2060 fathoms, and 403 separate observations of the temperature at the bottom were obtained. All the accidents, excepting two, happened to the sounding line in the first sixty of these soundings, namely before, and including, the 16th August, 1873. At the beginning of the voyage only one thermometer was used at each sounding and exactly sixty temperature observations correspond to these sixty soundings. The breakages of the sounding line which occurred during this period occasioned the loss of eleven thermometers, and this was due principally to the use of the inferior sounding line (No. 2) during the first months of the voyage.

At almost every one of the remaining soundings, 191 in number, two thermometers were used. The exact number of individual observations of temperature at the bottom was 343. We have seen that during this time, from August 1873 to the end of the voyage in May 1876, only two sounding lines were carried away, namely, those of the 14th and 16th June, 1874, entailing a loss of four thermometers. Therefore, in two years and nine months, 343 independent observations of the temperature at the bottom, in an average depth of 2060 fathoms, were made at an expenditure of four thermometers. This is at the rate of, in round numbers, 86 determinations per thermometer lost. After leaving New Zealand, the whole of the exploration of the Pacific Ocean, occupying eighteen months, was carried out without the loss of a sounding line or of a thermometer, or other instrument attached to it. One thermometer was lost by the parting of the temperature line on June 18, 1875. It may be added that, during the whole voyage, eleven thermometers collapsed under the high pressure, at very great depths.

We see then that, at a full station of average depth the length of sounding line handled was—

For the sounding	2060 fathoms
For intermediate temperatures	2500 ,,
And for intermediate waters	1875 ,,
Total . .	6435

When the depth was 3000 fathoms or more the length of sounding line handled in the day amounted to from 7500 to 8000 fathoms. In addition to this there would be the 3000 fathoms of dredge-line.

In dredging and trawling the *Challenger* was equally successful. From the date of sailing from England to October 3, 1873, or nine months, the dredge or trawl line parted six times. From October 3, 1873, till June 26, 1875, 113 stations, no line parted. In 1875 three lines, and in 1876, two lines parted. In all during three-and-a-half years, during which 354 stations were made, there were only eleven cases of the parting of the dredge or trawl line.

In the ordinary routine on board the *Challenger* the sounding was made with a sinker weighing 336 lbs. The "Baillie" tube which carried it weighed 25 lbs., and the water bottle weighed 20 lbs., so that the total weight at the end of the line was 381 lbs. in air. Excepting the water bottle which was of bronze, all this weight was of iron, and we find its weight in water by deducting one-eighth or 48 lbs. which leaves 333 lbs. as the effective sinking weight. The No. 1 line, which was in daily use, weighed in water 6 lbs. per 100 fathoms (in air it weighed 20 lbs.). Therefore, every 100 fathoms of line used in the sounding added 6 lbs. to the effective sinking weight, but at the same time by its friction it produced a retardation which depended on the velocity of descent. The retarding effect, except in very shallow water, is greater than the accelerating effect, therefore the net effect is one of retardation. With wire the opposite is the case. After the first 50 or 100 fathoms of line have run out, there is a continual and progressive retardation. The line was always allowed to run out free from coils on the deck and without any break or resistance.

Perhaps the greatest advantage which hemp line has over wire for sounding, and more particularly for dredging in deep water, is that it loses about 70 per cent. of its weight when immersed in water, whereas the wire loses only $13\frac{1}{2}$ per cent. Thus, the No. 1 line weighs in air 200 lbs. per thousand fathoms and only 60 lbs. in water. The same length of sounding wire weighs $14\frac{1}{2}$ lbs. in air and 12·6 lbs. in water. The breaking strain of the line is 14 cwt. or 1568 lbs., that of the wire is 210 lbs. The length of the line which weighs 1568 lbs. in water is 26,000 fathoms, while that of the wire which weighs 210 lbs. in water is 16,700 fathoms. Therefore, granting that we can sound in 16,700 fathoms with wire, there are nearly 10,000 fathoms more that can be explored only with hemp, and beyond 26,000 fathoms, if such depths existed, we should not be able to explore them at all.

Of course these limiting depths are purely theoretical, because, each being at its breaking strain, neither the line nor the wire could be hove up from them. They serve, however, to accentuate a very real advantage which the hemp line has over the metal wire. This advantage will make itself practically felt in dredging in the great depths of 4000 and even 5000 fathoms which are now known to exist. For instance, the wire rope used by Agassiz on board the *Blake* had a circumference of $1\frac{1}{8}$ inch. One fathom of it weighed 1·14 lb. in air and 1 lb. in water. Its breaking strain was 8750 lbs., so that its *breaking length* was 8750 fathoms in water and at rest. It is obvious that if it were to be used for dredging in 5000 fathoms the remaining 3750 lbs. would be quite inadequate to bear the weight of the dredge with its contents, and the strain which would have to be exerted in order to bring it up in a reasonable time, to say nothing of the margin for safety, or of the resistance while being dragged over the ground.

Much misunderstanding prevails about the relative rapidity with which sounding and dredging operations can be carried on with hemp and with wire. In well appointed cable ships, soundings in 2000 fathoms of water can be made in an hour from start to finish, but in order to do this the wire must be hove in as fast as possible and breakages frequently occur. When the wire carries deep-sea thermometers or other valuable instruments, it is impossible to work at this rate with any regard to the safety of the instruments. With hemp line the maximum rate of working can be observed whether the line carries instruments or not, because, if the line has been properly cared for, breakages do not occur.

The following short table shows the ordinary rate at which deep soundings were carried out on board the *Challenger* :—

Depth (d).		Time running out, with Sinker weighing		Time heaving in. Sinker slipped.	Average Rate of heaving in per minute.		Total Time of Sounding in Depth (d).
		3 cwt. or 150 kilos.	4 cwt. or 200 kilos.				
fathoms.	metres.	minutes.	minutes.	minutes.	fathoms.	metres.	minutes.
0	0	0	0	0	0	0	0
1000	1800	11·5	10·5	20	50	90	30·5
1500	2700	19·9	17·7	35	42·9	77·2	52·7
2000	3600	29·4	26·2	50	40	72	76·2
2500	4500	39·8	35·75	70	35·7	64·3	105·75
3000	5400	51·1	46·2	90	33·3	59·9	136·2
3500	6300	63·1	57·4	115	30·4	54·7	172·4
4000	7200	75·1	68·7	140	28·5	51·3	208·7
4500	8100	87·1	80·0	170	26·5	47·7	250·0

With regard to the dredge or trawl line, when clear of the bottom it was brought up from 3000 fathoms at the average rate of 1000 fathoms per hour.

In conclusion, I must thank you for the patient hearing which you have given me while I have been reciting facts which to me are old and to some of you will sound old-fashioned, but I am convinced that to the majority of you they are new. I look forward to an early date when I hope we shall meet again to hear the Prince's lecture which has been so unfortunately postponed.

[J. Y. B.]

WEEKLY EVENING MEETING,

Friday, June 5, 1903.

GEORGE MATTHEY, Esq., F.R.S., Vice-President, in the Chair.

PROFESSOR H. H. TURNER, D.Sc. F.R.S. Pres. R.A.S.

The New Star in Gemini.

DISCOVERY forms only a small part of the work of an astronomer; indeed, so engrossing is the study of the heavens under their normal aspect, that some astronomers are almost indifferent to new objects. It has been declared that the discovery of a new comet is merely a matter of money, i.e. of automatic searching for a given length of time. Those who note only the large comets, of which we have not seen one since 1882, may be surprised to hear that 100 faint comets have been discovered and watched since then. But during the same period only ten new stars have been found, as may be seen from the following complete list. (Harvard Circular No. 4, extended.)

LIST OF NEW STARS.

Ref. No.	Constellation.	Year.	Discoverer.
1	Cassiopeia . . .	1572	Tycho Brahé.
2	Cygnus . . .	1600	Janson.
3	Ophiuchus . . .	1604	Kepler.
4	Vulpecula . . .	1670	Anthelm.
5	Ophiuchus . . .	1848	Hind.
6	Scorpio . . .	1860	Auwers.
7	Corona Borealis . .	1866	Birmingham.
8	Cygnus . . .	1876	Schmidt.
9	Andromeda . . .	1885	Hartwig.
10	Perseus . . .	1887	Fleming.
11	Auriga . . .	1891	Anderson.
12	Norma . . .	1893	Fleming.
13	Carina . . .	1895	Fleming.
14	Centaurus . . .	1895	Fleming.
15	Sagittarius . .	1898	Fleming.
16	Aquila . . .	1899	Fleming.
17	Perseus . . .	1901	Anderson.
18	Gemini . . .	1903	At Oxford.

Of these no less than six were found by the scrutiny of photographic plates at Harvard College Observatory, most of them, however, only being recognised long after the original outburst. These were all found by Mrs. Fleming, who spends her life in examining the vast collections of photographs taken under the enterprising directorate of Professor Pickering.

Of the remaining four, two fall to Dr. Anderson, who watches the heavens assiduously from his quiet study window in Edinburgh. The first, Nova Aurigæ, had been shining for two months in the sky without attracting attention, as was found by scrutiny of the Harvard photographic records. If these could have been minutely examined as they were taken, the star would have been found two months earlier. But such immediate and comprehensive scrutiny is as impossible with the limited staff of the observatory as it would be to have all the books received at the British Museum read through as they came in. With the second discovery, Nova Persei, Dr. Anderson was more fortunate, catching the star within about twenty-four hours of its blazing up; and the news being flashed all over the world at once, the history of the star was well recorded from the very outset.

There remain two more of the ten discovered since 1882, the first and the last. The first, Nova Andromedæ, appeared in the midst of the great nebula in Andromeda, and its discovery was probably due to the special attention naturally paid to that wonderful object. It need not concern us further at present than to remark that there is a close connection between new stars and nebulæ, of which this is only one illustration.

The last of the ten, Nova Geminorum, was found at the University Observatory at Oxford on March 24, and is the reason of my giving this lecture. The discovery, although made in the course of regular and laborious work on the stars, was not the outcome of any systematic search. I will first briefly explain the work we are doing at the University Observatory. In conjunction with seventeen other observatories scattered over the world, we are making a great map of the whole sky. All the partners have instruments of similar pattern, and each of them has undertaken to take about 1200 photographs, which will cover twice over the portion of the sky assigned, and to measure carefully the positions of all the stars on these plates. There are about 400 stars on each plate, and each observatory must, therefore, measure nearly half-a-million star images. This is a considerable piece of work, and although at Oxford we have adopted the most rapid and economical methods of getting through it, it has occupied more than seven years. I am glad to say we are getting near the end ; about sixty or seventy plates only remain to be measured, and we hope to finish them before the end of the present year.

In this hope we have been making special efforts since the beginning of 1903 to get all the plates still required, so that none might be left over till 1904. Taking advantage of a spell of fine weather in February, Mr. Bellamy and I had secured about twenty or thirty plates, but on proceeding to develop them we found they were wofully deficient in faint stars. We had independent tests of developer and sky, and were driven to conclude that the plates were inferior ; and on referring to the makers (Messrs. Elliott and Son) they frankly and apologetically admitted an inferiority, and sent us some specially good ones in substitution. But all our work was to be done over again, with opportunities restricted by lapse of time; and working in this way under pressure, Mr. Bellamy was not very much surprised to find that one of the plates must be taken yet a third time, not because of any fault in the plate, but because the telescope had not been pointed to precisely the right region. This pointing, or "setting," is done by help of a star whose position is known—generally the brightest star in the region, since it is most easily identified; and if two stars are seen in the approximate place, it is generally safe to conclude that the brighter is the one required. In the case in question, two stars *were* seen, though the brighter did not seem to comply with the conditions. The brighter star was, in fact, the new star, Nova Geminorum, but Mr. Bellamy naturally had had no intimation of this, and when he found the plate wrongly " set " merely put it aside as defective.

A few days afterwards, some fortunate impulse led me to examine again the rejected plates, and on learning that this particular plate was wrongly " set," I took down another showing some of the same stars, to find out the reason. A mere glance was enough to show that there was a strange object on the rejected plate, and the exciting question arose, What was it? a planet, a variable star, or a new star? I need not trouble you with the details of the steps taken during that busy afternoon to discriminate between these three possibilities. It will suffice to say, that as night drew near it seemed fairly probable that the strange object could only be a new star, and if so, I was the first man to hold consciously in my hand the discovery photograph of a new star: for although six new stars had previously been found on photographs, they had all been found by a woman. The final test was still to be made if the weather was fine; was the object still there? For it was now March 24, and the photograph had been taken on March 16. The night was clear; the verification was made; and I may say, parenthetically, I have never been more surprised or delighted than when a differential transit and declination observation gave results agreeing precisely with my measures made during the afternoon on the photograph. And then it only remained to send postcards to such observers as would receive them before the next night, a letter to Dr. Anderson, and a telegram to the Central-stelle at Kiel.

It was certainly very pleasant when the answers to these communications began to come in—when there came a postcard from Mr. Newall saying that a spectroscopic observation supported the view that the object was a new star; and a note of congratulation from Dr. Anderson himself. And now we took a new view of the conduct of Messrs. Elliott in sending us that bad batch of plates; had they been good we should never have found the new star, for it appeared after the region had been taken the first time, and had the plate been successful we should not have taken another.

2 c 2

Nature of New Stars.

When I pass from this account (however imperfect) of the discovery of new stars to consider what they are, it is like leaving firm ground for quicksand. The inquiry is at present in that stage which is sometimes summed up by the daily papers in the phrase, " The police have a clue and are investigating the matter." The astronomical police think they have a clue, or rather several clues, to the origin of these outbursts, and various able members of the force are diligently investigating the matter; but they are as yet far from being able to prove their case against any suspected agency. The culprits to whom the evidence seems to point are the nebulæ— in almost every instance where it has been possible to collect information, a nebula appears if not definitely as the originator of the outburst, at least as an accessory after the fact. The new star of 1885 appeared in the midst of the nebula in Andromeda as we have seen; and one of those found at Harvard (Nova Centauri, 1895) also originated within a visible nebula.

The suggestion I wish to make to the jury is that in other cases also, although nebulæ could not be seen, *they were really there.* And I proceed to give evidence in support of this view.

Firstly, it is quite certain that some nebulæ are not seen merely because they are too faint for our present telescopes. For as telescopes have increased in power, and plates in sensitiveness, and exposures in duration, more and more nebulous matter has been revealed, and there is no reason to suppose that we have yet reached the limit. Illustrations will be given on the screen of successive advances in our knowledge of the nebulæ near the Pleiades.

But, secondly, it seems probable that there are nebulæ which are not shining at all, and of which we can only get evidence from their screening or obstructing the light of other bodies. Thus in the picture of the nebula near ζ Orionis there are many more stars on one side of the nebulous boundary than on the other. Do the stars really stop at this boundary? or is it not much more likely that the nebula here changes in character and becomes more obstructive to the light of stars beyond?

Again, there are dark patches of the sky apparently devoid of stars. The first thought that occurs to us in seeing one of them is that it is a black hole—*ein Loch im Himmel*, as Herschel said. But can it really be a gigantic tunnel piercing the whole universe and pointed directly towards us? Both these conditions are necessary for this explanation. A much simpler idea is that there is some obstructing body—which we may call a dark nebula—comparatively near us, hiding the light from stars beyond. Thus the idea of dark nebulæ was already in the minds of astronomers before the discovery of Nova Persei in February 1901, and we have now to notice a remarkable series of photographs which materially strengthen the case for the existence of such objects.

Nova Persei, like all new stars, rose to maximum brightness very suddenly and then faded away. The point which chiefly concerns us is that the star was only very bright, i.e. brighter than the second magnitude, for about a week. It remained visible to the naked eye for about two months, and there was a secondary outburst on March 26, but it is to the first week during which the star was conspicuously bright that I wish to direct your attention, for this great sudden flare had important consequences later in the star's history.

About six months after this flare, when the star had faded and became invisible to the naked eye, photographs taken of it with powerful instruments showed that it was surrounded by a nebula. As successive pictures were taken it was seen that this nebula was rapidly expanding in all directions, and the thought suggested was that of a great explosion driving fragments outwards from the centre. In comparing the phenomenon with any terrestrial explosion, let us say that of a rocket in mid air, allowance must of course be made for great differences of space and time. The fiery sparks from a rocket are scattered a few yards only within a second or two; the fragments of Nova Persei were being projected for millions of miles and the time was months instead of seconds. How *many* millions of miles we could not tell until we knew the distance from which we were viewing this magnificent explosion; and it presently became clear, by the method of parallax, that the star must be *at least* 10 billions of miles away, and might be at any greater distance. Thence it could be readily inferred that the velocity of the fragments must be at least 10,000 miles per second, and might have any value greater than this.

Now, although such a speed is too great, according to our experience, for material bodies, there is a well-known speed which is greater still—the velocity of light, which is nearly 200,000 miles per second. There was no reason whatever, why Nova Persei should not be 200 billion miles away from us, in which case the observed velocity of expansion would be that of light; and now a new interpretation of the facts was suggested, which perhaps an analogy from our solar system will help to elucidate.

The planets which circle round the Sun do not shine with light of their own, but with light received from the Sun and merely reflected by them; and the distances of the planets from the Sun are such that light, although travelling with enormous speed, takes an appreciable time to reach them. Thus the light we receive on the Earth takes eight minutes to reach us, and the light received by Saturn, which is ten times as far away as we are, takes 80 minutes. If the Sun were suddenly put out we should not know it for eight minutes, since we should still receive light during that time which had started on its journey before the extinction; and Saturn would receive light for 80 minutes, so that we should still see Saturn shining with reflected sunlight for more than an hour after the Sun had gone out. If the Sun were suddenly set ablaze again, we should be illuminated eight minutes afterwards, but it would be more than an hour before we saw Saturn.

Now suppose that we were at a great distance from the solar system, in a direction perpendicular to the planes of the planetary orbits, and that the Sun, originally dark, were to flame up suddenly to great brightness for one minute only, as Nova Persei did. Then after eight minutes we should see a speck of light appear where the Earth happened to be, or if there were, instead of a single Earth, a

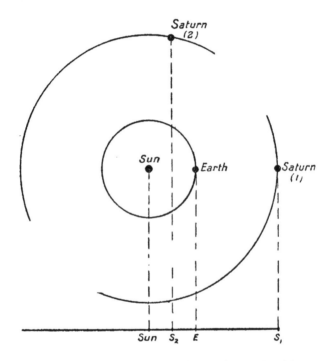

number of small earths scattered round its orbit, we should see a ring of light instead of a speck; and the illuminated ring would last just one minute and then disappear. After 80 minutes we should see Saturn as a speck of light, or again if we suppose him spread round his orbit, we should see this wider ring. If there were enough intermediate planets the appearance would be that of a continually expanding ring of light, similar to the expanding ripple when a stone is thrown into water. It is not the water which expands, but the wave motion, and so it is not, in the case of Nova Persei, actual matter which is exploded outwards, but *illumination* which lights up successively matter which was already there. The supposition is that all round Nova Persei there is one of those dark nebulæ of which we have spoken, and that the brilliant illumination of the star in the first week is successively lighting up for our inspection the parts of this nebula further and further from the centre.

There are some difficulties about this explanation which I do not wish to under-estimate, though I think they may be smoothed away. Some astronomers, for instance, feel it to be a difficulty that in the actual case of Nova Persei the expansion is at many points decidedly *not* outwards from the centre. But a little further consideration of our analogy will show that we must not expect this symmetrical expansion. We have previously supposed the solar system to be viewed from above, but let us now shift our imaginary view point to a position in the plane of the orbits. It is easy to see that Saturn may now appear either (1) on the side of the earth remote from the Sun, or (2) between the Sun and the Earth. In the former case the illumination which reaches the Earth before Saturn will appear to be travelling in the normal direction outwards, but in the latter case it will actually appear to travel *inwards*, exactly contrary to expectation. And it is easy to see that, by choosing our view point, we could obtain any anomaly in the perspective view, so that such anomalies in the case of the illumination of a vast irregular nebula need not trouble us.

On the other hand this explanation of the phenomena has received during the last few weeks a remarkable confirmation. If the light we are now receiving form the nebula is the light of the original flare of the star, we ought to be able to identify it by its colour, or by its spectrum which is a glorified name for colour. If, for instance, Nova Persei had flared up with a fierce red light which afterwards turned to green, and the nebula were to shine with a green light, we could not entertain the view that the nebula owed its illumination to the original red flare: on the other hand, a red light from the nebula would so far confirm that view of its origin. Now substituting for the terms red and green two varieties of spectrum, this has been found to be the case: the spectrum of the nebula agrees with the spectrum of the star at the outburst, and not with its subsequent spectrum.

This result is easily stated, but its attainment represents a truly marvellous achievement in astronomical photography. The light of the nebula round Nova Persei is so faint that an exposure of some hours is necessary to photograph it at all, even with a powerful telescope which would take a picture of the moon in the hundredth of a second. To get the spectrum, each point of light must be spread out into a line, which dilutes the already faint illumination very considerably, and no one could be blamed for regarding such an enterprise as hopeless. But at the Lick Observatory they have learnt by a long series of successes, to despair of nothing. With the help of his brother astronomers, Mr. Perrine devised a special instrument for the undertaking which would give him the best possible chance, and embarked on an exposure of 36 hours. Think of all that is implied! Such an exposure must of course be spread over several nights. After guiding the telescope with minute care for 9 long hours on one night and closing up the plate carefully before approaching dawn could impress it in the faintest degree, the astronomer cannot yet have the satisfaction of developing the plate and the relief of knowing that it is successful: he must return to his vigil on the next fine night and to his self-denial on its morrow; and so on for several nights. Mr. Perrine was at work altogether on four, and only after all this labour and suspense did he in the dark-room—

> " . . . . dare to put it to the touch
> To win or lose it all."

for it was quite possible, of course, that there would be nothing on the plate after all, if the brightness of the nebula had been over-estimated. How many men are there qualified by an equal experience to appreciate Mr. Perrine's anxiety while that plate was being developed? or his joy when it proved successful and a new fact was given to the world?

Turning now to the spectrum of the new star itself, we have learnt to ask by the spectroscope two questions: firstly, what is the source of light made of; and secondly, how quickly is it approaching us or receding from us?

There are two sets of similar lines side by side in the spectrum of new stars, one set bright, the other dark; indicating firstly a mass of hot hydrogen (for instance) itself shining; and secondly a mass of cooler hydrogen in front of a source of light, forming dark lines by absorption. From the relative position of the lines it would appear that the cool hydrogen is approaching us very rapidly— something like 1000 miles a second faster than the hot. And this suggests an explosion of some kind. We must not be alarmed by the magnitude of the velocity, nor even by its persistence, for Nova Persei is so far away, that even a velocity of 1000 miles per second continued for six months will look quite small: it would not suffice to carry a particle outside the patch which the star makes on a photographic plate.

Giving free scope to the possibilities as regards the scale of these magnificent phenomena, I will venture to collect the main facts enumerated in a single hypothesis for the origin of a new star; and I will ask my audience kindly to remember that the case is not yet considered ripe for judgment.

Far away from us—so far away that the winged-messenger Light only bring us news of the occurrence centuries afterwards—a star wandering through space, enters a vast "dark nebula." The friction of the encounter raises the temperature of the star enormously within a day or two, just as a meteor, on entering our tenuous upper atmosphere is set ablaze in a second or two. The hydrogen and helium and other gases which had been absorbed in the star, on this accession of temperature, are blown out of it in all directions with a velocity of 1000 or 2000 miles per second. The supply of them within the star is sufficient to keep up this eruption for months; and the velocity of the outer layers of gas, which might tend to suffer diminution from expansion or gravitational attraction, is maintained by the pressure of the new gas just issuing from within, until a huge volume of gas is produced which nevertheless at that immense distance would only subtend a few seconds of arc to our view. The eruption is moreover so violent that it rapidly disperses the whole materials of the star, the light dies down, and the mass of gas is left in that condition, not yet precisely understood, which we call a planetary nebula. To this must be added that the fierce light of the original outburst in some way illuminates the vast dark nebula which caused the catastrophe. Of the explanation thus sketched it may I think be said that there is no fact known as yet which is clearly irreconcilable with it. I cannot defend it in detail here for lack of time, and can only refer to one or two points of the general question which have not yet been noticed.

Firstly no special peculiarity has been assigned to the star which, wandering through space, meets the dark nebula; and the evidence collected both positively and negatively supports the view that the star may be just an ordinary star like the vast majority of those we know, including our own sun. When Nova Persei first blazed up, and before it exploded, its spectrum was just that of an ordinary star.

The name " new star " suggests that there was nothing seen in the place before, and this is generally true, but it only means that there was nothing bright enough to attract attention. Prior to the last 10 years there were no maps of stars fainter than those of magnitude $9\frac{1}{2}$; so that all that could be said of the previous history of a " new star " was that it must have been at least of the tenth magnitude. But it may have been precisely similar to our sun nevertheless. I drew your attention specially to the evidence that new stars were at a great distance, say 100 times or even 1000 times the distance of Sirius. A tenth magnitude star brought from 100 times the distance of Sirius and placed alongside him would not be much inferior in brightness, and if the distance were 1000 times as great would far outshine him.

Recently we have learnt a little more of the possibilities of prior existence from the magnificent records accumulated at Harvard; but even these only take us to magnitude twelve or thirteen. Nova Persei may have been shining before the outburst in 1901 as a star of the fourteenth magnitude and no one would have known it; in which case, if it had been brought up alongside Sirius, it would have been easily visible to the naked eye. We must make our surveys far more exhaustive before we can say with any certainty that the bodies which blaze up as new stars were previously " dark," or inferior in intrinsic lustre to other stars. In one case, that of

T Coronæ, the star was indeed previously known as an ordinary star. In this connection, our recent discovery at Oxford is of special interest. By great good fortune, it appeared in a region of which previous records were exceptionally good, for both Dr. Max Wolf at Heidelberg, and Mr. Parkhurst at the Yerkes Observatory, had taken photographs of the region within a month before the outburst, showing very faint stars, to the fifteenth or sixteenth magnitude; and a very faint object was seen to be in or near the position subsequently occupied by the Nova. With great kindness, Dr. Max Wolf immediately sent to us copies of his photograph of February 16, and we measured at Oxford with every care the positions of the Nova and the star suspected to be identical with it. Our verdict is that the positions are not identical, but that there is a difference of some seconds of arc. The photographic images are, however, rather indefinite, and this judgment may be reversed on appeal to Mr. Parkhurst's plate, which we have not yet seen. In any case, the existence of these two plates will carry the inquiry as to the previous history of Novæ one step forward, and our Oxford discovery will have made its contribution to this department of astronomy.

A second point worthy of notice is that all the new stars hitherto discovered lie in or near the Milky Way. So many facts concerning the distribution of the stars group themselves about the Milky Way that we must regard this wonderful belt as of fundamental importance in the structure of the visible universe. There is, for instance, another class of stars only found near the Milky Way, called Wolf-Rayet stars; and it is characteristic of the Harvard Observatory that nearly all of these objects have been discovered there by the examination of photographic plates. We have not time for more than a glance at the possibilities here opened up; but a mere glance is sufficient to indicate the important part played by these objects in the history of our stellar system. It is not too much to say that they may be the key which will unlock the approaches to stores of knowledge hitherto undreamt of.

And it follows that we must pay much more attention to the discovery of new stars in the future. It is practically certain that those hitherto discovered are only a few cases, and probably extreme cases, of events which are constantly happening in lesser degrees. The Harvard systematic search has produced about one new star per year, and we have every reason to believe this is only a fraction of the truth. Accepting it, however, as the whole truth, then during the millions of years which physicists, biologists and geologists concur in demanding for the past of our solar system, there must have been millions of such occurrences. If we are right in thinking that the subjects of such disturbances are ordinary stars like our Sun, may it not be that many of the stars we see have undergone one or more vicissitudes? May not our sun himself have been a "new star" at least once? The accession of heat cannot have been so great as in the case of Nova Persei for instance, or we should have been blown into gas again, but a moderate accession, from a similar cause, does not seem improbable. Geologists have long been asking for some such breach of continuity to explain the Glacial period and other observed facts. There is geological evidence to support for instance such a view as follows: that the Sun had sensibly cooled, producing a Glacial period which covered our Earth with ice; that in some way an accession of heat came to the Sun; the ice was melted and perhaps evaporated, causing a period of mighty torrents; since when the Sun has been gradually cooling to the present time. Direct evidence of such an occurrence we cannot obtain from astronomy, but neither can astronomy directly declare against it: we must listen to what the geologists have to say.

Astronomers can, however, throw more light on the possibilities by diligently investigating new and variable stars; and when issues such as the probable past and the possible future of our solar system are bound up with these researches, their importance must be obvious to everyone. I have little fear that they will be neglected, but as an Englishman I cannot help feeling a grave anxiety lest *we* should drop out of this important work for want of material resources, —for want of men and money, but especially men.

[H. H. T.]

GENERAL MONTHLY MEETING,

Monday, June 8, 1903.

Sir James Crichton-Browne, M.A.,LL.D. F.R.S., Treasurer and Vice-President, in the Chair.

Sir Walter Joseph Sendall, G.C.M.G.

was elected a Member of the Royal Institution.

The Presents received since the last Meeting were laid on the table, and the thanks of the Members returned for the same, viz.:—

FROM

The Secretary of State for India—Report on the Kodaikanal and Madras Observatories for 1902. 4to. 1903.
The French Government—Œuvres de Lavoisier, Tomes V. and VI. 4to. 1892-3.
Accademia dei Lincei, Reale, Roma—Classe di Scienze Fisiche, Matematiche e Naturali. Atti, Serie Quinta: Rendiconti. Vol. XII. 1º Semestre, Fasc. 7-9. 1903. 8vo.
   Classe di Scienze Morali, Storiche, etc. Serie Quinta, Vol. XII. Fasc. 1, 2. 8vo. 1903.
Allegheny Observatory—Miscellaneous Papers, N.S. No. 10. 8vo. 1903.
American Academy of Arts and Sciences—Proceedings, Vol. XXXVIII. No. 19. 8vo. 1903.

American Geographical Society—Bulletin, Vol. XXXV. No. 2. 8vo. 1903.
Asiatic Society of Bengal—Proceedings, 1902, Nos. 6–10. 8vo.
    Journal, Vol. LXXI. Part I. No. 1 and Extra No. 1; Part II. Nos. 2, 3;
    Part III. No. 2. 8vo. 1902.
Astronomical Society, Royal—Monthly Notices, Vol. LXIII. No. 6. 8vo. 1903.
Automobile Club—Journal for May. 4to. 1903.
Bankers, Institute of—Journal, Vol. XXIV. Nos. 5, 6. 8vo. 1903.
Belgium, Royal Academy of Sciences—Bulletin, 1903, Nos. 3, 4. 8vo.
Berlin Academy of Sciences—Sitzungsberichte, 1903, Nos. 1–24. 8vo.
Birmingham and Midland Institute—Meteorological Observations, 1902. 8vo.
    1903.
Borredon, G. Esq. (the Author)—Dell' Attrazione Planetaria. 8vo. 1903.
    La Legge del Sistema Planetario. 8vo. 1903.
Boston Public Library—Monthly Bulletin for May, 1903. 8vo.
British Architects, Royal Institute of—Journal, Third Series, Vol. X. Nos. 13, 14.
    4to. 1903.
British Astronomical Association—Journal, Vol. XIII. No. 7. 8vo. 1903.
Buenos Ayres, City—Monthly Bulletin of Municipal Statistics, March, 1903. 8vo.
California, University of—Publications, 1902. 8vo.
Chemical Industry, Society of—Journal, Vol. XXII. Nos. 9, 10. 8vo. 1903.
Chemical Society—Journal for June, 1903. 8vo.
    Proceedings, Vol. XIX. Nos. 266, 267. 8vo. 1903.
Chicago, John Crerar Library—Eighth Annual Report, 1902. 8vo. 1903.
Clarke, Ernest, M.D. F.R.C.S. M.R.I. (the Author)—Refraction of the Eye. 8vo.
    1903.
Crace, J. D. Esq. (the Author)—Francis Cranmer Penrose: A Memoir. 4to. 1903.
Cracovie, l'Académie des Sciences—Bulletins, Classe des Sciences Mathématiques
    et Naturelles, No. 3; Classe de Philologie, No. 3. 1903. 8vo.
Dax, Société de Borda—Bulletin, 1902, Nos. 3, 4. 8vo.
East India Association—Journal, N.S. Vol. XXXIV. No. 30. 8vo. 1903.
Editors—Analyst for May, 1903. 8vo.
    Astrophysical Journal for May, 1903.
    Athenæum for May, 1903. 4to.
    Author for June, 1903. 8vo.
    Board of Trade Journal for May, 1903. 8vo.
    Brewers' Journal for May, 1903. 8vo.
    Cambridge Appointments Gazette for May, 1903. 8vo.
    Chemical News for May, 1903. 4to.
    Chemist and Druggist for May, 1903. 8vo.
    Electrical Engineer for May, 1903. fol.
    Electrical Review for May, 1903. 4to.
    Electrical Times for May, 1903. 4to.
    Electricity for May, 1903. 8vo.
    Engineer for May, 1903. fol.
    Engineering for May, 1903. fol.
    Feilden's Magazine for May, 1903. 8vo.
    Homœopathic Review for May–June, 1903. 8vo.
    Horological Journal for June, 1903. 8vo.
    Journal of the British Dental Association for May, 1903. 8vo.
    Journal of Physical Chemistry for April, 1903. 8vo.
    Journal of State Medicine for May–June, 1903. 8vo.
    Law Journal for May, 1903. 8vo.
    London Technical Education Gazette for May, 1903. 8vo.
    Machinery Market for May, 1903. 8vo.
    Model Engineer for May, 1903. 8vo.
    Mois Scientifique for May, 1903. 8vo.
    Motor Car Journal for May, 1903. 8vo.
    Musical Times for May, 1903. 8vo.
    Nature for May, 1903. 4to.

Editors—continued.
New Church Magazine for June, 1903. 8vo.
Nuovo Cimento for March, 1903. 8vo.
Page's Magazine for May, 1903. 8vo.
Photographic News for May, 1903. 8vo.
Physical Review for May, 1903. 8vo.
Popular Astronomy for May, 1903. 8vo.
Public Health Engineer for May, 1903. 8vo.
Science Abstracts for April, 1903. 8vo.
Travel for May, 1903. 8vo.
Zoophilist for May–June, 1903. 4to.
Electrical Engineers, Institution of—Journal, Vol. XXXII. Part 3. 8vo. 1903.
Florence, Biblioteca Nazionale—Bulletin, May, 1903. 8vo.
Franklin Institute—Journal, Vol. CLV. No. 5. 8vo. 1903.
Geographical Society, Royal—Geographical Journal for June, 1903. 8vo.
Geological Society—Abstracts of Proceedings, Session 1902-3, Nos. 776-778. 8vo.
　　Quarterly Journal, Vol. LIX. Part 2. 8vo. 1903.
　　Geological Literature added to Library in 1902. 8vo. 1903.
Göttingen, Royal Academy of Sciences—Nachrichten : Mathematisch-physikalische
　　Klasse, 1903, Heft 2. 8vo.
Johns Hopkins University—American Journal of Philology, Vol. XXIV. No. 1.
　　8vo. 1903.
Layton, Messrs. C. and E. (the Publishers)—The Theory of Observations. By
　　F. N. Thiele. 4to. 1903.
Leighton, John, Esq. F.S.A. M.R.I.—Journal of the Ex-Libris Society, May, 1903.
　　8vo.
Life-Boat Institution, Royal National—Annual Report for 1903. 8vo.
Linnean Society—Journal, Zoology, Vol. XXVIII. No. 186. 8vo. 1903.
Manchester Geological Society—Transactions, Vol. XXVIII. Parts 4-7. 8vo.
　　1903.
Massachusetts Institute of Technology—Technology Quarterly, Vol. XVI. No. 1.
　　8vo. 1903.
Merck, E. Esq.—Annual Reports on the Advancement of Pharmaceutical
　　Chemistry, Vol. XVI. 1902. 8vo. 1903.
Meteorological Society—Journal, Vol. XXIX. April. 8vo. 1903.
Middlesex Hospital—Reports for the Year 1901. 8vo. 1903.
Moelans, J. D. Esq. (the Author)—Ballon-Parachute et Aerostat dirigeable. 8vo.
　　1902.
Montana University—Bulletin, Biological Series, Nos. 3, 4. 8vo. 1902.
Montpellier, Académie des Sciences—Mémoires, 2e Série, Tome III. No. 2. 8vo.
　　1902.
Munich, Royal Bavarian Academy of Sciences—Sitzungsberichte, 1903, Heft I.
　　8vo. 1903.
National Physical Laboratory—Reports for the Years 1901 and 1902. 4to. 1902-3.
Navy League—Navy League Journal for May–June, 1903. 8vo.
New South Wales, Agent-General for—Artesian Boring and Irrigation in America.
　　4to. 1902.
North of England Institute of Mining and Mechanical Engineers—Transactions,
　　Vol. LII. Nos. 4, 5; Vol. LIII. No. 1. 8vo. 1903.
Odontological Society—Transactions, Vol. XXXV. Nos. 5, 6. 8vo. 1903.
Paris, Société Française de Physique—Bulletin des Séances, 1902, Fasc. 3. 8vo.
　　1903.
Pennsylvania University—Contributions from the Zoological Laboratory, 1902.
　　8vo.
Pharmaceutical Society of Great Britain—Journal for May, 1903. 8vo.
Philadelphia Academy of Natural Sciences—Proceedings, Vol. LIV. Part 8. 8vo.
　　1903.
Photographic Society, Royal—Photographic Journal for April, 1903. 8vo.
Physical Society of London—Proceedings, Vol. XVIII. Part 4. 8vo. 1903.

Porter, A. W. Esq. M.R.I. (Joint Author)—Elementary Treatise on Electricity and Magnetism. Second edition. 8vo. 1903.

Queensland Government—North Queensland Ethnography: Bulletin, No. 5. fol. 1903.

Rennes University—Travaux Scientifiques, Tome I. Fasc. 1–3. 8vo. 1902.

Rochechouart, Sociéte les Amis des Sciences et Arts—Bulletin, Tome XII. Nos. 3–5. 8vo. 1902.

Rome, Ministry of Public Works—Giornale del Genio Civile for December, 1902. 8vo.

Royal Society of London—Philosophical Transactions, A, No. 336, B. No. 215. 4to. 1903.
Proceedings, Nos. 474, 475. 8vo. 1903.
International Catalogue of Scientific Literature; C. Physics, Part II. 8vo. 1903.

Selborne Society—Nature Notes for May, 1903. 8vo.

Smith, B. Leigh, Esq. M.R.I.—The Scottish Geographical Magazine, Vol. XIX. Nos. 5, 6. 8vo. 1903.

Society of Arts—Journal for May, 1903. 8vo.

Tacchini, Prof. P. Hon.Mem.R.I. (the Author)—Memorie della Società degli Spettroscopisti Italiani, Vol. XXXII. Disp. 4. 4to. 1903.

United Service Institution, Royal—Journal for May, 1903. 8vo.

United States Department of Agriculture—Monthly Weather Review, Feb. 1903. 4to.

United States Geological Survey—Twenty-second Annual Report 1900-1, Parts 1–4 Twenty-third Annual Report, 1901–2. 4to. 1901–2.

United States Patent Office—Official Gazette, Vol. CIII. No. 9; Vol. CIV. Nos. 1–4. 4to. 1903.

Verein zur Beförderung des Gewerbfleisses in Preussen—Verhandlungen, 1903 Heft 5. 4to.

Washington Philosophical Society—Bulletin, Vol. XIV. pp. 205–232. 8vo. 1903.

Welcker, Adair, Esq. (the Author)—A Dream of Worlds beyond us. Fifth edition. 8vo. 1903.

West, Mrs. Samuel, M.R.I. and Mrs. F. E. Colenso, M.R.I.—Sketches from the Life of Sir Edward Frankland, K.C.B. 8vo. 1902.

Western Society of Engineers—Journal, Vol. VIII. No. 2. 8vo. 1903.

Yerkes Observatory—Bulletin, No. 19. 8vo. 1903.

Zoological Society of London—Proceedings, 1902, Vol. II. Part 2. 8vo. 1903.
Transactions, Vol. XVI. Part 8. 4to. 1903.

WEEKLY EVENING MEETING,

Friday, June 19, 1903.

SIR WILLIAM CROOKES, F.R.S., Honorary Secretary and
Vice-President, in the Chair.

Professor PIERRE CURIE, Faculté des Sciences à la Sorbonne, Paris.

Le Radium.

I.

MR. BECQUEREL a découvert en 1896 que l'uranium et ses composés
émettent spontanément des radiations qui présentent des analogies
avec les rayons de Röntgen.   Ces rayons nouveaux impressionnent la
plaque photographique et rendent l'air qu'ils traversent conducteur
de l'électricité.   Ces rayons ne se réfléchissent pas, ne se réfractent
pas, ils peuvent traverser le papier noir et les lames métalliques
minces.*

Les composés du thorium émettent des radiations analogues et
d'une intensité comparable.†   On a appelé *rayons de Becquerel* les
rayons émis ainsi spontanément par certains corps, et nous avons
appelé *substances radio-actives* les substances susceptibles de les
émettre.

Nous avons découvert, Mme. Curie et moi, des substances radio-
actives nouvelles qui ne sont qu'à l'état de traces dans certains
minéraux, mais dont la radio-activité est très intense.   Nous avons
ainsi séparé le *polonium*, substance radio-active analogue au bismuth
par ses propriétés chimiques, et le *radium* ‡ qui est un corps voisin du
baryum.   Mr. Debierne a depuis séparé l'*actinium*, substance radio-
active que l'on peut rapprocher des terres rares. §

Le polonium, le radium, l'actinium émettent des radiations qui,
comme ordre de grandeur, sont un million de fois plus intenses que
celles émises par l'uranium et le thorium.   Avec des substances
aussi actives, les phénomènes de la radio-activité ont pu être étudiés
en détail et un grand nombre de recherches ont été exécutées sur ce

---

* *Becquerel*, C.R. de l'ac. des sciences, plusieurs notes, 1896 et 1897.
Rutherford, Phil. Mag., 1899.

† *Schmidt*, Wied. Ann., t. *65*, p. 141.   *Mme. Curie*, C.R. de l'ac. des sciences,
avril 1898.

‡ Découvert dans un travail fait en commun avec *Mr. Bémont.*

§ *P. Curie et Mme. Curie*, C.R. de l'ac. des sciences, juillet 1898.   *P. Curie
Mme. Curie et Mr. Bémont*, C.R. de l'ac. des sciences, déc. 1898.   *Debierne*, C.R.
de l'ac. des sciences, oct. 1899 et avril 1900.

sujet par divers physiciens dans ces dernières années. Nous ne parlerons ici que du *radium*, parce que nous sommes parvenus à prouver que ce corps constitue un élément nouveau et que nous avons pu l'isoler à l'état de sel pur.* Enfin, ce corps est celui qui a été le plus fréquemment utilisé dans les recherches de physique sur les propiiétés des substances radio-actives.

II.

Les rayons du radium impressionnent les plaques photographiques en un temps extrêmement court. L'action peut se produire à travers un écran quelconque. Les corps sont plus ou moins transparents, mais aucun écran n'est absolument opaque pour le rayonnement du radium.

Les rayons du radium provoquent la phosphorescence d'un très grand nombre de corps : sels alcalins, alcalino-terreux, matières organiques, peau, verre, papier, sels d'urane, etc. ; le diamant, le platinocyanure de baryum et le sulfure de zinc phosphorescent de Sidot sont particulièrement sensibles. Avec le sulfure de zinc phosphorescent la luminosité persiste assez longtemps quand on supprime l'action des rayons du radium.

Le rayonnement du radium est aussi intense quand le radium est placé dans l'air liquide (à — 180) que quand il est à la température ambiante. Voici une expérience qui montre les effets du rayonnement aux basses températures: on place au fond d'une éprouvette en verre une ampoule contenant un sel de radium et un petit écran au platinocyanure de baryum que le voisinage du radium rend lumineux. On plonge ensuite l'éprouvette dans l'air liquide et l'on constate que l'écran au platinocyanure de baryum est au moins aussi lumineux qu'avant l'immersion (expérience). Quand on répète cette expérience avec un écran au sulfure de zinc de Sidot, la luminosité de l'écran diminue fortement à la température de l'air liquide, mais cette diminution est due à la baisse du pouvoir phosphorescent du sulfure de zinc aux basses températures.

Les substances phosphorescentes sont, peu à peu, altérées par une action prolongée des rayons du radium, elles deviennent alors moins excitables et sont moins lumineuses sous l'action de ces rayons.

Les sels de radium sont spontanément lumineux ; on peut admettre qu'ils se rendent eux-mêmes phosphorescents par l'action des rayons de Becquerel qu'ils émettent. Le chlorure et le bromure de radium anhydres sont les sels qui donnent la luminosité la plus intense. On peut en obtenir d'assez lumineux pour que la lumière puisse se voir en plein jour. La lumière émise par les sels de radium rappelle alors comme teinte celle émise par le ver luisant (lampyre). La luminosité des sels de radium diminue avec le temps sans jamais disparaître complètement, et en même temps les sels d'abord incolores se colorent en gris, en jaune ou en violet.

---

* *Mme. Curie*, Thèse à la Faculté des sciences de Paris, 1903.

III.

Les rayons du radium rendent l'air qu'ils traversent conducteur de l'électricité. Quand on approche quelques décigrammes d'un sel de radium d'un électroscope chargé, celui-ci se décharge immédiatement. La décharge se produit encore, bien que plus lentement, lorsque l'on protège l'électroscope par une paroi solide épaisse. Le plomb, le platine absorbent fortement les radiations ; l'aluminium est le métal le plus transparent, les corps organiques absorbent relativement peu les rayons de Becquerel (expériences).

Les rayons du radium rendent également légèrement conducteurs les liquides diélectriques tels que l'éther de pétrole, le sulfure de carbone, la benzine, l'air liquide.*

Les rayons du radium dans certaines conditions facilitent le passage de l'étincelle entre deux conducteurs placés dans l'air. On peut faire l'expérience avec une bobine d'induction B (Fig. 1) ; les pôles du circuit induit P et P′ sont reliés par des fils métalliques à deux micromètres à étincelles M et M′ éloignés l'un de l'autre et offrant deux chemins distincts à peu près équivalents pour le passage de l'étincelle. On règle les micromètres de telle sorte que les étincelles passent à peu près aussi abondamment entre les boules de chacun

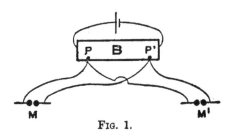

Fig. 1.

d'eux. Quand on approche le radium de l'un des deux micromètres, les étincelles cessent de passer à travers l'autre.

Les rayons les plus pénétrants semblent être les plus efficaces pour la production de ce phénomène ; car, en faisant agir le radium au travers d'une plaque de plomb de 2 centimètres d'épaisseur, l'action sur l'étincelle n'est pas fortement diminuée, alors que la plus grande partie du rayonnement est arrêtée par la plaque.

IV.

Les rayons du radium ne se réfléchissent pas, ne se réfractent pas. Ils forment un mélange hétérogène, et nous les diviserons en trois groupés, que nous désignerons par les lettres a, β et γ suivant la notation employée par Mr. Rutherford.

L'action d'un champ magnétique permet de les distinguer : dans un champ magnétique intense les rayons a sont légèrement déviés de leur trajet rectiligne et cela de la même manière que les "rayons

---

* *P. Curie,* C.R. de l'ac. des sciences, 17 fév. 1902.

canaux" des tubes à vide, tandis que les rayons β sont déviés comme des rayons cathodiques, et que les rayons γ ne sont pas déviés et se comportent comme des rayons de Röntgen.[*]

Le radium R (Fig. 2) est situé au fond d'une petite cavité cylindrique dans un bloc de plomb P. À l'abri de toute action magnétique, le rayonnement s'échappe de la cavité cylindrique sous la forme d'un pinceau rectiligne. Dans un champ magnétique uniforme normal au plan de la figure et dirigé vers l'arrière de ce plan, les

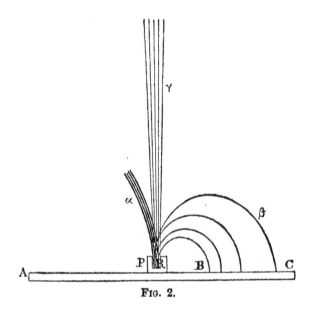

FIG. 2.

rayons β sont fortement déviés vers la droite et suivent un trajet circulaire, les rayons α sont à peine déviés vers la gauche, les rayons γ, de beaucoup les moins intenses, continuent à s'échapper rectilignement.

Les rayons α sont très peu pénétrants. Une lame d'aluminium de quelques centièmes de millimètres d'épaisseur les absorbent. Ces rayons ne sont que faiblement déviés par les champs magnétiques les plus intenses, et pour mettre en évidence cette déviation il faut en réalité employer un dispositif plus délicat qui celui de la figure 2 qui n'est qu'une figure schématique.[†] On peut assimiler ces rayons à des projectiles dont la masse serait comparable à celle des atomes;

---

[*] *Giesel*, Wied. Ann., 2 nov. 1899. *Meyer et Von Schweidler*, Akad. Anzeig. Wien, 3 et 9 nov. 1899. *Becquerel*, C.R., 11 déc. 1899, 26 jan. et 16 fév. 1903. *P. Curie*, C.R., 8 jan. 1900. *Villard*, C.R. de l'ac., t. *130*, p. 1010. *Rutherford*, Physik. Zeitsch., 15 jan. 1903.

[†] *Rutherford*, Phil. Mag., fév. 1903. *Becquerel*, C.R. de l'ac. des sciences, t. *136*, p. 199.

ces projectiles seraient chargés d'électricité positive et se déplaceraient avec une grande vitesse. En dehors de l'action du champs magnétique, les lois de l'absorption des rayons *a* par des écrans très minces superposés suffiraient pour caractériser ces rayons et en faire un groupe distinct.* En traversant des écrans successifs les rayons *a* deviennent en effet de moins en moins pénétrants (tandis que, dans les mêmes conditions, le pouvoir pénétrant des rayons de Röntgen va en augmentant). Il semble que l'énergie de chaque projectile diminue à la traversée de chaque écran.

Les rayons *a* sont ceux qui semblent actifs dans la très belle expérience réalisée dans le *spinthariscope* de Sir William Crookes. Dans cet appareil un fragment très petit d'un sel de radium (une fraction de milligramme) est maintenu par un fil métallique à une faible distance ($\frac{1}{2}$ millimètre) d'un écran au sulfure de zinc phosphorescent. En examinant dans l'obscurité avec une loupe la face de l'écran qui est tournée vers le radium, on aperçoit des points lumineux parsemés sur l'écran et faisant songer à un ciel étoilé ; ces points lumineux

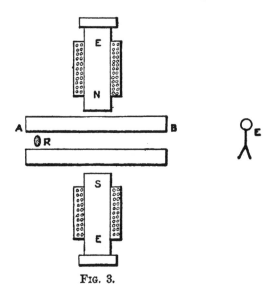

FIG. 3.

apparaissent et disparaissent continuellement. Dans la théorie balistique, on peut imaginer que chaque point lumineux qui apparaît résulte du choc d'un projectile. On aurait affaire pour la première fois à un phénomène permettant de distinguer l'action individuelle d'un atome.

Les *rayons* β sont analogues aux rayons cathodiques. Ils sont déviés par le champ magnétique de la même façon que ces derniers et

---

* *Mme. Curie,* C.R. de l'ac., 8 jan. 1900.

se comportent comme des projectiles chargés d'électricité négative qui s'échapperaient du radium avec une grande vitesse ; ces projectiles (électrons) auraient une masse environ 1000 fois plus petite que celle d'un atome d'hydrogène. L'expérience suivante donne la démonstration de la déviation magnétique des rayons β. Une ampoule en verre renfermant un sel de radium R est placée à l'une des extrémités d'un tube de plomb à parois très épaisses A B (Fig. 3, coupe de l'appareil). On place un électroscope E un peu au delà de l'autre extrémité du tube. Le pinceau de rayons issus du radium et limité par le tube provoque la décharge de l'électroscope. Le tube de plomb est situé entre les branches d'un électroaimant E E et orienté normalement à la ligne des pôles N S. Quand le courant circule dans le fil de l'électroaimant, les rayons β sont rejetés sur les parois du tube de plomb ; ils ne concourent plus à la décharge de l'électroscope, et cette décharge se fait lentement. Quand le courant est supprimé dans l'électroaimant, les rayons β agissent sur l'électroscope qui se décharge rapidement.

On peut prouver que les rayons β transportent de l'électricité négative, et ce résultat est en accord avec l'hypothèse dans laquelle on les considère comme des projectiles chargés d'électricité.* On peut employer pour cela le dispositif expérimental de la Fig. 4 ; le radium R R émet des rayons β ; parmi ces rayons ceux qui s'échappent vers la partie supérieure traversent successivement une feuille mince d'aluminium E E E E reliée électriquement à la terre et une couche isolante de paraffine $i\,i\,i\,i$; ils sont ensuite absorbés par un bloc de plomb M M qui est réuni à un électromètre au moyen d'un fil métallique isolé. On constate que le bloc de plomb M se charge continuellement d'électricité négative. Dans cette

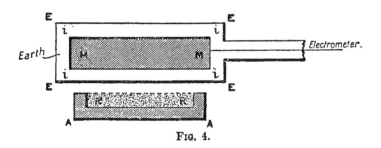

FIG. 4.

expérience les rayons a sont absorbés par la feuille d'aluminium en relation avec la terre. La couche de paraffine est nécessaire pour obtenir un isolement convenable du bloc de plomb M M ; cet isolement serait en effet tout à fait défectueux si le bloc de plomb était entouré d'air rendu conducteur par les rayons, et il serait alors

---

* *M. et Mme. Curie*, C.R. de l'ac. des sciences, 5 mars 1900.

impossible de constater à l'électromètre la charge électrique qui se dégage sur le morceau de plomb.

On peut faire l'expérience inverse : l'auge métallique A A (Fig. 5) est en relation avec l'électromètre et contient le radium R. Le tout est entouré de paraffine *i i i i* et d'une enveloppe métallique E E E en relation électrique avec la terre. Les rayons α très peu pénétrants

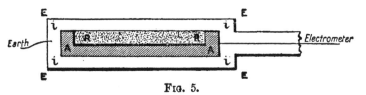

Fig. 5.

ne peuvent s'échapper; les rayons β traversent la paraffine et emportent de l'électricité négative, pendant que l'auge métallique se charge positivement.

Une ampoule de verre scellée et contenant un sel de radium se charge spontanément d'électricité comme une bouteille de Leyde. Si au bout d'un temps suffisant on fait avec un couteau à verre un trait sur les parois de l'ampoule, il part une étincelle qui perce le verre en un point où la paroi est amincie sous le couteau; en même temps l'opérateur éprouve une petite secousse dans les doigts, par suite du passage de la décharge.

· Le groupe des rayons β est constitué par la réunion de rayons qui diffèrent les uns des autres par leur pouvoir pénétrant. Certains rayons β sont absorbés par une lame de $\frac{1}{100}^{me}$ de millimètre d'épaisseur en aluminium, tandis que d'autres peuvent passer en se diffusant à travers une plaque de plomb de plusieurs millimètres d'épaisseur. Ou constate encore que les rayons β diffèrent les uns des autres par la courbure de la circonférence qu'ils décrivent dans un champ magnétique uniforme. Dans l'expérience représentée Fig. 2, les rayons β déviés par le champ magnétique impressionnent la plaque photographique A B C depuis B jusqu'en C. Les rayons les moins déviés impressionnent la plaque dans la région C, les rayons les plus déviés dans la région B. Sur la plaque on aura un véritable spectre produit par les rayons plus ou moins déviables séparés par le champ magnétique. En interposant une lame mince de métal sur le trajet des rayons contre la plaque photographique, on constate que les rayons les plus déviés sont supprimés. Les rayons les plus pénétrants sont donc les moins déviés.*

Dans la théorie balistique on suppose que les rayons β sont formés par des électrons animés d'une vitesse plus ou moins grande. Les rayons les plus pénétrants sont ceux dont la vitesse est la plus grande. Les recherches de Kaufmann interprétées dans la théorie des électrons (sous la forme que lui a donnée Mr. Abraham) conduisent

---

* *Becquerel*, C.R. de l'ac. des sciences, t. *130*, pp. 206, 372, 810.

à des conclusions d'une grande importance générale :* certains rayons β très pénétrants seraient constitués par des électrons animés d'une vitesse atteignant les $\frac{9}{10}$me de celle de la lumière ; la masse des électrons et peut-être celle de tous les corps serait la conséquence de réactions électromagnétiques ; l'énergie nécessaire pour donner à un corps chargé d'électricité une vitesse de plus en plus grande tendrait vers l'infini quand la vitesse du corps tendrait vers la vitesse de la lumière.

Les rayons γ non déviables et analogues aux rayons de Röntgen ne forment qu'une très faible partie du rayonnement total. Certains rayons γ sont extrêmement pénétrants et peuvent traverser plusieurs centimètres de plomb.

On peut utiliser les rayons de Becquerel pour faire des radiographies sans appareils spéciaux. Une petite ampoule en verre contenant quelques centigrammes d'un sel de radium remplace le tube de Crookes. On utilise les rayons β et γ. Les radiographies ainsi obtenues manquent de netteté par suite de la diffusion des rayons β par les corps qu'ils rencontrent. On obtient des radiographies bien nettes en faisant dévier les rayons β avec un électroaimant puissant et en utilisant seulement les rayons γ ; mais les rayons γ étant peu intenses, il faut alors plusieurs jours de pose pour obtenir une radiographie.

V.

Les sels de radium dégagent continuellement de la chaleur.† Ce dégagement est assez fort pour qu'on puisse le montrer par une

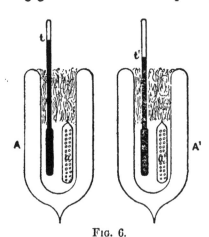

Fig. 6.

expérience grossière faite à l'aide de deux thermomètres à mercure ordinaires. On utilise deux vases isolateurs thermiques à vide, identiques entr'eux (A et A', Fig. 6). Dans l'un de ces vases A on place une ampoule de verre *a* contenant 7 décigrammes de bromure de radium pur ; dans le deuxième vase A' on place une ampoule de verre *a'* qui contient une substance inactive quelconque, par exemple du chlorure de baryum. La température de chaque enceinte est indiquée par un thermomètre dont le réservoir est placé au voisinage immédiat de l'ampoule. L'ouverture des isolateurs est fermée par du coton. Dans ces conditions le thermomètre *t* qui se trouve dans le

---

* *Kaufmann*, Nachrichten der k. Gesell. d. Wiss. zu Göttingen, 1901, Heft 2 ; C.R. de l'ac. des sciences, 13 oct. 1902.

† *Curie et Laborde* C.R. de l'ac. des sciences, 16 mars 1903.

même vase que le radium, indique constamment une température supérieure de 3° à celle indiquée par l'autre thermomètre *t'*.

On peut évaluer la quantité de chaleur dégagée par le radium à l'aide du calorimètre à glace de Bunsen. En plaçant dans ce calorimètre une ampoule de verre qui contient le sel de radium, on constate un apport continu de chaleur qui s'arrête dès que l'on éloigne le radium. La mesure faite avec un sel de radium préparé depuis long-temps indique que chaque gramme de radium dégage environ 80 petites calories pendant chaque heure. Le radium dégage donc pendant chaque heure une quantité de chaleur suffisante pour fondre son poids de glace. Cependant le sel de radium utilisé semble toujours rester dans le même état et, du reste, aucune réaction chimique ordinaire ne pourrait être invoquée pour expliquer un pareil dégagement continu de chaleur.

On constate encore qu'un sel de radium qui vient d'être préparé dégage une quantité de chaleur relativement faible. La chaleur dégagée en un temps donné augmente ensuite continuellement et tend vers une valeur déterminée qui n'est pas encore tout à fait atteinte au bout d'un mois.

Quand on dissout dans l'eau un sel de radium et que l'on enferme la solution dans un tube scellé, la quantité de chaleur dégagée par la solution est d'abord faible ; elle augmente ensuite et tend à devenir constante au bout d'un mois. Quand l'état limite est atteint le sel de radium enfermé en tube scellé dégage la même quantité de chaleur à l'état solide et à l'état de dissolution.

On peut encore évaluer la chaleur dégagée par le radium à diverses températures, en l'utilisant pour faire bouillir un gaz liquéfié et en mesurant le volume du gaz qui se dégage. On peut faire l'expérience avec le chlorure de methyle (à – 21°). L'expérience a été faite aussi par Mr. le professeur Dewar et Mr. Curie avec l'oxygène liquide (à — 180°) et *l'hydrogène liquide* (à – 292°). Ce dernier corps convient particulièrement bien pour réaliser l'expérience : un tube A, Fig. 7 (fermé à la partie inférieure et entouré d'un isolateur thermique à vide de Dewar), contient un peu d'hydrogène liquide H ; un tube de dégagement *t t* permet de recueillir le gaz dans une éprouvette graduée E remplie d'eau. Le tube A et son isolateur plongent tous deux dans un bain d'hydrogène liquide H'. Dans ces conditions

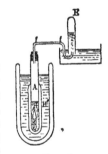

Fig. 7.

aucun dégagement gazeux ne se produit dans le tube A. Lorsque l'on place une ampoule *a* contenant 7 décigrammes de bromure de radium dans l'hydrogène du tube A, il se fait un dégagement continu de gaz hydrogène, et l'on recueille 73 centimètres cubes de gaz par minute.

VI.

Les rayons du radium provoquent diverses actions chimiques. Ils agissent sur les substances employées en photographie de la même façon que la lumière. Ils colorent le verre en violet ou en brun, les sels alcalins en jaune, en violet, en bleu ou en vert. Sous leur action la paraffine, le papier, le celluloïd jaunissent, le papier devient cassant, le phosphore ordinaire se transforme en phosphore rouge. D'une manière générale, les corps qui sont phosphorescents sous l'action des rayons du radium, subissent une transformation, et en même temps leur pouvoir phosphorescent tend à disparaître. Enfin, dans le voisinage des sels de radium on peut constater dans l'air la production d'ozone.

VII.

Les rayons du radium provoquent diverses actions physiologiques.

Un sel de radium, situé dans une boîte opaque en carton ou en métal, agit cependant sur l'œil et donne une sensation de lumière. Pour obtenir ce résultat on peut placer la boîte contenant le radium devant l'œil fermé ou contre la tempe. Dans ces expériences les milieux de l'œil deviennent lumineux par phosphorescence sous l'influence des rayons du radium et la lumière que l'on aperçoit a sa source dans l'œil lui-même.[*]

Les rayons du radium agissent sur l'épiderme ; si on tient pendant quelques minutes une ampoule contenant du radium sur la peau, on n'éprouve aucune sensation particulière, mais, 15 à 20 jours après, il se produit sur la peau une rougeur, puis une escharre dans la région où a été appliquée l'ampoule ; si l'action du radium a été assez longue, il se forme ensuite une plaie qui peut mettre plusieurs mois à guérir. L'action des rayons du radium sur l'épiderme est analogue à celle produite par les rayons de Röntgen. On essaye actuellement d'utiliser cette action dans le traitement des lupus et des cancers.[†]

Les rayons du radium agissent encore sur les centres nerveux et déterminent alors des paralysies et la mort. Ils semblent aussi agir d'une façon particulièrement intense sur les tissus vivants en voie d'évolution.[‡]

VIII.

Lorsque l'on place un corps solide quelconque dans le voisinage d'un sel de radium, on constate que ce corps acquiert les propriétés radiantes du radium ; il devient radio-actif. Cette *radio-activité*

---

[*] *Giesel*, Naturforscherversammlung München, 1899. *Himstedt et Nagel*, Ann. der Physik, t. *4*, 1901.

[†] *Walkhoff*, Phot. Rundschau, oct. 1900. *Giesel*, Berichte d. deutsch. chem. Gesell., t. *23*. *Becquerel et Curie*, C.R. de l'ac., t. *132*, p. 1289.

[‡] *Danysz*, C.R. de l'ac. des sciences, 16 fév. 1903. *G. Bohn*, C.R. de l'ac. des sciences, 27 avril 1903.

induite persiste encore un certain temps quand on éloigne le corps du radium, cependant elle s'affaiblit progressivement, elle diminue de moitié environ pendant chaque demi-heure et finit par s'éteindre.

Ce phénomène se produit d'une façon régulière et particulièrement intense si l'on enferme les corps avec un sel de radium dans une enceinte close. Il y a aussi grand avantage à placer dans l'enceinte une solution d'un sel de radium plutôt que le sel solide.*

Une solution d'un sel de radium est située en A, Fig. 8, dans un réservoir en verre qui communique par les tubes *t* et *t'* avec deux autres réservoirs en verre remplis d'air B et C.
On constate que les parois des réservoirs B et C sont radio-actifs, ils émettent des rayons de Becquerel analogues à ceux émis d'ordinaire par le radium lui-même, tandis que, au contraire, la solution du sel de radium émet très peu de rayons, la radio-activité est en quelque sorte extériorisée.

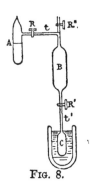

FIG. 8.

Les phénomènes qui viennent d'être décrits se produisent aussi bien dans un autre gaz que l'air et cela quelle que soit la pression du gaz. La radio-activité se communique de proche en proche par une sorte de conduction à travers les gaz ; elle peut même se propager d'un réservoir à un autre par un tube capillaire. Le gaz qui a séjourné près du radium a donc acquis la propriété de rendre les corps solides radio-actifs, le gaz lui-même est du reste radio-actif mais il n'émet que des rayons très peu pénétrants. (Les rayons émis par le gaz ne peuvent pas traverser les parois d'une réservoir en verre.) Lorsque le gaz ainsi modifié est entraîné loin du radium, il conserve assez longtemps ses propriétés ; il continue a émettre des rayons de Becquerel très peu pénétrants et à provoquer la radio-activité des corps solides. Son activité à ce double point de vue diminue cependant de moitié pendant chaque période de quatre jours et finit par s'éteindre.

Mr. Rutherford suppose que le radium dégage constamment une substance gazeuse radio-active qui se répand dans l'espace et provoque les phénomènes de la radio-activité induite. Il donne à cette substance hypothétique le nom d'*émanation du radium* et pense qu'elle se trouve à l'état de mélange dans les gaz qui ont séjourné dans le voisinage du radium. Sans admettre nécessairement la nature matérielle de l'émanation, on peut employer cette expression pour désigner l'énergie radio-active de forme spéciale emmagasinée dans le gaz.†

L'air chargé d'émanation provoque la phosphorescence des corps

---

* *Mr. et Mme. Curie*, C.R. de l'ac. des sciences, 6 nov. 1899. *Curie et Debierne*, C.R. de l'ac. des sciences, 4 mars 1901, 29 juillet 1901, 25 mars 1901.
† *Rutherford*, Phil. Mag., 1900, 1901, 1902, plusieurs mémoires. *Dorn*, Abh. Naturfrshgesel. Halle, juin 1900. *P. Curie*, C.R. de l'ac. des sciences, 17 nov. 1902, 26 jan. 1903.

qui se trouvent en sa présence ; le verre (plus particulièrement le
verre de Thuringe) donne une belle phosphorescence blanche ou verte.
Le sulfure de zinc de Sidot devient excessivement brillant sous
l'action de l'émanation.* On peut faire l'expérience avec l'appareil
représenté Fig. 8. Le robinet R étant fermé l'émanation radio-active
qui se dégage de la solution de sel de radium en A se répand dans
l'air au-dessus de la solution. Lorsque l'émanation s'est ainsi
accumulée en A pendant quelques jours, on fait le vide dans les
réservoirs B et C, dont les parois intérieures sont enduites de sulfure
de zinc phosphorescent. On ferme ensuite le robinet R″, et on ouvre
le robinet R. L'air chargé d'émanation est alors aspiré brusquement
dans les réservoirs B et C qui deviennent aussitôt lumineux.

L'émanation du radium se comporte comme un gaz à bien des
points de vue : elle se partage comme un gaz entre deux réservoirs
qui communiquent entre eux. Elle se diffuse dans l'air suivant la
loi de diffusion des gaz et possède un coëfficient de diffusion voisin
de celui de l'acide carbonique dans l'air.†

Mrs. Rutherford et Soddy ont découvert que l'émanation a la
propriété de se condenser à la température de l'air liquide.‡ On
peut montrer les effets de cette condensation en faisant encore usage
de l'appareil représenté Fig. 8. Le robinet R″ étant fermé et
l'émanation étant répandue dans tout l'appareil comme à la fin de
l'expérience précédemment décrite, les réservoirs B et C (couverts
intérieurement d'une couche de sulfure de zinc de Sidot) sont
lumineux. On ferme alors le robinet R et on plonge le réservoir C
dans l'air liquide. Au bout d'une demi-heure, on constate que le
réservoir B a perdu toute sa luminosité, tandis que le réservoir C
est encore lumineux. L'émanation a en effet quitté le réservoir B et
est venue se condenser en C dans la partie refroidie. Cependant le
réservoir C n'est pas très lumineux, parce que la phosphorescence du
sulfure du zinc est plus faible à la température de l'air liquide qu'à
la température ambiante. On ferme ensuite le robinet R′, ce qui
interrompt la communication entre les deux réservoirs B et C, on
retire le réservoir C de l'air liquide et on le laisse revenir à la
température ambiante. Le réservoir C est alors vivement illuminé
tandis que le réservoir B est toujours obscur ; l'émanation qui, au
début de l'expérience, était répandue dans les deux réservoirs se
trouve en effet, tout entière, maintenant dans le réservoir C.

Les expériences précédentes conduisent à assimiler l'émanation à
un gaz analogue à un gaz matériel. Cependant, jusqu'ici, l'hypothèse
de l'existence d'un pareil gaz est uniquement basée sur des mani-
festations radio-actives. Remarquons encore que, contrairement à
ce qui se passe pour la matière ordinaire, l'émanation disparaît spon-

* *Curie et Debierne*, C.R. de l'ac, 2 déc. 1901.
† *Curie et Danne*, C.R. de l'ac. des sciences, 1903. *Rutherford et Miss Brookes*,
Chemical News, 1902, 29 avril.
‡ *Rutherford et Soddy*, Phil. Mag., mai 1903.

tanément dans un tube scellé qui la renferme ; la quantité d'émanation diminue de moitié en quatre jours, et cette constante de temps est une donnée caractéristique de l'émanation du radium.

IX.

Après avoir énuméré les propriétés principales du radium, il convient de rappeler brièvement l'origine de sa découverte à laquelle Mme. Curie a pris une très grande part.*

L'étude des corps renfermant de l'uranium et du thorium avait montré que *la radio-activité est une propriété atomique* qui accompagne partout l'atome de ces deux corps simples ; la radio-activité d'une substance composée est en général d'autant plus forte, que la proportion du métal radio-actif contenue dans cette substance est elle-même plus grande. Certains minéraux d'uranium : la pechblende, la chalcolite, la carnotite, ont, cependant, une radio-activité plus forte que celle de l'uranium métallique. Nous nous sommes demandés si ces minéraux ne renfermaient pas, en petite proportion, quelques substances encore inconnues et fortement radio-actives, et nous avons recherché ces substances hypothétiques par les voies de l'analyse chimique, en nous guidant constamment par la radio-activité des matières traitées. Nos prévisions ont été vérifiées par l'expérience ; la pechblende contient des substances radio-actives nouvelles, mais ces substances sont dans le minerai dans une proportion excessivement faibles. Une tonne de pechblende, par exemple, contient une quantité de radium de l'ordre de grandeur de 1 décigramme. Dans ces conditions la préparation des sels de radium est pénible et coûteuse. Une tonne de minerai fournit quelques kilogrammes de bromure de baryum radifère, d'où l'on extrait ensuite le bromure de radium par une série de fractionnements.

Pendant la séparation du radium, Demarçay dont nous avons à déplorer la mort récente, a bien voulu examiner les spectres des produits que nous avions préparés. Ce concours nous a été précieux ; dès le début de nos recherches, l'analyse spectrale est venue confirmer nos prévisions, en nous apportant la preuve que le baryum radioactif que nous avions retiré de la pechblende contenait un élément nouveau. C'est à Demarçay que nous devons la première étude du spectre du radium.†

Le radium a une réaction spectrale très sensible, aussi sensible que celle du baryum, on peut reconnaître au spectroscope la présence du radium dans un sel de baryum radifère qui ne contient que $\frac{1}{10000}$me de radium. Mais la radio-activité du radium donne une réaction 10,000 fois plus sensible encore. Un électromètre ordinaire bien isolé permet de déceler facilement la présence du radium lorsqu'il est mélangé à des substances inactives dans la proportion de $\frac{1}{10^8}$.

---

* *Mme. Curie*, Thèse à la Faculté des sciences, Paris, 1903.
† *Demarçay*, C.R. de l'ac. des sciences, déc. 1898 et juillet 1900.

Le radium est l'homologue supérieur du baryum dans la série des métaux alcalino-terreux; son poids atomique égal à **225** a été déterminé par Mme Curie.

Bien que cet élément soit très voisin du baryum, il ne s'en trouve pas, même à l'état de trace, dans les minerais ordinaires de baryum. Le radium n'accompagne le baryum que dans les minerais d'urane, et ce fait a probablement une grande importance théorique.

X.

Le radium nous donne l'exemple d'un corps qui, tout en conservant le même état, donne lieu a un dégagement d'énergie continu et assez considérable. Ce fait paraît en désaccord avec les principes fondamentaux de l'énergétique et diverses hypothèses ont été proposées pour éviter cette contradiction.

Parmi ces hypothèses nous en retiendrons deux qui ont été émises dès le début des études sur la radio-activité.[*]

Dans la première hypothèse, ou suppose que le radium est un élément en voie d'évolution. On doit alors admettre que cette évolution est extrêmement lente de telle sorte qu'aucun changement d'état apréciable ne se fait sentir au bout de plusieurs années. L'énergie que le radium dégage pendant une année correspondrait donc à une transformation insignifiante de ce corps. Il semble d'ailleurs naturel de supposer que la quantité d'énergie mise en jeu dans la transformation des atomes est considérable.

La deuxième hypothèse consiste à supposer qu'il existe dans l'espace des rayonnements encore inconnues et inaccessibles à nos sens. Le radium serait capable d'absorber l'énergie de ces rayons hypothétiques et de la transformer en énergie radio-active.

Les deux hypothèses que nous venons énoncer ne sont pas du reste incompatibles.

---

[*] *Mme. Curie*, Révue générale des sciences, 30 jan. 1899.

GENERAL MONTHLY MEETING,

Monday, July 6, 1903.

SIR JAMES CRICHTON-BROWNE, M.D. LL.D. F.R.S., Treasurer and
Vice-President, in the Chair.

Lord Oxmantown,
R. L. B. Hammond Chambers, Esq. K.C.
Michael B. Field, Esq.
G. Marconi, Esq.
Henry D. McLaren, Esq. B.A.
Richard Pearce, Esq.
Ernest Percy Stuart Roupell, Esq.
W. Wavell, Esq.

were elected Members of the Royal Institution.

The Special Thanks of the Members were returned to the Misses
Gladstone for their present of a portrait of Dr. John Hall Gladstone,
F.R.S.

The PRESENTS received since the last Meeting were laid on the
table, and the thanks of the Members returned for the same, viz.:—

FROM

Astronomer Royal—Report on the Royal Observatory, Greenwich, 1903. 4to.
British Museum Trustees—Catalogue of Additions to the Manuscripts, 1894–99.
  8vo. 1903.
  Catalogue of Greek Coins: Parthia. 8vo. 1903.
The Secretary of State for India—An Introduction to the Grammar of the Kui
  Language. By L. Letchmajee. 8vo. 1903.
Abderhalden, E. Esq. (the Author)—Familiäre Cystindiathese. 8vo. 1903.
  Kochsalzsurrogates der Eingeborenen von Angoniland. 8vo. 1903.
Accademia dei Lincei, Reale, Roma—Classe di Scienze Fisiche, Matematiche e
  Naturali. Atti, Serie Quinta: Rendiconti. 1° Semestre, Vol. XII. Fasc. 1.
  Nos. 10, 11. 8vo. 1903.
Astronomical Society, Royal—Monthly Notices, Vol. LXIII. No. 7. 1903. 8vo.
Automobile Club—Journal for June, 1903.
Batavia, Royal Magnetical and Meteorological Observatory—Observations, Vol.
  XXIV. 1901. 4to. 1903.
Bell, A. Graham, Esq. (the Author)—The Tetrahedral Principle in Kite Structure.
  8vo. 1903.
Boston Public Library—Monthly Bulletin for June, 1903. 8vo.
British Architects, Royal Institute of—Journal, Third Series, Vol. X. Nos. 15, 16.
  4to. 1903.
British Association—Report of Seventy-second Meeting, 1902. 8vo. 1903.
British Astronomical Association—Journal, Vol. XIII. No. 8. 8vo. 1903.
Buenos Ayres, City—Monthly Bulletin of Municipal Statistics, April 1903. 4to.

Chemical Industry, Society of—Journal. Vol. XXII. Nos. 11, 12. 8vo. 1903.
Chemical Society—Proceedings, Vol. XIX. Nos. 268, 269. 8vo. 1903.
   Journal for July, 1903. 8vo.
Chicago, Field Columbian Museum—
   Publications—Botanical Series, Vol. III. No. 1. 8vo. 1903.
                    Report Series, Vol. II. No. 2. 8vo. 1902.
                    Zoological Series, Vol. III. Nos. 8, 9. 8vo. 1903.
Civil Engineers, Institution of—Proceedings, Vol. CLI. 8vo. 1903.
   Subject Index, Vols. CXIX. to CL. 8vo. 1903.
Cracovie, Academy of Sciences—Bulletin, Classe des Sciences Mathématiques.
   1903, No. 4; Classe de Philologie, 1903, No. 4. 8vo.
Editors—American Journal of Science for May–June, 1903. 8vo.
   Analyst for June, 1903. 8vo.
   Astrophysical Journal for June, 1903. 8vo.
   Athenæum for June, 1903. 4to.
   Author for July, 1903. 8vo.
   Board of Trade Journal for June, 1903. 8vo.
   Brewers' Journal for June, 1903. 8vo.
   Chemical News for June, 1903. 4to.
   Chemist and Druggist for June, 1903. 8vo.
   Electrical Engineer for June, 1903. fol.
   Electrical Review for June, 1903. 4to.
   Electrical Times for June, 1903. 4to.
   Electricity for June, 1903. 8vo.
   Engineer for June, 1903. fol.
   Engineering for June, 1903. fol.
   Feilden's Magazine for June, 1903. 8vo.
   Homœopathic Review for July, 1903. 8vo.
   Horological Journal for July, 1903. 8vo.
   Journal of the British Dental Association for June, 1903. 8vo.
   Journal of State Medicine for July, 1903. 8vo.
   Law Journal for June, 1903. 8vo.
   London Technical Education Gazette for June, 1903.
   London University Gazette for June, 1903. 4to.
   Machinery Market for June, 1903. 8vo.
   Machinery Users' Journal for May, 1903. 4to.
   Model Engineer for June, 1903. 8vo.
   Mois Scientifique for June, 1903. 8vo.
   Motor Car Journal for June, 1903. 8vo.
   Motor Car World for June, 1903. 4to.
   Musical Times for June, 1903. 8vo.
   Nature for June, 1903. 4to.
   New Church Magazine for July, 1903. 8vo.
   Page's Magazine for June, 1903. 8vo.
   Photographic News for June, 1903. 8vo.
   Physical Review for June, 1903. 8vo.
   Popular Astronomy for June–July, 1903. 8vo.
   Public Health Engineer for June, 1903. 8vo.
   Science Abstracts for May, 1903. 8vo.
   Terrestrial Magnetism for March, 1903. 8vo.
   Travel for June, 1903. 8vo.
   Zoophilist for July, 1903. 4to.
Florence, Biblioteca Nazionale—Bulletin for June, 1903. 8vo.
Florence, Reale Accademia dei Georgofili—Atti, Vol. XXV. Supplement; Vol.
   XXVI. Disp. 1, 2. 8vo. 1902-3.
Franklin Institute—Journal, Vol. CLV. No. 6. 8vo. 1903.
Geological Society—Abstracts of Proceedings, Nos. 779-781. 1903.
Johns Hopkins University—American Journal of Philology, Vol. XXIV. No. 2.
   8vo. 1903.

Leighton, John, Esq. M.R.I.—Ex-Libris Journal for June, 1903. 8vo.
London County Council—Annual Report of the Technical Education Board, 1902-3. 4to.
Longe, Francis D. Esq. (the Author)—Supplement to the "Fiction of the Ice Age." 8vo. 1903.
Manchester Literary and Philosophical Society—Proceedings, Vol. XLVII. Part 4. 8vo. 1903.
Mechanical Engineers Institution—Proceedings, 1902, No. 5. 8vo. 1903.
Mersey Conservancy—Report on the Navigation of the River Mersey, 1902. 8vo. 1903.
Microscopical Society, Royal—Journal for June, 1903. 8vo.
Musée Teyler—Archives, Série II. Vol. VIII. Parts 2, 3. 8vo. 1902-3.
Navy League—Navy League Journal for July, 1903. 8vo.
New Zealand, Registrar-General—Statistics of the Colony of New Zealand, 1901. 4to. 1903.
Nova Scotian Institute of Science—Proceedings, Vol. X. Part 4. 8vo. 1903.
Odontological Society—Transactions, Vol. XXXV. No. 7. 8vo. 1903.
Paris, Société Française de Physique—Bulletin des Séances, 1902, Fasc. 4; 1903, Fasc. 1. 8vo.
Pharmaceutical Society of Great Britain—Journal for June, 1903. 8vo.
Queensland Government—North Queensland Ethnography Bulletin, No. 6. fol. 1903.
Rome, Ministry of Public Works—Giornale del Genio Civile for Jan. 1903. 8vo.
Royal Society of London—Philosophical Transactions, A, Nos. 337-339. 4to. 1903. Proceedings, No. 476. 8vo. 1903.
Russell, Dr. W. J., M.R.I. (the Author)—The Formation of Definite Figures by the Deposition of Dust. 4to. 1903.
Saxon Society of Sciences, Royal—
    Abhandlungen, Band XX. No 6; Band XXI. No. 4; Band XXII. No. 1; Band XXVIII. Nos. 1-3. 4to. 1902-3.
    Berichte—
    Mathematisch-Physische Klasse, 1902, Nos. 6, 7; 1903, Nos. 1, 2. 8vo.
    Philologisch-Historische Klasse, 1902, No. 3; 1903, Nos. 1, 2. 8vo.
Selborne Society—Nature Notes for June–July, 1903. 8vo.
Society of Arts—Journal for June, 1903. 8vo.
Statistical Society, Royal—Journal, Vol. LXVI. No. 2. 8vo. 1903.
Tacchini, Prof. P., Hon. Mem. R.I. (the Author)—Memorie della Società degli Spettroscopisti Italiani, Vol. XXXII. Disp. 5. 4to. 1903.
Transvaal, Director of Agriculture—Transvaal Agricultural Journal, Vol. I. No. 3. 8vo. 1903.
United Service Institution, Royal—Journal for June, 1903. 8vo.
United States Department of Agriculture—Report of the Chief of the Weather Bureau, 1900-1, Vol. II. 4to. 1902.
    Monthly Weather Review for March, 1903. 4to.
United States Geological Survey—Monographs, XLII. and XLIII. 4to. 1903. Professional Papers, Nos. 1-8. 4to. 1902.
United States Patent Office—Official Gazette, Vol. CIV. Nos. 5-8. 8vo. 1903.
Verein zur Beförderung des Gewerbfleisses in Preussen—Verhandlungen, 1903, Heft 6. 8vo.
Vienna, Imperial Geological Institute—Verhandlungen, 1903, Nos. 5-8. 8vo.
Washington Academy of Sciences—Memoirs, Vol. VIII. (Seventh Memoir.) 4to. 1903.
Zeiss, Carl, Ltd. Messrs.—Das Zeisswerk und die Carl Zeiss Stiftung in Jena. 8vo. 1903.

GENERAL MONTHLY MEETING,

Monday, November 2, 1903.

Sir James Crichton-Browne, M.D. LL.D. F.R.S., Treasurer and Vice-President, in the Chair.

A. K. Huntingdon, Esq. F.C.S.
Sir Oliver Lodge, LL.D. Sc.D. F.R.S.
J. Francis Mason, Esq.
Edward H. Woods, Esq. M.Inst.C.E.

were elected Members of the Royal Institution.

The Presents received since the last Meeting were laid on the table, and the thanks of the Members returned for the same, viz. :—

FROM

The Secretary of State for India—
 Geological Survey of India—
  Palæontologia Indica, Ser. IX. Vol. III. No. 1. fol. 1903.
  Memoirs, Vol. XXXIV. Part 3. 8vo. 1903.
  Index to Vols. XXI.-XXX. of the Records. 8vo. 1903.
  General Report, 1902-3. 8vo. 1903.
 Mundari Grammar. By J. Hoffmann. 8vo. 1903.
 Tibetan-English Dictionary. By S. C. Das. 4to. 1903.
 Great Trigonometrical Survey of India, Vol. XVII. 4to. 1901.
 Report on Archæological Work in Burma for the year 1902-3. 4to. 1903.
 Archæological Survey: Punjab Circle, Progress Report for the year ending 31st March, 1903. 4to. 1903.
Accademia dei Lincei, Reale, Roma—Classe di Scienze Fisiche, Matematiche e Naturali. Atti, Serie Quinta: Rendiconti. Vol. XII. 1° Semestre, Fasc. 12; 2° Semestre, Fasc. 1-7. 1903. 8vo.
 Classe di Scienze Morali, Storiche, Serie Quinta, Vol. XII. Fasc. 3-6. 8vo. 1903.
African Society—Journal for July. 8vo. 1903.
Allegheny Observatory—Miscellaneous Papers, N.S. Nos. 11-14. 8vo. 1903.
American Academy of Arts and Sciences—Proceedings, Vol. XXXVIII. Nos. 20-26; Vol. XXXIX. Nos. 1-3. 8vo. 1903.
American Geographical Society—Bulletin, Vol. XXXV. No. 3. 8vo. 1903.
American Philosophical Society—Proceedings, Jan.-April, 1903. 8vo.
Amsterdam, Royal Academy of Sciences—Verhandelingen, 1° Sectie, Deel. VIII. Nos. 3-5; 2° Sectie, Deel. IX. Nos. 4-9.
 Zitingsverslagen, Vol. XI.
 Proceedings, Vol. V.
 Jaarbock, 1902. 8vo. 1902-3.
Asiatic Society, Royal—Journal, July-Oct. 1903. 8vo.
Asiatic Society of Bengal—Proceedings, 1902, No. 11, Extra No.; 1903, Nos. 1-5. 8vo. 1902-3.
 Journal, Vol. LXXI. Part I. No. 2 and Extra No. 2; Vol. LXXII. Part I. No. 1; Part III. No. 1. 8vo. 1902-3.
Astronomical Society, Royal—Monthly Notices, Vol. LXIII. Nos. 8, 9. 8vo. 1903.

Automobile Club—Journal for July–Oct. 4to. 1903.
Bankers, Institute of—Journal, Vol. XXIV. No. 7. 8vo. 1903.
Belgium, Royal Academy of Sciences—Bulletin, 1903, Nos. 5–8. 8vo.
Mém. Cour. et autres Mém. Tome LXIII. Fasc. 4–6. 8vo. 1903.
Mém. Cour. et des Savants étrang. Tomes LXI. and LXII. Fasc. 3. 4to 1902–3.
Berlin Academy of Sciences—Sitzungsberichte, 1903, Nos. 25–40. 8vo.
Berthelot, M. Daniel (the Author)—Sur les Thermomètres à Gaz. 4to. 1903.
Sur une Méthode pour la Mesure des Températures. 8vo. 1903.
Sur la notion des états correspondants. 8vo. 1903.
Borredon, G. Esq. (the Author)—La Luna é la Calamita del Mondo. 8vo. 1903.
Boston Public Library—Monthly Bulletin for July–Oct. 1903. 8vo.
51st Annual Report, 1902–3. 8vo.
Boston Society of Natural History—Proceedings, Vol. XXXI. Nos. 1 and 3–7. 8vo. 1903.
Memoirs, Vol. V. Nos. 8, 9. 4to. 1902–3.
British Architects, Royal Institute of—Journal, 3rd Series, Vol. X. Nos. 17–20. 4to. 1903.
Kalendar, 1903–4. 8vo. 1903.
British Astronomical Association—Journal, Vol. XIII. Nos. 9, 10. Memoirs, Vol. XI. Part III. 8vo. 1903.
Buenos Ayres, City—Monthly Bulletin of Municipal Statistics, May–June, 1903. 8vo.
Cambridge Philosophical Society—Proceedings, Vol. XII. Part 3. 8vo. 1903.
Cambridge University Library—Report of the Library Syndicate for 1902. 4to. 1903.
Canada, Geological Survey—Annual Report, New Series, Vol. XII. 1899. 8vo. 1903.
Catalogue of Canadian Birds, Part II. 8vo. 1903.
Canadian Government—Map of the Dominion of Canada. 1903.
Cassell & Co. Messrs.—Photograph of two Davy Lamps, from the R.I. Collection.
Chemical Industry, Society of—Journal, Vol. XXII. Nos. 13–19. 8vo. 1903.
Chemical Society—Journal for Aug.–Nov. 1903. 8vo.
Civil Engineers, Institution of—Proceedings, Vol. CLII. 8vo. 1903.
List of Members, 1903. 8vo.
Colonial Institute, Royal—Proceedings, Vol. XXXIV. 1902–3. 8vo.
Cornwall Polytechnic Society, Royal—Seventieth Annual Report. 8vo. 1903.
Cornwall, Royal Institution of—Journal, Vol. XV. Part 2. 8vo. 1903.
Coutts, John, Esq. (the Author)—The Method of Christ as traced in Chemistry Physics, and Spectrum Analysis. 16mo. 1903.
Cracovie, l'Académie des Sciences—Bulletin, Classes des Sciences Mathématiques et Naturelles, Nos. 5, 6; Classe de Philologie, No. 5. 1903. 8vo.
Crawford and Balcarres, The Earl of, K.T. M.P. F.R.S. M.R.I.—Bibliotheca-Lindesiana : Collations and Notes, No. VIII. Catalogue of Tracts by Luther and his Contemporaries, 1511–1598. 4to. 1903.
Dowson, W., M.D. (the Director)—The Wellcome Physiological Research Laboratories. 8vo. 1903.
East India Association—Journal, N.S. Vol. XXXIV. No. 31. 8vo. 1903.
Edinburgh Royal Society—Proceedings, Vol. XXIV. No. 5. 8vo. 1903.
Editors—Aeronautical Journal for July–Oct. 1903. 8vo.
American Journal of Science for July–Oct. 1903. 8vo.
Analyst for July–Oct. 1903. 8vo.
Astrophysical Journal for July–Oct. 1903.
Athenæum for July–Oct. 1903. 4to.
Author for July–Oct. 1903. 8vo.
Board of Trade Journal for July–Oct. 1903. 8vo.
Brewers' Journal for July–Oct. 1903. 8vo.
Cambridge Appointments Gazette for July–Oct. 1903. 8vo.
Chemical News for July–Oct. 1903. 4to.

Editors—continued.
Chemist and Druggist for July–Oct. 1903. 8vo.
Dioptric Review for June–Oct. 1903. 8vo.
Electrical Engineer for July–Oct. 1903. fol.
Electrical Review for July–Oct. 1903. 8vo.
Electrical Times for July–Oct. 1903. 4to.
Electricity for July–Oct. 1903. 8vo.
Electro-Chemist and Metallurgist for August, 1903   8vo.
Engineer for July–Oct. 1903. fol.
Engineering for July–Oct 1903   fol.
Feilden's Magazine for July–Oct. 1903. 8vo.
Focus for July, 1903. 8vo.
Homœopathic Review for Aug–Oct. 1903. 8vo.
Horological Journal for Aug.–Nov. 1903. 8vo.
Journal of the British Dental Association for July–Oct. 1903. 8vo.
Journal of Physical Chemistry for May–Oct. 1903. 8vo.
Journal of State Medicine for Aug.–Oct. 1903. 8vo.
Law Journal for July–Oct. 1903. 8vo.
London Technical Education Gazette for July–Oct. 1903. 8vo.
Machinery Market for July–Oct. 1903. 8vo.
Model Engineer for July–Oct. 1903   8vo.
Mois Scientifique for July–Oct. 1903. 8vo.
Motor Car Journal for July–Oct. 1903. 8vo.
Musical Times for July–Oct. 1903. 8vo.
Nature for July–Oct. 1903. 4to.
New Church Magazine for Aug.–Nov. 1903. 8vo.
Nuovo Cimento for April–May, 1903. 8vo.
Page's Magazine for July–Oct. 1903. 8vo.
Photographic News for July–Oct. 1903. 8vo.
Physical Review for July–Sept. 1903. 8vo.
Popular Astronomy for July–Oct. 1903. 8vo.
Public Health Engineer for July–Oct. 1903.   8vo.
Science Abstracts for June–Sept. 1903. 8vo.
Terrestrial Magnetism for June, 1903.
Travel for July–Oct. 1903. 8vo.
Zoophilist for Aug.–Oct. 1903. 4to.
Electrical Engineers, Institution of—Journal, Vol. XXXII. Parts 4, 5. 8vo. 1903.
Entomological Society—Transactions, 1903, Part II. 8vo.
Field Columbian Museum—Publications: Geological Series, Vol. II. No. 1;
    Zoological Series, Vol. III. Nos. 10, 11. 8vo. 1903.
Florence, Biblioteca Nazionale—Bulletin, July–Sept. 1903. 8vo.
Florence, Reale Accademia del Georgofili—Atti, Vol. XXVI. Disp. 3. 8vo. 1903.
Franklin Institute—Journal, Vol. CLVI. Nos. 1–3. 8vo. 1903.
Geographical Society, Royal—Geographical Journal for July–Oct. 1903. 8vo.
Geological Society—Quarterly Journal, Vol. LIX. Part 3. 8vo. 1903.
    Literature added to the Library in 1902. 8vo. 1903.
Geological Survey of the United Kingdom—Summary of Progress, 1902. 8vo.
    1903.
Glasgow, Royal Philosophical Society—Proceedings, Vol. XXXIV. 8vo. 1903.
Gordon, Mrs. Ogilvie (the Authoress)—The Geological Structure of Monzoni and
    Fassa   8vo   1903.
Göttingen, Royal Academy of Sciences—Nachrichten: Mathematisch-physikalische
    Klasse, 1903, Heft. 3, 4. 8vo.
    Geschaftliche Mittheilungen, 1903, Heft 1. 8vo.
Harlem, Société Hollandaise des Sciences—Archives Néerlandaises, Série II.
    Tome VIII. 8vo. 1903.
    Verhandelingen, Deel V. 4to. 1903.
Iron and Steel Institute—Journal, 1903, No. 1. 8vo. 1903.
    List of Members, 1903. 8vo.

Johns Hopkins University—University Circulars, Nos. 163, 164. 4to. 1903.
Junior Engineers, Institution of—Transactions, Vol. XII. 8vo. 1903.
Kansas University—Bulletin Vol. III. 8vo 1903.
Leicester, Borough of—Thirty-second Annual Report of the Public Libraries
    Committee. 8vo. 1902-3.
Leighton, John, Esq. F.S.A. M.R.I.—The Practical Photographer: No. 1, Bromide
    Printing. October 1903. 8vo.
    Journal of the Ex-Libris Society, July–Sept. 1903. 8vo.
    Portrait of the 6th Duke of Northumberland.
Life-Boat Institution, Royal National—Journal for August, 1903. 8vo.
Linnean Society—Transactions: Botany, Vol. VI. Parts 4–6; Zoology, Vol.
    VIII. Parts 9–12; Vol. IX. Parts 1, 2. 4to. 1902-3.
    Journal, Zoology, Vol. XXXVIII. No. 187; Botany, Vol. XXXVI. Nos. 246
    and 251. 8vo. 1903.
Literature, Royal Society of—Transactions, Vol. XXXI. Parts 1–3; Vols. XXII.-
    XXIII. and Vol. XXIV. Parts 1, 2. 8vo. 1899–1903.
Madrid, Royal Academy of Sciences—Memorias, Tome XVIII. Part 1; Tomes
    XX.-XXI. 4to. 1903.
    Annuario 1903. 16mo.
Manchester Literary and Philosophical Society –Memoirs and Proceedings, Vol.
    XLVII Parts 5, 6. 8vo. 1903.
Massachusetts Institute of Technology—Technology Quarterly, Vol. XVI. No 2.
    8vo. 1903.
Mechanical Engineers, Institution of—Proceedings, 1903, Nos. 1, 2. 8vo.
Meteorological Society—Journal, Vol. XXIX. No. 127, July. 8vo. 1903.
    Meteorological Record, Vol. XXII. No. 88. 8vo. 1903.
Metropolitan Asylums Board—Report for 1902. 8vo. 1903.
Microscopical Society, Royal—Journal, 1903, Parts 4, 5. 8vo.
Montana University—Bulletin, Geological Series, No. 1. 8vo. 1902.
Munich, Royal Bavarian Academy of Sciences—Sitzungsberichte, 1901, Heft II.
    8vo. 1903.
Musical Association - Proceedings, Twenty-ninth Session, 1902-3. 8vo.
Natal, Colony of—Report on the Mining Industry of Natal for 1902. 4to. 1903.
Navy League—Navy League Journal for Aug.–Oct. 1903. 8vo.
New South Wales, Comptroller of Prisons—Report for the Year 1902. fol. 1903.
New South Wales, Royal Society—Journal and Proceedings, Vol. XXXVI. 8vo.
    1903.
New Zealand, Agent-General for—Statistics relating to the Colony for 1901,
    Parts IV.-VII. fol. 1903.
    New Zealand Mines Record, Vol. V. Nos. 11, 12; Vol. VI. Nos. 1–9. 8vo.
    1902-3.
    Journal of the Department of Labour, May 1901. 8vo.
    New Zealand Handbook, No. 8, 1901. 8vo.
Norfolk and Norwich Naturalists' Society—Transactions, Vol. VII. Part 4. 8vo.
    1901.
North of England Institute of Mining and Mechanical Engineers—Transactions,
    Vol. LII. No. 6. 8vo. 1903.
    Report of the Committee on Mechanical Coal Cutting. 8vo. 1903.
    Annual Report, 1902. 8vo. 1903.
Nova Scotia, Agent-General for—Map of Nova Scotia. 1903.
Numismatic Society—Numismatic Chronicle, 1903, Part III. 8vo.
Odontological Society—Transactions, Vol. XXXV. No. 8. 8vo. 1903.
O'Halloran, George A. Esq. (Archivist)—Report on Canadian Archives, 1902.
    8vo. 1903.
Onnes, Dr. H. K.—Communications, Nos. 83, 84; Supplement, No. 6. 8vo. 1903.
Paris, Société Française de Physique—Bulletin des Séances, 1903, Fasc. 2. 8vo.
    1903.
Pennsylvania University—Miscellaneous Papers. 8vo and 4to. 1903.
Pharmaceutical Society of Great Britain—Journal for July–Oct. 1903. 8vo.

Philadelphia Academy of Natural Sciences—Proceedings, Vol. LV. Part 1. 8vo . 1903.

Photographic Society, Royal—Photographic Journal for May–July, 1903. 8vo.

Physical Society of London—Proceedings, Vol. XVIII. Part 5. 8vo. 1903.

Pomeroy, Hon. E.—National Education. By H. Spencer. (Reprinted from "Social Statics.") 8vo. 1903.

Rome, Ministry of Public Works—Giornale del Genio Civile for Feb.-April, 1903. 8vo.

Royal College of Surgeons—Calendar, 1903. 8vo.

Royal Commission for the Universal Exhibition, St. Louis, 1904—Speech by the Prince of Wales at the First Meeting. 8vo. 1903.

Royal Engineers, Corps of—Professional Papers, Vol. XXVIII. 8vo. 1903.

Royal Irish Academy—Proceedings, Vol. XXIV. Sec. A, Part 2; Sec. B, Part 3; Sec. C, Part 3. 8vo. 1903.

Transactions, Vol. XXXII. Sec. A, Part 6; Sec. C, Part 1. 4to. 1903.

Royal Society of London—Philosophical Transactions, A, Nos. 338–349, B, Nos. 216–219. 4to. 1903.

Proceedings, Nos. 477–482. 8vo. 1903.

Reports of the Sleeping Sickness Commission, No. 1. 8vo. 1903.

Reports to the Malaria Committee, 8th Series. 8vo. 1903.

St. Bartholomew's Hospital—Statistical Tables for 1902. 8vo. 1903.

St. Pétersbourg, L'Académie Impériale des Sciences—Mémoires: VIII. Série, Classe Physico-Mathématique, Vols. XI.-XII. and XIII. Nos. 1–5 and 7. 4to. 1900-3.

Bulletin, Tome XIII. Nos. 4, 5; XIV. XVI. XVII. Nos. 1–4. 8vo. 1900-2.

Comptes Rendus de la Commission Sismique Permanente, Tome I. Liv. 2. 4to. 1903.

Sanitary Institute—Journal, Vol. XXIV. Parts II.-III. and Supplements. 8vo. 1903.

Selborne Society—Nature Notes for Aug.-Oct. 1903. 8vo.

Sennett, A. R. Esq. M.R.I. (the Author)—Fragments from Continental Journeyings. 8vo. 1903.

Smith, B. Leigh, Esq. M.A. M.R.I.—Transactions of the Institute of Naval Architects, Vol. XLV. 4to. 1903.

The Scottish Geographical Magazine, Vol. XIX. Nos. 7–10. 8vo. 1903.

Smithsonian Institution—Miscellaneous Collections, 1372 and 1376. 8vo. 1902-3.

Contributions to Knowledge, 1373. 4to. 1903.

Society of Arts—Journal for July-Oct. 1903. 8vo.

Swedish Academy of Sciences—Handlingar, Band XXXVI. and XXXVII. Nos. 1, 2. 4to. 1902-3.

Bihang, Band XXVIII Nos. 1–4. 8vo. 1903.

Lefnadsteckningar, Band IV. Heft 3. 8vo. 1903.

Arkiv för Botanique, Band I. Háfte 1–3. 8vo. 1903.

„ Kemi, Band I. Háfte 1. 8vo. 1903.

„ Matematik, Band I. Háfte 1, 2. 8vo. 1903.

„ Zoologi, Band I. Háfte 1, 2. 8vo. 1903.

Ärsbok för ar, 1903. 8vo.

Tacchini, Prof. P. Hon. Mem. R.I. (the Author)—Memorie della Società degli Spettroscopisti Italiani, Vol. XXXII. Disp. 6–10. 4to. 1903.

Tasmania, Royal Society—Proceedings, 1902. 8vo 1903.

Toronto University—Studies, Physical Science Series, Nos. 1, 2. 8vo. 1903.

Transvaal Department of Agriculture—Journal, Vol. I. No. 4. 8vo. 1903.

United Service Institution, Royal—Journal for July-Oct. 1903. 8vo.

United States Department of Agriculture—Monthly Weather Review, April-July, 1903. 4to.

Experiment Station Record, Jan.-July, 1903. 8vo.

United States Geological Survey—Mineral Resources of U.S. 1901. 8vo. 1902.

Bulletin, Nos. 191, 195–207. 8vo. 1902.

Geologic Atlas of U.S. Folios 72–90. fol. 1901-3.

United States Geological Survey—Water Supply Papers, Nos. 65–79. 8vo. 1902–3.
Twenty-second and Twenty-third Annual Reports. 14 vols. 4to. 1901–2.
United States Patent Office—Official Gazette, Vol. CIV. No. 9; Vol. CV.–CVI.
Nos. 1–5. 4to. 1903.
Annual Report of the Commissioner of Patents, 1902. 8vo. 1903.
United States Surgeon-General's Office—Index Catalogue of the Library, Series II.
Vol. VIII. 4to. 1903.
Upsal, L'Observatoire Météorologique—Bulletin Mensuel, Vol. XXXIV. 1902. 4to.
1903.
Verein zur Beförderung des Gewerbfleisses in Preussen—Verhandlungen, 1903.
Heft 7, 8. 4to.
Victoria Institute—Journal, Vol. XXXV. 8vo. 1903.
Vienna, Imperial Geological Institute—Jahrbuch Jahrgang, 1902. 8vo. 1903.
Abhandlungen, Band XX. Heft 7. 4to. 1903.
Walter, Miss M. (the Translator)—Notes from a Diary in Asiatic Turkey. By
Earl Percy. (In German.) MS. fol. 1902.
Washington, Academy of Sciences—Proceedings, Vol. V. pp. 99–229. 8vo. 1903.
Wells, Henry M. Esq. (the Author)--Cylinder Oil and Cylinder Lubrication. 4to.
1903.
Western Society of Engineers—Journal, Vol. VIII. Nos. 3, 4. 8vo. 1803.
Yorkshire Archæological Society—Journal, Part 67. 8vo. 1903.
Yorkshire Philosophical Society—Annual Report for 1902. 8vo. 1903.
Zoological Society of London—Proceedings, 1903, Vol. I. Parts 1, 2; Vol. II.
Part 1. 8vo. 1903.
Transactions, Vol. XVII. Parts 1, 2, 4to. 1903.
Zurich Naturforschende Gesellschaft—Vierteljahrsschrift Jahrg XLVII. Heft. 3, 4.
8vo. 1903.
Die Elektrische Wellen von A. Weilenmann. 4to. 1903.

GENERAL MONTHLY MEETING,

Monday, December 7, 1903.

Sir James Crichton-Browne, M.D. LL.D. F.R.S., Treasurer and Vice-President, in the Chair.

Andrew Carnegie, Esq. LL.D.
John Sadler Curgenven, Esq. M.R.C.S. L.R.C.P.
J. Emerson Reynolds, M.D. Sc.D. F.R.S.
Rudolf Wissmann, Esq.

were elected Members of the Royal Institution.

The Chairman announced the decease, on the 30th of November, of Sir Frederick Bramwell, Bart., and the following letter from His Grace the President was read:—

Alnwick Castle, *2nd December,* 1903.

My Dear Sir William Crookes—

It is with the utmost regret that I find an important engagement in Newcastle will unavoidably prevent my attending the Monthly Meeting of the Managers of the Royal Institution next Monday, and the subsequent General Meeting of Members. I am especially grieved, because it would have been a melancholy satisfaction to have had the opportunity of expressing my deep sense of the loss the Institution has sustained in the death of our respected and valued friend, Sir Frederick Bramwell.

His great talents and high professional reputation are known to all the world —the deep interest he always displayed in the work of the Royal Institution, and the conspicuous services he rendered it, cannot, I think, fail to be within the cognisance of every member. But it is only those who have taken part as his colleagues in the management, and who have had the privilege of enjoying his personal friendship, who can be fully aware of his genial kindliness, his wise counsels, and his indefatigable energy.

I am glad to think that we were able to do something in his life-time to convey to him a sense of our esteem when we added his bust to our collection. To us it is a reminder of a loss which will long be felt.

Believe me, Yours very truly,
(Signed) Northumberland.

The Chairman said: In the absence of His Grace, the President, it devolves upon me to submit to you the Resolution which he would have proposed had he been here, and in doing so I would say that we miss this afternoon a distinguished figure that was with us at our last monthly meeting, and that has been familiar in this theatre for seven and twenty years, for ever since Sir Frederick Bramwell joined the Royal Institution in 1876 he has taken an active interest in its affairs and has been seldom absent from its meetings.

I am sure that our members, and especially our old members, during the coming session will be conscious of a void, a want, when they look round and see him no longer in his accustomed place, and as for us who have been his fellow workers here, I am sure I may affirm that until we ourselves are memories of the past, kind thoughts of him, grateful reminiscences of his genial presence, will haunt this place. In him the Royal Institution had a wise counsellor, a vigilant guardian, a generous friend, and it is fitting that it should bear its share in the wide-spread mourning for his death.

I say wide-spread mourning, because many other public institutions which he benefited by his services are suffering from the loss of him, and indeed I might say that London and the whole country are the poorer because he has gone. One of the great band of Engineers, of Engineering Sculptors, if I might so call them, who during the last century have been remodelling the face of the earth, perhaps with some sacrifice of pristine beauty, but with a vast augmentation of expression and meaning, one of that great band, he has left his mark upon his times. Identified with no one monumental undertaking, he has in scores of places done something to promote the public health, to facilitate human intercourse, to improve our industries.

In this theatre he spoke of himself not long ago, with characteristic modesty, as one of the mediocrities who carry on the work of the world. There was no mediocrity about him. He was a man of commanding intellect, of excellent attainments, of quick and deep insight, of choice humour, of unique personality; and it was by virtue of this combination of qualities, that he carried on his work in the world in such a manner as to be always far in advance of his fellow-labourers in the rank and file, and to secure on all hands honourable recognition.

But, gentlemen, the life is larger than the work of the man—and it seems to me that Sir Frederick Bramwell's life was memorable and beneficent. He always diffused around him a wholesome, genial, hopeful, righteous influence, that has radiated away through realms subtler than ether—through the thoughts and emotions of the fellow-men with whom he was brought into contact, and that may continue to reverberate when the structures of stone and of steel he erected have crumbled away, for who can limit the range of the psychons that we are each of us momentarily emitting from our brains?

It is consolatory to reflect that with Sir Frederick Bramwell there was no rusting from rest or indolence, no piteous interval of bodily enfeeblement or mental decay. He died with his harness on his back. Far advanced in years, he carried into old age much of the elasticity and vigour and charm of youth, and was always happiest when he was most busily employed. I recollect hearing him describe the premonitory attack of illness that carried him off—a slight stroke which happened three years ago—as "the melancholy result of a fortnight's holiday." "I was quite well and very busy," he said; "but my family and my friends and doctors insisted on my having a change, so I went away and was idle for a fortnight, and this is the consequence."

His age was as a lusty winter, frosty but kindly, full of briskness and usefulness.

I need not recount Sir Frederick Bramwell's services to the Royal Institution. These were suitably and gracefully acknowledged by our President on the occasion of the presentation of Mr. Onslow Ford's bust of him, which stands in the anteroom, and which will keep alive the memory of him when those who knew him have passed away. In its finances, in its administration, in its scientific department, he gave it invaluable assistance, and it will be long before it finds a better friend.

A typical Englishman, bold, upright, straightforward, practical, an eminent engineer, enterprising yet prudent, with a mastery of figures and mechanical problems rarely equalled, an able scientist, skilled in his own branch, but interested in all extension of natural knowledge—a faithful friend, a good man, Sir Frederick Bramwell was one whom we loved and revered when he was amongst us, and to whom we would wish to pay all possible posthumous honour. I beg to propose the following Resolution, passed by the Managers :—

Resolved, That the Managers of the Royal Institution of Great Britain desire to record their deep sense of the loss sustained by the Institution in the decease of Sir Frederick Bramwell, Bart., D.C.L. LL.D. F.R.S. M.Inst.C.E.

His personality, and the signal services he rendered to the Institution by his unvarying devotion to and deep interest in its welfare and in the advancement of experimental research for twenty-seven years as Manager, Hon. Secretary, and Vice-President, will be long remembered.

The Managers desire to offer Lady Bramwell and her family the expression of the most sincere sympathy with them in their bereavement.

This was unanimously adopted, all present standing.

The Managers reported that at their Meeting held that day they had elected Professor Louis Compton Miall, F.R.S., Fullerian Professor of Physiology for three years (the appointment dating from January 14, 1904).

The PRESENTS received since the last Meeting were laid on the table, and the thanks of the Members returned for the same, viz. :—

FROM

The Secretary of State for India—Archæological Survey of Western India: Progress Report, 1903. fol.
  Report on the Madras Government Museum and Connemara Library, 1902-3. fol. 1903.
Accademia dei Lincei, Reale, Roma—Classe di Scienze Fisiche, Matematiche e Naturali. Atti, Serie Quinta: Rendiconti. Vol. XII. 2o Semestre, Fasc. 1. Nos. 8, 9. 8vo. 1903.
American Academy of Arts and Sciences—Proceedings, Vol. XXXIX. No. 4. 8vo. 1903.
American Philosophical Society—Proceedings, Vol. XLII. April-May. 8vo. 1903.
Automobile Club—Journal for Nov. 1903.
Bankers, Institute of—Journal, Vol. XXIV. Part 8. 8vo. 1903.
Belgium, Royal Academy of Sciences—Mém. Cour. et des Savants étrang. Tome LXII. Fasc. 4. 4to. 1903.
  Mém. Cour. et autres Méms. Tome LXIII. Fasc. 7. 8vo. 1903.

Boston Public Library—Monthly Bulletin for Nov. 1903. 8vo.
Botanic Society of London Royal—Quarterly Record, Vol. VIII. No. 95 (July–Sept). 8vo. 1903.
British Architects, Royal Institute of—Journal, Third Series, Vol. XI. Nos. 1–3. 4to. 1903.
British Astronomical Association—Journal, Vol. XIV. No. 1. 8vo. 1903.
British Fire Prevention Committee—Official Report of the International Fire Prevention Congress, 1903. fol. 1903.
Brooklyn Institute of Arts and Sciences—Cold Spring Harbour Monographs, I.-II. 8vo. 1903.
Bureau International des Poids et Mesures—Procès-Verbaux, Série 2, Tome II. 8vo. 1903.
Canada, Meteorological Service—Report for 1901. 4to. 1903.
Chemical Industry, Society of—Journal, Vol. XXII. Nos. 20–22. 8vo. 1903.
Chemical Society—Proceedings, Vol. XIX. Nos. 270, 271. 8vo. 1903.
 Journal for Dec. 1903. 8vo.
City of London College—The Calendar, 1903-4. 8vo. 1903.
Clinical Society—Transactions, Vol. XXXVI. 8vo. 1903.
Cracovie, Academy of Sciences—Bulletin, Classe des Sciences Mathématiques, 1903, No. 7; Classe de Philologie, 1903, Nos. 6, 7. 8vo.
Dax, Société de Borda—Bulletin, 1903, 1–2eme Trimestre. 8vo.
East India Association—Journal, Vol. XXXV. No. 32. 8vo. 1903.
Editors—American Journal of Science for Nov.–Dec. 1903. 8vo.
 Analyst for Nov.. 1903. 8vo.
 Archivio di Fiscologia for Nov. 1903. 8vo.
 Astrophysical Journal for Nov. 1903. 8vo.
 Athenæum for Nov. 1903. 4to.
 Author for Nov.–Dec. 1903. 8vo.
 Board of Trade Journal for Nov. 1903. 8vo.
 Brewers' Journal for Nov. 1903. 8vo.
 Chemical News for Nov. 1903. 4to.
 Chemist and Druggist for Nov. 1903. 8vo.
 Dioptric Review for Nov. 1903. 8vo.
 Electrical Engineer for Nov. 1903. fol.
 Electrical Review for Nov. 1903. 8vo.
 Electrical Times for Nov. 1903. 4to.
 Electricity for Nov. 1903. 8vo.
 Electro-Chemist and Metallurgist for Oct.–Nov. 1903. 8vo.
 Engineer for Nov. 1903. fol.
 Engineering for Nov. 1903. fol.
 Feilden's Magazine for Nov. 1903. 8vo.
 Homœopathic Review for Nov.–Dec. 1903. 8vo.
 Horological Journal for Dec. 1903. 8vo.
 Humane Review for Oct. 1903. 8vo.
 Journal of the British Dental Association for Nov. 1903. 8vo
 Journal of Physical Chemistry for Nov. 1903. 8vo.
 Journal of State Medicine for Nov. 1903. 8vo.
 Law Journal for Nov. 1903. 8vo.
 London Technical Education Gazette for Nov. 1903. 4to.
 London University Gazette for Oct.–Nov. 1903. 4to.
 Machinery Market for Nov. 1903. 8vo.
 Model Engineer for Nov. 1903. 8vo.
 Mois Scientifique for Nov. 1903. 8vo.
 Motor Car Journal for Nov. 1903. 8vo.
 Motor Car World for Nov. 1903. 4to.
 Musical Times for Nov. 1903. 8vo.
 Nature for Nov. 1903. 4to.
 New Church Magazine for Dec. 1903. 8vo.
 Nuovo Cimento for June, 1903. 8vo.
 Page's Magazine for Nov.–Dec. 1903. 8vo.

Editors—continued.
Photographic News for Nov. 1903. 8vo.
Physical Review for Oct. 1903. 8vo.
Popular Astronomy for Nov. 1903. 8vo.
Public Health Engineer for Nov. 1903. 8vo.
Science Abstracts for Oct.–Nov. 1903. 8vo.
Travel for Nov. 1903. 8vo.
Zoophilist for Nov.–Dec. 1903. 4to.
Entomological Society—Transactions, 1903, Part III. 8vo.
Fleming, Prof. J. A., M.A. D.Sc. F.R.S. M.R.I. (the Author)—Handbook for the
Electrical Laboratory, Vol. II. 8vo. 1903.
Florence, Biblioteca Nazionale—Bulletin for Oct.–Nov. 1903. 8vo.
Franklin Institute—Journal, Vol. CLVI. Nos. 4–5. 8vo. 1903.
Geographical Society, Royal—Journal, Vol. XXII. Nos. 5, 6. 8vo. 1903.
Geological Society—Abstracts of Proceedings, Nos. 782, 783. 1903.
Quarterly Journal, Vol. LIX. No. 4. 8vo. 1903.
List of Fellows, 1903. 8vo.
Horticultural Society, Royal—Journal, Vol. XXVIII. Parts 1, 2. 8vo. 1903.
Johns Hopkins University—American Journal of Philology, Vol. XXIV. No. 3.
8vo. 1903.
Leighton, John, Esq., M.R.I.—Ex-Libris Journal for Oct. 1903. 8vo.
Liége, Université de—Les Installations et les Programmes de l'Institut Électro-
technique Montefiore 4to. 1903.
Linnean Society—Proceedings, Oct. 1903. 8vo.
Journal: Botany, Vol. XXXV. No. 247, Vol. XXXVI. No. 252; Zoology,
Vol. XXIX. No. 188 8vo. 1903.
List of Members, 1903–4. 8vo.
Manchester Geological Society—Transactions, Vol. XXVIII. Parts 8, 9. 8vo. 1903.
Mathematical Society—Index to Proceedings, Vols. I.–XXX. 8vo. 1903.
Mather & Crowther, Messrs. (the Publishers)—Practical Advertising, 1903–4. 4to.
Medical and Chirurgical Society, Royal—Transactions, Vol. LXXXVI. 8vo. 1903.
Meteorological Society, Royal—Quarterly Journal, Oct. 1903. 8vo.
Meteorological Record, Vol. XXIII. No. 89. 8vo. 1903.
Mexico Sociedad Cientifica "Antonio Alzate"—Memorias y Revista, Tome XVIII.
3–5, Tome XIX. 2–4. 8vo. 1903.
Navy League—Navy League Journal for Nov.–Dec. 1903. 8vo.
Odontological Society—Transactions, Vol. XXXVI. No. 1. 8vo. 1903.
Pharmaceutical Society of Great Britain—Journal for Nov. 1903. 8vo.
Photographic Society, Royal—Journal, Vol. XLIII. No. 8. 8vo. 1903.
Physical Society—Proceedings, Vol. XVIII. Part 6. 8vo. 1903.
Quekett Microscopical Club—Journal, Series 2, Vol. VIII. No. 53. 8vo. 1903.
Nov.
Rennes, Université de—Travaux Scientifiques, Tome II. Fasc. 1, 2. 8vo. 1903.
Reynolds, A. R., Esq. (Commissioner of Health, Chicago)—Report of Streams
Examination, Sanitary District of Chicago. 8vo. 1903.
Rome, Ministry of Public Works—Giornale del Genio Civile for May–June, 1903.
8vo.
Royal Society of London—Philosophical Transactions, A, Nos. 350–353; B, Nos.
220–222. 4to. 1903.
Proceedings, Nos. 483, 484. 8vo 1903.
Reports of the Sleeping Sickness Commission, Nos. 2–4. 8vo. 1903.
Rouchechouart, La Société des Amis des Sciences et Arts—Bulletin, Tome XII.
No. 6. 8vo. 1902.
Russell, T. H., Esq. (the Author)—Chemical and Physical Laboratories. 8vo. 1903.
Saxon Society of Sciences, Royal—
Mathematisch-Physische Klasse—
Abhandlungen, Band XXVIII. Nos. 4, 5. 4to. 1903.
Berichte, Band LV. Nos. 3–5. 8vo. 1903.
Philologisch-Historische Klasse—
Abhandlungen, Band XXI. No. 3; Band XXII. Nos. 2, 3. 4to. 1903.

Selborne Society—Nature Notes for Nov.-Dec. 1903. 8vo.

Smith, B. Leigh, Esq., M.R.I.—Scottish Geographical Magazine, Vol. XIX. Nos. 11, 12. 8vo. 1903.

Société Archéologique du Midi de la France—Bulletin, Nouvelle Série, Nos. 29, 30. 8vo. 1903.

Society of Accountants and Auditors—Incorporated Accountants' Year Book, 1903–4. 8vo.

Society of Arts—Journal for Nov. 1903. 8vo.

Solvay, E., Esq., Hon.M.R.I. (the Author)—Le Procédé de Fabrication de la Soude à l'ammoniaque. 8vo. 1903.

South Australian School of Mines—Annual Report, 1902. 8vo. 1903.

Swithinbank, H., Esq., J.P. M.R.I., and Newman, G., M.D. (the Authors)—Bacteriology of Milk. 8vo. 1903.

Tacchini, Prof. P., Hon. M.R.I. (the Author)—Memorie della Società degli Spettroscopisti Italiani, Vol. XXXII. Disp. 11. 4to. 1903.

Transvaal, Director of Agriculture—Transvaal Agricultural Journal, Vol. II. No. 5. 8vo. 1903.

United Service Institution, Royal—Journal for Nov. 1903. 8vo.

United States Department of Agriculture—Monthly Weather Review for August, 1903. 4to.

United States Patent Office—Official Gazette, Vol. CVI. No. 9; Vol. CVII. Nos. 1–4. 8vo. 1903.

Verein zur Beförderung des Gewerbfleisses in Preussen—Verhandlungen, 1903. Heft 9. 8vo.

Western Society of Engineers—Journal, Vol. VIII. No. 5. 8vo. 1903.

WEEKLY EVENING MEETING,

Friday, January 16, 1903.

Sir James Crichton-Browne, M.D. LL.D. F.R.S., Treasurer and
Vice-President, in the Chair.

Professor Sir James Dewar, M.A. LL.D. D.Sc. F.R.S. *M.R.I.*

Low Temperature Investigations.

In the Friday Evening Discourse delivered in the year 1896, entitled " New Researches on Liquid Air" (Proc. Roy. Inst.), it was
shown that seven substances, having very different coefficients of
expansion, viz. cadmium, lead, copper, silver, calc spar, rock crystal,
silver iodide, all gave the same density for liquid oxygen, when
used to determine the weight displacement in the liquid, provided
the correcting factor used in each case was the calculated mean coefficient of cubical expansion found by extending the values of Fizeau
to low temperatures.   The fact of the uniformity in the resulting
oxygen density proved that the parabolic law of Fizeau may safely
be used for extrapolation at low temperatures as far as the boiling
point of air, especially in the case of the metals.

The determination of the densities of substances at the boiling
point of oxygen—and hence of their mean coefficients of expansion
between that temperature and ordinary temperatures—opens out
a very large field of investigation, from which, if a sufficiently
large number of observations were available, valuable deductions might be drawn.   On account, however, of the expense and
trouble of producing quantities of liquid oxygen, its use for this
purpose is not likely to become general, although, when available, it
is the easiest body to use in conducting such experiments, especially
when the vacuum vessel containing it is immersed in a larger vessel
containing the same fluid or well evaporated air.   The ease with
which liquid air can now be obtained in many laboratories suggests
that its application to work of this kind would be a convenience.
The use of a mixture of varying composition and density like liquid
air necessitates a determination of its density with accuracy and
rapidity before and during the course of the experiments.   For this
purpose, liquid air that had been allowed to evaporate for twenty-four
hours in advance was used in large silver-coated vacuum vessels of
some 3 litres capacity.   In order to ascertain the density of the

liquid, a polished silver ball, which had been weighed once for all in liquid oxygen, was weighed in the sample of liquid air, and from the relative weights thus found the density of the liquid air could be approximately determined, that of liquid oxygen being 1·137. To prevent any disturbing ebullition in the liquid-air flask in which the weighings took place, and to reduce the rate of its evaporation to a minimum during the course of an experiment, the substance to be used was previously cooled in a supplementary vessel containing liquid air and then transferred to the large flask. Substances like solid carbonic acid and ice were weighed in the cool, gaseous air of the vacuum vessel, and their weights subsequently corrected for buoyancy. The temperatures of the densest and lightest samples of liquid air were ascertained by the hydrogen thermometer, and that of the others deduced by graphic interpolation. As the entire range of temperature through which the bodies were cooled amounted to about 200°, a degree or two up or down has no real influence on the results; the extreme range of temperature in the air samples was from 83·8° to 86·1° absolute.

Salts were employed in the form of compressed blocks. The salt, previously reduced to a fine powder, was moistened with water and compressed in a cylindrical steel mould under great hydraulic pressure. During compression the saturated salt solution drained away, and finally a cylindrical block of some 50 grammes of the salt was obtained free from porosity and hard enough to allow its surface to be polished. In this form salts and other materials similarly treated are especially adapted for accurate specific gravity determinations. After such treatment it was found that all the mechanically attached water was got rid of in the case of hydrated salts, and also in such as did not combine with water. In order to get cylindrical blocks of the salts showing no porosity, the presence of water, or rather the saturated salt solution, was found to be essential during the application of pressure. In the same way it was found to be an advantage in compressing such a substance as solid carbonic acid, to moisten it with a fluid like ether before applying the hydraulic pressure.

Recalling the work of Playfair and Joule,[*] which originated in a suggestion of Dalton's that the volume of a hydrated salt in solution was simply the volume of the water of crystallisation as ice, some hydrated salts were selected, as well as some other bodies whose coefficients of expansion they had determined. Substances of special interest included in the list, were ice, mercury, sulphur, iodine, and solid carbonic acid, the latter being particularly important as an example of a solidified gas.

The specific gravity of the actual portion of the substance weighed in the liquid air was, with one or two exceptions, determined also at

---

[*] " Researches on Atomic Volume and Specific Gravity," Chem. Soc. Jour., vol. i., 121.

the temperature of the laboratory, about 17° C. From the two sets
of observations the value of the mean coefficient of cubical expansion
between 17° C. and the temperature of liquid air, was calculated, and
whenever the expression coefficient of expansion is used, the volume
coefficient is meant.

Ice at Low Temperatures.

The actual density at the temperature of liquid air of pieces of ice
cut from large blocks, gave the value 0·92999. The density at
0° being 0·91599, this gives for the mean cubical coefficient
0·00008099.

We may take 0·0001551 as the mean coefficient of expansion of
ice between 0° and −20° C. Thus the mean coefficient of expansion
between 0° and −188° C. is about half of that between 0° and
−20° C. The mean coefficient of expansion of water in passing
from 4° to −10° C. is −0·000362, and from 4° to 40° C. it is
0·0002155. Hence the mean coefficient of expansion of ice between
0° and −188° C., is about one-fourth of that of water between 0°
and 10° C., and half of that between 4° and 100° C.

If the densities of ice at still lower temperatures could be determined, the values of the coefficient of expansion thence deduced would,
we have every reason to believe, be less than the value given above.
We shall therefore not be overstraining the case if we use the
value just found to determine an upper limit to the density of ice at
the absolute zero. The result is 0·9368, corresponding to a specific
volume 1·0675. Now the density of water at the boiling-point, is
0·9586 (corresponding to the specific volume 1·0432), so that ice can
never be cooled low enough to reduce its volume to that of the liquid
taken at any temperature under one atmosphere pressure. In other
words, ice molecules can never be so closely packed by thermal contraction as the water molecules are in the ordinary liquid condition,
or the volume of ice at or near the absolute zero is not the minimum
volume of the water molecules. It has been observed by Professor
Poynting * that if we suppose water could be cooled without freezing,
then taking Brunner's coefficient for ice, and Hallstrom's formula for
the volume of water at temperatures below 4° C., it follows that ice
and water would have the same specific volume at some temperature
between −120° and −130° C. Applying then the ordinary thermo-
dynamic relation, no change of state between ice and water could be
brought about below this temperature. Clausius has shown that the
latent heat of fusion of ice must be lowered with the temperature of
fusion some 0·603 of a unit per degree. If such a decrement is
assumed to be constant, then about −130° C. the latent heat of fluidity
would vanish. Thus under a pressure of about 16,000 atmospheres

---

* "Change of State, Solid, Liquid," Phil. Mag., 1881.

at this low temperature there ought theoretically, if the extrapolation pressure were legitimate, to be no distinction between the solid and liquid forms of water. At temperatures below this limit no amount of pressure would transform ice into water.

In inferring at what temperature this kind of critical point of the possible transition of ice into water takes place, no consideration of the change in the specific heats of ice and water under the greatly increased pressures have been included. If this is done, then it appears that ice under 50 tons and a corresponding temperature of about $-50°$ C., would be all transformed into water, so that no lower temperature could be reached by any forced transition. All such speculations based on imperfect data are cleared away by the important experiment made by Tamman, who has shown that ice under a pressure of 20 to 30 tons on the square inch and a temperature of $-22°$ or $-23°$ becomes transformed into a new variety of ice, which under the specified conditions of temperature and pressure is denser than water. This new ice has a density greater than water, viz. $1 \cdot 11$, so that no amount of pressure on ice below the temperature of $-23°$, can cause any transition. All the theoretical anticipations of the relations of ice and water at very low temperatures and high pressures have been entirely falsified by the results of Tamman.

Ice near its melting point can easily be squeezed into the form of wire when forced by hydraulic pressure from a steel cylinder having a small aperture in the bottom. If the temperature of the ice is lowered to -80 C. by embedding the steel cylinder and plunger in solid carbonic acid under a pressure of 50 tons, the flow still takes place. The ice wire was now made up of what looked like a set of disc-like scales, which contrasted strongly in appearance with the transparent clear ice wire got when the experiment was made at $0°$ C. On cooling the ice and its accessories as above to a temperature approaching that of liquid air, no pressure the apparatus would stand caused any flow, but only intermittent explosive ejections.

Water was frozen in a steel cylinder in successive portions so as to include lead shot in the middle and upper portions of the ice, and the whole cooled to $-80°$ C. It was now subjected to 100 tons pressure in order to see if the lead spheres had fallen partly through the ice. After this great pressure the ice was clear, the lead spheres, however, had now very irregular shapes, but no motion of the kind anticipated had taken place. Thus ice under the pressure of 100 tons, or 15,000 atmospheres, which by theory ought to have lowered the melting point below $-80°$, shows no such action. The transparent ice after the experiment showed no increase of density, but during the gradual heating up from $-80°$, it became milk-white from some new crystalline arrangement. Such high pressure experiments are greatly favoured by the increased strength of the steel dies and pistons at the lowest temperatures.

Ice is a highly expansive substance, and as a result transparent blocks dropped into liquid air crack in all directions from the sudden cooling. That the expansibility is being diminished at very low temperatures may be inferred from the fact that clear pieces of ice that have been slowly cooled to the temperature of liquid air when dropped into liquid hydrogen do not crack, although the relatively sudden drop in temperature is actually in this experiment greater than in the similar experiment with liquid air. The limiting density of ice at low temperatures may be determined from observing that it floats upon the surface of liquid oxygen, while it sinks in liquid nitrogen, the density of the former liquid being 1·13 and the latter 0·81.

Solid Carbonic Acid at Low Temperatures.

The density of solid carbonic acid at its boiling-point was formerly given as 1·5,* but the mean of my results came to 1·53. Recently the same value has been found by Behn. Taking this value and 1·6267, the density found at −188·8° C., the mean coefficient of expansion is found to be 0·0005704. This is a very large coefficient of expansion, being greater than that of any substance recorded in Table I. on page 423, and comparable with that of sulphur between 80° and 100°, which, according to Kopp, is 0·00062. The coefficient of expansion of liquid carbonic acid at its melting-point taken from the recent observations of Behn† is 0·002989, so that the rate of expansion of the liquid at its minimum value is very nearly five times that of the solid. When solid carbonic acid was subjected to pressure in the same steel cylinder as was used in the ice experiments, a wire of the solid, composed of a series of adhering disc-like plates, was easily formed.

Coefficients of Expansion; Hydrated Salts; Organic Bodies, etc., at Low Temperatures.

A general summary of the values found for the coefficients of expansion between 17° and −188° of a number of substances is given in Table I. on page 423.

In the solid state mercury has a coefficient about half of that in the fluid state, while sodium has about the same value as that of mercury at the ordinary temperature. The coefficient for sulphur is about half of that between 0° and 100° C.; and that of iodine is not far removed from the value given for the solid at ordinary temperatures. The rate of expansion of liquid iodine is about three times this value. The value found for naphthalin is about half that of the liquid near its melting-point.

With the exception of carbonate of soda and chrome alum, hydrated salts have a coefficient of expansion not differing greatly from that of ice at low temperatures. It will be noted that iodoform is a highly expansive body like iodine, and that oxalate of methyl has

---

* See Proc. Roy. Inst., 1878, " The Liquefaction of Gases."
† Chem. Jour., 1901.

nearly as great a coefficient as paraffin, which is one of the most expansive solids.

It will be possible by cooling the moulds with liquid air during the process of hydraulic compression, to produce cylindrical blocks of solid bodies of lower melting-points, like alcohol, ether, nitrous oxide, ammonia, chlorine, etc., and to ascertain their coefficients of expansion in the solid state between the individual melting-points and the boiling-point of liquid air.

This method, which works well with liquid oxygen or air, fails with liquid hydrogen, as the density of the liquid is too small to give accurate values. For temperatures about 20° absolute, recourse must be had to measurements of the coefficient of linear expansion, and such observations at present could only be applied with any accuracy to metallic bodies and alloys.

TABLE I.

	Density at −188°.	Density at 17°.	Mean Coefficient Cubical Expansion. a. $0 \cdot 000$
Sulphate of aluminium (18)[1]	1·7194	1·6913	0811
Biborate of soda (10)	1·7284	1·6937	1000
Chloride of calcium (6)	1·7187	1·6775	1191
Chloride of magnesium (6)	1·6039	1·5693	1072
Potash alum (24)	1·6414	1·6144	0813
Chrome alum (24), large crystal	1·8335	1·8199	0365
,, ,, ...	1·7842	1·7669	0478
Carbonate of soda (10).	1·4926	1·4460	1563
Phosphate of soda (12)	1·5446	1·5200	0787
Hyposulphate of soda (5)	1·7635	1·7290	0969
Ferrycyanide of potassium (3)	1·8988	1·8533	1195
Ferricyanide of potassium	1·8944	1·8109	2244
Nitro-prusside of sodium (4)	1·7196	1·6803	1138
Chloride of ammonium, sample i. . . .	1·5757	1·5188	1820
,, ,, sample ii. . . .	1·5809	1·5216	1893
Oxalic acid (2)	1·7024	1·6145	2643
Oxalate of methyl	1·5278	1·4260	3482
Paraffin	0·9770	0·9103	3567
Naphthalin	1·2355	1·1589	3200
Chloral hydrate	1·9744	1·9151	1482
Urea	1·3617	1·3190	1579
Iodoform	4·4459	4·1955	2930
Iodine	4·8943[2]	4·6631	2510
Sulphur	2·0989	2·0522	1152
Mercury	14·382	14·193	0887
Sodium	1·0066	0·972	1865
Graphite (Cumberland)	2·1302	2·0990	0733

[1] The figures ( ) refer to the number of molecules of water of crystallisation.   [2] At −38·85°.

The Action of Low Temperatures on Metals and Alloys.

In a former discourse it was shown that all the chief metals and alloys acquired a greatly increased cohesive attraction at low temperatures as measured by the breaking stress. The increase in the breaking stress may reach from 30 to 50, or even 100 per cent. It was further shown that in some metals, before rupture took place at the temperature of liquid air, no diminution of the extension under stress had taken place as compared with similar tests made at the ordinary temperature. This led to testing the flow of a metal into wire about the temperature of liquid air. The only metal that could be examined in this way was lead. At the ordinary temperature in the apparatus used lead flowed into wire under a pressure of $7\frac{1}{2}$ tons, but at $-170°$ C. it was necessary to apply the pressure of $67\frac{1}{2}$ tons, or nine times the pressure, to cause any flow. In the same manner solder flowed into wire at the ordinary temperature when 35 tons was applied, but at the temperature of $-170°$ C. the application of 125 tons pressure caused no motion of the alloy through the aperture. This is the greatest pressure that any of the dies used in the experiments would stand without explosive rupture.

Ccoled Rubber Films.

One of the most interesting illustrations of the increased strength and elasticity of a body at the lowest temperatures is to take the case of a very thin transparent film of indiarubber. The film is stretched over one end of a short glass tube about the diameter of a good wide test tube, the other end being contracted and sealed on to a long, narrow tube that after being bent twice at right angles, has still one limb more than 30 in. long. The film end of the test tube can now be immersed in liquid air, while the end of the long tube is placed in a vessel containing mercury, in order to observe the diminution of pressure in inches of mercury. When the whole test tube part covered by the film is cooled in liquid air, a diminution of from 9 to 10 in. of mercury may be observed. Under such conditions the film is perfectly tight, provided it has been tied on to the glass after a little coating of melted rubber has been applied to the surface. No liquid oxygen seems able to diffuse through the film, which is indeed remarkable considering the rapidity with which gaseous oxygen is known to pass. But the most remarkable fact of all is that the liquid air surrounding the film may be replaced by a vessel containing liquid hydrogen, which instantly solidifies all the air in the film-enclosed space, giving almost a perfect vacuum, as proved by the mercury column rising to the height of the barometer at the time, and yet the film stands the pressure when cooled to $-252°$ C., and further resists the passage of hydrogen by molecular diffusion. In the cooled

state such films, when struck with a cork hammer, give out a clear
metallic ring, and if the striking is continued during the heating up
of the film, a complete gamut of notes is produced from the varying
elasticity.   After returning to the ordinary temperature the film
recovers all its ordinary properties.

MOLECULAR VOLUMES AT THE ZERO OF TEMPERATURE.

Theoretical formulæ enable an estimate to be formed of the
volume of the gram-molecule of many bodies at the zero of temperature.   The direct experimental method is to ascertain the densities
of bodies as near the zero of temperature as possible.   By means of
the use of liquid hydrogen as a cooling agent instead of liquid air
densities might be determined within 20° of the zero.   In the meantime the limiting densities of oxygen, nitrogen and hydrogen have
been found, together with the coefficients of expansion about their
boiling points.   The approximate results are given in the following
table:—

TABLE II.

	Density at Boiling Point.	Density at 62·5° Ab.	Density at 20° Ab.	Density at 15° Ab.	Coefficient of Expansion.
Oxygen	1·12	1·24	1·42	..	·004
Nitrogen	0·80	0·88	1·03	..	·006
Hydrogen	0·07	..	..	0·076	·013

Thus solid oxygen and nitrogen are respectively some 18 and 14
times denser than solid hydrogen, while the expansion coefficients of
oxygen, nitrogen and hydrogen are roughly in the ratio of 1, 1½ and 3.
With these values the molecular volumes at the absolute zero can be
inferred, if we assume the general application of what is called the
Matthias Law of the rectilinear semi-diameter.   The values which
result for oxygen, nitrogen and hydrogen are respectively 21·2, 25·5,
24·2.   The volume in cubic centimetres of the gram-molecule of
these three elements does not differ much from the mean value 23·6 c.c.
The experiments already described on the density of ice and solid
carbonic acid, about 90° absolute, enables an approximate estimate
to be made of their zero volumes, which results in the values of 19·2
for the ice molecule and 25·7 for the carbonic acid one.   From these
values along with the molecular volumes given above for solid
hydrogen and oxygen, we can ascertain the volume change that would
result in the formation of the compound molecules of ice and solid
carbonic acid—provided they could be formed by an imaginary combination taking place, at the zero of the solid hydrogen and oxygen
on the one hand, and the solid oxygen and carbon on the other.   Thus

it appears that 100 volumes of mixed hydrogen and oxygen in the
solid state would after combination produce 55·2 volumes of ice, or
the contraction would amount to 45 per cent. of the initial volume
of the mixture. This value is of the order of magnitude of the contraction which results from the combination of oxygen (solid) with
metallic bodies like lithium and sodium, which is about 60 per cent.
On the other hand, the production of solid carbonic acid from the
diamond or graphite and solid oxygen, would in the former case
involve an expansion of $4\frac{1}{2}$ per cent., while in the latter the contraction
would not exceed some 3 per cent. Such considerations confirm the
view that what we call the molecular volume at zero is not the real
volume of the molecules, but includes a considerable amount of
unoccupied free space.

[J. D.]

LONDON   PRINTED BY WILLIAM CLOWES AND SONS, LIMITED
GREAT WINDMILL STREET, W , AND DUKE STREET, STAMFORD STREET, S.E

Royal Institution of Great Britain.

WEEKLY EVENING MEETING,

Friday, January 15, 1904.

Sir Benjamin Baker, K.C.B. K.C.M.G. LL.D. F.R.S. M.Inst.C.E.,
Vice-President, in the Chair.

The Right Hon. Lord Rayleigh, O.M. M.A. D.C.L. LL.D. Sc.D.
F.R.S. *M.R.I.*, Professor of Natural Philosophy, R.I.

Shadows.

My subject is shadows, in the literal sense of the word—shadows thrown by light, and shadows thrown by sound. The ordinary shadow thrown by light is familiar to all. When a fairly large obstacle is placed between a small source of light and a white screen, a well-defined shadow of the obstacle is thrown on the screen. This is a simple consequence of the approximately rectilinear path of light. Optical shadows may be thrown over great distances, if the light is of sufficient intensity : in a lunar eclipse the shadow of the earth is thrown on the moon : in a solar eclipse the shadow of the moon is thrown on the earth. Acoustic shadows, or shadows thrown by sound, are not so familiar to most people; they are less perfect than optical shadows, although their imperfections are usually over-estimated in ordinary observations. The ear is able to adjust its sensitiveness over a wide range, so that, unless an acoustic shadow is very complete, it often escapes detection by the unaided ear, the sound being sufficiently well heard in all positions. In certain circumstances, however, acoustic shadows may be very pronounced, and capable of easy observation.

The difference between acoustic and optical shadows was considered of so much importance by Newton, that it prevented him from accepting the wave theory of light. How, he argued, can light and sound be essentially similar in their physical characteristics, when light casts definite shadows, while sound shadows are imperfect or non-existent ? This difficulty disappears when due weight is given to the consideration that the lengths of light waves and sound waves are of different orders of magnitude. Visible light consists of waves of which the average length is about one forty-thousandth of an inch. Audible sound consists of waves ranging in length from about an inch to nearly forty feet : the wave length corresponding to the middle C of the musical scale is roughly equal to four feet. It is, therefore, no matter for wonder that the effects produced by sound waves and by light waves differ in important particulars.

Moreover, the wave length is not the only magnitude on which the perfection of the shadow depends; the size of the obstacle, and the distance across which the shadow is thrown, must also be taken into consideration. The optical shadow of a small object, thrown across a considerable distance, partakes of the imperfections generally observed in connection with sound shadows.

It was calculated by the French mathematician, Poisson, that, according to the wave theory of light, there should be a bright spot in the middle of the shadow of a small circular disc—a result that was thought to disprove the wave theory by a *reductio ad absurdum*. Although unknown to Poisson, this very phenomenon had actually been observed some years earlier, and was easily verified when a suitably arranged experiment was made.

Under suitable conditions a bright spot can be observed at the centre of the shadow of a three-penny bit. The coin may be supported by three or four very fine wires, and its shadow thrown by

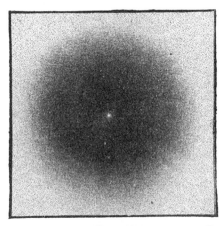

Fig. 1.—Reproduction of a Photograph of the Shadow of a Silver Penny Piece.

sunlight admitted at a pin-hole aperture placed in the shutter of a darkened room. The coin may be at a distance of about fifteen feet from the aperture, and the screen at about fifteen feet beyond the coin. To obtain a more convenient illumination, a larger aperture in the shutter may be filled by a short focus lens, which forms a diminutive image of the sun; this image serves as a point source of light. A smaller disc has some advantages. Fig. 1 is reproduced from a photograph of the shadow of a silver *penny* piece, struck at the time of the Coronation. The shadow, formed in the manner just described, was allowed to fall directly on a photographic plate; after development a negative was obtained, in which the dark parts of the shadow were represented by transparent gelatine, while the bright parts were represented by opaque deposits of silver. To obtain a correct representation, a contact print was formed from the negative in the usual

way, upon a lantern plate; and from this Fig. 1 has been reproduced.

It is at once evident that at the centre of the shadow, where one
would expect the darkness to be most complete, there is a distinct
bright spot. This result has always been considered a valuable
confirmation of the wave theory of light.

I now propose to speak of acoustic shadows—shadows thrown by
sound. The most suitable source of sound for the following experiments is the bird-call,* which emits a note of high pitch—so high,
indeed, that it is inaudible to most elderly people. The sound emitted
has two characteristics, valuable for our purpose—the wave length
is very short; and the sound is thrown forward, without too much
tendency to spread, thus differing from sounds produced by most other
means.

Since the sound emitted is nearly inaudible, some objective method
of observing it is required. For this purpose we may utilise the
discovery of Barrett and Tyndall, that a gas flame issuing under somewhat high pressure from a pin-hole burner *flares* when sound waves
impinge on it, but recovers and burns steadily when the sound ceases.
The sensitiveness of the flame depends on the pressure of the gas,
which should be adjusted so that flaring just does not occur in the
absence of sound. If the bird-call is directed towards the sensitive
flame, the latter flares so long as the call is sounded and no obstacle
intervenes. On interposing the hand about midway between the two,
the flame recovers and burns steadily. Thus the sound emitted by
the bird-call casts a shadow, and to this extent resembles light.

It will now be shown that the sensitive flame flares when it is
placed at the centre of the acoustic shadow thrown from a circular
disc, but recovers in any other position within the shadow; thus
proving that there is sound at the centre of the shadow, although at
a small distance from this point there is silence. The part of the
flame which is sensitive to sound is that just above the pin-hole orifice,
so that it is necessary to arrange the bird-call, the centre of the disc,
and the pin-hole orifice in a straight line. For the disc, it is convenient to use a circular plate of glass about 18 inches in diameter
with a piece of black paper pasted over its middle portion, a small
hole being cut in the *paper* exactly at the centre of the disc. The
glass disc is hung by two wires, and the positions of the bird-call and
sensitive flame can be adjusted by sighting through the hole in the
paper. If the disc is caused to oscillate in its own plane, the flame
flares every time that the disc passes through its position of equilibrium, and recovers whenever the disc is not in that position. The
analogy between this experiment, and that in which a bright spot is
formed at the centre of the optical shadow of a small disc, is sufficiently obvious.

The approximate theory of the shadow of the circular disc is

---

* See 'Proc. Roy. Inst.,' Jan. 17, 1902.

easily given, and it explains the leading features of the phenomenon. But, even in the simpler case of sound, an exact calculation which shall take full account of the conditions to be satisfied
at the edge, has so far baffled the efforts of mathematicians. When
the obstacle is a sphere, the problem is more tractable, and, in a
recent memoir in the 'Philosophical Transactions' a solution is given,
embracing the cases where the circumference of the sphere is as
great as two or even ten wave-lengths. When the sphere is small
relatively to the wave-length, the calculation is easy, but the difficulty rapidly increases as the diameter rises. The diagram gives
the intensity in various positions on the surface of the sphere when
plane waves of sound, i.e. waves proceeding from a distant source,
impinge upon it. The intensity is a maximum at the point 0°
nearest to the source, which may be called the pole. From the pole
to the equator, distant 90° from it, the intensity falls off, and the
fall continues as we enter the hinder hemisphere. But at an
angular distance from the pole of about 135° in one case and 165°
in the other, the intensity reaches a minimum and thence increases
towards the antipole at 180°.

In private experiments the distribution of sound over the surface
of the sphere may be explored with the aid of a small Helmholtz resonator and a flexible tube, and in this way evidence may be obtained
of the rise of sound in the neighbourhood of the antipole. A more
satisfactory demonstration is obtained by the method already employed in the case of the disc, the disc being replaced by a globe
(about 12 inches in diameter), or by a croquet-ball of about $3\frac{1}{2}$
inches diameter. In the former case the burner may be situated
behind the sphere at such a distance as 5 inches. In the latter a
distance of $1\frac{1}{2}$ inches (from the surface) suffices. By a suitable
adjustment of the flame, flaring ensues when everything is exactly
in line, but the flame recovers when the ball is displaced slightly
in a transverse direction. Since the wave-length of the sound is
3 cm. and the circumference of the croquet-ball is about 30 cm., this
case corresponds to the curve B of our diagram.

In connection with the mathematical investigation which led to
the results represented graphically in Fig. 2, there is a point of interest which I should like to mention. The investigation was carried
out upon the supposition that the source of sound is at a considerable
distance, so that the waves reaching the sphere are plane; and that
the receiver, by which the sound is detected, is situated on the
surface of the sphere. At any given position on the surface of the
sphere, the receiver will indicate the reception of sound of a certain
intensity, which may be read off from Fig. 2. Now the final results
assume a form which shows that, if the positions of the source and
the receiver are interchanged, the latter will indicate the reception
of sound of the same intensity as in the original arrangement. Thus
each of the curves in Fig. 2 represents the solution of two distinct
problems: the intensity of the sound derived from a distant source

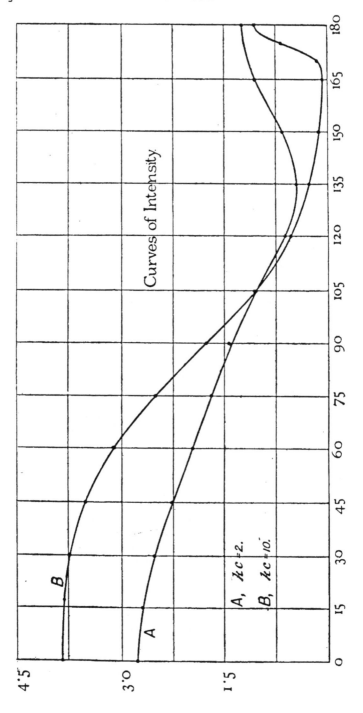

Fig. 2.

and detected at any point on the surface of the sphere; and the intensity of the sound derived from a source on the surface of the sphere, and observed at a distant point. This result forms an interesting example of a principle of very wide application, which I have termed the *Principle of Reciprocity*. Some special cases were given many years ago by Helmholtz.

It is a matter of common observation that if one person can see another, either directly or by means of any number of reflections in mirrors, then the second person can equally well see the first. The same law applies to hearing, apparent exceptions being easily explained. For instance, such is the case of a lady sitting in a closed carriage, listening to a gentleman talking to her through the open window. If the street is noisy, the lady can hear what the gentleman says very much more distinctly than he can hear what she replies. This is due to the fact that the gentleman's ears are assailed by noises of the street from which the lady's ears are shielded by the walls of the carriage.

Another instance may be mentioned, which will appeal to electricians. In the arrangement known as Wheatstone's bridge, resistances are joined in the form of a lozenge, a galvanometer being connected between two opposite angles of the lozenge, while a battery is connected between the other two angles. When the resistances are suitably adjusted, no current flows through the galvanometer; but a slight want of adjustment produces a deflection of the galvanometer, thus indicating the passage of a small current. Now, if the positions of the battery and the galvanometer are interchanged, without alteration of resistance, the same current as before will flow through the galvanometer, and therefore the deflection will be the same as before. Thus with a given cell, galvanometer and set of resistances, the sensitiveness of the Wheatstone's bridge arrangement is the same whichever pair of opposite angles of the lozenge are joined by the galvanometer. If a source of alternating E.M.F. is used instead of the battery, and a telephone is substituted for the galvanometer, then the principle of reciprocity still applies, whether the resistances are inductive or non-inductive.

A simple illustration, of a mechanical nature, is now shown. Fig. 3 represents a straightened piece of watch-spring clamped at one end to a firm support. A weight can be hung at either of the points A or B of the spring, when it may be observed that the deflection at B due to the suspension of the weight at A, is exactly equal to the deflection at A due to the suspension of the weight at B. This result is equally true wherever the points A and B may be situated; it applies not only to a loaded spring, which has been chosen as suitable for a simple lantern demonstration, but also to any sort of beam or girder.

It will have become clear, from what has been said, that waves encounter considerable difficulty in passing round the outside of a curved surface. I wish now to refer to a complementary phenomenon—the ease with which waves travel round the *inside* of a curved surface. This is the case of the whispering gallery, of which there is a good example in St. Paul's Cathedral. The late Sir George Airy considered that the effect could be explained as an instance of concentrated echo, the sound being concentrated by the curved walls, just as light may be brought to a focus by a concave mirror. From

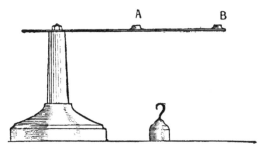

FIG. 3.—To Illustrate a Simple Mechanical Application of
the Principle of Reciprocity.

my own observations, made in St. Paul's Cathedral, I think that Airy's explanation is not the true one; for it is not necessary, in order to observe the effect, that the whisperer and the listener should occupy particular positions in the gallery. Any positions will do equally well. Again, whispering is heard more distinctly than ordinary conversation, especially if the whisperer's face is directed

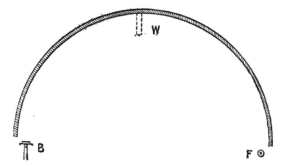

FIG. 4.—Model Illustrating the Peculiarities of a
Whispering Gallery.

along the gallery towards the listener. It is known that a whisper has less tendency to spread than the full-spoken voice; thus a whisper, heard easily in front of the whisperer, is inaudible behind that person's head. These considerations led me to form a fairly satisfactory theory of the whispering gallery, nearly twenty-five years ago.* The phenomenon may be illustrated experimentally

---

* 'Theory of Sound,' § 287.

by the small scale arrangement represented diagrammatically in Fig. 4. A strip of zinc, about 2 feet wide and 12 feet long, is bent into the form of a semicircle; this forms the model of the whispering gallery. The bird-call B is adjusted so that it throws the sound tangentially against the inner surface of the zinc: it thus takes the place of the whisperer. The sensitive flame F takes the place of the listener. A flame is always more sensitive to sound reaching it in one direction than in others; the flame F is therefore adjusted so that it is sensitive to sounds leaving the gallery tangentially. The flaring of the flame shows that sound is reaching it: if an obstacle is interposed in the straight line F B the flame flares as before; but if a lath of wood W, which need not be more than 2 inches wide, is placed against the inner surface of the zinc, the flame recovers, showing that the sound has been intercepted. Thus the sound creeps round the inside surface of the zinc, and there is no disturbance except at points within a limited distance from that surface.

[R.]

WEEKLY EVENING MEETING,

Friday, January 22, 1904.

His Grace The Duke of Northumberland, K.G. D.C.L. F.R.S.,
President, in the Chair.

The Rev. Walter Sidgreaves, S.J. F.R.A.S., Director.

*Spectroscopic Studies of Astrophysical Problems
at Stonyhurst College Observatory.*

I hold in my hand two papers, prints of the Royal Institution,
kindly sent me by my friend Professor Dewar. The first bears
the date 24th May, 1889, and is a discourse by my predecessor,
Fr. Perry. In it he says: " For the last ten years I have been
anxiously endeavouring to make Stonyhurst as efficient an observatory for solar physics as the means at my disposal would admit."
It was the last time he addressed you in this hall, for the close of
that year was the close of his life, and found him struggling in the
discharge of a duty he had accepted, which pressed upon his sensitive
nature so far as to hide from him the real danger in which he was on
the eve of the Solar Eclipse of December 1889.

It would be right on this occasion to take up the narrative where
he left it ; and this was made more easy for me by the second paper,
bearing a later date, April 11th, 1902, by Professor Dewar, 'On
Problems of the Atmosphere.' In his concluding paragraph he
draws attention to a definite and rational answer of Arrhenius to the
question: "What is the cause of the electric discharges which are
generally believed to occasion auroras?" There is much in the
answer to favour an alternative hypothesis : and it would have been
congenial to a combative nature to have placed before you my
reasons for preferring a theory of my own to that of Arrhenius.
And I feel that I owe an apology to Professor Dewar for not accepting in full the suggestion contained in his presentation of the two
papers. My apology is this: In October last two great spots crossed
the solar disc. The greater spot was associated with a smaller
magnetic disturbance, and the smaller spot with the greatest magnetic
storm recorded by the Stonyhurst magnetographs. This contrast
was all in my favour; but the facts brought out so much sun-spot
literature, that I was frightened by the thought that the subject could
not be divested of its worn-out attire, and would not afford that
interest which is the one encouragement of a stranger in my position
this evening.

Nevertheless, I am able to take up the narrative in another
direction. Fr. Perry had provided the Observatory with two costly
spectrometers : a Rowland grating on a massive stand, with circles

and micrometers, for the prosecution of his studies in solar physics, and a well equipped prismatic instrument for the stars.

With the first of these I had hopes of carrying on Fr. Perry's work on spot-spectra, with the acknowledged advantage of the grating dispersion. It was a promising field for work with an excellent instrument; and my disappointment was great when I turned from the text-books to the instrument, and tried to see what others had seen and sketched. I photographed the spectra of many spots in the rich year of 1893, only to learn more clearly that any conclusions based on the spectra of sun-spots must remain outside my sphere of influence. My last attempt was an imitation of a spot-spectrum for comparison with the real thing. A thread drawn across the slit, and kept in a state of irregular vibration, gave me all that could be desired, except the much-needed help to read the real spectrum correctly. The imitation (projection on screen) seems to me perfect; so perfect that I am unable, by the widening of the lines alone, to say which of the two is the counterfeit. I know them only by the diamond marks on the negatives, and by the companion smaller spots on the unpretended photograph. Others can differentiate between widened and unaffected lines, and they see all the lines widened in the imitation. This is as it should be; for the widening is the result of the well-known photographic effect of light encroaching on darkness: the dark line is thinned by the neighbouring brightness of the con- tinuous spectrum, off the spot, and hardly at all in its shaded hand. Allowance, therefore, had to be made for this photographic effect, and the difference between the real and the pretended spectrum was so small that nothing remained on my photographs to deal with. This experience cast a doubt over the reality of spot-spectra, which has only quite recently been removed by the transparency now on the lantern. It is from a photograph taken at the Yerkes Observatory, with a solar image of 7 inches diameter, by the great telescope. The spectrum shows widened lines, which I do not think could be imitated; and Professor Hale is probably right in saying that a much larger image of the sun is wanted than has been found possible with the light supplied by the small heliostat at Stonyhurst.

On the whole, you will not be surprised that I gave up the spectroscopic study of the sun-spot problem, and have since then left it to better eyes.

But in other ways the solar spectrograph has written its teaching in clearer characters. It has given the answer to a question modestly expressed in the discourse already mentioned: "Might not the last- mentioned observations suggest the question, whether absorbent vapours may not sometimes be cast up from the seething mass beneath, although a down-rush be the prevailing feature of a sun- spot?" The photograph projected on the screen is the spectrum of a spot, in the H K region; and if I read correctly the language of light, it says, by these bright reversals of calcium vapour, that the vapour *has* been cast up from the hotter lower regions high into the solar atmosphere, and in drifting away has fallen down like the lava of a volcanic eruption down the slopes of the crater. High in the solar atmosphere there is no absorbing vapour between it and our spectrograph to dull its glow. Falling down, it loses its central part by the absorption of attenuated vapour above it, and shows this forked appearance of the extremity at a distance from the spot.

The neighbouring hydrogen line, H_ϵ, is also clearly reversed, and probably quite similarly, but it is much weaker and shows only the form of the brighter parts of the Ca line. It appeared on four negatives. exposed within the same half hour, on September 9th, 1893, to the spot on the eastern limb. Possibly these are the only photographs of a reversed H_ϵ on the solar surface.

I will now pass to the second instrument left to me by Fr. Perry, the stellar spectrometer. It was designed for visual work, fitted with every convenience for mapping faint spectra in the dark. I spent hours with it before accepting the conclusion that I could have no confidence in my direct eye-observations. Photography was a necessity for me; and photography was impossible without alterations, which would have exposed me to the reproach of having ruined an excellent visual spectroscope. I had to make use of tools of my own, for I could not give up the stars. These had always been to me my wonder-land; and to gain ever so little knowledge of themselves and their movements, through the delicate touches of light, was more attractive than any discoveries on the surface of a star so near at home as the sun. I will not detain you with a description of my tools further than to say that I have never been ashamed to show and explain the inexpensive additions to an excellent 4-inch refractor which have enabled a small visual prismatic camera to furnish the photographic material for a spectroscopic study of the problematic variable star of the Lyre.

The material is not so abundant as could be desired. The phenomenally cloudy state of our night skies during the past two years is enough to account for the deficiency, without stopping to enumerate and explain all the adversities which have contributed to reduce the collection of photographic plates to the 54 which go to form what has been called at the Observatory the spectrographic chart of β Lyræ.

The star is a well-known variable, and its light-changes have attracted the attention of observers for over a century. Its characteristic, as a variable, is an alternately greater and less loss of light between successive equal maxima, the cycle of changes being complete in nearly thirteen days. The periodic changes are represented by the curve of one cycle projected on the screen, in which it will be observed that the four extremes, two maxima and two minima, divide the cycle into approximately equal parts. This might have suggested a binary star of short period, in a circular orbit lying in our line of sight. The two stars would then alternately eclipse one another, and one being brighter than the other the consequent darkenings would be unequal. It was reserved to Professor Pickering to draw this conclusion from Mrs. Fleming's discovery, on the Harvard spectrograms, of both bright and dark hydrogen lines in juxtaposition. These were found to exchange places periodically, and synchronously with the light changes, the bright lines being found on the red sides of their dark companions during the first half of the light cycle, and on the violet sides in the second half. This would be the result of the relative velocity of the components of a close binary, revolving on the orbit described, one giving the bright lines, the other the dark lines ; and Pickering's measure of the greatest separations of the pairs of lines gave the relative velocity at about 300 miles in the second. This explanation was simple and attractive ; but the simplicity did not last long. Professor Vogel's examination of a goodly series of photographs at Potsdam resulted in perplexing variations ; and he pointed out the uncertainty of a velocity deduced from the separation of centres which cannot be accurately pointed, on account of their unknown positions on broad and overlapping lines.

It was to facilitate the study of these new difficulties in their relation to the light curve that the spectrographic chart was compiled in the order now shown on the screen. The enlarged photographs were originally mounted in one column on a long cardboard, here in two columns, in the order of their time-intervals from the preceding chief minimum, regardless of their absolute dates. These intervals, called periodic dates, were written, in days and hours, opposite the spectrograms in one margin, and the civil dates of the preceding chief minima in the other margin. The phases of the light variations are thus followed throughout the entire cycle by corresponding phases of the changing spectrum, at intervals averaging about six hours.

Mounted in this order, the enlargements were found to be a great help in the examination of the original negatives. Each of these was examined on the micrometer stage in connection with its enlarged positive, and then the details, better seen on the negatives, could be safely recognised on the positives without the aid of a magnifier. Comparisons then were easily made ; and the chart became a guide, directing attention to the parts of the spectrum which called for closer examination on the original plates.

I will first draw your attention to some general features in the progressive changes of the spectrum. And the one which claims the first mention is the fidelity with which the same spectral phase recurs at the same phase of the light cycle. This was important to me, because my arrangement of the photographs on the chart had not found favour with eminent German astronomers ; and it is the corner-stone of my superstructure. I do not pretend that no exceptions will be found in the future ; indeed, the regularity of a pendulum would be too precise for some of my propositions. But it is worthy of notice that amongst the 54 plates which fill in the light cycle there is no case of going back upon the continuity of succession in the spectral changes, notwithstanding the long time-interval between some of the consecutive periodic dates on the chart.

Secondly. The bright lines of hydrogen and helium lie on the less refractive sides of their dark companions during the first half of the period, and on the opposite sides in the second half, while about the two epochs of minimum-light the pairs of lines are superposed. And, so far, the chart is a confirmation, in a general way, of Pickering's explanation. Neglecting details, we might say that at the two minima of light, when the stars are in conjunction, with partial eclipses, they are crossing our line of sight, and there is no displacement of the spectral lines. Midway between the two eclipses the stars are travelling along our sight-line in opposite directions, one receding, the other approaching; the bright-line-star receding in the first half and approaching in the second half of the cycle of changes. Orbital motion in a plane through our line of sight, or only a little inclined to it, is thus indicated by the general aspect of the chart; and the circular orbit is certainly most in accord with the light variations.

Thirdly. There is a general decline of the bright lines throughout the light-period, and they are nearly extinct at the end. There is here an obvious suggestion of higher temperature at the beginning, and a progressive cooling during the rest of the cycle—a condition of things not easily explained on any other supposition than that of a tidal disturbance at the periodic returns of a near approach of two stars.

Collating these three general views of the chart, I consider myself entitled by the first, to treat the spectrographic chart as, in general characters, representative of any single cycle of spectral changes; and by the second, to assert that these changes must ultimately be attributed to orbital motion. The third is a suggestion, strong in itself, but inconsistent with a circular orbit. And it remains to seek in the details of the chart either confirmation or refutation of elliptical motion.

To begin with the greater eclipse, we note that just before, at, and just after the principal minimum, the line marked H_ζ appears as a weak broad bright band divided centrally by a fine dark line, and the other lines are single and dark. There is no displacement. The bright and dark lines are superposed centrally, and the motion of the stars must be across our line of sight. This is the first of the two epochs of spectral conjunction, and it coincides with the stellar eclipsing conjunction. Then comes the outburst of bright lines, all on one side of their dark companions, to your right on the screen, which is the red end of the spectrum; and they remain there until well after the second minimum, as seen on the 26 successive periodic dates, including the date of 6–20, which is 9 hours after the second minimum. It is not until 17 hours after this minimum that we again find the spectrum as it was at the first eclipse, viz., a broad H_ζ bright line divided by a fine dark line, and the other hydrogen and helium lines single and dark. This is the second spectral conjunction, and, unlike the first, it is retarded on the stellar conjunction, which of necessity occurs at the time of least light; and the retardation is about 24 hours, between the periodic dates of 7–4 and 7–19.

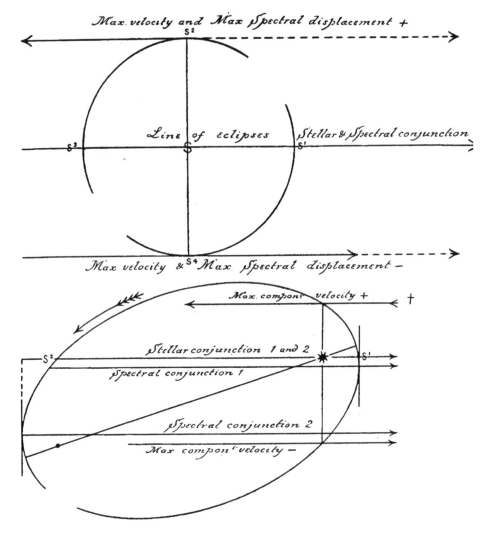

We have here direct indication of elliptical motion, as may be seen in the diagram projected on the screen. Suppose an elliptic orbit of a star S about another star occupying the focus nearer to Earth,

† Rambaut, 'On the Parallax of the Double Stars,' *Monthly Notices, R.A.S.*, vol. 50, p. 305.

and the major axis inclined, in the plane of the orbit, to our line of sight, represented by the arrow-headed parallel lines. The line $S_1 S_2$ is the line of eclipses, or stellar conjunctions, and the two lines of sight drawn normals to the curve are the two lines of spectral-conjunction, when the star is moving in the tangent at right angles to the sight-line. It is obvious that the excentricity of the ellipse and the inclination of the axis may be such as to leave no measurable time-interval between the stellar and spectral conjunctions at periastron, while the latter is considerably retarded at apastron.

The importance of this retardation led me to look for some independent confirmation of it, and I found it both in H_ϵ and in H_ζ, but only in such manner as to show that with more photographs clustering about the two epochs there might be confirmation amounting to certainty.

The line H_ϵ has a dark line at a little distance on the more refrangible side, which gives it the appearance of a double dark line gradually closing up and reopening in the progress of the light period. This appearance was found to be partly true by their measured length-intervals, as may be seen in the tabulated results projected on the screen. In this table the two dark lines are denoted by H_ϵ for the hydrogen, and c for its companion. There are 14 measures of the length-interval $H_\epsilon - c$ in the first half of the cycle, and 12 in the second half. All the plates were measured, but only these could be measured accurately. On many plates the line c is very weak, but these confirm the tabulated numbers inasmuch as all belonging to the first half of the light period gave readings below the general mean, and those of the second half gave readings above the mean. The mean difference between the two sets of readings, treated as a Doppler-effect, would give a relative velocity of the origins at about 100 km/secs. I have no doubt that Vogel's objection applies here, and that the cause of the apparent shifts of the lines is the one-side-thinning of dark H_ϵ by its bright companion. This would throw the apparent centre nearer to the fiducial line c in the first half, and further away in the second half of the period. The range of the readings could not be expected to give a reliable amplitude of the oscillating displacement of the bright line, from which the velocity could be deduced; but the mean of the measures should give the zero of displacement, and this is the reading of the plate belonging to the periodic date 7–4. Unfortunately it is the only plate available for the measurement between the dates 5–9 and 9–1; but on it the two lines are very fine and clear, and its date separates all the lower readings which precede it, from all the higher readings which follow it. In this way the result is a striking confirmation of the one-day retardation of the second spectral conjunction. It would be more satisfactory if the coincidence of the first conjunctions had also been indicated by the readings at the principal minimum. These indicate correctly a preceding spectral conjunction, but occurring about eight hours too soon. But on the plates at this epoch the two lines are neither sharp nor thin, but wide and ill defined, and we are more dependent upon a mean value from many plates which are wanting.

A more complete result is offered by the similar relative oscillations of the dark component H_ζ. These measures were ready in the tabulated scale readings of the prominent lines; and the results are shown in the second part of the table on the screen, in measures referred to the centre of dark H_γ. There is a clear relative oscillation of the line between a well-marked minimum and a well-marked maximum distance from H_γ; and the mean of the two occurs twice in the light period, on the dates 0–0, 0–5, 0–6, and again on 7–4, 7–6, indicating the spectral conjunctions *at* the first minimum and about 18 hours *after* the second minimum.

It is only necessary to add that the apparent oscillation of dark H_ζ is the indication of the real oscillation of bright H_γ. But it may be well to point out the cause of the oscillation appearing in H_ζ and not in the other hydrogen lines. In the first half of the period, the fiducial point of the scale readings is the centre of that part of dark H_γ which is seen outside the bright line towards the violet end of the spectrum; in the second half, the fiducial point is shifted to another part of the same dark line, viz., the centre of the red side part. The other hydrogen lines, H_δ and H_ϵ are affected in the same manner as H_γ. Not so H_ζ; this dark line appears in the second half of the period, with *both* edges visible, and its centre (omitting a smaller change to be considered later) is practically the same at both epochs; its apparent shift in one direction is the real shift of the zero point in the opposite direction.

We have, therefore, I venture to say, a good probability in favour of the elliptical orbit; and we must now look to other details of the chart, to see in what way they may or may not support the conclusion.

And first, for the greatest development of the bright lines, we find them at their best on the plates of 1–11 and 1–14, periodic dates, and again on a more recent photograph, too late to find its place on the chart, on 1–9 date; while at 0–17 they are still quite weak. They come out, therefore, it may be quite suddenly, between these dates, or about one day after the principal minimum. It has long been known that this excessive strength is associated with the epoch of chief minimum, and the effect has been attributed by Dr. Vogel to *contrast* with the weaker continuous spectrum at the minimum of light. Contrast, of itself, is certainly enough to account for much; but it does not answer in this case to the chart. The chart shows the lines strong, not *at* the chief minimum, but well after it. Now contrast should be the same before and after, for the continuous spectrum is equally weak on both sides of the minimum; and yet on the fourteen dates between the second maximum and the chief minimum, following one another at short intervals of the series up to within seven hours of the minimum, there are no strong bright lines, but on the contrary they are weak and dying out. I see no answer to this, and, *contrast* failing, there remains only *greater intensity* to account for the temporary strength of the lines.

To me, therefore, the more probable interpretation of the chart is : that shortly after the chief minimum there is a more or less sudden evolution of heated gases, which cool down during the rest of the period. The decline of the bright lines is in strict harmony with this fall of temperature, allowance being made for an apparent revival after the second spectral conjunction during which they are hidden by the masking effect of the dark lines. The periodic rise of temperature is quick and its fall is slow, as might be expected from a tidal disturbance at the near approach on an elliptical orbit.

Secondly. The highest temperature, brought out by tidal action, may be expected shortly after the near approach of the stars ; and the whole-day-delay in the manifestation of the bright lines is possibly too long. But the greatest heat and greatest light need not go hand in hand together. Heat may have done its work, and the light display may be delayed until the masking effect of the superposition has been removed by the separation of the bright and dark lines, which is always the condition of this bright evolution.

Thirdly. At this first appearance of separated bright and dark lines, the relative displacement is already well advanced as early as 33 hours after conjunction at the minimum. It is so far advanced that it may already have reached its maximum, for the measured width of the bright part differs so little from the subsequent measures in the series that it cannot be asserted confidently that there is any greater. And this early indication of the maximum velocity in the sight-line is in keeping with the quick swing back on an elliptical orbit, and impossible in the slower rounding of the circle.

Fourthly. To consider a hitherto neglected peculiarity in the behaviour of the line marked H_ζ, you will see on the spectrogram at six hours after the chief minimum that this line has anticipated the other lines of the spectrum in appearing with its one-sided bright line. It is quite 24 hours earlier ; and it has lost its more refracted bright part without any gain of width on the opposite side ; the more refracted part seems to have simply vanished. I can only think of two possible explanations. Either the two bright lines, as seen on the previous date 0–5, are parts of one broad hydrogen line which has contracted asymmetrically, so far as to show nothing on the more refractive side; or they are separate lines, and one of them has vanished. Of these suppositions the first has against it serious difficulties, which must be passed by here as requiring too long a discussion, and may be omitted without loss: for the periodic enlarging and thinning of the line cannot well be accounted for otherwise than as an effect of disturbances set up at the near approaches of two stars, a state of things which implies elliptical motion, and is the condition wanted for the simpler explanation of the one of two lines disappearing.

Supposing, therefore, that at the principal minimum there are two bright lines x and y, the more refracted x has vanished at about six hours after this minimum, and does not reappear for some days in the period. It may, therefore, have disappeared through the increment of temperature, brought about at the tidal disturbance of the periastron, as the same line manifestly disappeared in the spectrum of the great Nova of Perseus. The changes in the Nova's spectrum are now projected on the screen. There are four spectra of two very different types following alternately. The dates are written in the margin; and you will observe that the change from the earlier type to the new type of March 25th and 28th was very rapid. It occurred within the 24 hours from the 27th to the 28th, and also between the 21st and 22nd, for on March 22nd a very good photograph was obtained, precisely the same as on the 25th, but the clear sky break in the clouds was of short duration, and the trail on the plate gave too narrow a spectrum for a good reproduction. Confining our attention to the H_ζ line, we find in it the same changes as in the spectrum of β Lyræ. It has a broad and bright extension on the more refractive side, on March 25th and 28th, covering the whole interval between H_ζ and H_η ; and on the 21st and 27th the extension has vanished. There can be no doubt about the apparent extension; it cannot be of hydrogen origin. This is manifest, first, by its position in the spectrum, and, second, by the distinctly weaker part of the whole band, which is the true H_ζ line, known by its position and by its relative weakness in the series of hydrogen lines.

Comparing these spectral changes of the Nova with its light curve, as plotted from the excellent series of measures made at the Radcliffe Observatory, Oxford, the apparent extension occurs at the minima of the light. And taking these measured magnitudes for our thermometric indications, the extension vanished at the higher temperatures and appeared at the lower temperatures. It was a new bright band which appeared for the first time on March 22nd, when the temperature of the star had fallen below the dissociation temperature of the origin, probably cyanogen. The subsequent oscillations of heat were, for a time, crossing the dissociation point, until the temperature of the star had fallen permanently below it, when the earlier spectral type was finally lost. This interpretation you will find confirmed in a remarkable manner by the tabulated comparisons of the star's light-magnitudes with the two types of its spectrum. The table shows (projection on screen), without any exception, that the new bright band appeared only when the magnitude of the star was below 4·57; and the temperature indicated by this magnitude is probably the dissociation temperature of the origin; for the plate referred to in the table, as exposed on the night of this magnitude of the star, shows an undecided spectrum, belonging partly to both types, a notice of which was sent to the R.A.S. before the magnitudes were published.

The oscillations of heat and light may well be explained in the Nova by the oscillating volume of gaseous matter evolved at the great catastrophe, which in its dilatation and contraction was alternately cooler and hotter. And to return to the same line, x, in the spectrum of β Lyræ, is it not possible that we have here also an oscillating temperature brought about, not as in the Nova by compression and rarefaction, but by the periodic approach and recession of two stars, and, in consequence, a temporary extinction of the line at the higher temperature of pericentre?

This was the thought that led me to a careful and often repeated measurement of the total width of the composite line H_ζ on all the plates of the series. For the line x might be present whenever the total width was greatest, although possibly hidden by the dark line; and it would be absent when the whole band was distinctly narrower. The measures went very well in favour of the hypothesis until the last quarter of the period was reached, as is now exhibited in the abridged tabulations on the screen. At the principal minimum on five dates, from 10 hours before it to 5 hours after it, the mean width is 20·8 micrometer divisions between extremes 22 and 18. On eleven dates between the periodic dates 0–6 and 2–13 the mean is 15·5 between extremes 17 and 14; and on the twenty-three dates from 2–20 to 9–14 the mean is up again to 22 divisions between extremes of 24 and 18. But in the last quarter of the period the mean width is down again to 16·5 divisions. Except for this fall-off the demonstration would be complete. Omitting it for the moment, we have the narrower width only for about two and a half days after the principal minimum, indicating a temporary extinction of the line x during the period of highest temperature above the dissociation point. But the same indication recurs towards the end of the cycle, when the temperature should be lowest. Is this a clear negation of the hypothesis? I think not; for all the bright lines are weaker at this time, and the line x is a weaker line than its companion line y, and fails more on its more refractive side; the loss of width at this time may therefore be owing only to general loss of light.

But a more important objection has yet to be met. We have seen that dark H_ζ bisects the broad bright line at the principal minimum; and I offered this, together with the apparent concentric positions of the other pairs of bright and dark lines, as evidence of spectral conjunction at the principal minimum, a proposition which is incompatible with a foreign origin of the part I have called x. The answer is found in the table of figures showing the apparent oscillation of dark H_ζ. The mean position of this line is not the true position of the hydrogen line. It is quite three micrometer divisions to the more refractive side. This might be accepted as a probable error of a single reading, but not as the mean of 14 measures on different plates. The true hydrogen centre is well within the red side part of the bright line, supposing it to overlap the dark line; and the dark line may be, what its appearance suggests, a stranger, for it has not the family likeness of the other hydrogen lines. It is natural to ask: What has become of the line?

The only answer I can give is by the same question on another star-spectrum, the spectrum of o Ceti. You see on the screen those two brilliant lines, H_δ and H_γ : What has become of H_ϵ, H_β, and H_α ? The latter could not appear in the photograph, but it has been searched for in vain as a feature of the visual spectrum.

I have now put before you all the parts of my spectroscopic study bearing upon the interesting problem of the form of the orbital motion of β Lyræ. The velocity of this motion is another problem, the interest of which is enhanced by the difficulties attending its evaluation. It will have been already observed that an eclipsing pair of stars must have their motions at some part of their orbit, very nearly in our line of sight. And, if the orbit is circular, the maximum measured velocity, on the Doppler principle of line displacement, must be the uniform velocity of the motion, and must be found at the times when the stars are to us in quadrature; whereas, in an elliptical orbit the maximum velocity must occur nearer the epoch of pericentre. If, then, we could measure the displacement-separations of the bright and dark companion lines, and in the process fix also the epochs of greatest and zero separations, we should be able to decide the question between circular and elliptical motion, as well as quote the relative velocity of the pair of stars, independently of any motion of the system in space. But here we meet with the obstacle already alluded to, which throws doubt on all direct measures of the relative displacements of broad lines, viz., the unknown positions of their spectral centres—the positions in which the lines would be found when reduced to the thinness characteristic of the more perfect gaseous state of the origin.

An indirect method of solving the problem was suggested by the tabulated widths of the components of the composite band at H_ζ. For simplicity of expression we may transfer all the relative motion to one of the pair of stars about the other at rest. The method may then be described as the measurement of the arithmetical sum of the opposite displacements of the same spectral line at two opposite epochs, in place of the relative displacement of the two lines at the same epoch. The method was possible with the H_ζ line, because its red side bright edge was never absent, never masked by the dark line. The measures could be taken from the centre of the dark line to the edge of the bright line. And, supposing the bright line star to be the moving one, the measures would be greater when the motion was from us; and half the difference of the greatest and least measures would give the oscillation amplitude of the bright line, and consequently the maximum relative velocity of the stars. The epochs of the greatest and least measures would determine the points on the orbit of greatest positive and negative velocities, and those of the mean measure would settle the points of spectral conjunction.

The method supposes constancy of width of the lines; and this may be incorrect physically, or apparently falsified by atmospheric conditions. But both these variations might be expected to be balanced in a large number of measures on different civil dates, unless we suppose a periodic widening of the bright line, which could only be explained by greater pressure at a pericentre; and the admission of a pericentre is the acceptance of an elliptic orbit. The only systematic error remaining is the fraction of a measurement due to a small shift of the apparent centre of the dark line. The dark line is measurably thinner when the bright line is seen on both sides of it than when seen on one side only, and half the difference appears, in my trial of the method, to be a satisfactory correction.

It is obvious, from this description of the suggested method of dealing with the velocity problem, that many photographs are wanted on different civil dates for the elimination of errors; and at least four times the number would be necessary for a determination of the four epochs of maximum and mean displacements. But the need of a great multitude of observations has never yet blocked the way of a zealous astronomer, nor is he now deterred by the weight of photographic plates to be packed in his store room. You could hardly expect me to place before you any result obtained from the small number of plates in my present collection. But the trial was made, and I will venture to show you the figures, confident that you will not condemn my method of studying the problem on account of its present shortcomings. The tabulations on the screen give the mean greatest length-interval from the centre of the dark line to the edge of the bright line at 16·3 divisions of the micrometer head, from 18 dates, between extreme readings of 18·5 and 14·0 divisions; and the mean least length at 7·5 divisions, without variations on four successive dates. The difference, 8·8, is a little too great, owing to the encroachment of the bright line already mentioned, and the correction is shown in the lower half of the table. The resulting maximum relative velocity of the stars is 120 km./secs.; and this is encouraging when taken in conjunction with Belopolsky's direct measures at the Pulkowa Observatory: for, on the supposition of equal velocities, the maximum sight-line-velocity of either star with reference to the sun would be 60 km./secs., to compare with the 67 kms. given by the light power of the great Russian telescope.

In the next table (projection on screen) we have the analysis of the measures which would point out the four epochs wanted, if the number of dates were great enough. At present the table offers no certainties. Of the four epochs indicated in it two are in favour of elliptical motion and two are against it, while all four are impossible in a circular orbit. The two epochs which favour the ellipse are well marked in the tabulations, one within 15 hours, the other within 24 hours intervals; but the number of dates within each interval is too small. And for the unfavourable epochs, the number of dates might be enough, but they spread over long intervals of the cycle, one over two and a half days, the other quite four days.

Seen in this light, the necessity of long and patient work is more clear. We are fortunate in having so many as four plates within the 15-hour interval; for it should be remembered that the plates must be of different civil dates, in order to eliminate variations dependent on atmospheric conditions, and consequently the plates must be exposed in different light cycles. It might be very long before one succeeded in acquiring a dozen plates within any one 24-hour interval of a cycle.

The hour is nearly ended, and I promised myself not to exceed it. It is, therefore, with some regret that I must leave my subject unfinished. It would have been a satisfaction to have laid the whole case of β Lyræ before you, as it has come before me in my spectroscopic study of its problems. I must not, however, leave unanswered a question which I imagine you are mentally asking : What do I consider truly demonstrated? My answer is simple : Only this, that the method of studying the variable spectrum of β Lyræ, as I have illustrated this evening, is full of promise, and worthy of being carried out with a telespectrograph more capable than mine of writing with accuracy the history of a star which has been rightly named by Miss Clerke the "Problem Star."

[W. S.]

WEEKLY EVENING MEETING,

Friday, January 29, 1904.

Sir James Crichton Browne, M.D. LL.D. F.R.S., Treasurer and Vice-President, in the Chair.

David George Hogarth, Esq., M.A.

The Marshes of the Nile Delta.

[Abstract.]

There is a part of Egypt never visited by tourists, but not without considerable interest owing to the singularity of both its conformation and its population. This is the seaward belt of the central Delta. The southern part of this belt has few inhabitants except those of recent origin, mostly derived from refugees who migrated thither during the 18th and early 19th centuries; but there are abundant traces of a large ancient population, and it was to study these traces that the lecturer visited the locality in the spring of 1903.

Leaving the Berari railway one soon passes northward beyond the margin of cultivation and the last mud-hamlets, built in fantastic pepper-pot forms, on to great salt flats singularly devoid of all life, and rendered treacherous by the overflowing of old drains and lost arms of the Nile. These are being reclaimed gradually from the south by the *Société Anonyme du Behéra*, officered by Britons, which offers a fine example of how western capital, judiciously applied, may improve an eastern land and the society upon it, without spoiling the native character and rendering it hybrid. The process of reclamation and the inducements offered to new settlers were described. Considering what the unreclaimed portion of these flats is like now, it is amazing that signs of ancient inhabitation and of the culture of cereals, olives and vines should be so abundant. So difficult is the land to drain that one cannot but think it must lie lower now than it did in antiquity. But to a certain extent it was always a fen, as we know from many recorded facts and especially a description by Heliodorus.

The character of the ancient mounds was described, and their disposition along the lines of two ancient Nile arms, the Thermuthiac and the Athribitic; and the lecturer then passed to the northern or littoral belt, which is mostly lagoon and sand, but somewhat thickly inhabited by a fishing and gardening population. The surplus fish, out of the enormous piscatorial wealth of Lake Burlos, is roughly cured, or taken a short way up the canals inland and sold at auction. The type of the fishermen, which shows traces of survival from a type known on the ancient monuments, was commented upon, and the life of the lake-shore settlements described. The most primæval society left in Egypt, it is, however, certain to suffer great modification in the near future; for the increase of births and social security, resultant on the present state of things in the Nile valley, is causing a steady flow of surplus population into the northern vacuum.

[D. G. H.]

GENERAL MONTHLY MEETING,

Monday, February 1, 1904.

Sir James Crichton-Browne, M.A. LL.D. F.R.S., Treasurer and Vice-President, in the Chair.

William Booth Bryan, Esq., M.Inst.C.E.
Charles Edward Challis, Esq.
Alexander F. H. Dick, Esq.
William Gowland, Esq., F.C.S. F.S.A.
Sir Charles Hartley, K.C.M.G. M.Inst.C.E. F.R.S.E.
Miss C. Jones,
Cyril Frederick Lan-Davis, Esq.
Charles Edwin Layton, Esq.
Henrik Loeffler, Esq., M.A.
Brigade-Surgeon Lieut.-Col. C. W. MacRury, I.M.S. (retired).
William M. Ogilvie, Esq.
Capel George Pitt Pownall, Esq.
Lady Priestley,
Charles Frederick Rousselet, Esq., F.R.M.S.
Ralph Edward Tatham, Esq.
James Weir, Esq.

were elected Members of the Royal Institution.

The Special Thanks of the Members were returned to the following gentlemen for their Donations to the Fund for the Promotion of Experimental Research at Low Temperatures:—

Sir Andrew Noble, Bart., K.C.B. F.R.S.	.. £100
Frank McClean, Esq., M.A. LL.D. F.R.S.	.. 50
J. B. Broün-Morison, Esq., J.P. D.L. ..	.. 5
Lord Greenock, J.P. D.L.	.. 5

The following letter from Lady Horsley, in acknowledgment of the Resolution of Condolence on the death of Sir Frederick Bramwell, Bart., passed at the Meeting of the Members on December 7, 1903, was read:—

25 Cavendish Square, W., *December* 14, 1903.

Dear Sir William,

May I ask you to express on behalf of my mother, my sister and myself, our gratitude to the President and Managers of the Royal Institution for their message of sympathy in the loss which has befallen us.

Believe me,

Very gratefully yours,

Sir William Crookes, F.R.S.　　　　　(Signed) Eldred Horsley.

The PRESENTS received since the last Meeting were laid on the table, and the thanks of the Members returned for the same, viz.:—

FROM

The Secretary of State for India—
 Geological Survey of India—
  Memoirs, Vol. XXXV. Part 2. 8vo. 1903.
*The Astronomer Royal—*Greenwich Observations, 1900. 8vo. 1903.
  Greenwich Spectroscopic and Photographic Results, 1900. 4to. 1903.
  Annals of the Cape Observatory, Vol. XI. Part 1. 4to. 1903.
  Annals of the Royal Observatory, Cape of Good Hope, Vol. II. Part 3. 4to. 1903.
*Accademia dei Lincei, Reale, Roma—*Classe di Scienze Fisiche, Matematiche e Naturali. Atti. Serie Quinta: Rendiconti Vol. XII. 2° Semestre, Fasc. 10–12; Vol. XIII. Fasc. 1, No. 1. 8vo. 1903–4.
  Classe di Scienze Morali, Storiche, Serie Quinta, Vol. XII. Fasc. 7–10. 8vo. 1903.
*Allegheny Observatory—*Miscellaneous Papers, N.S. Nos. 15, 16. 8vo. 1903.
*American Academy of Arts and Sciences—*Proceedings, Vol. XXXIX. No. 5. 8vo. 1903.
*American Geographical Society—*Bulletin, Vol. XXXV. No. 4. 8vo. 1903.
*Ångstrom, K. Esq. (the Author)—*Energy in the Visible Spectrum of the Hefner Standard. 8vo. 1903.
*Astronomical Society, Royal—*Monthly Notices, Vol. LXIV. Nos. 1, 2. 8vo. 1903.
*Automobile Club—*Journal for Dec. 1903 and Jan. 1904. 4to.
*Bankers, Institute of—*Journal, Vol. XXIV. No. 9; Vol. XXV. No. 1. 8vo. 1903–4.
*Beckenhaupt, C. Esq. (the Author)—*Bedurfnisse und Fortschritte des Menschengeschlechtes. 8vo. 1903.
*Belgium, Royal Academy of Sciences—*Bulletin, 1903, Nos. 9, 10. 8vo.
*Berlin Academy of Sciences—*Sitzungsberichte, 1903, Nos. 41–53. 8vo.
*Boston Public Library—*Monthly Bulletin for Dec. 1903 and Jan. 1904. 8vo.
*British Architects, Royal Institute of—*Journal, Third Series, Vol. XI. Nos. 4–6. 4to. 1903–4.
*British Astronomical Association—*Journal, Vol. XIV. Nos. 2, 3. Memoirs, Vol. XI. Part 4. 8vo. 1903.
*Brown, Henry J. Esq.—*Memoir of Benjamin Franklin Stevens. By G. Manville Fenn. 8vo. 1903.
*Buenos Ayres, City—*Monthly Bulletin of Municipal Statistics, Oct.–Nov. 1903. 8vo.
*Busk, Miss Elinor and Miss Frances (M.R.I.)—*Descriptive Sociology. By Herbert Spencer. In 8 Parts. fol. 1873–1881.
*Cambridge Philosophical Society—*Proceedings, Vol. XII. Part 4. 8vo. 1903.
*Canada, Geological Survey—*Maps, Nos. 593, 598, 600, 608–610, 633, 635–637. fol. 1903.
*Canada, Royal Society of—*Proceedings and Transactions, Second Series, Vol. VIII. 8vo. 1902.
*Canadian Government (Department of the Interior)—*Maps of Assiniboia, Saskatchewan, Alberta, Lake Louise and Banff. 1903.
*Chemical Industry, Society of—*Journal, Vol. XXII. No. 23; Vol. XXIII. No. 1. 8vo. 1903–4.
*Chemical Society—*Proceedings, Vol. XIX. Nos. 272–274. 8vo. 1903.
*Civil Engineers, Institution of—*Proceedings, Vol. CLIV. 8vo. 1903.
  Engineering Conference, 1903. 8vo.
*Cracovie, l'Académie des Sciences—*Bulletin, Classes des Sciences Mathématiques et Naturelles, Nos. 8, 9; Classe de Philologie, Nos. 8, 9. 1903. 8vo.
*Devonshire Association—*Transactions, Vol. XXXV. 8vo. 1903.
  Devonshire Wills, Part 5. 8vo. 1903.

Edinburgh Royal Society—Proceedings, Vol. XXIV. No. 6. 8vo. 1903.
　　Transactions, Vol. XL. Part 3. 4to. 1903.
Editors—Aeronautical Journal for Jan. 1904. 8vo.
　　American Journal of Science for Jan. 1904. 8vo.
　　Analyst for Dec. 1903 and Jan. 1904. 8vo.
　　Astrophysical Journal for Dec. 1903 and Jan. 1904.
　　Athenæum for Dec. 1903 and Jan. 1904. 4to.
　　Board of Trade Journal for Dec. 1903 and Jan. 1904. 8vo.
　　Brewers' Journal for Dec. 1903 and Jan. 1904. 8vo.
　　Chemical News for Dec. 1903 and Jan. 1904. 4to.
　　Chemist and Druggist for Dec. 1903 and Jan. 1904. 8vo.
　　Electrical Engineer for Dec. 1903 and Jan. 1904. fol.
　　Electrical Review for Dec. 1903 and Jan. 1904. 8vo.
　　Electrical Times for Dec. 1903 and Jan. 1904. 4to.
　　Electricity for Dec. 1903 and Jan. 1904. 8vo.
　　Electro-Chemist and Metallurgist for Dec. 1903 and Jan. 1904. 8vo.
　　Engineer for Dec. 1903 and Jan. 1904. fol.
　　Engineering for Dec. 1903 and Jan. 1904. fol.
　　Feilden's Magazine for Dec. 1903 and Jan. 1904. 8vo.
　　Homœopathic Review for Jan. 1904. 8vo.
　　Horological Journal for Jan. 1904. 8vo.
　　Journal of the British Dental Association for Dec. 1903 and Jan. 1904. 8vo.
　　Journal of Physical Chemistry for Dec. 1903. 8vo.
　　Journal of State Medicine for Dec. 1903 and Jan. 1904. 8vo.
　　Law Journal for Dec. 1903 and Jan 1904. 8vo.
　　London Technical Education Gazette for Dec. 1903 and Jan. 1904. fol.
　　London University Gazette for Dec. 1903 and Jan. 1904. fol.
　　Machinery Market for Dec. 1903 and Jan. 1904. 8vo.
　　Model Engineer for Dec. 1903 and Jan. 1904. 8vo.
　　Mois Scientifique for Dec. 1903. 8vo
　　Motor Car Journal for Dec. 1903 and Jan. 1904. 8vo.
　　Musical Times for Dec. 1903 and Jan. 1904. 8vo.
　　Nature for Dec. 1903 and Jan. 1904. 4to.
　　New Church Magazine for Jan.-Feb. 1904. 8vo.
　　Nuovo Cimento for July-Aug. 1903. 8vo.
　　Page's Magazine for Dec. 1903 and Jan. 1904. 8vo.
　　Photographic News for Dec. 1903 and Jan. 1904. 8vo.
　　Physical Review for Nov.-Dec. 1903. 8vo.
　　Physiologiste Russe for May, 1903. 8vo.
　　Popular Astronomy for Dec. 1903 and Jan. 1904. 8vo.
　　Public Health Engineer for Dec. 1903 and Jan. 1904. 8vo.
　　Science Abstracts for Jan. 1904. 8vo.
　　Terrestrial Magnetism for Sept. 1903.
　　Travel for Dec. 1903 and Jan 1904. 8vo.
　　Zoophilist for Dec. 1903 and Jan. 1904. 4to.
Florence, Biblioteca Nazionale—Bulletin, Dec. 1903. 8vo.
Franklin Institute—Journal, Vol. CLVI. No. 6; Vol. CLVII. No. 1. 8vo. 1903-4.
Geographical Society, Royal—Geographical Journal for Jan. 1904. 8vo.
Geological Society—Abstracts of Proceedings, Nos. 784-786. 8vo. 1903.
Gottingen, Royal Academy of Sciences—Nachrichten ; Mathematisch-physikalische
　　Klasse, 1903, Heft 5. 8vo.
Hallock-Greenewalt, Miss M. (the Author)—Pulse and Rhythm. 8vo. 1903.
Harlem, Société Hollandaise des Sciences—Archives Néerlandaises, Serie II.
　　Tome VIII. 5e Livraison. 8vo. 1903.
Hess, Adolfo, Esq. (the Translator)—Il Radio. By W. J. Hammer. 8vo. 1903.
Johns Hopkins University—University Circulars, No. 165. 4to. 1903.
Lagrange, Charles, Esq. (the Author)—La Machine à mouvement perpetuel et la
　　question du Radium. 8vo. 1903.
　　　Newton et le principe de la limite. 8vo. 1903.

Leighton, John, Esq., F.S A. M.R.I—Journal of the Ex-Libris Society, Nov.-Dec 1803. 8vo.
Life-Boat Institution, Royal National—Journal for Feb. 1904. 8vo.
Manchester Literary and Philosophical Society—Memoirs and Proceedings, Vol. XLVIII. Part 1. 8vo. 1903.
Massachusetts Institute of Technology—Technology Quarterly, Vol. XVI. No 3 8vo. 1903.
Mechanical Engineers, Institution of—Proceedings, 1903, No. 3. 8vo.
Meteorological Society—Meteorological Record, Vol. XXIII. No. 90. 8vo. 1903.
Mexico, Secretaria de Comunicaciones y Obras Publicas—Anales, Num. 1-8. 8vo. 1902-3.
Microscopical Society, Royal—Journal, 1903, Part 6. 8vo.
Molteno, P. A. Esq , LL.B. M.R.I. (the Author)—A Federal South Africa. 8vo. 1896.
   The Life and Times of Sir John Charles Molteno. 2 vols. 8vo. 1900.
Montana University—Bulletin, Biological Series, No. 5. 8vo. 1903.
Munich, Royal Bavarian Academy of Sciences—Sitzungsberichte, 1901, Heft. III. 8vo. 1903.
Musée Teyler—Archives, Série II. Vol. VIII. 4me partie. 8vo. 1903.
Navy League—Navy League Journal for Jan. 1904. 8vo.
New South Wales, Agent-General for—Year Book of New South Wales, 1904. 8vo.
New York Academy of Sciences—Annals, Vol. XIV. Part 3; Vol. XV. Part 1. 8vo. 1903.
New Zealand, Agent-General for—New Zealand Official Year Book, 1903. 8vo.
North of England Institute of Mining and Mechanical Engineers—Transactions, Vol. LII. No. 7; Vol. LIV. No. 1. 8vo. 1903.
Odontological Society—Transactions, Vol. XXXVI. No. 2. 8vo. 1903.
Peru, Cuerpo de Ingenieros de Minas—Boletin, No. 3. 8vo. 1903.
Pharmaceutical Society of Great Britain—Journal for Dec. 1903 and Jan. 1904. 8vo.
   The Calendar for 1904. 8vo.
Philadelphia Academy of Natural Sciences—Proceedings, Vol. LV. Part 2. 8vo. 1903.
Photographic Society, Royal—Photographic Journal for Oct.-Dec. 1903. 8vo.
Physical Society of London—Proceedings, Vol. XVIII. Part 7. 8vo. 1903.
Pollock, Sir Montagu, Bart., M.R.I. (the Author)—Light and Water. 8vo. 1903.
Rome, Ministry of Public Works—Giornale del Genio Civile for July-Aug. 1903. 8vo.
Royal Dublin Society—Transactions, Vol. VIII. Parts 2-5. 4to. 1903.
   Proceedings, Vol. X. Part 1. 8vo. 1903.
   Economic Proceedings, Vol. I. Part 4. 8vo. 1903.
Royal Irish Academy—Proceedings, Vol. XXIV. Sec. A, Part 3; Sec. B, Part 4 ; Sec. C, Part 4. 8vo. 1903.
   Transactions, Vol. XXXII. Sec. A, Parts 7-9; Sec. B, Parts 3, 4; Sec. C, Parts 2, 3. 4to. 1903.
Royal Society of London—Philosophical Transactions, A, Nos. 354-359.
   Proceedings, Nos. 485, 486. 8vo. 1903.
   Ceylon Pearl Oyster Fisheries and Marine Biology. Report to the Colonial Government by Professor W. A. Herdman and others. Part 1. 4to. 1903.
Salford, Borough of—Fifty-fifth Report of the Museum and Libraries Committee. 8vo. 1903.
Sanitary Institute—Journal, Vol. XXIV. Part IV. and Supplement. 8vo. 1904.
Scottish Meteorological Society—Journal, Third Series, Nos. 18, 19. 8vo. 1903.
Scottish Society of Arts, Royal—Transactions, Vol. XVI. Part I. 8vo. 1903.
Selborne Society—Nature Notes for Jan. 1904. 8vo.
Smith, B. Leigh, Esq., M.A. M.R.I.—The Scottish Geographical Magazine, Vol. XX. No. 1. 8vo. 1904.
Smithsonian Institution—Annual Report, 1902. 8vo. 1903.
   Contribution to Knowledge, No. 1413. 4to. 1903.

Society of Arts—Journal for Dec. 1903 and Jan. 1904. 8vo.

Statistical Society, Royal—Journal, Vol. LXVI. Part 4. 8vo. 1903.

Stoney, G. Johnstone, Esq., F.R S. M.R.I. (the Author)—How to Simplify British Weights and Measures. 8vo. 1903.

Tacchini, Prof. P., Hon. Mem. R.I. (the Author)—Memorie della Società degli Spettroscopisti Italiani, Vol. XXXII. Disp. 12ª. 4to. 1903.

Thompson, S. P. Esq., F.R.S. M.R I. (the Author)—Gilbert of Colchester, The Father of Electrical Science. 8vo. 1903.

    Gilbert: Physician. 8vo. 1903.

United Service Institution, Royal—Journal for Dec. 1903 and Jan. 1904. 8vo.

United States Department of Agriculture—Monthly Weather Review for Sept. 1903. 4to.

    Atmospheric Radiation. By F. W. Very. 4to. 1903.

    Studies among Snow Crystals. By W. A. Bentley. 4to. 1903.

    Semi-diurnal Tides in Northern Part of Indian Ocean. By R. A. Harris. 4to. 1903.

    Experiment Station Record, Aug.-Dec. 1903. 8vo.

United States Geological Survey—Bulletin, Nos. 209-217. 8vo. 1903.

    Monographs, XLIV. XLV. and Atlas. 4to. 1903.

United States Patent Office—Official Gazette, Vol. CVII. Nos. 5-9; Vol. CVIII. Nos. 1-3. 4to. 1903-4.

Verein zur Beförderung des Gewerbfleisses in Preussen—Verhandlungen, 1903, Heft 10. 4to.

Vienna, Imperial Geological Institute—Verhandlungen, 1903, Nos. 12-15. 8vo. Jahrbuch, 1903, Heft 1, 2. 8vo.

Western Society of Engineers—Journal, Vol VIII. No. 6. 8vo. 1903.

Yorkshire Archæological Society—Journal, Part 68. 8vo. 1903.

WEEKLY EVENING MEETING,

Friday, February 5, 1904.

HIS GRACE THE DUKE OF NORTHUMBERLAND, K.G. D.C.L. F.R.S., President, in the Chair.

ALFRED AUSTIN, Esq., B.A., Poet Laureate.

The Growing Distaste for the Higher Kinds of Poetry.

[A report of this Discourse appears in the " Fortnightly Review," March, 1904.]

WEEKLY EVENING MEETING,

Friday, February 12, 1904.

His Grace The Duke of Northumberland, K.G. D.C.L. F.R.S.,
President, in the Chair.

W. N. Shaw, Esq., Sc.D. F.R.S. *M.R.I.*, Secretary of the
Meteorological Council.

Some Aspects of Modern Weather Forecasting.

[ABSTRACT.]

After referring to the circumstances under which he was called upon to deliver the Evening discourse in the absence of the Dean of Westminster, the lecturer explained that he had chosen the subject, not because he regarded weather forecasting as the only, or, from the scientific point of view, the most important practical branch of Meteorology, but because, in a general sense, the possibility of its application to forecasting—the deduction of effects from given causes —was the touchstone of scientific knowledge.

The process of modern forecasting was illustrated by the daily weather charts of the period from February 1, 1904, up to the evening of the 12th, which exhibited the passage over the British Isles of a remarkable sequence of cyclonic depressions, reaching a climax in a very deep and stormy one on the evening of the lecture. It was thus pointed out that the barometric distribution and its changes were the key to the situation as regards the weather, and this was supported by exhibiting the sequence of weather accompanying recognised types of barometric changes, as shown in the self-recording instruments at the observatories in connection with the Meteorological Office.

Some cases of difficulty in the quantitative association of rainfall or temperature changes with barometric variations were then illustrated. The barometric distributions in the weather maps for April 8 and April 16, 1903, were shown to be almost identical, and yet the weather on the later date was 10° colder than on the earlier. The observatory records for June 22, 1900, showed that a barometric disturbance of about the fiftieth of an inch, too small to be noticed on the scale of the daily charts, passed across the country from Valencia to Kew, over Falmouth, in about twenty-four hours, and produced at each observatory characteristic changes of temperature and wind, and also in each case about a fifth of an inch of rainfall.

Some examples of the irregularity of motion of the centres of depressions were also given, including one which travelled up the western coasts of the British Isles on October 14 and 15, and down the eastern coasts on October 16 and 17, 1903, one which developed from scarcely visible indications into a gale on December 30, 1900, and one which disappeared, or "filled up," as it is technically called, on February 6, 1904. The conclusion was drawn that the suggested extension of the area of observation by means of wireless telegraphy from ships crossing the Atlantic would not immediately place forecasting in the position of an exact science, but would add greatly to the facilities for studying the life-history of depressions.

The irregularities and uncertainties illustrated by the examples given, might be attributed in part to the complexities of pressure due to the irregular distribution of land and sea in the Northern hemisphere. Charts of the mean isobars for the World for January and July showed greater simplicity of arrangement in the Southern hemisphere, where the ocean was almost uninterrupted, than in the Northern hemisphere, where there were alternately large areas of sea and land. The comparative simplicity of the South as compared with the North was also illustrated by a chart representing an attempt at a synoptic barometric chart for the World for September 21, 1901.

The simplification of the barometric distribution at successively higher layers of the atmosphere, as illustrated by Teisserenc de Bort's chart of mean isobars at the 4000 metre level, was pointed out, and illustrations were also given of the method of computing the barometric distribution at high levels from observations at the surface, using data obtained from observations at high level observatories, or those made with balloons and kites.

Some indication of the connection between the complexity of the surface and the simplicity of the upper strata might be established by means of careful observations of the actual course of air upon the surface and the accompanying weather conditions.

The actual course of air along the surface was often misunderstood. The conventional S-shaped curves representing the stream lines from anticyclonic to cyclonic regions were shown to be quite incorrect as a representation of the actual paths of air along the surface. A diagram contributed to the Quarterly Journal, Royal Meteorological Society,* showed the computed paths for special case of a storm of circular isobars and uniform winds, travelling without change of type at a speed equal to that of its winds. An instrument made by the Cambridge Scientific Instrument Company to draw the actual paths of air for a number of different assumptions as to relative speed of wind and centre, and of incurvature of wind from isobars, was also shown, and the general character of the differences of path exhibited under different conditions was discussed.

---

* The Meteorological Aspects of the Storm of February 26–27, 1903. Q. J. R. Met. Soc. vol. xxix. p. 233, 1903.

In illustration of the application of these considerations to practical meteorology, it was noted that rainfall is an indication of the existence of rising air, and conversely the disappearance of cloud may be an indication of descending air. It was further noted that if the ascent and descent of air extended from or to the surface, the actual paths of air along the surface, as traced from the direction and speed of the winds, ought to show convergence in the case of rising air and divergence in the case of descending air.

The chart for April 16, 1903, was referred to for an obvious case of dilatation or divergence of air from a centre corresponding with fine weather, the centre of the area of divergence being specially marked "no rain," and the actual trajectories or paths of air for two different travelling storms were contrasted, to show how the rainfall might be related to the convergence of the paths of air. The two occasions selected were (1) the rapid travelling storm of March 24–25, 1902, and (2) the slow travelling storm of November 11–13, 1901.* The trajectories or actual paths of air for these two storms had been constructed from two hourly maps drawn for the purpose from a collection of records of self-recording barographs, etc. Those for March 24–25 showed the paths to be looped curves with very little convergence, whereas those for the storm of November 11–13 showed very great convergence; so much so that if four puffs of smoke could be imagined starting at the same time from Aberdeen, Blacksod Point, Brest and Yarmouth respectively, and travelling for 24 hours, they would find themselves at the end of the time enclosing a very small area in the neighbourhood of London.

Corresponding to this difference of convergence as shown by the paths, was the difference of rainfall as illustrated by two maps showing the distribution of the rain deposited from the two storms. The first, with little convergence, gave hardly anywhere more than half-an-inch; the second, with its great convergence, gave four inches of rain in some parts of its area.

[W. N. S.]

---

* *See* Pilot Charts for the North Atlantic and Mediterranean, issued by the Meteorological Office. February 1904.

WEEKLY EVENING MEETING,

Friday, February 19, 1904,

His Grace The Duke of Northumberland, K.G. D.C.L. F.R.S.,
President, in the Chair.

C. T. R. Wilson, Esq., M.A. F.R.S.,
Fellow of Sidney Sussex College, Cambridge.

Condensation Nuclei.

A familiar experiment was first shown illustrating the action of
ordinary dust particles as condensation nuclei. From a large globe,
which had been allowed to stand for some hours, some of the air was
removed by opening communication with an exhausted vessel. Only
a very few drops were formed as a result of the expansion. On
allowing air to enter the globe through a cotton-wool filter, so that
the pressure was brought back to its original value (that of the
atmosphere), and allowing the air to expand as before, the drops
formed were again very few. The ordinary air of the room was now
admitted; an expansion of the air in this case resulted in the production of a thick fog.

When air has been freed from dust by filtering, or by repeatedly
forming a cloud by expansion, and allowing it to settle, the vapour
which, in the presence of the nuclei would have separated out in
drops, must be in the "supersaturated" condition immediately after
the expansion is completed.

Another method of producing clouds was now shown. Air was
allowed to escape through a fine orifice into an atmosphere of steam;
the mixed air and steam were then passed through a Liebig's condenser, where the greater part of the steam was condensed, and then
into a large glass globe, where the clouds were observed. From this
vessel the air was drawn off by a pump which maintained the
pressure in the globe and condenser at a considerable number of cms.
of mercury below that of the atmosphere. Before reaching the jet
the air of the room had to pass through a cotton-wool filter, and then
through a long tube containing water; finally it was led through an
aluminium tube to the orifice. The latter was about half a mm. wide.
The fall of pressure in passing through the orifice was about 15 or
20 cms. In the absence of the filter, the air being admitted
directly to the water tube through a tap turned just sufficiently to
give the same flow as with the filter, a dense fog poured out from the
end of the condenser tube; on closing the tap and letting the air

enter through the filter the fog rapidly cleared, and only a fine rain continued to be produced. While the apparatus was in this condition an X-ray tube was set in action near the aluminium tube; the rain was succeeded by fog, which continued to pour out from the end of the condenser so long as the X-rays were kept in action. Condensation nuclei are, as this experiment proves, produced in air exposed to Röntgen rays. Later experiments will, however, show that they have entirely different properties from the ordinary dust nuclei.

When air has been completely freed from dust particles, so that a slight expansion of the air (initially saturated with water vapour) does not result in the formation of any drops, it is found that quite a high degree of supersaturation may be brought about without the appearance of a single drop. There is, however, a limit to the supersaturation which can exist without condensation of the vapour in drops resulting. To study this condensation in dust-free air, and to measure the expansion required to produce the necessary degree of supersaturation, a special form of expansion apparatus is required. The lantern slide thrown on the screen shows the construction and mode of working of the apparatus. The second slide is a photograph of the machine in action, the exposure having been made immediately after an expansion; the cloud formed (in this case on nuclei produced by the action of radium) is plainly visible along the path of a concentrated beam of light from a lantern.

Let us now try an actual experiment with the expansion apparatus. On making a slight expansion a cloud forms on the dust particles which are present; this slowly settles to the bottom of the vessel. The air is allowed to contract to its original volume, and a second expansion of the same amount made. The drops formed are on this occasion comparatively few, and they fall rapidly; the dust particles have nearly all been carried down with the drops formed by the previous expansion. The fewer the nuclei on which water condenses the larger will be the share of water available for each drop, and the more rapid will be the fall. The next expansion produces no drops. While the air is in the expanded condition, the piston being at the bottom of the expansion cylinder, air is removed from the cloud chamber by opening the connexion to the air-pump until the pressure is about 13 or 14 cms. of mercury below that of the atmosphere; the piston is again allowed to rise by putting the air space below it in communication with the atmosphere. The next expansion is thus comparatively large, the pressure after the expansion has taken place and the temperature has risen to its original value being 13 cms. or more below the initial pressure. Yet, in spite of the high degree of supersaturation reached, not a drop of water is seen. Making the fall of pressure 16 cms., however, we see on expansion a shower of drops. And although these drops are few and large, falling therefore rapidly, yet, however often the same expansion be repeated, the drops produced on expansion show no diminution in number. Thus the nuclei removed with the drops are continually replaced by others manufactured within the apparatus itself.

To produce the necessary supersaturation to cause condensation in the form of drops in dust-free air, the air must be allowed to expand suddenly till the final volume is 1·25 times the initial volume. The condensation is rain-like in form, and moreover the number of drops remains small although the expansion considerably exceeds this lower limit. Expansions exceeding the limit, v_2/v_1 = 1·38, however, give fogs, which increase rapidly in density, i.e. in the number of the drops, as the expansion is increased beyond this second limit. The expansions required for the rain-like and cloud-like condensations correspond to a fourfold and eightfold super-saturation respectively.

A further experiment will throw light on the nature of the nuclei associated with the rain-like condensation. Let us expose the moist air to the action of X-rays before causing it to expand. First let us try an expansion very slightly less than that required to give the rain-like condensation without the rays. You observe, no drops are formed. Now let the expansion be slightly greater than the critical value, 1·25. A fog is seen on expansion. Thus the X-rays produce in the air immense numbers of nuclei having the same properties, so far as their power of assisting condensation goes, as the compara-tively few nuclei which the rain-like condensation makes visible. Now, a gas exposed to X-rays conducts electricity, and the otherwise complicated ·phenomena of this conduction are all reduced to com-parative simplicity by the theory that under the action of the rays equal numbers of freely moving positively and negatively electrified bodies (the ions) are produced from the originally neutral gas. It is at once suggested that the condensation nuclei produced by X-rays are simply these ions.

Let us now impart conducting power to the gas by exposing it to the action of the radiation from radium. Again we have the same result, no drops produced if the expansion be less than 1·25, fog if the expansion exceed this limit.

If we substitute for the glass shade, which has thus far formed the cloud-chamber, a glass cylinder with a horizontal metal top, we have the means of testing whether the condensation nuclei produced by Röntgen or radium rays are really electrically charged, whether in fact it is the ions themselves which act as condensation nuclei or other particles produced by the rays. If, for example, the roof of the cloud chamber be kept positively charged and the floor negatively, the negatively charged ions will travel upwards and the positively charged ones downwards. In the absence of an electric field the positive and negative ions produced by the action of the rays will go on increasing in number until as many are neutralised by recom-bination with ions of the opposite kind, or by coming in contact with the walls of the vessel, in each second, as are set free in that time by the rays. If the rays be cut off, the removal of ions by recombiuation and diffusion will continue, and the number of ions in the vessel will diminish rapidly.

Experiment shows that, while in the absence of an electric field, quite a considerable fog is formed when an expansion, slightly exceeding 1·25, is effected 10 seconds after the rays have been cut off, with 200 volts between the upper and lower plates the same expansion, allowed to take place 3 or 4 seconds after the stopping of the rays, produces only a very slight shower. Or, again, if the rays be kept on all the time the resulting fog is very much less dense with the electric field acting than without it. These results are easily explained if we assume that the condensation nuclei are the ions, and apply the result obtained by purely electrical methods, that the ions travel about 1·6 cm. per second in a field of 1 volt per cm. The nuclei causing the rain-like condensation without exposure to Röntgen or radium rays are also removed by the action of an electric field; we have thus the direct proof that they also are ions. Recent experiments have proved that a charged conductor suspended within a closed space loses its charge by leakage through the air, and that the conduction shows all the peculiarities of that met with in an ionised gas. And, indeed, it appears that this ionisation is due to the action of radiation of the radium type from the walls of the vessel and from outside the vessel. The condensation method of detecting ions is, it may be pointed out, a very delicate one; a single ion if present in the vessel will be detected.

The positive and negative ions are not alike in their power of acting as condensation nuclei. In most of the experiments shown to-night the negative ions alone have in fact come into action. The positive require a considerably greater expansion in order that water may condense upon them. The final volume must for the positive ions be about 1·31 times the initial instead of only 1·25, corresponding to a six-fold instead of a four-fold supersaturation.

To demonstrate the difference between the positive and negative ions the same form of apparatus is used as in the previous experiment. Instead, however, of a difference of potential of 200 volts, only two or three volts are applied between the plates. And in this experiment only a thin layer close to the lower plate is exposed to the action of the rays. Under these conditions, if the upper plate is the positive one, the negative ions will be attracted upwards out of the ionised layer, and will occupy the greater part of the volume of the vessel, while the positive ones will have only a short distance to travel before reaching the lower plate. If the rays be cut off before the expansion is made it is easy to arrange the interval to be of such a duration that all the positive ions have been removed, while only a small fraction of the negative ions have reached the upper plate before the expansion takes place. Thus we can try the effect of expansion when the vessel is charged with practically negative ions only. By reversing the electrical field the action of positive ions, almost free from negative ions, can be studied. When the expansion is between 1·25 and 1·31 a fog or a mere shower is obtained, according as the direction of the field is such as to drive negative or positive ions upward.

The ions are by no means the only nuclei which can be produced within moist air from which the dust particles have been removed. Among the most interesting of such apparently uncharged nuclei are those produced in moist air exposed to ultra-violet light. It is impossible in the time available to do more than allude to them here.

[C. T. R. W.]

WEEKLY EVENING MEETING,

Friday, February 26, 1904.

DONALD W. C. HOOD, ESQ., C.V.O. M.D. F.R.C.P.,
Vice-President, in the Chair.

ALEXANDER SIEMENS, ESQ., M. Inst. C.E. *M.R.I.*

New Developments in Electric Railways.

THERE is no doubt that during the nineteenth century greater changes were wrought in the mode of living, not only in Europe but all over the world, than in any of the preceding centuries; and a common feature of nearly all of these changes has been the saving of time effected by them.

An illustration of the difference between the beginning of the nineteenth century and the present time, is furnished by some of the newspapers which publish extracts from their own pages a hundred years ago.

News travelled slowly then; while nowadays the newspapers are expected to contain every day a record of the most important events that happened on the preceding day, or even on the same day, in any part of the globe.

In other walks of life the saving of time has likewise been the dominant factor by which the introduction of changes has been secured.

Historically, the application of the steam engine for industrial purposes preceded the utilisation of electricity for the propagation of news, and the fundamental changes, which took place in consequence, were so important that for the greater part of its duration the nineteenth century might be called the age of steam.

During the last thirty years, however, gas engines and electric motors have taken the place of the steam engines in many different ways; but it would not be possible to enter into a general discussion of this subject in the short time allowed for to-night's lecture. One of the most important fields in which the steam engine has been of the greatest service to mankind, is the transportation of goods and people by land and by sea.

By its aid the intercommunication between the various parts of each country and between all places, where civilised people desire to trade, has become rapid, frequent and reliable; and this, coupled with the facilities for obtaining messages by means of electricity, has absolutely changed the old order of things in the intercourse of nations.

The principal effect of this change has been the equalisation of the chances of success for workers in all parts of the civilised world, whatever their occupation may be; provided they keep themselves well informed, as to what is being done elsewhere, or they will not be able to keep abreast of the times.

Naturally there are other factors, which influence the competition that takes place everywhere in life, but the importance of rapid, frequent and reliable intercommunication cannot be overrated, and that must be the excuse for bringing to your notice to-night the endeavours that have been made to introduce electric motors on main lines of railways, with a view of accelerating the speed and increasing the opportunities of travelling without diminishing its safety.

It is exactly twenty-five years ago that the first electric loco-motive ran on a short line in an exhibition at Berlin, drawing a few carriages on which people were taken round the exhibition.

How primitive all the arrangements were can best be seen by the picture now thrown on the screen.

The two rails, insulated from each other and from earth, served as lead and return, while the switches were of the simplest form.

Some progress was made, when in 1881 a short line from Lichter-felde to the Cadettenhaus was opened, although on this line also the rails served as conductors. In the same year a short line with overhead conductors was run in connection with the Paris electrical exhibition.

Two years later, in the autumn of 1883, the line from Portrush to Bushmills was opened, in which the conductor was about eighteen inches above the ground by the side of the rails, while the rails served as return circuit.

From these small beginnings the electric tramways of modern times have developed, of which examples can be seen in nearly every town.

Their success naturally led to attempts being made to introduce electricity as the motive power on railways; but the totally different conditions obtaining there, made so many and such intricate demands on the electrical engineer, that the satisfactory solution of the pro-blem has been long delayed—and even now opinions differ as to the best method for universal adoption.

At first, short lines of railways have been carried out, partaking more of the character of tramlines with fixed stopping places, such for instance, as the elevated railway along the Mersey docks or the tube railways in London, of which the City and South London line was the pioneer not only in this country but for all other countries as well.

On these lines continuous currents at the comparatively low pressures not exceeding 500 volts are in use, and they are conveyed to the motors from a third rail, while the running rails serve as the return for the currents.

Where the distances from the power stations have demanded it, recourse is had to higher pressures at the generators, which is converted into the working pressure either by rotary converters and transformers, when the generators produce three-phase currents, or by the three-wire system with modifications, when the generators produce continuous currents of high pressure.

A fair example of a modern railway on these lines is the elevated railway of Berlin, which, however, in some parts dips under the level of the streets.

Messrs. Siemens and Halske have kindly lent a number of slides, which show the endeavours that have been made to mitigate the innate hideousness of an iron structure in the streets.

Some other features of this railway also deserve attention: one is the method of building a square channel immediately under the carriage-way of the streets, which is supported by transverse iron girders. The shallow tunnel, thus formed, can be readily ventilated, and is easily accessible.

In New York a similar construction has been adopted for the new underground line, which will soon be ready for opening.

Another feature of the Berlin line is the way of connecting the main line with the short branch leading at right angles from it to the Potsdamer Platz. The lines form a triangle, and are rising or dipping from the corners of the triangle in such a way that no down line crosses an up line at the same level.

All the points for directing the trains to and from the various lines are moved electrically, on the lock-and-block system, from one signal cabin placed at a high level, so that the signalman has a good view of all trains approaching the triangle.

Another attempt to transport passengers without interfering with other traffic has been made by Messrs. Schuckert of Nürnberg, who have built a line in Elberfeld and Barmen, on the monorail system, invented by the late Mr. Eugen Langen of Cologne.

The permanent way of this line consists of two girders supported either by arches or by " A " frames, and placed at such a level above the street that the passenger carriages, hanging from wheels, which run along the girders, are well above the traffic in the street.

Ample precautions are adopted to prevent the wheels running off the rails, and the current is taken by a trolley from a wire close to the girder. The space between the girders is partly covered in so that a footway is provided, which makes the trolley line and the rails accessible at all times even when carriages are running.

Getting in and out of the train, formed by a motor carriage and trailers, is only possible at the stations, which resemble the ordinary high level stations.

For the greater part of its length the railway is suspended over the river Wupper, and the slides clearly show that no beautiful prospect is injured by the presence of the superstructure over the river.

At the ends of the line and also at one intermediate station, loops are provided for transferring the car from one line of girders to the other.

In all these railways continuous current motors have been employed, as these can be easily regulated to run at various speeds, and they automatically adapt themselves to their load. It is, however, difficult to construct such motors for high voltages or to transform high tension continuous currents into low tension currents.

When the question of applying electric motors to the traffic of main lines was seriously taken up, it became at once evident that the necessary energy could only be supplied if high tension currents were employed.

Owing to the facility with which alternate currents can be transformed by means of stationary contrivances, which require no supervision or manipulation, it is natural that most attempts to solve the problem are based on the use of alternate currents. For this purpose single-phase as well as multi-phase alternate currents have been employed, and either has been used direct or transformed into continuous current.

As long ago as 1884, Dr. John Hopkinson and Prof. Adams read papers before the Institution of Electrical Engineers on alternate current motors, and in the discussion experiments were mentioned, which had been carried out at the works of Siemens Brothers at Woolwich, in which a continuous current dynamo, with laminated poles, had been driven as a motor from a single-phase alternate current generator. The dynamo even turned round when the alternate current was only sent through the armature; in other words, it worked as an induction motor.

These experiments did not give very promising results, as the periodicity of the alternate currents was too high; they were, therefore, dropped in favour of more promising work.

A few years later, when three-phase currents found general favour for transmission of electrical energy to great distances, Mr. Wilhelm von Siemens came to the conclusion that electricity could only be applied to main lines by means of overhead conductors and in the form of alternate currents. He, therefore, had a short line laid down in the works of Siemens and Halske in Charlottenburg, on which a small three-phase locomotive was shown to the German railway authorities in December 1892.

The experiment was to demonstrate to them the feasibility of using overhead conductors for main lines, and to induce them to place a suitable length of the State railways at the disposal of the firm, to make their trials on a sufficiently large scale. Unfortunately, the railway officials could not be convinced that practical results could be attained on these lines, and the request of the firm was refused.

In order to demonstrate in a practical manner the feasibility of their proposals, the firm constructed an experimental line of full size in 1898, near Gross-Lichterfelde, which was visited in May 1901 by the Institution of Electrical Engineers, when a three-phase electric locomotive was shown in operation, which utilised currents of a pressure of 10,000 volts and could attain a speed of sixty miles per hour.

This line was, however, only about a mile long, and the trials were obviously not exhaustive enough to decide definitely how far the electrical appliances in use were adapted for high speed traffic on main lines. It served, all the same, the useful purpose of settling a practical form of conductor and of collector.

The picture of the locomotive, as shown on the screen, was taken when the president of the Imperial German Railway Office and Dr. Georg Siemens, director of the Deutsche Bank, visited the line on the 29th March, 1900.

Eventually it was decided to make experiments on a large scale on the military railway between Marienfelde and Zossen, and with a view of obtaining the co-operation of all the most competent engineers, a syndicate was formed, called " Studiengesellschaft für elektrische Schnellbahnen," by the Deutsche Bank, Delbrück Leo & Co., National Bank, Jakob S. H. Stern, A. Borsig, Phil. Holzmann, Friedr. Krupp, v. d. Zypen and Charlier, Allgemeine Elektricitäts Gesellschaft, and Siemens and Halske.

The military authorities placed the line at the disposal of the syndicate, and the President of the Imperial German Railway Office became its chairman.

While the problem was attacked in this fashion in the north of Germany, Messrs. Brown, Boveri & Co., had opened in July 1899, a service on the main line between Burgdorf and Thun, employing three-phase currents of 16,000 volts, which are transformed to 750 volts for use in the motors.

On the other side of the Alps, Messrs. Ganz & Co. fitted up the Lecco-Sondrio railway with three-phase motors, in which their system of " cascade " connection served for regulating the starting torque of the motors.

By this system the 3000-volt current of the line is connected direct to the stator of the first motor, while the rotor of the first motor is connected to the rotor of the second motor, both electrically and mechanically; while a resistance is introduced in the circuit of the second stator.

As the train accelerates this resistance is cut out until at half-speed, the stator of the second motor is short circuited. The second motor is then cut out altogether, and the resistance inserted into the circuit of the first rotor; again it is gradually diminished until at full-speed the first rotor is short circuited.

Just about two years ago Messrs. Mordey and Jenkin read a paper on " Electric Traction on Railways," before the Institution of Civil Engineers, here in London, and came to the conclusion that a single-phase alternating current system was the best for the generation, transmission and distribution of power, with one overhead conductor and a return conductor on the ground, which can be either the rails or an insulated conductor at practically the earth's potential.

Their idea of utilising the single-phase currents was to have a single-phase motor driving a large continuous current generator, and a smaller one for exciting the fields of the generator and of the train motor.

All this machinery is carried on the locomotive, which, therefore, resembles a Heilmann locomotive, except that the steam engine is replaced by the single-phase motor.

In this system the regulation of speed is effected on the Ward-Leonard plan by adjusting the strength of the magnetic field of the continuous current generator.

A very similar system has been carried out by the Oerlikon Company for the electric railway between Seebach and Wettingen. And they have kindly placed at my disposal some photographs of the railway and of the locomotive, from which the slides have been prepared.

One feature which deserves attention is the form of the collector, which is a curved bar turning round a horizontal axle on the carriage and pressed against the conductor by a spiral spring. This arrangement allows the collector to adapt itself to all the various positions which the conducting wire may successively take up relative to the moving carriage.

It is, however, possible to use single-phase motors direct for the propulsion of vehicles by reverting to the induction-motor, and this has been done by Mr. Lamme, an engineer of the Westinghouse Company in America, and by Dr. Finzi in Milan, and by the Union Company of Berlin.

This latter company is running such electric motor cars on an experimental track about 4 km. long between Johannisthal and Spindlersfeld near Berlin.

Their working current is 6000 volts, which passes direct into the stator of the motors, while about the sixth part of the current is transformed down for lighting and regulating purposes.

The working conductor is supported by being connected at intervals of 3 metres to a carrying wire, which has not much strain on it, the object being that, in the case of the wire breaking, its ends cannot touch the ground.

The Union Company have kindly sent me some photographs of the line, which fully explain the arrangements.

Reverting now to the Studiengesellschaft, which was formed in November 1899, they resolved to carry out experiments for the purpose of attaining a speed of 200 km. (125 miles) per hour, and to utilise three-phase currents of at least 10,000 volts.

The Allgemeine Elektricitäts Gesellschaft and Siemens and Halske were requested each to design a motor car capable of accomplishing this speed in a satisfactory manner, and Mr. Lasche, an engineer of the Allgemeine Elektricitäts Gesellschaft, described to the Engineering Conference in Glasgow, in September 1901, the car which had been built by his company.

Before, however, the design of the car could be decided on, some preliminary experiments had to be carried out, to determine the power necessary for propelling the car at the unprecedented speed of 200 km. per hour.

It was very evident that the resistance of the air would be the most important factor, but the usual formulæ relating to the air resistance of railway trains gave such high values that a confirmation of these results by a special experiment appeared very desirable.

Messrs. Siemens and Halske, therefore, fixed up a 200 horse-power motor, with its axle in a vertical position, and an ordinary tramcar wheel was pressed on it on top. To this tramcar wheel a wooden plank was bolted 85 mm. thick, 500 mm. broad and 6350 mm. long. On the ends of this plank two plain boards were fixed in the first instance, to represent the surface of a car, and the power supplied to the motor was measured while it was revolving at various speeds.

Afterwards, instead of the plain surface, a broken surface, roughly parabolic in plan, was substituted, and the air resistance dropped very considerably to about 90 kg. per square metre at a speed of 200 km. per hour.

Thus it was found that for air resistance and for mechanical friction, the motors had to develop 950 horse-power to drive a car weighing 96 tons at a speed of 200 km. per hour.

For safety's sake the motors were, however, made capable of developing about 50 per cent. more power as their normal output, and the actual experience afterwards proved the wisdom of this increase.

In any case it was necessary to provide nearly 3000 horse-power for the period of acceleration, and allowing an overload of 100 per cent. for this period, the normal output of the four motors together was fixed at about 1500 horse-power.

There were a number of other investigations to be made in order to determine the safest way of collecting the high tension current, transforming it to the working pressure, and regulating the motors, so that the actual experiments did not begin until the autumn of the year 1901.

The line from Marienfelde to Zossen is about 23 km. (14½ statute miles) long, and the smallest radius of a curve is 2000 m., while the ruling gradient is 1 : 200.

Along this line the three conductors are carried in the way settled by experiment for the trial line at Gross-Lichterfelde.

The three wires are supported by insulators at a vertical distance of 1 m. from each other, and the insulators in turn are elastically supported, as shown by the slide, so that the collectors, in passing the supporting posts, do not receive violent blows.

Each conductor is connected at the insulators to a loop, which passes round the support of the insulators, and also round a vertical wire, which is connected to the running rails, and through them to earth.

This device is added for the case that a high tension wire breaks, as the loop then comes in contact with the earthed wire, and prevents any danger from high tension currents.

When the Allgemeine Elektricitäts Gesellschaft car ran at its highest speed one of the conductors broke and struck a man who happened to stand near, but luckily the earth connection proved to be effective, and no harm was done.

Excepting this accident, which was caused by the abnormal swaying of the car, the method of suspending the conductors gave every satisfaction, and the experiment on the Zossen line thus confirmed the previous favourable experience gained at Gross-Lichterfelde.

The current for the experiments was supplied from the power station of the Berlin Electricity Works, at Oberspree, through a feeder about 13 km. (8½ statute miles) long, and it was generated by a 3000 kw. three-phase alternator, driven direct by a steam engine.

It was possible to regulate the output so as to give from 25 to 50 periods per second, which corresponded, after transforming up, with a difference of potential of from 6000 to 14,000 volts at the station end of the feeder.

As the set is designed to give its full output at 50 periods, it was rather over-loaded at the lower speeds, so that the motors could not develop their full power during the acceleration period.

The connection between the periods of the generator and the speed of the car is determined by the diameter of the car wheels and the number of poles of the motors.

In this case, six-pole motors were employed, driving the axle direct, and the diameter of the wheels is 1·25 metre.

Consequently, the kilometres per hour are found by multiplying the periods per second by 4·714.

Thus :—30 periods correspond to 141·42 km. per hour.

       40   ,,   ,,   188·56 ,,   ,,
       50   ,,   ,,   235·70 ,,   ,,

In other words, the speed of the cars, neglecting the slip, depends on the generator, and not on the motor; in this way the differences in speed on different days and on the two cars are explained.

One of the most difficult problems was to transfer the electric current from the stationary conductor to the moving car at the very high speeds which were eventually attained.

Here again, the experience of Gross-Lichterfelde proved to be of service, as the comparison between the collectors on the Lichterfelde locomotive and on the Zossen line clearly shows. In fact, the syndicate prescribed the adoption of this collector for both cars.

During the first experiments the Lichterfelde collector proved satisfactory, but at speeds of 175 km. and more per hour the mass of the part touching the conductor proved to be too great, and a further spring was added, so that the contact arm, constructed of brass and aluminium, which is pressed by the spring against the wire, weighs only 600 gr. (1⅓ lb. avoirdupois).

The spring again is carried by a hinged frame, which in its turn is fastened to a vertical axle.

To counterbalance the influence of the air pressure on all this structure a square board is fixed to the frame on the other side of the vertical axis, and its size is such that the pressure of air on it just balances the air pressure on the other parts of the collector.

It was therefore possible to adjust the pressure between collector and conductor to remain the same within narrow limits at any speed.

The official report of the Studiengesellschaft, for the year 1901, points out that the permanent way of the military line between Marienfelde and Zossen was not in a very good condition, but that it was deemed expedient to gain some experience before relaying it in a substantial manner.

After describing the preliminary tests at the works of the two manufacturing firms, the report gives the official regulations, drawn up by commissioners deputed by the Royal Railway Direction, Berlin, the Royal Government at Potsdam, and the Royal Post Office at Berlin. They fill five printed pages of the report.

In this year four sets of trials were run :—

During the first set the generating station supplied current at 25 periods, and the motors were connected only long enough to attain a speed of 100 km. per hour. As soon as this speed was reached the current was cut off and the brakes applied.

During the second set, and with currents of the same periodicity, the cars were run the whole distance at a speed not exceeding 100 km. per hour.

As soon as these trials had been run, and all the machinery had been found to work smoothly, the third set was commenced, in which the speed was increased to 130 km.

Up to this speed no serious defects had shown themselves, but when, during the fourth set of trials, the speed was increased still more, the permanent way gave out, and the cars began to sway in a dangerous manner.

The highest speed during this year was 160 km. per hour, and the conclusions arrived at were that the conductors and collectors had proved quite satisfactory at these speeds, that three-phase motors were suitable for rapid transit, that the brakes were not quite efficient enough, and that the determination of the resistance of the air was not accurate enough.

The report further mentioned that the permanent way of the military line was not safe for speeds exceeding 120 km. per hour, and that further observations were required to determine the amount of energy supplied by the power station and the durability of all the machinery and apparatus employed.

During the year 1902 the speeds of the cars were not allowed to exceed 125 km. per hour, as it had not been possible to arrange for the relaying of the permanent way, but observations were made on air resistance at various speeds, and on tractive force for the cars alone or when drawing other vehicles.

Exact measurements were also taken of the electric losses in feeders and conductors, and further trials were made with the brakes.

Some trials were also run with an electric locomotive, constructed by Siemens and Halske, for utilising three-phase currents of 12,000 volts direct in the motors, and dispensing with the heavy transformers used on the motor cars.

The locomotive succeeded in drawing a train of 11 carriages, weighing 153 tons, at a speed of 51 km. per hour, when the current had 45 periods and 12,500 volts pressure.

At the time of the experiments it had been fitted with two motors only, but eventually it will have four, and then it will be capable of exerting about double the power.

As the motors gave no trouble, and no difficulty was experienced in handling these high tension currents, these trials show the possibility of a very simple construction of main line electric locomotives.

The result of the 1902 experiments was that some alterations in the construction of the bogies of the motor cars were undertaken— notably, the wheel-base was lengthened from 3·8 to 5·0 m.; this change involved a modification in the brake mechanism, which had again failed to give satisfaction, and it also very much improved the steady running of the cars at the very high speeds.

In the beginning of last year the relaying of the permanent way was commenced, and it was made equal to the standard permanent way of the Prussian State Main lines for express traffic, with the addition of guard rails.

When the alterations to the motor cars had been finished, further trial runs were made, during which the Siemens car attained, on the 15th of September, a speed of 145 km., and in the following week, on the 23rd of September, a speed of 175 km.

During the latter run, as mentioned before, the collector did not work satisfactorily, and a slight alteration was made in its construction.

On the 26th of September the speed of the Siemens car reached 189 km., and on the 6th of October the speed of 201 km., or 125 miles per hour, was first accomplished.

It is notable that the electrical appliances and connections of this car remained exactly the same from 1901 to the present time.

The car of the Allgemeine Elektricitäts Gesellschaft commenced to run a few days later, and attained, on the 28th of October, the maximum speed of 210·2 km. per hour, with a periodicity of 47·5 per second. During this run the conductor was damaged, and the generating machinery rather strained.

In consequence of this, the speed of further runs was somewhat diminished, the next best being made on the 11th of November, when the periodicity was 45·3 per second, the speed of the Allgemeine Elektricitäts Gesellschaft car 196 km., corresponding to a slip of 8·4 per cent., and the speed of the Siemens car 208 km., corresponding to a slip of 2·8 per cent.

Again, on the 25th November, with a periodicity of $45\frac{1}{4}$ per sec. the Allgemeine Elektricitäts Gesellschaft car attained 204·9 km., a slip of 3·9 per cent., and the Siemens car 207·3 km. per hour, a slip of $2\frac{3}{4}$ per cent.

The difference in the slip of the two motor cars is probably due to the fact that at full speed the Siemens motors are metallically short circuited, while in the circuit of the Allgemeine Elektricitäts Gesellschaft motors the liquid resistance is not quite removed.

I am indebted to the Allgemeine Elektricitäts Gesellschaft and to Siemens and Halske for the photographs and the information about last year's trials, which I have been able to place before you to-night.

In any case, the results of these trials clearly establish the fact that it is possible to run up to a speed of 200 km., or 125 statute miles, per hour on the same permanent way as is now used on well-built main lines for express traffic, and that at such a speed the high tension electric current can pass from the stationary conductors along the line to the moving car without difficulty.

These two facts are quite independent of the kind of motor employed, and are equally applicable to the other railway motors which have been described during the lecture.

The electrical engineers can claim, therefore, that they are quite ready to run main line traffic by means of electrical motors, and that they are prepared to do so at about double the speed of the present express trains—in fact, that technically, the problem has been solved in a satisfactory manner.

It would be rash, however, to assume from these successes that steam locomotives will disappear in the near future from the main lines of railways, to be replaced by fast running motor cars, following each other in quick succession.

No innovation in this material life has any chance of general and permanent introduction, unless in replacing existing means it introduces a saving of money.

Applying this rule to the present case it is at once apparent that the existing main lines, although their permanent way has been proved to be sufficiently strong, do not lend themselves to the immediate use for rapid transit on account of the curves which occur in them.

As stated before, the worst curve on the Zossen line has a radius of 2000 metres, which is, for ordinary traffic, almost negligible; but during the rapid transit experiments the speed of the cars had to be lowered to 160–170 km. per hour when passing through these curves, in order to make their running reasonably safe.

It will, therefore, be necessary to build special main lines for this fast traffic with no intermediate stations, and with no level crossings of any kind. They should not have any curves with less than 2000 metre radius, except, perhaps, near their stations, where the speed of the car must of necessity be slow.

In order to justify the capital outlay for such lines, and for their equipment with frequently running cars, an enormous traffic is required, and it is not likely that many places will be found where this condition is fulfilled.

On the other hand, increased facilities for communication create traffic, and the possibility of reaching Liverpool, for instance, from London in less than two hours may attract a sufficient number of passengers to warrant the making of a new bee-line railway between the two towns.

Another question is, whether it would not stimulate the traffic between two towns, if the service between them were run at very frequent intervals without increasing the speed of the individual car or train of cars.

No answer, applicable everywhere, can be given; each case has to be tested on its own merits, and in every case it is the economical result which will decide the issue; but in estimating the merits of any particular scheme due consideration should be given to the well established principle, "Time is Money."

[A. S.]

WEEKLY EVENING MEETING,

Friday, March 4, 1904.

SIR JAMES CRICHTON-BROWNE, M.D. LL.D. F.R.S., Treasurer
and Vice-President, in the Chair.

WILLIAM STIRLING, M.D. D.Sc. LL.D., Professor of Physiology, and
Dean of the Medical Faculty of the University of Manchester.

Breathing, in Living Beings.

(ABSTRACT.)

MAY I crave your lenient judgment during a passing hour? The
subject is so vast, and the time so short, that I hardly feel equal to
the task.

I hope you will season your just criticisms with a sprinkling of
the sweet waters of mercy. If so, I am encouraged to proceed at
once to give a short account of some of the mechanisms related to the
process of breathing in living beings.

It has been said that the most striking facts connected with
respiration are its universality and its continuity. In popular language
"the breath is the life." Breathing is not only a sign of life, it is a
condition of its existence. Permanent cessation of breathing is regarded as a sign of death. Link up with this the icy coldness of death
and you have two significant facts.

Respiration and calorification are therefore intimately related; in
fact, calorification is one form of expression of the results of respiratory activity.

The popular view of respiration is an inference from what is
observed in man and animals. During life the rise and fall of the
chest goes on rhythmically from the beginning to the end. The
respiratory exchanges effected in the breathing organs—lungs or
gills—constitute "external respiration." This, however, scarcely
touches the main problem, viz. what is called "internal respiration,"
or tissue respiration—i.e. the actual breathing by the living cells and
tissues which make up a complex organism.

We are told that man does not live by bread alone. We know
he requires, in addition, solids, fluids, and air. Taking these to
represent the three graces, then air is of all the graces best.

The higher animals have practically no reserve stores of air—
unlike what happens with the storage of fats and proteids—and hence
the necessity for mechanisms by which air is continually supplied to
the living tissues, and also by which the waste product of combustion,

viz. carbon dioxide, is got rid of. Closure of the wind-pipe, even for a few minutes, brings death with it from suffocation. The entrance of oxygen is prevented and the escape of carbon dioxide is arrested.

The process of breathing is common to all living beings—to plants and animals alike. It consists essentially in the consumption of oxygen by the tissues and the giving out of carbon dioxide. It is immaterial whether the animals or plants live in water or air, the principle is the same in both cases. Living active protoplasm demands a supply of oxygen.

All the world's a stage. The human body is at once a stage and a tabernacle—a vast theatre—and the myriads of diverse cells of which it is composed, the players.

The cells or players, as active living entities, not only require food but they require energy. The respiratory exchanges in and by the living cells provide the energy for the organism. This breathing by the cells is called "internal respiration." In a complex organism, therefore, the respiratory exchanges represent the algebraic sum of the respiratory activity of the several tissues that make up the organism. The various tissues, however, breathe at very unequal rates.

In one of his charming " contes philosophiques," Voltaire describes the visit of a giant of Sirius to our planet. Before reaching his journey's end he would have to traverse an aerial medium ; and on arriving would see before him a fluid medium in continual movement, and tracts of solid land. After investigation—or no doubt he would be told, even though he was not personally conducted—that the water surface of this our globe is two and a half times greater than the land surface. He would discover that there are animals that live in air, others in water, and again others on land. Our visitor would find out that the respirable media are two—water and air—and that there are 210 parts of free oxygen in a litre of air, while there are only 3–10 dissolved in a litre of water.

Had Voltaire's friend paid us another visit during the present century, we should be able to tell him that the water of the Thames above London contains 7·40 c.c. of O per litre; at Woolwich only 0·25; the decrease being due to the pollution of the river. Putting it broadly, water contains only 3–10 parts per litre, while air contains 210. Water-breathers under good conditions have twenty times less O than air-breathers. It is as if air-breathers on land had the percentage of O_2 reduced to 1.

He would also be told that carbon dioxide—CO_2—is also remarkably soluble in water, and readily combines with certain bases present in water; thus water forms an admirable medium into which an animal may discharge its effete and poisonous irrespirable CO_2.

He would also be told that our blood contains 60 volumes per cent. of gases, and that there is more O and less CO_2 in arterial blood than in venous blood.

Perhaps the name of Sir H. Davy might be whispered to him, for he was one of the first to detect the presence of gases O and CO_2 in blood.

In story, one has heard of the " Quest of the Holy Grail." I have even listened with rapt attention to an entrancing lecture on the " Quest of the Ideal." For the cell, the quest is the " quest of oxygen," and it is not happy till it gets it.

We speak of a distinction between air-breathers and water-breathers. If, however, we push the matter to its ultimate issue, we find that all our tissues—and equally those of plants—live in a watery medium ; in us the fluid lymph which exudes from our capillary blood-vessels, and in plants in the sap. Thus we come upon what at first seems a paradox, but is not so; all our cells not only live in water, but they live in running water. They are bathed everywhere by the lymph which is the ;real nutrient fluid for our cells. Thus, in its final form, all respiration is actually aquatic. The process of internal respiration, besides other conditions, requires the presence of a certain amount of water. In fact, all vital phenomena require the presence of water.

The unity and identity of the process in animal and vegetable cells, as the theatre of combustion, is the striking fact. The means by which the necessary oxygen is brought to the cells is as varied as the forms of animated organisms themselves. This function exists for the cells, and not the cells for the function.

If the mountain will not go to Mohammed, Mohammed must go to the mountain. There are, at least, two principles on which animal cells obtain oxygen.

The air, or water containing air, is carried to the cells. This is the principle adopted in the lower invertebrates, as in sponges and with regard to certain air-breathers such as insects.

The other principle is this, that an intermediary carries the respiratory oxygen from some more or less central localised or diffuse surface to the cells. This intermediary is the blood—an internal medium of exchange. The fluid part of the blood may carry the oxygen supply and remove the carbonic dioxide waste. This is the case in many of the invertebrates, and it reaches its highest development in the vertebrates. Hence in them the circulating and respiratory systems reach their fullest development.

In most invertebrates the fluid part of the blood contains the nutritive substances and also the oxygen and carbonic acid. In the vertebrates, the hæmoglobin of the red blood corpuscles carries the oxygen from the gills or lungs to the tissues, whilst the CO_2 is contained in and carried chiefly by the blood plasma from the tissues to the gills or lungs.

It is singular that in the cephalopods, such as the squid and cuttle-fish, the blood is bluish in tint ; and this is due to the presence in the plasma of a respiratory pigment called hæmocyanin. This body has a composition like that of hæmoglobin, but copper is substituted for the iron of the hæmoglobin. Copper also exists in organic combination in the red part of the feathers of the Plantain-eater or Turaco.

The real aristocracy with genuine blue blood are the crabs, lobsters, squids, and cuttle-fishes.

Perhaps one of the most striking ways of dissociating this accessory mechanism from the activity of the cell itself is by the use of a poison. When a person is poisoned by coal gas, what happens? The coal gas contains carbon monoxide. This gas does not poison invertebrate animals or plants. Still it kills vertebrate animals. Why? It does not kill by acting on the living cells, only by depriving them of oxygen and asphyxiating them. It combines with the respiratory pigment hæmoglobin. Chloroform, ether, and similar drugs destroy the actual life of the cell elements by destroying their irritability.

As this year of grace marks the centenary of the death of Joseph Priestley, I may be permitted to refer to his early discovery of the action of green plants.

In 1771, Priestley found that air vitiated by combustion of a candle, or by the breathing of animals—such as mice—could be made pure or respirable again by the action of green plants.

Under certain conditions, however, Priestley found that plants gave off carbonic acid, and the air did not support combustion or animal life. He regarded these as " bad experiments," and he selected what he was pleased to regard as " good experiments," i.e. those in which the air, rendered impure by the respiration of animals, was rendered respirable by the action of green plants.

In 1779 John Ingen-Housz published his " Experiments on Vegetables, discovering their great power of purifying the common air in sunshine, and of injuring it in the shade and at night."

He confirmed Priestley's observations that green plants thrive in putrid air; and that vegetables could convert air fouled by burning of a candle, and restore it again to its former purity and fitness for supporting flame, and for the respiration of animals—or, as he puts it " plants correct bad air."

In 1787 Ingen-Housz, an English physician at the Austrian court, found that only in daylight did green plants give off oxygen. In darkness, or where there was little light, they behaved like animals so far as exchange of gases is concerned, i.e. they used up oxygen and exhaled carbonic acid. He found also that all roots, when left out of the ground, yielded by day and by night foul air, i.e. carbonic acid.

In the same year, 1804—the year of Priestley's death—Nicolas Theodore de Saussure, a Swiss naturalist and chemist, published his " Recherches Chimiques sur la Végétation " (Paris, 1804), a veritable encyclopædia of experiments of the effects of air on flowers, fruits, plants and vegetation generally, and on the effects of these on atmospheric acid.

It is an old adage—the exception proves the rule. The exception "probes" the rule as the surgeon's probe probes a wound. The tactus eruditus of the surgeon, by his probe—indeed an elongated tactile sense —enables him to discover the presence or absence of a body in a wound. Had Priestley used the probe of a bad experiment, he in all probability would have anticipated the discovery of Ingen-Housz.

Some of you, no doubt, recollect the words of Goldsmith's famous description of his own bedroom and of the furniture of the inn—

> " The house where nut-brown draughts inspired."

And how his imagination stooped to trace the story of—

> "The chest that contrived a double debt to pay,
> A bed by night, a chest of drawers by day."

As to himself he tells us how—

> " A night-cap decked his brows instead of bay,
> A cap by night—a stocking all the day."

Green plants contrive a double debt to pay : they give off oxygen by day, and at night exhale CO_2.

How do the vast number of plants, the microbes, the bacteria without chlorophyll get oxygen? Most of them get it as we get it. Some, however, cannot live in pure oxygen and are anaerobic, such as the micro-organisms that cause tetanus, malignant œdema, and those that set up butyric acid fermentation.

Pushing the matter still further, it is extremely probable that the oxidation processes in our tissues are largely due to the presence of oxydases.

This raises the question as to the part played by the nucleus of a cell in its respiratory processes.

Is the source of muscular energy to be sought in oxidation or cleavage processes in tissues? In some animals there is not a direct relation between the muscular work and oxygen consumed, though there is to heat production. Bunge, on this ground, thought that the intestinal parasites of warm-blooded animals must have their oxygen at a minimum. In the intestinal contents there is no estimable oxygen, there active reduction processes go on. Entozoa might get oxygen from O_2 diffusing from blood-vessels.

Bunge found that intestinal worms of the cat and pike can live in an alkaline solution of common salt, free from gases, under Hg, for four to six days. They made active movements, and gave off much CO_2.

Ascaris lumbricoides from the intestine of the pig, lived four to six days in 1 per cent. boiled NaCl solution. It made little difference whether oxygen or hydrogen was passed through the fluid. They lived seven to nine days if fluid was saturated with carbon dioxide, so that they have accommodated themselves to high percentages of carbon dioxide.

They give off to the fluid valerianic acid, an acid with a characteristic butyric acid odour. These worms contain a very large quantity of glycogen, the dry body yielding 20 to 34 per cent. of this carbohydrate.

100 grammes Ascaris, placed in boiled normal saline solution, used per day—

> 0·7 gramme glycogen.
> 0·1   „   sugar,
> No fat;

and yielded—

> 0·4 gramme CO_2
> 0·3   „   valerianic acid.

It would seem that glycogen had split into CO_2 and valerianic acid—

$$4\ C_6H_{12}O_6 = 9\ CO_2 + 3\ C_5H_{10}O_2 + 9\ H_2$$
$$720\ \ =\ \ 396\ \ +\ \ 306\ \ +\ \ 18.$$

Is it a genuine fermentation?

Weinland found that he could express by Buchner's method a substance "zymase," which could split glycogen into CO_2 and valerianic acid.

Turning now to respiration in invertebrate animals, and dealing first with those which live in water, let us see some of the contrivances by which this end is achieved. The mechanisms are but means to an end. The ultimate union of oxygen, and the discharge of carbon dioxide with the liberation of energy, occur in the protoplasm of the cell itself.

There are two distinct processes, and it may be that the oxygen is introduced by one portal and the carbon dioxide got rid of by another, or it may be that one portal may do for both processes—the letting in of oxygen and the giving off of carbon dioxide.

Although the principle itself is simple, the variety of mechanisms adopted by nature to secure this double function is remarkable. Let us glance at some of the mechanisms proceeding from the simple to the complex, and first with regard to those animals that live in water.

Consider the oceanic fauna. It is immense both from the point of view of number and variety. Save insects and certain groups of molluscs, all invertebrates are aquatic. Amongst vertebrates, fishes have aquatic respiration; and some mammals, *e.g.* cetaceans or whales, have water as their sphere of existence, though they depend on the air for their respiratory oxygen.

The evolution from an aquatic to an aerial mode of existence can be traced in the animal kingdom, and may even be seen within limits in the history of certain species.

Every living cell, animal or vegetable, requires for its continued existence a supply of oxygen, and every living cell exhales carbon dioxide. The exchange of these two gases between the fluids of the body and the outer medium is the process of respiration. The simplest form of respiratory exchange occurs where there is no specially differentiated organ or mechanism for this purpose, so-called diffuse respiration. The whole surface of the organism in a watery medium may be concerned in this respiratory exchange. This is only possible, however, as long as the boundary surface, skin, or otherwise is permeable to gases and no great respiratory exchanges are necessary.

Before showing you some lantern slides, I should like to point out how one process is made to aid another.

Motion associated with respiratory processes.

Ciliary motion with respiration and the capture of prey for food.

The old idea of one function for an organ is exploded. One speaks of one man one vote. One man one value. It is not really so.

With Shelley we may say—

"Nothing in this world is single;
All things, by a law Divine,
In each other's being mingle."

As regards the surfaces for these respiratory exchanges for diffuse respiration, it may take place through the inner surface of the body cavity of cœlenterates, the under surface of the bell of a medusa, the tentacles of an echinus, the respiratory tree at the hind gut of the sea cucumber, or the intestine of the young of the dragon fly, or by the intestinal mucous membrane of the mites which have no lungs or other directly respiratory organ. In the higher animals we have tracheæ, gills and lungs.

In some animals, the respiratory mechanism is closely related to the motor apparatus, as in some crustacea. In some mollusca the nutritive and respiratory mechanisms are closely related. In the highest of all there is central apparatus—gills or lungs—for the respiratory exchange between the blood and the air, and a circulatory apparatus for carrying the blood to and from the respiratory organs. The adaptivity of insects to varied conditions of oxygen supply is marvellous.

Before showing some classical experiments and illustrating the principles already laid down, I should like again to draw your attention to the association of several processes with respiratory mechanisms.

[The lecture was illustrated by means of lantern slides, showing the respiratory mechanisms from the lowest to the highest animals, and also by a number of experiments dealing with the chemical exchanges in the process of respiration. Lastly, the classical experiment of John Hunter, on the pneumaticity of the bones of birds, was shown in the duck. A candle flame was extinguished when held in front of the divided trachea, when air was blown into the divided humerus bone of the wing.]

[W. S.]

GENERAL MONTHLY MEETING,

Monday, March 7, 1904.

Sir James Crichton-Browne, M.D. LL.D. F.R.S., Treasurer and
Vice-President, in the Chair.

Miss Lilian Annie Black,
Henry Thomas Davidge, Esq., B.Sc.
Henry Fielding Dickens, Esq., K.C.
Frederic S. Eve, Esq., F.R.C.S.
Francis L. M. Forster, Esq.
Miss Charlotte E. M. Gibbons,
Thomas Grove Hull, Esq.
Mrs. Gerard Leigh,
Vivian Byam Lewes, Esq., F.C.S.
The Hon. Mary Portman,
Henry Ralph Prendergast, Esq.
William Napier Shaw, Esq., M.A. Sc.D. F.R.S.
James Sorley, Esq., F.R.S.E.
Robert Henry Scanes Spicer, M.D. B.Sc.
The Hon. Sir Joseph Walton,
Edward Wormald, Esq.

were elected Members of the Royal Institution.

The Presents received since the last Meeting were laid on the
table, and the thanks of the Members returned for the same, viz.:—

FROM

The British Museum Trustees—Catalogue of English Pottery. 4to. 1903.
   Supplementary Catalogue of Chinese Books and MSS. 4to. 1903.
   Catalogue of the Franks Collection of Bookplates, Vol. I. 8vo. 1903.
   Catalogue of German and Flemish Woodcuts, Vol. I. 8vo. 1903.
The British Museum Trustees (Natural History)—Catalogue of Madreporarian
   Corals, Vol. IV. 4to. 1903.
   Catalogue of the Library, Vol. I. 4to. 1903.
   Catalogue of Lepidoptera Phalænæ, Vol. IV. and Plates. 8vo. 1903.
   First Report on Economic Zoology. By F. V. Theobald. 8vo. 1903.
   Hand List of Birds, Vol. IV. 8vo. 1903.
Accademia dei Lincei, Reale, Roma—Classe di Scienze Fisiche, Matematiche e
   Naturali. Atti, Serie Quinta: Rendiconti. Vol. XIII. 1° Semestre, Fasc. 2-3,
   8vo. 1904.
Allegheny Observatory—Miscellaneous Papers, New Series, No. 17. 8vo. 1904.
American Academy of Arts and Sciences—Proceedings, Vol. XXXIX. Nos. 6-12.
   8vo. 1903.
American Geographical Society—Bulletin, Vol. XXXV. No. 5. 8vo. 1903.
Antolik, Karl, Esq. (the Author)—Uber Klangfiguren gespannter Membranen
   und starrer Platten. 8vo. 1904.

Astronomical Society, Royal—Monthlv Notices, Vol. LXIV. No. 3. 8vo. 1904.
Automobile Club—Journal for Feb. 1904.
Bankers, Institute of—Journal, Vol. XXV. Part 2. 8vo. 1904.
  List of Members, 1904. 8vo.
Basel, Naturforschenden Gesellschaft—Verhandlungen, Band XV. Heft 2. 8vo. 1904.
Belgium, Royal Academy of Sciences—Bulletin, 1903, Nos. 11, 12. 8vo.
  Annuaire, 1904. 8vo.
Birmingham and Midland Institute Scientific Society—Meteorological Observations, 1903. 8vo. 1904.
Boston Public Library—Annual List of Books, 1902-3. 8vo. 1904.
  Monthly Bulletin for Feb. 1904. 8vo.
Botanic Society of London, Royal—Quarterly Record, Vol. VIII. No. 96 (Oct.-Dec.). 8vo. 1903.
British Architects, Royal Institute of—Journal, Third Series, Vol. XI. Nos. 7-9. 4to. 1904.
British Astronomical Association—Journal, Vol. XIV. No. 4. 8vo. 1904.
  Memoirs, Vol. XII. Part 2. 8vo 1904.
Brough, H. Bennett, Esq. (the Author)—The Mining of Non-Metallic Minerals 8vo. 1904.
Buenos Ayres—Monthly Bulletin of Municipal Statistics for Dec. 1903. 4to.
Chemical Industry, Society of—Journal, Vol. XXIII. Nos. 2-4. 8vo. 1904.
Chemical Society—Proceedings, Vol. XIX. No. 274; Vol. XX. Nos. 275-276. 8vo. 1904.
  Journal for Jan.-Feb. 1904. 8vo.
Chicago, University of—Publications of the Yerkes Observatory, Vol. III. Part 1 4to. 1903.
  Report of the Director of the Yerkes Observatory, 1899-1902. 4to. 1903.
  Decennial Publications: The Spectra of Stars of Secchi's Fourth Type. By G. E. Hale and others. 4to. 1903.
East India Association—Journal, Vol. XXXV. No. 33. 8vo. 1904.
Editors—American Journal of Science for Feb. 1904. 8vo.
  Analyst for Feb. 1904. 8vo.
  Astrophysical Journal for Feb. 1904. 8vo.
  Athenæum for Feb. 1904. 4to.
  Author for Jan.-March 1904. 8vo.
  Board of Trade Journal for Feb. 1904. 8vo.
  Brewers' Journal for Feb. 1904. 8vo.
  Chemical News for Feb. 1904. 4to.
  Chemist and Druggist for Feb. 1904. 8vo.
  Electrical Engineer for Feb. 1904. 4to.
  Electrical Review for Feb. 1904. 4to.
  Electrical Times for Feb. 1904. 4to.
  Electricity for Feb. 1904. 8vo.
  Electro-Chemist and Metallurgist for Feb. 1904. 8vo.
  Engineer for Feb. 1904. fol.
  Engineering for Feb. 1904. fol.
  Engineering Review for Feb. 1904. 8vo.
  Homœopathic Review for Feb. 1904. 8vo.
  Horological Journal for Feb. 1904. 8vo.
  Journal of the British Dental Association for Feb. 1904. 8vo.
  Journal of Physical Chemistry for Jan. 1904. 8vo.
  Journal of State Medicine for Feb. 1904. 8vo.
  Law Journal for Feb. 1904. 8vo.
  London Technical Education Gazette for Feb. 1904. 4to.
  London University Gazette for Feb. 1904. 4to.
  Machinery Market for Feb. 1904. 8vo.
  Model Engineer for Feb. 1904. 8vo.
  Mois Scientifique for Feb. 1904. 8vo.

Editors—continued.
Motor Car Journal for Feb. 1904. 8vo.
Motor Car World for Feb. 1904. 4to.
Musical Times for Feb. 1904. 8vo.
Nature for Feb. 1904. 4to.
New Church Magazine for March, 1904. 8vo.
Nuovo Cimento for Sept.-Oct. 1903. 8vo.
Page's Magazine for Feb. 1904. 8vo.
Photographic News for Feb. 1904. 8vo.
Physical Review for Jan.-Feb. 1904. 8vo.
Public Health Engineer for Feb. 1904. 8vo.
Science Abstracts for Feb. 1904. 8vo.
Zoophilist for Feb. 1904. 4to.
Electrical Engineers, Institution of—Journal, Vol. XXXIII. Part 1. 8vo. 1904.
Florence, Biblioteca Nazionale—Bulletin for Jan.-Feb. 1904. 8vo.
Franklin Institute—Journal, Vol. CLVII. No. 2. 8vo. 1904.
Genève, Société de Physique—Compte Rendu, XX. 1903. 8vo.
Geographical Society, Royal—Journal, Vol. XXIII. No. 2. 8vo. 1904.
Geological Society—Abstracts of Proceedings, Nos. 787-790. 1904.
Quarterly Journal, Vol. LX. No. 1. 8vo. 1904.
Göttingen, Royal Academy of Sciences—Nachrichten-Geschäftliche Mittheilungen, 1903, Heft 2. 8vo.
Mathematisch-physikalische Klasse, 1903, Heft 6. 8vo.
Hickson Ward & Co., Messrs. (the Publishers)—Report of the Select Committee on Ventilation appointed by the House of Commons. 8vo. 1904.
Iron and Steel Institute—Journal, 1903, No. 2. 8vo.
Leighton, John, Esq., M.R.I.—Ex-Libris Journal for Jan. 1904. 8vo.
Literature, Royal Society of—Chronicon Adae de Usk, A.D. 1377-1421. Edited by Sir E. Maunde Thompson. 8vo. 1904.
Madrid, Royal Academy of Sciences—Anuario, 1904. 16mo.
Meteorological Society, Royal—Quarterly Journal for Jan. 1904. 8vo.
List of Members, 1904. 8vo.
Mexico, Secretaria de Comunicaciones y Obras Publicas—Anales, No. 9. 8vo. 1904.
Microscopical Society, Royal—Journal, 1904, Part 1. 8vo.
Mimir—Icelandic Institutions and Addresses, 1903. 12mo.
Mitchell & Co., Messrs. (the Publishers)—Newspaper Press Directory, 1904. 4to.
Munich, Royal Academy of Sciences—Sitzungsberichte, 1903, Heft IV. 8vo. 1904.
Navy League—Navy League Journal for Feb.-March, 1904. 8vo.
The British Navy, Past and Present. By Rear-Admiral S. Eardley-Wilmot. 8vo. 1904.
North of England Institute of Mining Engineers—Transactions, Vol. LIV. Part 2, 8vo. 1904.
Odontological Society—Transactions, Vol. XXXVI. No. 3. 8vo. 1904.
Pharmaceutical Society of Great Britain—Journal for Feb. 1904. 8vo.
Photographic Society, Royal—Journal, Vol. XLIV. No. 1. 8vo. 1903.
Radcliffe Library, Oxford—Catalogue of Books, 1903. 4to. 1904.
Righi, Professor A. (the Author)—La Moderna Teoria dei Fenomeni Fisici. 8vo. 1904.
Rio de Janeiro, Observatorio—Bolletin Mensal, April-June, 1903. 8vo.
Royal College of Physicians—List of Fellows, etc., 1904. 8vo.
Royal Society of London—Philosophical Transactions, A, Nos. 360, 361; B, No. 223. 4to. 1904.
Proceedings, Nos. 487-490. 8vo. 1904.
Selborne Society—Nature Notes for Feb. 1904. 8vo.
Smith, B. Leigh, Esq. M.R.I.—Scottish Geographical Magazine, Vols. XX. Nos. 2, 3. 8vo. 1904.
Society of Arts—Journal for Feb. 1904. 8vo.

St. Bartholomew's Hospital—Reports, Vol. **XXXIX**. 1903.   8vo.   1904.
Tacchini, Prof. P. Hon. Mem. R.I. (the Author)—Memorie della Società degli
      Spettroscopisti Italiani, Vol. **XXXIII**. Disp. 1.   4to.   1904.
Torroja, E., Esq. (the Author)—Teoria Geometrica de las Lineas Alabaedas.
      8vo.   1904.
United Service Institution, Royal—Journal for Feb. 1904.   8vo.
United States Department of Agriculture—Monthly Weather Review for Oct.-Nov.
      1903.   4to.
      Report of the Chief of the Weather Bureau for 1903.   8vo.
      Weather Folk-Lore.   By E. B. Garriott.   8vo.   1903.
      Experiment Station Record for Jan. 1904.   8vo.
United States Patent Office—Official Gazette, Vol. **CVIII**. Nos. 4–8.   8vo.   1904.
Verein zur Beförderung des Gewerbfleisses in Preussen—Verhandlungen, 1904,
      Heft 1, 2.   8vo.
Vienna, Imperial Geological Institute—Verhandlungen, 1904, No. 1.   8vo.
Washington Academy of Sciences—Proceedings, Vol. **V**. pp. 231–429.   8vo.   1904
Washington, Philosophical Society of—Bulletin, Vol. **XIV**. pp. 233–246.   8vo.
      1903.

WEEKLY EVENING MEETING,

Friday, March 11, 1904.

His Grace The Duke of Northumberland, K.G. D.C.L. F.R.S.
President, in the Chair.

Professor Frederick T. Trouton, M.A. D.Sc. F.R.S. *M.R.I.*

The Motion of Viscous Substances.

Viscosity is one of the more familiar phenomena. An excellent example of the behaviour of a viscous substance can be afforded by treacle or honey. For instance, on helping oneself to honey, the slow and leisurely way in which the substance leaves the spoon is only too patent. This is in marked contrast to liquids such as water which flow freely.

If we stir such a liquid as treacle, we experience a resistance, a force opposing the movement; on diluting the treacle with water less resistance is afforded by it to stirring—the liquid does not feel so thick.

Liquids of any desired thickness or viscosity intermediate between treacle and water may be prepared by mixing them in the proper proportions.

Another convenient series of substances can be easily made, beginning with benzine, which is a particularly mobile liquid, and ending with pitch, ordinarily viewed as a solid.

The substance next to benzine is just a little less easily stirred than it. Next to this we have a material a little thicker, and so on by slight increases until we arrive at pitch.

In considering this set of substances we see the difficulty there is in defining exactly what a liquid is. The members of the series, at one end, would be said by all to be liquid, while at the other end they appear at first sight equally entitled to be considered solid; but if we examine them carefully we find that as we go up the series each one can flow just like its neighbour only not quite so fast. Even the seemingly solid pitch flows when given time.

The question naturally suggests itself, How much is each one thicker or more viscous than the one next it in the series? This at once leads us to the further question, What is the precise meaning we propose to attach to the term viscosity?

Comparative measurements might be made of the rate at which the substances flowed under similar circumstances, and these made the basis of a scale of viscosity, but a clearer and more precise definition is got by having recourse to the more mechanical ideas used in the well-known definition of viscosity.

A good idea of the lines on which this is done may be obtained in this way. Imagine a cube of the viscous material standing on a

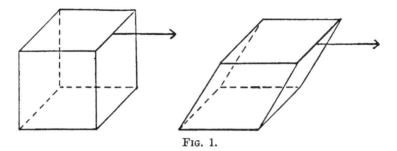

FIG. 1.

table (Fig. 1) to be gradually twisted over to the right by a force applied along the top. Then the slower the block twists the more viscous it is said to be.

Tube Method.

The viscosity of the substances at the more liquid end of the benzine-pitch series, can be determined by observing the rate at which they can be forced to move through tubes under recognised conditions. Fig. 2 shows how a layer in such a tube lying between two cross

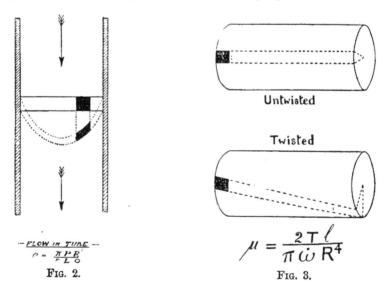

Untwisted

Twisted

$$\mu = \frac{2T\ell}{\pi\dot\omega R^4}$$

FIG. 2.   FIG. 3.

sections at any moment, becomes bent forward at its centre as the flow proceeds. The little square shown in the first position has taken up a skewed shape in the second. This helps us to see that we have in this case the same kind of motion as contemplated in the definition above referred to. The subjoined table exhibits the determinations made in this way for five of the substances in the series.

Torsion Method.

The viscosity in the case of the more viscous materials of the series was in the first instance determined by a method in which a column or rod was twisted round and round, one end being held. From the amount of the twist required to force it to turn at a given rate, the viscosity can be calculated. The table shows the value found in this way for several substances.

The apparatus employed in these determinations consisted in a horizontal shaft turning on anti-friction wheels, round which a cord

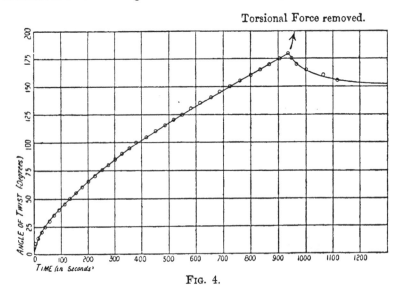

Fig. 4.

carrying a weight is wound. The weight unwinding the cord turns the shaft. A rod of the material to be tested is fixed in line between the end of the shaft and a stationary clamp. The rate at which the shaft turns is measured from a divided dial face. A knowledge of this and of the weight enables the viscosity to be calculated.

Fig. 3 will help us to understand that there is a direct connection between the simple alteration in shape dealt with in the definition of viscosity and the effect produced on the substance of the rod when twisting. The small square is seen in the second figure to have become bent in consequence of the twist imparted to the rod.

The results obtained with this apparatus are shown in Fig. 4. The curve shows the angle through which the rod had twisted at any moment after the application of the force.

"It will be seen that just at first it turns more rapidly than afterwards when it settles down to a uniform rate of twisting. It is apparently possible to keep on turning the rod as long as one pleases. By painting a white line down a rod, and then twisting it, a beautiful spiral line is produced. On removing the twisting force the rod turns back a short distance, at first rapidly, but slows down gradually to rest.

Traction Method.

A very interesting question arises as to how the material moves in a rod of pitch or such-like substance when drawn out. If we subject a rod of pitch to traction, say by suspending it from one end and hanging weights from the other end, we find that it draws out at an approximate uniform rate. The rate is at first a little faster than it is later on; it however finally settles down to a uniform rate, provided the tension is kept the same. This is similar to the effect observed in the case of torsion. Also, as in the case of torsion, we get a slight recovery on removal of the force.

The exact way in which the particles move in a rod as it is drawn out is not at all clear. In the case of the flow through a tube, we know that the centre flows faster than the outer parts because the sides are held back. Nothing in the case of the rod corresponds to this. To try and observe the character of the flow in a rod the plan was tried of drawing out a rod made up of two shorter rods of different colours, but otherwise alike, joined end to end. The junction was made as sharp as possible and lay at right angles to the axis of the rod. Rods of shoemaker's wax and of glass were used. Difficulty was found in getting two differently coloured glasses of exactly the same fusibility, and also in forming a really sharp line of junction between them. As far as the observations go, they show at least that the particles lying in a plane do not move so as to lie on a curved surface, such as occurs in the flow through a tube.

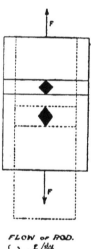

FLOW OF ROD.

$$\begin{cases} \lambda - \xi \Big/ \frac{du}{dt} \\ \mu - \tfrac{1}{3}\lambda \end{cases}$$

Fig. 5.

Fig. 5 is intended to show how a small quantity of the material, included between two near planes, lies after a short time. A small cube of the material is shown in the initial stage and its subsequent shape when drawn out. The connection will thus be appreciated between the kind of movement here taking place, and that consistent with the definition of viscosity as illustrated by the model.

The rate of elongation of the rod divided by the tension gives us a coefficient of viscous traction. We have every reason to expect this to be about three times the viscosity of the material, and experiment supports this view. The value of the viscosity of pitch

found in this way is shown in the table. The agreement with the
values found by other methods is satisfactory for experiments of this
character.

Bending of Bars.

When a bar of pitch is laid horizontally, supported only at its
ends, it sags in the middle. In doing so the upper parts of the

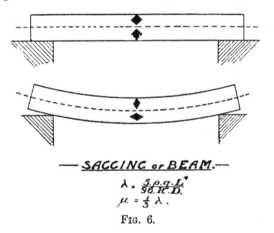

— *SAGGING of BEAM.* —

$$\lambda = \frac{5 \rho g L^4}{96 \cdot h^2 D}.$$

$$\mu = \frac{4}{3} \lambda.$$

Fig. 6.

material become compressed, while the lower are drawn out. This
is shown in Fig. 6. Two little squares are put in to help us to under-

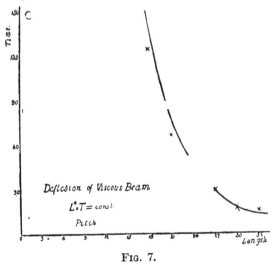

Fig. 7.

stand the movement of the material; these are subsequently transformed into the skew shape we have seen to be associated with
viscous flow.

The upper square is compressed horizontally, the lower is drawn out in the same direction.

Experiment shows that the rate at which the bar falls at its centre is approximately that given by theory based on similar suppositions to those which were made in the case of traction.

Fig. 7 shows the time which, according to the theory, should be taken by rods of different lengths in falling at their centres through the same distance. The shorter the bar the slower it falls. The marked points are those actually observed. The agreement is, under the circumstances, satisfactory.

The results obtained in this way for the viscosity of several materials are exhibited in the tables. When the material of the bar is so soft that it sags too quickly to be easily observed, it may be immersed in a liquid of very nearly its own density. In this way the forces bending it can be made as small as we please and consequently the rate at which it sags also.

Shape of Falling Stream.

The shape of a falling stream of a viscous liquid is interesting to observe. It can be well seen when helping oneself to honey. If the liquid is a thick one we have practically a case of simple viscous traction of a rod continued until extreme thinning of the rod is produced. In Fig. 8 is shown the shape taken by such a liquid when falling from a circular hole in the bottom of the containing vessel.

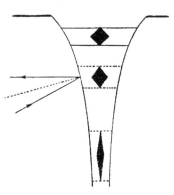

$$\lambda = \frac{W.g.y}{2.J.\frac{dy}{ds}}$$

$$\mu = \tfrac{1}{3}\lambda$$

Fig. 8.

There is an interesting point in connection with the shape or outline of the falling stream which is at first surprising. It comes out that, when the flow is slow, the same shape ought to be assumed by all substances under the same conditions as to size of orifice and height in containing vessel. This certainly appeared to be so in the case of the substances in the series examined. How this comes about will be understood by considering that if the material is removed slowly below owing to high viscosity, it is fed in at the top equally slowly.

A short length such as marked above draws out as it falls so that it occupies a greater length at subsequent positions. Now if we know the rate at which it is drawn out and the force acting upon it, we can calculate the viscosity just as in the direct experiments on traction.

In order to know the rate at which the material is being drawn out, two things are required. The amount of material passing down per second—this is easily found—and the slope of the surface to the vertical at the place in question. We can observe this on reflecting a beam of light from the surface of the column so as to pass out horizontally, by measuring the angle between the incident beam and the reflected one. So much for the rate of drawing out; now the

VISCOSITY in C. G. S. UNITS.

METHOD	SODA GLASS 575°	SODA GLASS 660°	SODA GLASS 710°	PITCH	PITCH TAR 7/8 : 1/8	PITCH TAR 3/4 : 1/4	PITCH TAR 1/2 : 1/2	TAR	TAR BENZ. 1/2 : 1/2	BENZINE
TORSION	1.1×10^{13}	2.3×10^{11}	4.5×10^{10}	1.04×10^{10}						
TRACTION				1.55×10^{10}						
BENDING BEAM				1.09×10^{10}	3.2×10^{7}	2.6×10^{5}				
FALLING STREAM					3.6×10^{7}	3.7×10^{5}				
TUBE						2.3×10^{5}	8.3×10^{3}	520	.24	.007

other quantity required, the force of traction, is simply the weight of the suspended column situated below the point. By cutting off the column, collecting and weighing, we get its amount. It is true that the value of this varies slightly, for if a stream is watched it will be seen to grow longer and then break off when it gets too heavy to be supported, somewhere higher up. However, this goes on only at the end of the stream where it is extremely thin, and the alteration in weight thus produced may be safely neglected.

From the table it will be seen that the value of the viscosity so obtained is, on the whole, in satisfactory agreement with that obtained by other methods. It will be noticed that in the case of the thinner liquid, the inertia term which was neglected has probably made itself felt.

[F. T. T.]

WEEKLY EVENING MEETING,

Friday, March 18, 1904.

SIR WILLIAM CROOKES, F.R.S., Honorary Secretary and Vice-President, in the Chair.

HENRY ARTHUR JONES, Esq.

The Foundations of a National Drama.

[ABSTRACT.]

AT the present moment we seem to be urged and beckoned on every hand to overhaul and reorganize our national resources, to set every room of our house in order.   There is a general instinct of alarm and uneasiness, and whatever may be the result of the present search into the causes and conditions of our national prosperity, it will not be without some effect in every sphere of English thought and action. Now, in whatever spheres it may be decided to abandon the doctrine and policy of *laissez-faire*, I hope the English drama may put in a claim to be rescued from its present state of national neglect and national contempt.   In that reorganisation of our national means and resources, in that refixing of our national aims and goals towards which we seem to be summoned, not merely by the warnings of statesmen and the shrill cries of contending politicians, but by those threatening, hovering portents—those pillars of cloud and fire that daily and nightly guide our nation to its destiny—in that awakening of new national hopes and ambitions and ideals, I hope I may put in a very urgent claim that the drama shall be recognized as a great civilizing and humanizing force, a great potential influence in our community, a great potential educator.

I use the word "educator" with much reluctance, knowing well that I shall be misunderstood and misrepresented by all those whose business and interest it is to keep the drama on its present level.   But in the widest and truest sense I claim that in a closely-packed democracy such as ours the drama is and must be an increasingly-powerful teacher, either of bad manners or of good manners, of bad literature or of good literature, of bad habits or of good habits.   Potentially it is the cheapest, the easiest, the most winning, the most powerful teacher of that great science which it so much concerns every one of us to know through and through, I mean the science of wise living.   In that supreme science, the drama is or should be a supreme teacher, a supreme educator.

I will beg leave then to affirm, on behalf of the myriads of amusement-seekers, that it is desirable to have a national English drama ; wisely regulated, wisely encouraged, thoroughly organized, suitably housed, recognized and honoured as one of the fine arts.

2 L 2

Clearly the first function of drama is to represent life and character by means of a story in action; its second and higher function is to interpret life by the same means. But the first and fundamental purpose of the drama is to represent life.

I think, if you will carefully listen to the remarks and judgments upon plays that come within your earshot during the next few months, even from cultivated men and women—I think you will come to the conclusion that the English playgoing public have for the most part lost all sense that the drama is the art of representing life, and that there is a keen and high pleasure to be got out of it on that level.

By the representation of life I do not mean that the drama should copy the crude actualities of the street and the home. Very often the highest truths of life and character cannot be brought into a realistic scheme. The drama must always remain, like sculpture, a highly conventional art; and its greatest achievements will always be wrought under wide, and large, and astounding conventions. Shakespeare's plays are not untrue to life because they do not perpetually phonograph the actual conversations of actual persons.

I have not time here to do more than explain in the briefest way that I am not contending for a realistic drama. In the past the greatest examples of drama have been set in frankly poetic, fantastic and unrealistic schemes. But whether a play is poetic, realistic or fantastic, its first purpose should be the representation of life, and the implicit enforcement of the great plain simple truths of life. Realistically, or poetically, or fantastically, it should show you the lives and characters of men and women; and it should do this by means of a carefully-chosen, carefully-planned, and always moving story.

Ten years ago, in the years 1893 and 1894, we seemed to be advancing towards a serious drama of English life; we began to gather round us a public who came to the theatre prepared to judge a modern play by a higher standard than the number of jokes, tricks, antics and songs it contained. To-day the English dramatist, who pays his countrymen the compliment of writing a play in which he attempts to paint their daily life for them in a serious straightforward way, finds that he is not generally judged upon this ground at all; he is not generally judged and rewarded according to his ability to paint life and character; he is generally judged according to his ability to amuse the audience without troubling them to think. And I believe that this tendency on the part of the English playgoers to demand mere tit-bits of amusement, and to reject all study of life and character in the theatre, I believe these tendencies and tastes have largely increased during the past ten years, and are still increasing. Insomuch we may say that the legitimate purpose of the drama, which is to paint life and character in a story; and the legitimate pleasure to be gained from the drama, that is to say, the keen and intellectual delight in watching a faithful representation of life and character and passion—this legitimate purpose and this legitimate pleasure of play-writing and playgoing are to-day swallowed up and lost sight of in the demand for mere thoughtless entertainment, whose one purpose is not to show the people their lives, but to provide them with a means of escape from their lives. That is to say, the purpose of the entertainments provided in our most successful theatres is indeed the very opposite to the legitimate purpose of the drama, the very negation and suffocation of any serious or thoughtful drama whatever.

I do not say that one or two of us may not get in an occasional success of a hundred and fifty nights with a comedy, or even with a play of serious interest, if by a miraculous chance one can get it suitably played. But any play of great serious interest, such as would meet with instant and great recognition and reward in France or Germany, is most likely to be condemned and censured by the mass of English playgoers as "unpleasant."

I am aware it is useless to condemn a man for not paying to be bored or disgusted. But the fact that he is bored and disgusted raises the further question: "Why is he bored and disgusted?" "What are the things that bore and disgust him?"

I question whether any subject has recently gathered around it such a thick fungus of cant and ignorance as that of the "problem play." For a number of years past the parrot-phrase "problem play" has been applied to almost every play that attempts to paint sincerely any great passion, any great reality of human life. No doubt great extravagances and absurdities were committed by the swarm of foolish doctrinaire playwrights who tried to imitate Ibsen. But the stream of just contempt that was poured upon these absurdities has run over its bounds, and has almost swamped all sincere and serious play-writing in England.

What are the necessary foundations of a national English drama?

Speaking through you to the great body of English playgoers, I would say to them: If we are to have an English drama at all, it is necessary—

1. To distinguish and separate our drama from popular amusement; to affirm and reaffirm that popular amusement and the art of the drama are totally different things; and that there is a higher and greater pleasure to be obtained from the drama than from popular amusement.

2. To found a national or répertoire theatre where high and severe literary and artistic standards may be set ; where great traditions may be gradually established and maintained amongst authors, actors, critics and audiences.

3. To insure so far as possible that the dramatist shall be recognized and rewarded when and in so far as he has painted life and character, and not when and in so far as he has merely tickled and bemused the populace.

4. To bring our acted drama again into living relation with English literature; to dissolve the foolish prejudice and contempt that literature now shows for the acted drama; to win from literature

the avowal that the drama is the most live, the most subtle, the most difficult form of literature; to beg that plays shall be read and judged by literary men who are also judges of the acted drama. To bring about a general habit of reading plays, such as prevails in France.

5. To inform our drama with a broad, sane, and profound morality; a morality that neither dreads nor wishes to escape from the permanent facts of human life, and the permanent passions of men and women; a morality equally apart from the morality that is practised amongst wax dolls; and from the morality that allows the present sniggering, veiled indecencies of popular farce and musical comedy.

6. To give our actors and actresses a constant and thorough training in widely varied characters, and in the difficult and intricate technique of their art; so that in place of our present crowd of intelligent amateurs, we may have a large body of competent artists to interpret and vitalize great characters and great emotions in such a way as to render them credible, and interesting, and satisfying to the public.

7. To break down so far as possible, and at any rate in some theatres, the present system of long runs with its attendant ill-effects on our performers; to establish throughout the country répertoire theatres and companies, to the end that our actors may get constant practice in different parts, and to the end that the author may see his play interpreted by different companies and in different ways.

8. To distinguish between the play that has failed because it has been inadequately or unsuitably interpreted, and the play that has failed on its own demerits; to distinguish between the play that has failed from the low aims or mistaken workmanship of the playwright, and the play that has failed from the low tastes of the public, or from the mistakes of casting or production.

9. To bring the drama into relation with the other arts; to cut it asunder from all flaring advertisements, and big capital letters, and from all tawdry and trumpery accessories; to establish it as a fine art.

You will have noticed that many of these proposals overlap and include each other. Virtually they are all contained in the one pressing necessity for our drama that it shall be recognized as something distinct from popular amusement. And this one pressing necessity can be best and most effectually met by the fostering of the drama as a national art in a national theatre.

[H. A. J.]

GENERAL MONTHLY MEETING,

Monday, April 11, 1904.

Sir JAMES CRICHTON-BROWNE, M.D. LL.D. F.R.S., Treasurer and
Vice-President, in the Chair.

William Baker Anderson, Esq.
Joseph Benson, Esq.
Mrs. Douglas Cow,
Mrs. J. Mackenzie Davidson,
John Archibald Watt Dollar, Esq.
Bayntun Hippisley, Esq., J.P.
Edward William Linging, Esq.
Mrs. Gilbert Master,
Mrs. Guy E. Broün-Morison,
Julius C. Prince, Esq.
Ernest Angelo Short, Esq.
W. A. Watson-Taylor, Esq.
Henry Letheby Tidy, Esq.
Charles Selby Whitehead, Esq.

were elected Members of the Royal Institution.

The Special Thanks of the Members were returned to Francis
Gaskell, Esq., M.A. *M.R.I.*, for his Donation of £50 to the Fund for
the Promotion of Experimental Research at Low Temperatures.

The PRESENTS received since the last Meeting were laid on the
table, and the thanks of the Members returned for the same, viz.:—

FROM

The Lords of the Admiralty—Nautical Almanac for 1907. 8vo.
The Secretary of State for India—Kurukh (Orää) -English Dictionary. By F.
  Hahn. Part I. 8vo. 1903.
  Report on the Kodaikanal and Madras Observatories for 1903. fol. 1904.
  Progress Report for the Archæological Survey Circle, United Provinces. (With
  Photographs and Drawings.) fol. 1903.
Accademia dei Lincei, Reale, Roma—Classe di Scienze Fisiche, Matematiche e
  Naturali. Atti, Serie Quinta: Rendiconti. 1° Semestre, Vol. XIII. Nos. 4-6.
  8vo. 1904.
Agricultural Society of England, Royal—Journal. Vol. LXIV. 8vo. 1904.
American Academy of Arts and Sciences—Proceedings, Vol. XXXIX. Nos. 13-15.
  8vo. 1904.
  Memoirs, Vol. XIII. No. 1. 4to. 1904.
American Geographical Society—Bulletin, Vol. XXXVI. No. 1. 8vo. 1904.
Antiquaries, Society of—Archæologia, Vol. LVIII. Part 2. 4to. 1903.
  Proceedings, Vol. XIX. No. 2. 8vo. 1903.
Astronomical Society, Royal—Monthly Notices, Vol. LXIV. No. 4. 8vo. 1904.
Automobile Club—Journal for March, 1904. 4to.

Bankers, Institute of—Journal, Vol. XXV. Part 3. 8vo. 1904.
Belgium, Royal Academy of Sciences—Bulletin, 1904, Nos. 1-2. 8vo.
Boston Public Library—Monthly Bulletin for March, 1904. 8vo.
British Architects, Royal Institute of—Journal, Third Series, Vol. XI. Nos. 10-11.
    4to. 1904.
British Astronomical Association—Journal, Vol. XIV. No. 5. Memoirs, Vol. XIII.
    Part 1. 8vo. 1904.
Buenos Ayres, City—Monthly Bulletin of Municipal Statistics, Jan. 1904. fol.
Canadian Government (Department of the Interior)—Maps of South-Eastern
    Alaska, North-West Territories and Manitoba. 1904.
Chemical Industry, Society of—Journal, Vol. XXIII. No. 5. 8vo. 1904.
Chemical Society—Journal for March, 1904. 8vo.
    Proceedings, Vol. XX. Nos. 277-278. 8vo. 1904.
Dax, Société de Borda—Bulletin, 1903, 3e Trimestre. 8vo.
Edinburgh Royal Society—Proceedings, Vol. XXV. No. 1. 8vo. 1904.
Editors—American Journal of Science for March-April, 1904. 8vo.
    Analyst for March-April, 1904. 8vo.
    Astrophysical Journal for March, 1904.
    Athenæum for March, 1904. 4to.
    Board of Trade Journal for March, 1904. 8vo.
    Brewers' Journal for March, 1904. 8vo.
    Chemical News for March, 1904. 4to.
    Chemist and Druggist for March, 1904. 8vo.
    Electrical Engineer for March, 1904. fol.
    Electrical Review for March, 1904. 4to.
    Electrical Times for March, 1904. 4to.
    Electricity for March, 1904. 8vo.
    Electro-Chemist and Metallurgist for March, 1904. 8vo.
    Engineer for March, 1904. fol.
    Engineering for March, 1904. fol.
    Engineering Review for March, 1904. 8vo.
    Homœopathic Review for March, 1904. 8vo.
    Horological Journal for March, 1904. 8vo.
    Journal of the British Dental Association for March, 1904. 8vo.
    Journal of Physical Chemistry for Feb. 1904. 8vo.
    Journal of State Medicine for March, 1904. 8vo.
    Law Journal for March, 1904. 8vo.
    London Technical Education Gazette for March, 1904. fol.
    London University Gazette for March, 1904. fol.
    Machinery Market for March, 1904. 8vo.
    Model Engineer for March, 1904. 8vo.
    Mois Scientifique for March, 1904. 8vo.
    Motor Car Journal for March, 1904. 8vo.
    Musical Times for March, 1904. 8vo.
    Nature for March, 1904. 4to.
    New Church Magazine for March, 1904. 8vo.
    Nuovo Cimento for Nov.-Dec. 1903 and Jan. 1904. 8vo.
    Page's Magazine for March, 1904. 8vo.
    Photographic News for March, 1904. 8vo.
    Physical Review for March, 1904. 8vo.
    Public Health Engineer for March, 1904. 8vo.
    Science Abstracts for March, 1904. 8vo.
    Terrestrial Magnetism for Dec. 1903. 8vo
    Zoophilist for March, 1904. 4to.
Florence, Biblioteca Nazionale—Bulletin, March, 1904. 8vo.
Franklin Institute—Journal, Vol. CLVII. No. 3. 8vo. 1904.
Geneva, Société de Physique—Mémoires, Vol. XXXIV. Fasc. 4 4to. 1904.
Geographical Society, Royal—Geographical Journal, Vol. XXIII. Nos. 3-4. 8vo.
    1904.

Geological Society—Abstracts of Proceedings, Nos. 791-792. 8vo 1904.
Harlem, Société Hollandaise des Sciences—Archives Néerlandaises, Série II. Tome XIX. 1e et 2e Livraisons. 8vo. 1904.
Harper & Bros., Messrs (the Publishers)—The Rise and Progress of the Standard Oil Company. By G. H. Montague. 8vo. 1904.
Manchester Geological Society—Transactions, Vol. XXVIII. Parts 10-12. 8vo. 1904.
Massachusetts Institute of Technology—Technology Quarterly, Vol. XVI. No. 4. 8vo. 1903.
Mexico, Instituto Geologico—Parergones, Tom I. No. 1. 8vo. 1904.
Mill, H. R., Esq , D Sc. (the Author)—Mean and Extreme Annual Rainfall over British Isles. 8vo. 1904.
    The Great Dustfall of February, 1903, and its Origin. 8vo. 1904.
Montana University—Bulletin, Biological Series, No. 6. 8vo. 1903.
    President's Report, 1902-3. 8vo.
Munich, Royal Bavarian Academy of Sciences—Abhandlungen, Vol. XXI. Part 1. 4to. 1904.
Navy League—Navy League Journal for April, 1904. 8vo.
North of England Institute of Mining and Mechanical Engineers—Transactions Vol. LIII. No. 2. 8vo. 1904.
Odontological Society—Transactions, Vol. XXXVI. No. 4. 8vo. 1904.
Paris, Société Française de Physique—Bulletin des Séances, 1903, Fasc. 3-4. 8vo.
Pennsylvania, University of—Catalogue, 1903-4. 8vo.
Pharmaceutical Society of Great Britain—Journal for March, 1904. 8vo.
Photographic Society. Royal—Photographic Journal, Vol. XLIV. No. 2. 8vo. 1904.
Rome, Ministry of Public Works—Giornale del Genio Civile for Sept.-Oct. 1903. 8vo.
Royal College of Physicians—Catalogue of Accessions to the Library. 8vo. 1903.
Royal Irish Academy—Transactions, Vol. XXXII. Sec. A, Part 10. 4to. 1904.
Royal Society of London—Philosophical Transactions, A, Nos. 362-363; B. Nos. 224-226. 4to. 1904.
    Proceedings, No. 491. 8vo. 1904.
Selborne Society—Nature Notes for March, 1904. 8vo.
Smith, B. Leigh, Esq., M.A. M.R.I.—The Scottish Geographical Magazine, Vol. XX. No. 4. 8vo. 1904.
Smithsonian Institution—Miscellaneous Collection, Vol. XLIV. No. 1374; Vol. XLV. (Quarterly Issue), Parts 1-2. 8vo. 1903-4.
Society of Arts—Journal for March, 1904. 8vo.
Statistical Society, Royal—Journal, Vol. LXVII. Part 1. 8vo. 1904.
Tacchini, Prof. P., Hon. Mem. R.I. (the Author)—Memorie della Società degli Spettroscopisti Italiani, Vol. XXXIII. Disp. 2. 4to. 1904.
Toronto University—Studies: Physical Science Series, No. 3; Physiological Series, No. 4. 8vo. 1903.
Transvaal, Agricultural Department—Journal for Jan. 1904. 8vo.
United Service Institution, Royal—Journal for March, 1904. 8vo.
United States Department of Agriculture—Monthly Weather Review for Dec. 1903. 4to.
    Weather Bureau, Bulletin L., Climatology of California. By A. G. McAdie. 4to. 1903.
    Experiment Station Record, Feb. 1904. 8vo.
United States Patent Office—Official Gazette, Vol. CIX. Nos. 1-3. 4to. 1904.
Verein zur Beförderung des Gewerbfleisses in Preussen—Verhandlungen, 1904, Heft 3. 4to.
Vincent, Miss E.—Haydn's Dictionary of Dates. 23rd ed. 8vo. 1904.
Western Society of Engineers—Journal, Vol. XIX. No. 1. 8vo. 1904.
Woodhouse, A. J., Esq , M.R.I.—Transactions of the New Zealand Institute Vols. I.-XXXV. 1868-1903. 8vo.
Zoological Society—Proceedings, 1903, Vol. II. Part 2. 8vo. 1904.

WEEKLY EVENING MEETING,
Friday, April 15, 1904.

Sir JAMES CRICHTON-BROWNE, M.D. LL.D. F.R.S., Treasurer
and Vice-President, in the Chair.

The Right Hon. and Right Rev. Monsignor
THE COUNT VAY DE VAYA AND LUSKOD, D.P.H.H. K.C.I.C.

First Impressions of Seoul.

(FRAGMENT OF LECTURE ON "KOREA AND THE KOREANS.")

. . . . At last I arrive safely in Seoul. It is eventide and the
moon is just appearing. In the dimness the most desolate imperial
residence in the world seems still more desolate, more wretched,
miserable and deserted.

My sedan chair is being carried through a long street, or rather
road.

Small houses stand on either hand, but houses they cannot be
called—those I have seen up to the present can at the best be termed
"hovels"—at last we reach the walls of the inner city; for till now
we have been merely in the outer town. The wall is ragged and
thorny. In front stand a number of roofed and painted gates, I
almost imagine myself back in Peking, for the picture is a replica,
but in miniature. However, I am unable in the dusk to see how
much smaller it is. The general effect is the same, imprinted with
the familiar Chinese characteristics.

The moon is now shining brightly, but it shows nothing new in
the aspect of the road within the walls. The main street of Seoul is
as deep in clay and mud as it was at the Creation, when the "waters
dried up." Its houses have not altered; they are no more than the
clay huts of prehistoric man, his protection against cold or heat.

I requested the bearers of my chair to walk slowly; I did not
wish to lose my first impression. The first sight of an unknown
country stamps itself on our minds in a manner unique. There is a
fascination in the unknown—a wonderful interest attached to the
unexpected. Our wanderings amongst strange peoples in the streets
of a strange city are not for the pen to describe.

Everything that is uncommon is mysterious until reality tears
aside the veil; and as long as it is built up by our imagination and
peopled by her fantastic creations, so long does it remain a City of
Dreams.

The streets are getting broader and the clay huts grow even more

RECEPTION HALL IN THE OLD PALACE.

insignificant.   I stop for a moment in the great square ; it may be the centre of the city, but is little more than a cross-road leading into a few side-streets.

It is scarcely seven o'clock and yet over all broods a death-like silence, a peaceful calm, as complete as one can imagine.   The broad streets seem an immense cemetery and the mean little flat-roofed houses graves.   One might think it is All Saints' Day for on each grave a little lamp is burning.   A lantern hangs from each eave, showing a yellowish flame.

But the people themselves—like ghosts they are returning to their homes, each robed in white, each and all mute.   Without a sound they flit over the roads of this endless graveyard, until they disappear into the depths of some one of the illuminated tombs.

I have never been so impressed by any other city I have seen as I was by my first sight of Seoul.   As I saw the city just now, by the light of a November moon, dark, dumb, desolate and ghostly, it resembled more some fairy city than reality.   Almost like those storied places sung of in the poetry of almost every people, whose tale is listened to with such rapture by the little folk of the nursery who know nothing as yet of life's seamy side.

Such a town was Seoul to me, the first few hours after my arrival.

The Dawn.

I am aroused by the sound of drums and trumpets.   Of whom ? Of ghosts ?   What can have happened that the house of silence should have been disturbed by such an awful uproar ?

I hasten to my window.   The long street, the square, every inch of ground is occupied by soldiers.   These are short and yellow, wearing a black uniform, the black cloth set off by a broad red collar. The black coats, red collars and yellow faces make a motley colour-scheme, almost as though they were checkered.   The men seem to like it.

If the mixture serves no other purpose it offers an excellent target for an enemy, which was probably the idea of its inventors.

The din continues.   The trumpets blare and these black, red and yellow little people, like tin soldiers, keep moving before me.   To and fro, up one street and down another they go, like property soldiers, now appearing on and again disappearing from the stage— always the same supers, but one would think they were a mighty army.

And all the time the bayonets flash on the rifle barrels, whose weight seems rather too much for the little men.   The drums still beat and fanfares ring out on the frosty morning.

What has happened ?   Has the coronation not been postponed after all ?   Is the Emperor at last inaugurating the festivities so long looked forward to ?

I ring the bell, and a servant wearing a pig-tail wound up in a knot and dressed in white enters. His long coat is of linen, his head covered by a bell-shaped hat of horse-hair, which resembles in shape the glass lid used to protect preserves from flies.

This quaint servant seems more surprised at my question than I at his livery.

"But the army has been reorganised by European officers. It has been taught, in the Western style, to march, manœuvre and kill, and for the performance of this gay farce, seats have been erected, and now you, a European, coming from the West ask with obvious irony, what does all this mean ? "

I can see how amusing the whole situation is, and what a ludicrous side it has. The fact of the collar being a few inches deeper, or of the colour of the uniform, does not alter the character of that uniform ; it is still a distinctive mark, even in its best edition.

The rifle always destroys, whether the mechanism is new or old that projects the ball, and whether or no a soldier is a couple of feet taller—with a yellow or a white complexion—his calling is always a gloomy one. For do we not consider that soldier most efficient who destroys the greatest number of lives ?

Dawn now turns out into morning and the doors of the shops open one by one. Most of them are only protected for the night by mats or a few planks. Later on the customers begin to arrive, all of them dressed in white. Men and women alike wear long linen coats (kaftans), and their lined foot-gear is also of linen—in fact they are white from top to toe, excepting the black hat of horsehair.

Now and again I see a sedan chair, which however is not larger than a good-sized box—its occupant huddled up inside. I cannot see any carriage, trap or horse, in spite of the growing traffic, which however, is perfectly noiseless. Perhaps this may account for the fact of my still being under the impression of being in a deserted city.

My First Walk through the City.

It is generally on the first day that we catch the most characteristic traits, or at any rate the most salient features strike our imagination. When our perceptive powers are still fresh, they are most influenced by little peculiarities.

After breakfast I go out for a stroll, and find the palace gate in front of me, outside which some soldiers are standing. Beyond it stretches a long street towards which I turn. This is the same street which yesterday resembled a vast graveyard. The houses now stand open, as the wooden wall, looking on the street, has been removed. There are a considerable number of shops, but small and mean,. displaying no wares that attract my attention. The cabinet makers make the best show, consisting of small chests, with brass ornamentation, having large polished locks. These are no less quaint than they are tasteful.

They seem to sell well, for in a whole row I can see nothing else.

South City Gate, Seoul

the King.

SEOUL.

There is also no lack of seeds, but the baskets do not offer a quarter of the variety of a Chinese grocery. I do not think I saw any more shops, at least any that I remarked. They seemed small and empty, never more than a couple of customers in the shop.

What attracted especially my attention was the large number of sentry-boxes. Every five or ten yards you came across a box with a stubby black, red and yellow soldier inside, armed!

No matter where I turn there are sentry-boxes everywhere, to the right, to the left, in front and behind me. Can it be a fact that a whole army is required to keep this little folk in order?

Street Life.

No sooner had I put this question to myself than I perceive a disturbance. Some coolies carrying vegetables engaged in a battle royal, and two boys pitching into each other. But the private stands there unmoved. His look seems rather to approve than condemn. He is evidently not intended to keep the peace; this does not seem part of his duties. So the coolies may fight as much as they like among the cabbages. (The group, by the way, forms an interesting picture, the coolies in white with the green loads on their backs in the thick of the fray.) The smaller of the boys commences to cry as blood is dripping from his forehead, but the soldier is not affected by the sight of this either. I wonder if what he just muttered was that the "Red Cross" was not his business?

As I went on I heard more screaming and quarrelling, and witnessed a few more little skirmishes. It was not until now that I realised how unaccustomed I was to quarrels and fights, as in China I never saw one man fighting another. They have there a civilisation of thousands of years to thank for that.

The Emperor's New Hall.

I now approach a hall which is being repaired; it has a pointed roof and broad eaves, similar to those of the Palace at Pekin.

Quite a forest of wood is stored up there in the shape of beams. As I see with what precision the workmen make the various parts fit together without the use of nails, I am delighted that the traditions of ancient architecture are not yet extinct.

I am now in the neighbourhood of the Royal Palace. In front of the main gate is a large square which further on turns into a street, with public buildings on either side. These are the ministerial offices, where is spun the web of the Korean government.

The Old Palace.

Externally, the Palace has little to distinguish it. The façade is rather low and the walls are mud coated, while the gates are not much better, in the Chinese style and crowned by tiles. The gates, which are wide open, lead into a large inner courtyard, where there

are a number of ordinary and state sedan-chairs. Crowds of servants,
attendants and coolies are warming themselves in the sun ; others are
playing at ball, which they kick off and catch with their legs.

In the middle of the street you meet mandarins hurrying to their
offices, magistrates and other men of consequence, most of them in
chairs, or rather boxes. These are carried by two servants. The
vehicle is covered with a cloth, that of the better class matching in
colour the servants' liveries. I have seen grey and yellow ones also.
These belong to the Korean aristocracy.

The most attractive of all was the "carriage" of a noble in
mourning. His chair had quite recently been covered with cloth
of a yellowish hue, which the two servants also wore, their coats
reaching nearly to the ground. In order to give their limbs free
play these had been split up as far as the waist. But this can be
nothing more than fashion, for not even the whip would make a
Korean hurry. They also wear a broad girdle tied up in a bow
round the waist.

When in mourning they wear straw hats, but not black, which
are shaped like a fair sized orthodox bread-basket. These have wide
brims reaching the shoulders and entirely concealing the face. In
such a weird costume they strongly resemble a yellow mushroom
sprung up on a summer's day. Straw sandals complete the costume.

In spite of these strange details and absurd combinations the
general effect is good ; the colours, the silk-covered chair, straw hat
and sandals blend harmoniously together. Seen from a distance they
all have the appearance of ivory knick-knacks, such as you see exhibited
for sale in Japanese curio shops.

A Korean Wedding.

But I hear a noise in the distance, and from the direction of the
western gate a motley crowd comes towards me. It must be either a
funeral or a wedding. So far I cannot distinguish which. The next
moment two children detach themselves from the crowd. They seem
to lead the procession. Their dress is glaring, of green, purple and
scarlet silk, with their dark hair encircling their foreheads in gleam-
ing plaits. They are also decked out with flowers and butterflies.

Behind them a large box painted red and polished is carried.
This is evidently the dowry. Now follow the dancers, in pairs, but
widely apart from each other. Their costume—I cannot describe it !
Almost shapeless, it consisted of skirt over skirt, kerchiefs, veils, all
pell-mell, and of every colour of the rainbow.

I take note of many things which to-morrow might escape me.

Street life is one ever flowing stream. In Seoul, I observe, everybody lives on the thoroughfares. That is probably the reason why
its streets are so wide and the dwellings so cramped. In this trait
the Korean is like the Spaniard or Italian. He is never so happy as
when out of doors. There he stands on his threshold, or basking in
the sunny courtyard, or he lights his pipe and strolls up and down

COREAN PEASANTS.

for hours. His carriage is slow and stately. I wonder where he is going to and what he is thinking of—nowhere and of nothing—"il flâne." There is no suitable word in another language for this aimless meandering. "Loitering" indicates only physical slowness, and moral vacuum is not simultaneously connoted by it.

The Korean T. Atkins.

Now and again a private comes by. He is the coming man! If he learns nothing in the barrack-yard, he does learn how to walk.

He has had his pigtail shorn. At first he bemoaned it; for this head-dress of his embodied a general principle. With its departure he was cut adrift from all his old associations and traditions.

But like the child he is at heart, he soon forgets his pigtail and its traditions along with it, and to-day is proud of the metamorphosis.

As the man of progress and of the future he scorns the white coats, sandals and hats of his countrymen.

A Korean School.

From a small house at the corner a very babel of sound issues forth. It is the inarticulate mechanical repetition of one chapter— exactly the same method our own schoolmasters used to employ for instilling knowledge.

As the door in the courtyard is open I enter. In front of me I find a room, not more than 10 feet square, in which ten or more youngsters are crowded together. There they sit on the floor, dressed in green instead of white, and their long hair hanging down in fine plaits.

Each has a big A B C book in his hand. Every word has a different letter. These they repeat, and in this way knowledge is driven into them. They pronounce everything out loud, moving all the time the upper part of their body to right and left, backwards and forwards.

The dominie is seated in front; he also is squatting on the floor. His eyes are shielded by goggles of enormous size, and he wears on his head a horse-hair crown.

He is wisdom personified, outwardly at any rate. His thoughts seem to be ranging far away in the distance, and from his olympic seat he casts an indifferent eye on his perspiring pupils.

But, as a famous Chinese pedagogue says, Chinese spelling and writing can only be mastered mechanically. His best scholar is the jackass.

The R. C. Mission.

From there to the Mission is but a few yards. As I enter its iron grilled gate my surprise is as great as agreeable. For I see before me a grand cathedral, and on either side spacious buildings standing in their own wooded grounds.

It was built on the model of one of the old cathedrals in the Netherlands—red brick, Gothic, a style I do not like to find in the

East. But this is only a shortcoming in my artistic sense; as a building nothing can be said against it. In its way it is perfect. But what struck me most was its cleanliness. The stone floor was as bright as a mirror.

The bishop is away on circuit and will return only in ten days, so the vicar received me. He showed me over the whole little colony, the school, and convent and orphanage. But of these I will speak more fully elsewhere.

As I take my leave the sun was setting. The peaks of the engirdling hills were reflected in purple tints on the topaz sky. The Mission down below in the dell appeared in a bluish mist, only the cathedral cresting the hill.

Returning home by a circuitous route I found the streets even more thronged than in the morning. I glanced into a few shops, but there is not much worth seeing. The furriers seem to be the busiest. They are cutting out and sewing a number of tunics, capes and fur coats. There are also a good many jackets, and still more without sleeves to protect the chest and back. Over these they wear thin white linen kaftans. No wonder they look like walking eider-downs.

To the right I noticed a tavern, much like the Chinese roadside inn. In the large open stable a row of small rough-haired horses were standing with straw rugs on their backs. A coolie was carrying water from the well in two brass vessels hanging on the ends of a long pole.

The pole does not, however, rest upon his shoulders, but is fastened crosswise to his back. Man and load have the appearance of a living pair of scales.

Next came some unpretentious little barracks, which in their smallness are after the pattern of the soldiers, a number of whom are looking out of the windows. In the absence of any better occupation they are chewing pumpkin seeds.

Now we arrive at the curiosity shops—several porcelain, a few bronze articles, many tiles and a farrago of rubbish.

On the cross-road some more barracks—a long low building. The little men in front of it were wearing not only red collars but also red dolmans. Here the cavalry are garrisoned. A little scrap of a hussar was just galloping home. This warrior is not a whit taller than Hop-o'-my-Thumb—his charger scarce larger than a well-developed calf of two months. By the side of this toy hussar rattled a formidable sabre which seemed in danger of pulling him down from his horse. His seat is without that poor enough. On his coming nearer I saw that the murderous instrument is an ordinary cavalry sword. His uniform is the most checkered I ever saw, though in this respect all European nations are conspicuous enough.

The dolman of the Korean hussar is of a cinnamon colour, his collar and cuffs emerald green, and his breeches' stripes saffron. If it was the plumage of a parrot that served the model they have attained it most effectually.

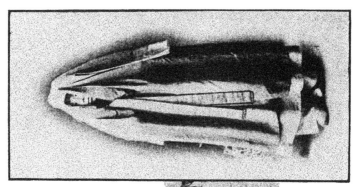

STREET SCENE.

Canine Street Police.

I was wandering further on, when in front of a gate some dogs nearly knocked me down.

The streets of Seoul, like those of Pekin and Constantinople, are full of them, but the dogs here are well kept and strong. If a single one of these starts barking, this signal of some approaching danger is in a minute responded to throughout a whole quarter. It was so in my case. As I came too near the threshold the guardian on duty there was under the impression that I intended to encroach on his domain. His attitude towards me was anything but friendly, and not being armed with either stick or umbrella I instinctively stooped down to pick up a stone. This movement on my part, however, was sufficient to make him drive me summarily into his own courtyard.

Apropos of the Korean canine race, the subject is worthy of a few words, because they are some of the most typical figures in the streets of Seoul. I must confess I never have seen better trained dogs than these. In the streets they are the meekest of quadrupeds and as quiet as lambs.

A single word is quite sufficient for the Seoul dog to make him scamper home to his doorway. He knows that it is his duty to be there. He will lie in the little yard for hours and hours, but prefers, best of all, to take his ease on the doorstep with his head in the street, so as not to lose sight of anyone approaching. He hardly takes any notice of you as long as you walk in the middle of the road. The farthest he would go is to stare at dark-clothed people with other than yellow faces, to the sight of whom he is not accustomed, as ever since he came into this world he has seen none other than white kaftans.

But the moment we direct our steps towards the house he gives a growl or two, and upon your approach barks as loud as he can. He reserves his attack until you are about a yard within his range. By that time the auxiliary forces from the neighbourhood have concentrated, and you have the whole brigade snarling and yapping at your heels. This fearsome pandemonium brings the master of the house, or a member of his family, to the seat of the disturbance, and a single word or merely a sign suffices for the Cerberus to retire to a corner wagging his tail.

The Evening.

Darkness has set in. Calmness reigns supreme. The fresh autumnal night is silently spreading its grey veil of mist over the white city. But behold. Is it not the northern light that breaks through the dark? In the direction of Puk-Han it begins to dawn. The sky unexpectedly flashes up; its grave red light is getting more and more acute. Now flames of hundreds of torches illuminate the atmosphere. Another surprise. As if the many strange phenomena of the day had not yet reached their climax. A torchlight procession, the like of which I have not seen before. Pedestrians, sedan-chairs, men on horseback, are coming forward in an endless row. And what a pageant this is! What an effective group! The minutest detail has been carried out with artistic taste. The smallest traits are wonderfully grouped together to enhance the general effect.

The procession is headed by children, dressed in white from top to toe, wearing bell-shaped headgear. Then follow bearers of torch-lights and banners, servants carrying texts attached to poles, others dangling lanterns, and behind these another group burning straw plaits.

The next section of the procession consists of riders, of whom eight are entirely covered in white cloaks. You would imagine they were phantoms, if it were not that they wept bitterly. These are the paid mourners, like the moaning women of ancient Rome. It is a native funeral. A member of the Min family is being taken to his last resting place. He is a descendant of a famous clan, a relative of the late Empress of Korea, so regal pomp is awarded him. And the funeral procession is really grand, although all dresses worn therein are of unbleached linen. The trimmings are for the most part of paper, but in such striking combinations, and designed and finished so perfectly, that we disregard the details and only admire the general effect. The group of moaning women is followed by monsters dressed as guys, such as gruesome fables are peopled with. One wears a red masque, another a yellow, this a green and that a blue one. The appearance of all is awe-inspiring, their heads being adorned with horns, cocks-combs and crowns. Now more and more new groups follow, approaching stately and disappearing slowly in the darkness of the night.

A Princely Funeral.

How long the procession lasted I know not, but some thousand persons must have marched by ere the two gilt catafalques appeared on the scene. Both were alike, resembling monumental pagodas, gabled in many places, designed with the quaint originality of this people, and ornamented with all the fulness of its fancy. The two coffins, prescribed by ancient traditions, rest on pedestals in the shadow of high baldachinos. Behind the coffin walks a person wrapped in sackcloth, suggestive of the cloth worn over their uniforms by mem-bers of the Society of Misericordia in Italy. The catafalques and coffins are carried on their shoulders by thirty-two mourners, proceed-ing slowly and rhythmically.

But the pageant is not yet at an end. On a number of sedan-chairs are heaped up the personal belongings of the defunct. His clothes, household furniture, horses and cows—all follow him so that they may be consumed as a burnt-offering by his graveside; all in *effigy*, for they are but of paper. It is in such cheap edition that the ancient traditions are being preserved by the more practical progeny of the present day. The silver coins, thrown by the riding *weepers* amongst the crowd, are likewise make-believe, representing nothing

MARBLE TOWER.

FERRY BOAT.

but small discs of paper. One sedan-chair follows another; hosts of carriers and servants accompany the members of the family. There is the whole tribe; a whole brigade is riding behind the gabled catafalque. All are covered with sackcloth, even the mendicant is dressed in white—the whole procession is white. And as they turn round on top of the hill the effect of the picture is unique. The weeping women, the monsters, the mourners and attendants, the gigantic catafalques and the immense crowd were one of the strangest sights I ever contemplated. The furled banners, dangling texts, open sunshades, lanterns with dim lights in the darkness of the night formed the quaintest setting. The light of torches, the burning fascicles of bulrushes and straw are tinting in a vibrating red the long, white and ghostly procession. The beating of drums, the droning of bagpipes furnish the music, and the weeping women the proper chorus; this strange funeral, in fact, is the most perfect "danse macabre."

The full moon is rising slowly and stately behind the hills, fuller than usually, as though anxious to light up the weird procession; her melancholy light filters through the night, and her silvery rays intensify the ghostliness of the scene.

The first day spent in the capital of Korea is nearing its end. Quietness penetrates the night, such profound quietness as can only be enjoyed in Seoul. As I am walking homeward, the alley leading to the Legation is dark and deserted, and I try to recall to my memory all that I have perceived and heard, all that was new to me and striking; all the contrasts and the incoherency of the earliest perceptions.

I will write it down forthwith, ere *knowledge* spoils the glamour of *first impressions*, whilst every tint is shining in glaring colour, whilst every detail can be observed through the microscope of novelty.

On the last day of my sojourn here I will look through these short notes, and correct, in red ink, any mistakes that may be found therein. Town and people will thus be better known, but the charm of the first day will vanish for ever.

WEEKLY EVENING MEETING,

Friday, April 22, 1904.

GEORGE MATTHEY, Esq., F.R.S., Vice-President, in the Chair.

COLONEL DAVID BRUCE, R.A.M.C. F.R.S.

Sleeping Sickness in Uganda.

FIRST allow me to remind you of the general position of Uganda in Central Africa, which is represented in the following map (Fig. 1). The port of entry into the country is Mombasa, the Uganda railway running from here to Victoria Nyanza. On the north-west shore of the lake is Entebbe, the seat of the English Government in the Uganda Protectorate. Kampala, or Mengo, the native capital, lies some 20 miles to the north-east. Uganda proper lies to the north-west of the lake, Ankole and Unyoro to the west of Uganda, and Busoga to the east. The other lakes, Albert Nyanza, Lake Albert Elward and Lake Tanganyika, form the boundary between the Uganda Protectorate and the Congo State. To the east of Busoga is British East Africa.

The portion of the map which is shaded with horizontal lines represents the part of the country in which sleeping sickness is raging—that is the sleeping sickness area. And first let us consider how the disease was introduced into the country. There are various theories in regard to this. It is quite impossible, in my opinion, that the disease could have been indigenous in the country. None of the chiefs or missionaries, who have been many years in the country, ever saw a case of the disease before the year 1901. In April of that year the Drs. Cook, Medical Missionaries at Kampala, reported the first case.

It first broke out in the part of the country lying to the east, called Busoga. Dr. Moffatt, C.M.G., the Principal Medical Officer of Uganda, is of opinion that the disease was introduced into this part of the country when Emin Pasha's Soudanese and their wives and followers, numbering some 10,000, were brought into and settled in Busoga. These natives were brought from the edge of the Congo territory lying to the west, and therefore from a country in which

---

NOTE.—The maps, tables and illustrations in this paper are taken, with the permission of the Royal Society, from the Further Report on Sleeping Sickness in Uganda, by Lieut.-Col. David Bruce, R.A.M.C. F.R.S., David Nabarro, M.D., and Capt. E. D. W. Greig, I.M.S. (Harrison & Sons, London).

The illustrations showing the parasites have been kindly lent by the 'British Medical Journal.'

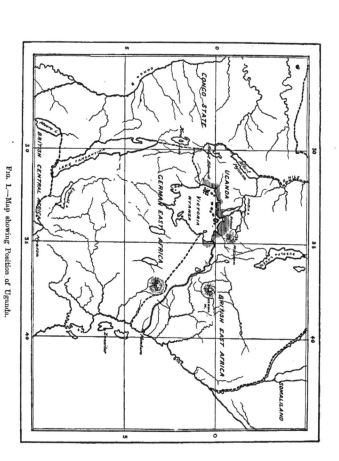

Fig. 1.—Map showing Position of Uganda.

sleeping sickness has been endemic for an unknown time. It seems, then, quite probable that some of these natives, brought in with the remains of Emin Pasha's expedition, may have brought the disease into Busoga, and that from this focus it slowly spread to the neighbouring population. Be that as it may, the disease broke out in this part of the country some time between 1896, according to Dr. Hodges, and 1901, when the disease was definitely diagnosed, and in a short time reduced a populous and richly cultivated country to a depopulated wilderness.

Now, having discussed the introduction of sleeping sickness into Uganda, let me for a few minutes draw your attention to the disease itself. Sleeping sickness is a curious disease, and is essentially a disturbance of the functions of the brain. A slow chronic inflammatory process takes place in the brain substance, which after a time gives rise to the peculiar symptoms of the disease. But for a long time, sometimes years, the preliminary symptoms of sleeping sickness may be of so slight a character that no one suspects there is anything wrong. That is to say, the sleeping sickness patient may go about doing his ordinary work for years without his friends noticing there is anything the matter. But gradually a slight change in his demeanour becomes evident; he is less inclined to exert himself; he lies about more during the day, and at last his intimates see that he has the first symptoms of this absolutely fatal malady.

The face is sad, heavy, dull-eyed and apathetic. The man is, however, well nourished, and this is the rule if the patients are well nursed and fed. If you examine the man's pulse, you find it rapid and weak. If you ask him to hold out his hands, you find that they are weak and tremulous. When asked to walk, his gait is weak and uncertain. When he answers a question, his voice is weak, indistinct and monotonous. The symptoms gradually deepen, and after several months the patient is unable to walk, unable to speak, and unable to feed himself. He is then, of course, altogether confined to his bed, lying in an absolutely lethargic condition all day long. It is in this stage that the sick are often neglected by their friends: they remain unfed, and become emaciated.

In regard to other symptoms, it may be mentioned that during the illness the temperature has shown some elevation of an irregular character, often normal in the morning and rising to 102° or so in the evening. (Fig. 2.)

Here you see the irregular course of the fever, and also that during the last few weeks of life the temperature falls several degrees below the normal line, showing the gradual extinction of the vital forces.

This, then, is a short description of this peculiar disease called sleeping sickness; and now the question arises, what is it that causes this peculiar disease, and gives rise to these curious symptoms? I may pass over without notice the various theories which have been held up to the present time to account for this disease, and ask your attention to what is revealed on a careful microscopical examination of the blood of these cases. If the blood from a case of sleeping sickness is examined under a high-power microscope, an active, wriggling parasite may be seen, which is known by the name of trypanosome.

Here (Fig. 3) is a representation of the trypanosome found in sleeping sickness.

These blood parasites belong to the lowest group in the animal kingdom, viz. the protozoa. The trypanosome consists of a single cell, and in its best known form is a sinuous, worm-like creature, provided with a macronucleus and a micronucleus, a long terminal flagellum, and a narrow fin-like membrane, continuous with the flagellum and running the whole length of the body. When alive it is extremely rapid in its motions, constantly dashing about, and lashing the red blood corpuscles into motion with its flagellum. It swims equally well with either extremity in front.

Among the first to draw attention to these blood parasites was the late Surgeon-Major Timothy Lewis, F.R.S. R.A.M.C.; he discovered a trypanosome, in 1888, in the blood of rats in India, to which was afterwards given the name of *Trypanosoma Lewisi.* This rat trypanosome is found all over the world, and even in Uganda the blood of the ordinary common wild field rat was often found to contain myriads of these creatures. This trypanosome does not appear to do any great harm, or to have any effect on the health of the rats. The next important trypanosome was found also in India, in the blood of horses suffering from surra. This disease, surra, is closely related to the tsetse-fly disease of South Africa, or, as it is called by the natives, nagana.

The trypanosome which causes tsetse-fly disease lives in the blood of the wild animals, such as the buffalo and various antelopes, without evidently interfering with their health, but when transferred by the tsetse fly from the blood of these animals to that of the domestic animals, it causes the death of the latter. Almost all the domestic animals are highly susceptible to nagana, especially horses, dogs and cattle, and even monkeys, but curiously enough man himself is insusceptible.

But now let us return to our examination of the blood of cases of sleeping sickness (Fig. 4). The method of examination is simple: 10 c.c. of blood are drawn, by means of a hollow needle, from one of the veins of the arm, and this is then centrifuged to get rid, as far as possible, of the red blood corpuscles. When this has been done the clear fluid is decanted off and again centrifuged, and the sediment now resulting is subjected to microscopical examination.

I draw your attention to this table giving the result of the examination of sixteen cases, and here you find that in every case, with the exception of one, this trypanosome is found. In all probability it would have been found in this case if there had been an opportunity for further examination, but the man

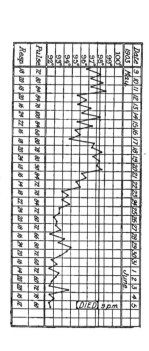

Fig. 2.

Date.	Name.	Sex.	Age.	Duration of Disease.	Trypanosoma
1903					
April 16	Benjamini	Male	28	1st stage	Present
„ 18	Esaka	„	28	1st „	„
„ 18	Waiswa	„	10	1st „	„
„ 18	Kidorme	„	20	2nd „	„
„ 18	Zebuganza	„	40	1st „	„
„ 18	Budara	„	22	2nd „	„
„ 20	Kimbra	„	30	2nd „	„
„ 20	Matasa	„	24	1st „	„
„ 21	Seera	„	25	2nd „	Absent
„ 22	Warosansa	„	32	2nd „	Present
„ 22	Katola	„	25	1st „	„
„ 27	Koagoffa	„	30	1st „	„
„ 27	Kitaroma	„	20	1st „	„
May 12	Nakaiba	Female	8	1st „	„
„ 12	Musa	Male	20	1st „	„
„ 12	Diwarana	„	14	1st „	„

unfortunately died before the trypanosomes had been discovered in his blood.

But there is another fluid in the body which is more easily examined than blood for such a small parasite, and that is the cerebro-spinal fluid.

This cerebro-spinal fluid is a clear transparent fluid, exactly resembling water in outward appearance, which fills the various cavities of the brain, and surrounds the spinal cord so as to prevent damage to these delicate organs. It is easily obtained by introducing a hollow needle between the vertebræ in the lumbar region. Ten to fifteen cubic centimetres of the fluid are drawn off, which is then centrifuged and the sediment examined. As there are few or no red corpuscles in the fluid to interfere with the vision, naturally the actively moving trypanosomes are more easily detected.

I now give a table showing the result of the examination of the cerebro-spinal fluid cases of sleeping sickness.

Here, as you see, forty cases have been examined and the trypanosomes found in every case. This is rather a suggestive fact, and it begins to appear probable that these parasites may have some causal relationship to the disease.

But it may be that this trypanosome is a mere accidental con- comitant of the disease, living in the blood and cerebro-spinal fluid without affecting the health, much in the same way as the rat trypanosome lives in rats, or the nagana trypanosome in the wild animals. So it may be that natives suffering from other diseases also harbour these trypanosomes in their cerebro-spinal fluid. To find if this were so we must examine the cerebro-spinal fluid of natives who come into hospital for other complaints than sleeping sickness. This was done, with the result that none of the patients in

Date.	Name.	Sex.	Age.	Duration of Case.	No. of Examination.	Trypanosoma.
1903						
May 14	Kaperi	Male	8	3rd stage	1	Present
March 26	Seera	,,	25	1st ,,	1	,,
,, 26	Budara	,,	22	2nd ,,	1	,,
,, 26	Kimbra	,,	30	2nd ,,	1	,,
,, 26	Kagoya	Female	20	3rd ,,	1	,,
,, 27	Zeboganza	Male	40	1st ,,	2	,,
,, 27	Yakubu	,,	12	2nd ,,	1	,,
,, 28	Kidorme	,,	20	2nd ,,	1	,,
,, 28	Leobeni	,,	25	3rd ,,	1	,,
,, 29	Waiswa	,,	10	1st ,,	1	,,
,, 31	Dekodemo	,,	25	3rd ,,	1	,,
April 1	Fatoma	Female	18	1st ,,	2	,,
,, 6	Katola	Male	25	1st ,,	1	,,
,, 6	Esaka	,,	28	1st ,,	1	,,
,, 6	Nakaiba	Female	10	1st ,,	1	,,
,, 6	Zakibu	Male	20	2nd ,,	1	,,
,, 6	Warosansa	,,	32	2nd ,,	1	,,
,, 8	Jansi	,,	25	1st ,,	1	,,
,, 9	Feragi	,,	12	1st ,,	1	,,
,, 10	Katoola	,,	20	1st ,,	1	,,
,, 10	Donah	,,	38	1st ,,	1	,,
,, 10	Asumani..	,,	25	1st ,,	2	,,
,, 10	Kainavidi	,,	20	1st ,,	1	,,
,, 10	Moosura Madunga	,,	30	1st ,,	2	,,
,, 10	Msabwa	,,	30	1st ,,	1	,,
,, 10	Adam	,,	30	2nd ,,	1	,,
,, 13	Nonbi	Female	30	1st ,,	2	,,
,, 13	Benjamini	Male	24	1st ,,	5	,,
,, 13	Kiagoffu	,,	30	1st ,,	1	,,
,, 13	Kitaroma	,,	20	1st ,,	1	,,
,, 14	Nateneri	,,	25	1st ,,	1	,,
,, 14	Mutaisa	,,	15	1st ,,	1	,,
,, 14	Erissa	,,	20	1st ,,	1	,,
,, 14	Bagwibwa	Female	18	1st ,,	1	,,
,, 14	Johana	Male	20	1st ,,	1	,,
,, 14	Mwasa	Female	18	1st ,,	1	,,
,, 14	Rukina	Male	25	2nd ,,	1	,,
,, 20	Matasa	,,	24	1st ,,	4	,,
May 4	Kiagabidoia	,,	50	1st ,,	5	,,
,, 14	Divarana	,,	14	1st ,,	1	,,

hospital suffering from other diseases were found to harbour these parasites.

Now, having seen that this trypanosome is found in the blood and cerebro-spinal fluid of all cases of sleeping sickness, and that it is not found in the cerebro-spinal fluid of natives suffering from other diseases, let me ask you to consider that in a slow and chronic disease such as this, sometimes taking years to develop, there must be many natives living in the sleeping sickness area who have these trypanosomes in their blood without as yet showing

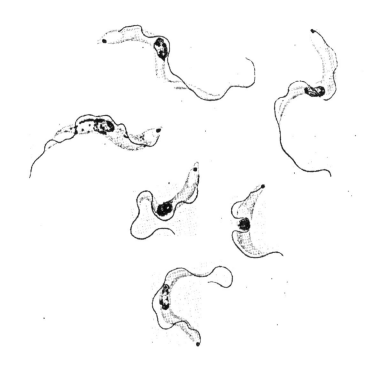

Fig. 3.—Blood Parasites.

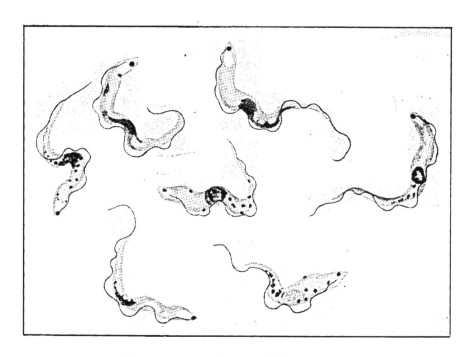

Fig. 4.—Blood of Sleeping Sickness Cases.

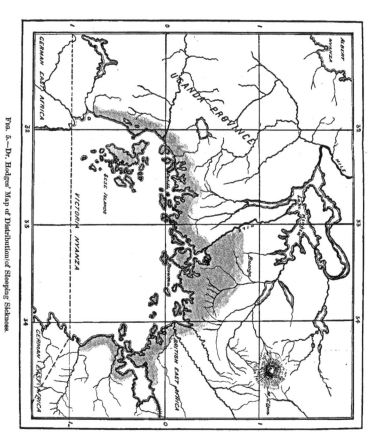

Fig. 5.—Dr. Hodges' Map of Distribution of Sleeping Sickness.

any manifest symptoms of the disease. This seems to be an important point, because if this trypanosome is in reality the cause of sleeping sickness, a certain proportion of the natives inhabiting the sleeping sickness area ought to harbour these parasites in their blood. On the other hand, if this parasite is the true cause of this disease, then no native living in a non-sleeping sickness area ought to harbour a single trypanosome in his blood. It will, therefore, be interesting to examine the blood of natives in the sleeping sickness area and the non-sleeping sickness area of Uganda. Further, it will act as a check if we examine natives living in a non-sleeping area, say in Nairobi, in British East Africa, which is some hundreds of miles away from any infected place.

Eighty natives from the sleeping sickness area were examined, with the result that twenty-three were found to have trypanosomes in their blood, giving a percentage of 28·7. One hundred and seventeen were examined from non-sleeping sickness areas, but not a single trypanosome was found.

You will all agree with me that these results make it very highly probable that the trypanosome under discussion is the real cause of this disease; but there are other methods of adding to this proof, for example, by experiments on animals. If this trypanosome gives rise to symptoms of sleeping sickness in one of the lower animals, this will be a great addition to the proof that this parasite is the cause of sleeping sickness.

The best animals procurable in Entebbe for the purpose of animal experimentation are monkeys. The infective material is injected under the skin, into the spinal canal, and also into the cavity of the brain. The animals show no symptoms for a long time; their temperature remains absolutely normal, and they appear to be in perfect health, but after some months fever of an irregular type sets in and the animals begin to show symptoms of lethargy, sitting about all day, and taking very little interest in their surroundings. Towards the end they sit all day long with their heads bent on their chests, apparently asleep, and show a strong resemblance to the later symptoms of the disease in man.

During this time the monkeys show, constantly, trypanosomes in their blood, sometimes in fairly large numbers.

Therefore it is shown that the trypanosomes derived from cases of sleeping sickness give rise to a long chronic disease in the monkey with symptoms closely resembling those seen in man. From these animal experiments, taken in connection with the other observations, we may now assert that these trypanosomes are the cause of sleeping sickness.

I now pass on to the "distribution of sleeping sickness in Uganda."

This has been investigated by Dr. Hodges, one of the Uganda Colonial Surgeons, and he has prepared this map (Fig. 5), which discloses a remarkable fact. Sleeping sickness is found to have a very peculiar distribution. It is found to be restricted to the numerous islands which dot the northern part of the lake, and to a narrow belt of country a few miles wide skirting the shores of the lake. In no part of Uganda can a single case be found more than a few miles from the lake shore. This part of the country—the islands and the shore of the lake—is, however, the most thickly populated, there being here a population of more than 100 to the square mile. In this area since 1901 the disease has raged, and many places have become depopulated.

In Busoga, where, as we saw, the disease first broke out, cases are found further inland than in Uganda, but here also the same rule holds good. As Mr. Cubbitt, Assistant Collector in Busoga, wrote: "It would seem to be a fairly accurate statement to make, that sleeping sickness confines itself to the territories adjoining the lake, roughly speaking, from a ten to twenty mile radius of the coast." The Uganda Prime Minister, Apolo, also gives it as his opinion that a strip along the lake shore, ten miles broad, would cover the infected area, and that any cases found further inland are always imported. The islands have been specially affected by the disease. For example, the Island of Buvuma in 1901 had a population of 22,000; in 1903 only 8000 remained alive.

Now there must be some cause for this peculiar distribution. Sleeping sickness, evidently, cannot be due to a food poison, as has been suggested, since the people living outside the sleeping sickness strip eat the same food, and have the same habits as those living on the lake shore.

Then again, we have found that the cause of the disease is a trypanosome, a blood parasite, which is not likely to be conveyed in food or clothes, or directly from man to man, but most probably must be carried by some blood-sucking insect.

This leads to the question: "Does the distribution of sleeping sickness in Uganda coincide with the distribution of any particular biting insect?"

Knowing that we are dealing with a trypanosome, and knowing that the trypanosome of nagana is carried in South Africa by a tsetse fly (*G. morsitans*), naturally we will suspect that the trypanosome of this disease is also carried by a tsetse fly. Now on the lake shore near Entebbe a tsetse fly (*G. palpalis*, Fig. 6) is found in large numbers. This may be the insect carrier we are in search of. The Prime Minister and Regents, on being consulted, recognised the fly as one known to the Muganda as the kivu, and said it was found along the shores of the lake. They were supplied with several dozen nets, killing bottles and boxes, and on their part promised to have the distribution of this fly and of sleeping sickness worked out. The bishops, missionaries, and Government officials also promised their assistance.

During June, July and August of last year some 460 collections of biting flies were sent in from all parts of Uganda. As each

FIG. 6.—*Glossina palpalis*, Rob Desv., ♂. (× 3¾)*

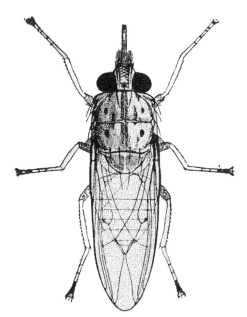

FIG. 7.—A Tsetse fly (*Glossina longipennis*, Corti, from Somaliland) in resting attitude, showing position of wings. (× 3½)*

---

* These two illustrations are taken from "A Monograph of the Tsetse flies," by Ernest Edward Austen, British Museum (Natural History), Cromwell Road, S.W.

package came in it was examined for tsetse flies. If the parcel contained one or more tsetse flies, a red disc was stuck on a large map over the locality from which the flies had been sent. If, on the other hand, no tsetse flies were found, a blue disc was fixed over the spot.

In the same way and at the same time a second map was prepared, to show the distribution of sleeping sickness. That is to say, if the note accompanying the collection of flies stated that sleeping sickness was prevalent, then a red disc was placed over the locality, and if, on the contrary, no cases of sleeping sickness were reported, a blue disc was affixed.

It is evident that two maps so prepared should show at a glance whether the distribution of sleeping sickness and this tsetse fly correspond or not.

The accompanying maps are prepared from these two maps. On comparing them the similarity of the distribution of sleeping sickness and *Glossina palpalis* is self-evident.

In order to work out more minutely the habits of the *Glossina palpalis*, the peninsula on which Entebbe stands was taken in detail and carefully searched for the fly.

The result of this showed that the fly is only found on the shore of the lake where there is forest. This forest is thick jungle with high trees and dense undergrowth. The fly is never found on open sandy beaches backed by grass plains, even although there may be some small scrub near the water's edge. It is never found in the grass of the grassy plains, even though the grass be long and tangled. It has not been found by us in banana plantations, and not at any time far from the lake shore.

The habitat, then, of this fly is the shore of the lake where there is forest. In Busoga, on the other hand, it appears to be found further inland, but what the physical characters of this province are which would account for this I have not learned. The fly also passes down the Nile as far as Kakoge Ferry, some fifty miles north of the Ripon Falls, and it has even been received from Fajao on the Somerset Nile, and from Tengri and the Achwa River, still further north, and near Wadelai, and also from Lake Albert.

It is important that the distribution of this fly should be fully worked up, but enough has been done to show that the distribution of this species of tsetse fly is, like sleeping sickness, confined to the shores of the lake and the islands. It is on the densely-wooded shore of the lake that the half-naked natives of the mainland and islands meet in thousands to trade in fish, bananas, earthenware, etc. If the *Glossina palpalis* can act as a carrier of the trypanosome of sleeping sickness, the circumstances could not be made more favourable than they are for the spread of the disease.

The next point, therefore, to solve is: " Can this tsetse fly carry this trypanosome from persons suffering from sleeping sickness to healthy animals? "

The best animal to carry out these experiments on, of course, is the monkey. The method used is simply to feed tsetse flies on a sleeping sickness case, and, at varying intervals of time, to place the same cage of flies on a monkey. The sleeping sickness patients do not seem to feel the bites of the flies, as they make no complaints or other signs of inconvenience. It is convenient to have, as a rule, about 30 flies in each cage, but only those which fill themselves are to be reckoned as having fed.

As the result of many experiments, several of which were thrown on the screen, it was shown that the tsetse fly is capable of conveying the virus of sleeping sickness from the sick natives to healthy monkeys. These feeding experiments were made at various intervals of time, and it was found that the tsetse fly can still give rise to the disease at the end of 48 hours, but not longer. That is to say, a fly which has fed on a sleeping sickness case, and then kept in a cage without further feeding for 48 hours, is thus capable of transmitting the disease to a healthy monkey, but if kept for three days is no longer capable.

This proves that this tsetse fly can convey the infection from the sick to the healthy. But as 28 per cent. of the natives of the sleeping sickness area have this trypanosome in their blood, doubtless the tsetse flies caught in this area, which feed on these natives, will be able to convey the disease to a healthy animal without any artificial feeding.

A further set of experiments was therefore made to show that the tsetse flies caught on the lake shore were already infective, from having fed on the natives living along the shore. Cages full of the freshly caught flies were straightway placed on healthy monkeys, and after some days the examination of these monkeys showed that they had become infected with sleeping sickness.

This then concludes the story of sleeping sickness in Uganda. We have seen that probably this disease was introduced from the Congo on account of the greater movement of natives under the march of civilisation and the Pax Britannica. We have seen that the disease is caused by the entrance into the blood of a protozoal parasite, and that the infection is carried from the sick to the healthy by a species of tsetse fly. We have seen that the distribution of this fly corresponds with the distribution of the disease. Where there is no fly there is no sleeping sickness. In other words, we are dealing with a human tsetse fly disease.

[D. B.]

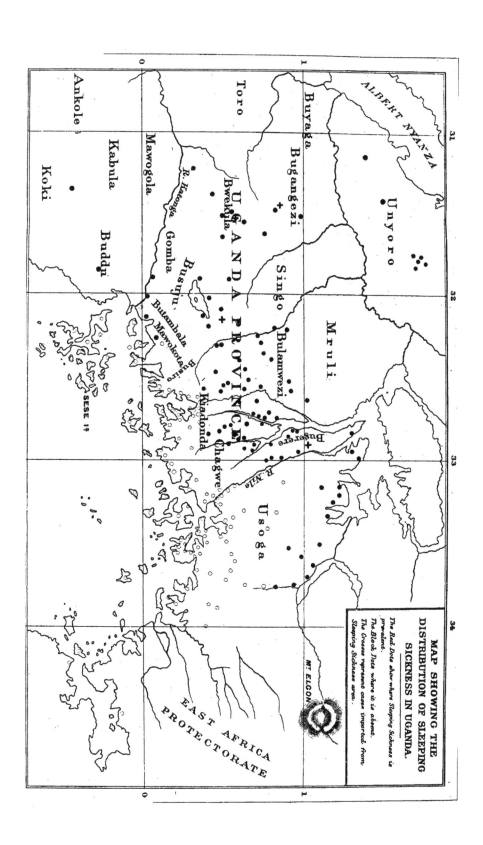

MAP SHOWING THE
DISTRIBUTION OF SLEEPING
SICKNESS IN UGANDA.

The Red Dots show where Sleeping Sickness is
prevalent.
The Black Dots where it is absent.
The Crosses represent cases imported from
Sleeping Sickness area.

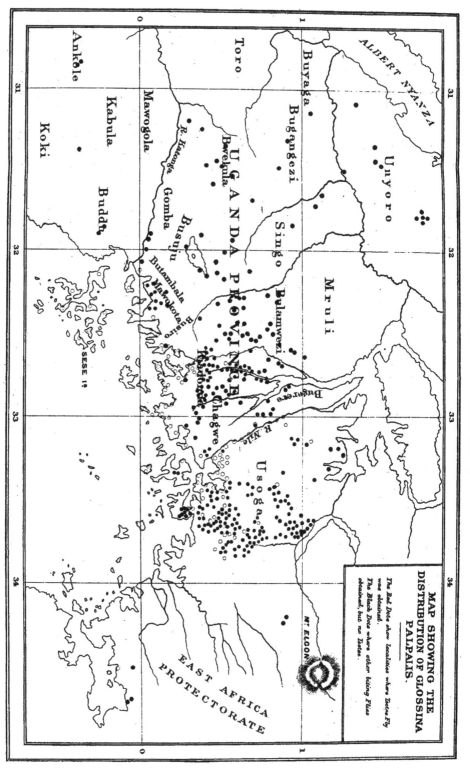

MAP SHOWING THE
DISTRIBUTION OF GLOSSINA
PALPALIS.

The Red Dots show localities where Tsetse Fly
was obtained.
The Black Dots where other biting Flies
obtained, but no Tsetse.

Scale of Miles.
0   10   20   30   40   50   60   70

ALBERT NYANZA

Unyoro

Buyaga

Bugangezi

Toro

UGANDA PROVINCE

Bwekula

R. Katonga

Gomba

Busuju

Butambala

Mawokota

Kisiro

Mawogola

Kabula

Budda

Koki

Ankole

SESE IS

Singo

Mruli

Bulamwezi

Kyagwe

Chagwe

Bugerere

R. Nile

Usoga

Mt ELGON

EAST AFRICA
PROTECTORATE

WEEKLY EVENING MEETING,

Friday, April 29, 1904.

Sir William Crookes, F.R.S., Honorary Secretary and
Vice-President, in the Chair.

Westminster Abbey in the early part of the Seventeenth Century.

By the Dean of Westminster.

The most conspicuous of the churchmen who helped to make English
history in the first half of the seventeenth century had nearly
all been connected at some period of their lives with Westminster
Abbey. Andrewes and Neile were successively Deans; Williams
and Laud were together as Dean and Prebendary; Heylyn, Laud's
biographer, who wrote the Church history of the period as we still
read it to-day, was a Prebendary; and Hacket, the biographer of
Williams, who wrote it from the opposite point of view, as no one
cares to read it now, had gone up to Cambridge as a Westminster
scholar in 1608, together with his schoolfellow, the delightful
George Herbert. The story of their times has been often told, and I
have no claim to tell it afresh. The story of the Abbey, too, is
accessible to all in Stanley's brilliant pages, and in the more recent
' Annals ' by which our late Dean's daughter has admirably supple-
mented that unique book. I have gleaned where they have already
reaped: but the gleanings of such a history as ours are not to be
neglected; and it has seemed worth while to piece together a number
of hitherto unnoticed facts, inserting a well-known story now and
then in the hope of giving a little life to what might otherwise be a
dull picture.

We must of necessity begin by seeking an introduction to the
early Deans. In the first year of the century the Dean was a very
aged man. The Right Worshipful Mr. Gabriel Goodman, Doctor of
Divinity (to give him his title as it appears in contemporary documents), had been Dean for forty years. He had in 1561 succeeded
Dr. Bill, Queen Elizabeth's first Dean, who had succumbed after a
year's endeavour to preside simultaneously over three of the greatest
educational establishments in England, as Master of Trinity, Provost
of Eton, and Dean of Westminster. Dr. Bill's memory was cherished
in Westminster School for many generations; for in 1622, sixty-one
years after his death, the Treasurer's accounts contain the entry of a
sum ' paid to Thomas Brering for mending the scholer's coverletts
given by Doctor Bill sometyme Deane of this Colledge.'

But we are now concerned with Dr. Goodman, who specially

interests me as a member of my own College, having been a Fellow
of Christ's from 1552 to 1554; who moreover owed his place as a
Prebendary, as I owed mine, to the head of the great house of Cecil;
and who afterwards as Dean had Lord Burleigh as his Lord High
Steward, as it is my good fortune to have the Marquis of Salisbury
to-day. I should wish to linger on Dean Goodman, who must, I
think, have been a lovable man. The Abbey owed to him its new
organisation as a College of Prebendaries, including a famous school:
and I cannot refrain from quoting an extract from a letter which he
wrote to Lord Burleigh in reference to the project of the new
statutes in 1577: 'I beseech your honour that there may be that
moderation used, which may be most convenient for all in respect.
Hitherto I and the company, I thank God, have agreed very brotherly
with great quietness, as any such company, I hope. I would be
sorry, if by seeking to better things, dissension should grow to un-
quietness.'

But we must leave Dean Goodman who is really a survival from
the last century, and who will be buried in St. Benet's Chapel, near
Dean Bill, before the year has half run its course. His successor will
again be sought among the Prebendaries, and will again be a
Cambridge man, as indeed all the Deans were for the first hundred
years. Lancelot Andrewes, master of fifteen languages, the witty
courtier, the prince of preachers, and one of the saintliest names of
the English Church, had been a Prebendary for four years, when, on
the nomination of the Lord High Steward, as was the custom of
those days, he was appointed Dean. I find by certain lists of
preachers at the Chapel Royal, which have somehow strayed into our
muniment room, that with Andrewes began the tradition, which still
remains in force, in accordance with which the Good Friday sermon at
St. James's is preached by the Dean of Westminster. His memory
is preserved at the Deanery by a curious old portrait on wood, by
the coloured glass in the Jericho parlour, and by the wainscot there
and in the room above. In 1605, after giving us nine of the best years
of his life, he became Bishop of Chichester; but he never forgot to
pray for τὸ ἐπιζεφύριον μοναστήριον, ' the West Monastery,' as he calls
us in his famous book of devotions.

Our new Dean, Richard Neile, was installed on that notorious day,
the original 'Fifth of November.' He presented a striking contrast
to the refined and graceful scholar whom he succeeded; he had
forced his own way up, and was a clumsy, though a powerful courtier.
We see the two ex-Deans at a later period standing by the chair of
King James, who knew the good points of each and trusted both.
' My Lord,' said the King, ' cannot I take my subjects' money without
all this formality in Parliament?' Neile, now Bishop of Durham,
replies: ' God forbid, Sir, but you should; you are the breath of our
nostrils.' The King turns to Andrewes, the Bishop of Winchester,
' Well, my Lord, what say you?' ' Sir, I have no skill to judge of
parliamentary causes.' ' No put-offs, my Lord: answer me presently.'

' Then, Sir, I think it lawful for you to take my brother Neile's money, for he offers it.'

Richard Neile was a Westminster boy, when the famous William Camden was our second master. His father was a tallow-chandler in King Street. Dean Goodman saw the lad's abilities, and sent him to Cambridge in 1580. An unknown benefactor (it was the Lady Mildred, Lord Burleigh's learned wife) had just given scholarships to St. John's College, the holders of which were to be called Dr. Goodman's scholars. But for this bounty, says Neile, ' I thinke I shoulde never have bin sent to the Universitie, but that the best of my Fortune would have bin to have become some Bookesellers apprentice in Paules Churcheyard : To which Trade of life Mr. Graute then Schoolemaster here persuaded my Mother to have disposed of mee.' His gratitude, when he returned to live here as Dean, was expressed by his sending up two or three boys to the University every year.

He made an admirable Dean, busy and business-like, putting the college estates and accounts into order, repairing Henry VII.'s Chapel, even mending the wax effigies of the kings. He must have kept more state than his predecessors, for he found his house too small, and built a quaint little chamber on the top of the long wooden gallery. He also built ' for the Deanes use a large Stable sufficient to receave 14 or 16 Geldings,' which with the coachhouse and other rooms cost the immense sum, for those days, of 100*l.* But though he spent boldly in every direction his good management largely increased our revenues, which not long before had been exceedingly scanty. He made William Neile, his elder brother, a kind of *factotum*, giving him various lay posts in the college. Their father Paul had died six years before Richard went up to Cambridge, and Sibill, their mother, after less than a twelvemonth became Mrs. Newell. Her son Robert Newell was presented to the Abbey living of Islip in 1609, and he became a Prebendary in 1620. So the good Dean, like all others of his time, ' provided for his own.' He arranged moreover that his mother should lie in the great north porch : and at his very last Chapter meeting he secured a remarkable testimonial for himself and his wife in the shape of a grant of a small pew behind the pulpit for Mrs. Neile's use when she might happen to be in Westminster, and a key for himself to the seat where the lessons were read in the choir. He was evidently reluctant to sever his connection with the great church which he had served so well ; and it is interesting to find him back again two years after he had ceased to be Dean, with a special mandate from the King for the removal to the Abbey of the body of Mary Queen of Scots.

He was with us five years, in the last two of which he was Bishop of Rochester as well. He then left us to climb the ladder of preferment as Bishop successively of Lichfield, of Lincoln, of Durham, of Winchester ; dying at seventy-eight as Archbishop of York, just a couple of days before the Long Parliament met, and the deluge began. He was a good churchman, and we shall hear of him again, for he was the making of his chaplain Laud, for whom he obtained the promise of a prebend in the year that he ceased to be Dean. The two men had both risen from the ranks, and the son of the tallow-chandler of Westminster was the steadfast patron and the lifelong friend of the son of the cloth-merchant of Reading. The next Dean was George Montaigne (or Mountain), who came to us at the end of 1610. He left us for the Bisphoric of Lincoln in 1617, but his affection for the neighbourhood of the Court seems to have led him to rent a prebendal house and spend a large sum of money upon it, as we shall see later on. This feature of his somewhat unattractive character reappears in the story by which he is best remembered. He had come back to be Bishop of London in 1620. Eight years later King Charles wished to transfer him to Durham, in order to bring Laud to London. But the mountain refused to be moved, and yielded at last only on the understanding that the utmost of his removal should be 'from London House in the City to Durham House in the Strand.' As a matter of fact, the death of the Archbishop of York provided him with a yet more honourable and less remote see.

When Dr. Mountain left us in 1617, that curious adventurer the Archbishop of Spalato hoped to have got his place, but it was given to Dr. Tolson (or Tounson).* In those days the Dean was accustomed to grant dispensations to parishioners of St. Margaret's, which was then under his exclusive jurisdiction. Dr. Tounson used to hand the fees received for these dispensations to the parish overseers for distribution to the poor. Their receipts for 1618 include such items as this :—

Of the right worll. Mr. Dctr. Tounson, Deane of Westm., for license by him made to eate Fleshe in the Lent season, videlicet:
Of the Right honorable Lord Pagett for a license . . . xxvi^s viii^d .
[Knights and Ladies paid 13s. 4d. and commoners 6s. 8d.—*Westm. Records*, p. 94.]

One of his latest exploits as Dean was to forbid ' ladies in yellow ruffs to be admitted into his Church.' It appears that he had misunderstood a wish expressed by King James in this regard. A fortnight later he left us for the Bishopric of Salisbury ; but he died the next year, and the case of Mrs. Tounson and her fifteen children was so piteous that her brother, Dr. Davenant, was appointed to the vacant see. Yet Mrs. Tonuson cannot have been penniless, for in 1624 I find a lease granted to ' Margaret Touuson of Sarum, widow,' of the Ancre's House, called after the old anchorite of Westminster, which abutted on the south side of St. Margaret's chancel. There is a Chapter Order, however, which forbids her to eject the curate, the famous Dr. Isaac Bargrave, afterwards Dean of Canterbury.

For the next twenty years Westminster Abbey was to play a notable part in English history, for John Williams came as Dean in

---

* The uncertainty of the name is explained when we discover that his father's name was Toulnesonn (Chester ' Registers,' p. 117).

July 1620, and William Laud as Prebendary in Jan. 1621. I concern myself only with them as they appear as figures on our domestic stage. At the age of sixteen Williams had entered St. John's College, Cambridge; at the same age, but nine years before, Laud had entered St. John's College, Oxford. Williams's friend and biographer was a Cambridge man, John Hacket: Laud's was an Oxford man, Peter Heylyn. Williams was promoted early; he was Dean of Westminster at thirty-nine. Laud was promoted late; and embittered thereby; he was Prebendary of Westminster at forty-eight, and there is a trace of dissatisfaction in the entry in his Diary: 'Having had the advowson of it ten years the November before.' The rivalry of these two able men came to be an important element in the history of their time: they both sought Canterbury, and they ended one at Canterbury and the other at York. There is one piece of paper, and perhaps only one, which contains their autographs together, before they had arrived at episcopal signatures. Laud's first Chapter Meeting was on May 4, 1621, and the following Order is signed at the top by 'John Williams' as Dean, and at the bottom by 'William Laud' as the junior of the Prebendaries:—

It is also consented unto in full chapter and now ordered and decreed that Mr. Deane of Westminster and the Præbendaries resident here or as many as he shall call to the nomber of six shall have full powre . . . to the altering of a lease now taken in trust for the good of the colledg. in the name of Mr. Ellis Wynn; and any other thinge or thinges, concerning the quieting of a controversye like to prove a suite in law, about an house for Mr. Dr. Laud, Dean of Glocester, belonging to him as Præbendary of this churche, and the which he is dispossessed of, which we hope to end in peaceable and quiet manner. . .

Before Laud signed another Chapter Order he had become Bishop of St. Davids, and wrote himself *Guil: Menevens.* His rival had attained far greater distinction. When Lord Keeper Francis Bacon was ejected from his office for scandalous practices, King James protested that 'he would have a clergyman: he would have no more lawyers, for they are all so nursed in corruption that they cannot get away from it.' So he chose the Dean of Westminster, who had been Chaplain to Lord Keeper Egerton. When the Lords objected that they were to have over them a man who was not one of themselves, the King made him Bishop of Lincoln, with a special license to retain the Deanery of Westminster.

We are now ready to read the Chapter Order of December 4, 1622:—

It is ordered and decreed by common consent of the Right reverend Father in God, the Lord Bishop of Lincolne. Lord Keeper of the great seale of England. and Deane of the Collegiate churche of St. Peter in Westm., and the chapter of the same, whereas for the regaining of one of the Prebendaries houses, situate in St. Margaret's churche yard, lately in the possession of Dr. Bulkeley Prebendary, and since demised by lease unto William Man Gent., to the use of the Lord Bishop of London then Deane of Westm . the sum of 200*li.*, was to be repaid to the said Bishop of London in consideration of his charges expended in repairing the said house, that the sum of 100*li.* should be paid by the Lord Bishop of St. Davids in regard of his enioing the said house, so repaired, and that 100*li.* more should be paied by the Deane and chapter: of which sum the Deane and chapter have presently laid downe 30*li.* in part: and the remainder of the said 100*li.* my Lord Keeper is pleased to lay downe for the present, and to receive the same at the two chapiters next ensuinge, by even portions. And it is farther decreed by the Deane and chapter aforesaid, that the said Lord Bishop of St. Davids in consideration of his charges, and monye disbursed, shall receive by himself, or his Executors, the sum of 80*li.*, to be repaid to him, by his next successor in the said house and Prebend, . . .

From these Orders it would appear that the junior Prebendary was somewhat of a stormy petrel at his first entrance amongst us. To understand the position we must study some hitherto unwritten pages of our domestic history. As late as 1740, when by the aid of a grant from Parliament, the site was cleared, there were two prebendal houses on the north side of the nave, and three smaller tenements between them. The artists have always disapproved of them, and every picture that I have been able to find shows the site as clear as it is to-day. But the one which interests us most, the house nearest the north porch, was occupied by a Prebendary in the time of King Edward VI. An Act of Parliament of 1552, defining the precincts of the church, speaks of ' the Prebendaries house now in the possession of Barnard Sandyforth,* Clarke, one of the Prebendaries of the said Churche, and the grounde and other howses adioyninge to the same howse on the northe parte of the said Churche.' In 1590 we find Dr. Bulkeley in this house; for ' a tenement or lodg' next to him was then let to a notable citizen of Westminster, Maurice Pickering, who with Joan his wife had presented the burgesses with their now famous cup two years before. We first come across Maurice Pickering as a verger in 1572, and then as allowed to have a deputy in that office. He was for a long time, keeper of the Gatehouse. In 1592, he and ' Joan his wife ' are allowed to have ' a quill of water at their own costs to their new house ' from the house of Dr. Bulkeley.

So far then, we have two houses, Dr. Bulkeley, Prebendary, and Maurice Pickering, Gent., living side by side. We shall presently find a stable, and then a house, further west close by the tower. In 1604 Mr. Pickering is gone, and Hugh Parlor and Edmund, his son, are Dr. Bulkeley's neighbours. Then the house is let to Sir Edw. Zouche (1608), and then Dudley Norton, soon to be knighted, comes in 1610.

In this year we get our first sight of the stable next to this house. It is let to William Man and William Neale. The former was son of William Man and collector to the Dean and Chapter. His mother, Widow Man, had married again, and we have the account of what it cost William Meredith, Dr. Goodman's Secretary, to woo and secure this lady. The latter was elder brother of Dean Neile. When Neile

---

* He was Prebendary from 1546 to 1554, going out when the monks came back.

was Bishop of Lichfield or of Durham, it suited him to have this stable. But later the site was used to build another prebendal house. Here Dr. Durant de Brevalt lived in 1674, and Dr. Barker from 1716 till it was cleared away about 1740.

This story is put together out of old leases and Chapter Orders, and more might be said of Maurice Pickering's house, as *e. g.* that it was commonly called ' Mason's Lodge,' and that its place was taken in 1668 by three small houses, built by one John Shorter, who undertook to remove certain insanitary arrangements hard by the Abbey buttresses.

But we must return to Dr. Bulkeley, who long survived his neighbour, Mr. Pickering. He ceased, however, to live in this house, and it was let for forty years to William Man in 1613, when Dr. Mountain was Dean. Dr. Bulkeley had surrendered it into the hands of the Dean and Chapter, and in 1616 he was assigned in lieu of it ' a great stone house ' within the Close. This was the house long called 'the Dean's house ' because the Dean had occupied it during the ten unhappy years in which a Bishop of Westminster was in the Deanery. In 1662 it was rebuilt by Inigo Jones for Colonel Ashburnham, who bought the site, which only returned to the Chapter in 1741, when the houses outside in the churchyard were pulled down.

When this house was assigned to Dr. Bulkeley, it was further stipulated that it should go to his successor, ' if he be Dr. Nuel only.' But Dr. Newell did not succeed Dr. Bulkeley. He was Neile's half-brother, and doubtless through him had obtained the promise of a prebend. His hopes, however, were realised some months before Dr. Bulkeley died. Laud succeeded Bulkeley. What house was he to have? William Man had sublet the house outside to Dr. Mountain, formerly Dean. But Dr. Mountain was just now migrating to London House ; and so Dr. Bulkeley's old house could be got by arrangement with William Man. The large sum of 200*l.* was due to the new Bishop of London for improvements ; and the two Chapter Orders show (1) that Laud claimed this house as of right, and (2) that Williams's liberality helped to smooth the difficulty about the money.

So the sad little Prebendary got his house on the sunless side of the Abbey, with power to come in and out by what tradition calls ' the Demons' door ' ; while the cheerful Dean, his junior, was on the south side, enjoying in the sunshine the preferments in Church and State which are indicated by his signature ' Jo: Lincoln, C.S. et Dec: Westm.' As a matter of fact, it was several years before Laud moved into his house, for he preferred to reside with Bishop Neile at Durham House in the Strand. Only when that house was required for the ambassador-extraordinary of the King of France, who was coming over with the Royal bride, Henrietta Maria, on January 3, 1626, the move was hurriedly made. His books were hardly on the shelves when he found himself practically responsible for a task which involved much historical inquiry, the drawing up of the Coronation Service for the new king Charles. Laud's star was now in the ascendant, and Williams, no longer Lord Keeper, was prohibited from coming to Westminster. Laud acted as Deputy-Dean at the well-ordered coronation, the one thing (so far as I can find) for which we at Westminster have special cause to remember him.

There is a curious sequel to the story of Laud's house. Dr. Richard Steward became a Prebendary in 1638, and in 1640 he was Prolocutor of the famous Convocation which continued its sittings in Henry VII.'s Chapel after the Short Parliament had been dissolved. We learn from Peter Heylyn's Life of Laud (pp. 423, 438) that a certain committee ' was desired by the Prolocutor to hold their meetings in his house, situate on the North-side of the Abbey-Church and therefore most convenient both to himself and to them.' The Long Parliament soon afterwards swept away Dean and Prebendaries, and they assigned Dr. Steward's house to their Serjeant-at-Arms. Dr. Steward was nominally Dean after Williams became Archbishop of York, but he died at Paris during the Interregnum. When the King came to his own again, a mandate was sent to the new Dean, Dr. Earles, to exhume the bodies of the Commonwealth leaders who had been buried at the east end of Henry VII.'s Chapel. The remains of Pym and others were thrown into a pit in the churchyard ' near the back-door of one of the Prebendaries.' It is the grim irony of history: for this had been Laud's back door.

When we turn from Deans and Prebendaries and try to picture the interior of the church during our period, we are at a loss for guidance. The pictures of successive coronations are useless, partly because the ordinary arrangement of the church was upset, and yet more because the artists took no trouble to give a correct idea of the building. We must pick up what we can from our Chapter books, accounts, and muniments.

In the time of Dean Neile the altar was well cared for. The great sum of 58*l.* was spent on a large ' backe Front of Cloathe of gold and blue velvett.' Out of the palls offered by King James and his Queen at their coronation was made a splendid altar cloth, and another was provided at the cost of 22*l.* for daily use.

Westminster was conservative in its ritual, and had maintained throughout our period the use of copes and wafer breads, which the Puritanism of other places had abandoned. The wafers were, no doubt, of the larger Protestant form ordered in Queen Elizabeth's time, and never since disallowed. Cosin says,* 'Though there was no necessity, yet there was a liberty still reserved of using wafer-bread, which has continued in divers churches of the kingdom, and Westminster for one, till the 17th of King Charles.' In 1614 the Parliament decided that the whole House was to receive the Communion, not at Westminster Abbey, 'for feare of copes and wafer

---

* ' Works,' v. 481, cf. 518 f.

cakes,' but at St. Margaret's. When the Parliament next met, seven
years later, Williams was the Dean, and a note is preserved to the
effect that 'the Speaker of the Commons acquainted the House that
the Dean and Chapter of Westminster refuse to permit them to
receive the Communion there, because they were not first asked, and
because the preacher was not one of themselves; but that if they would
appoint a canon preacher, they might receive the Communion with
ordinary bread; and that the House rejected the offer, and chose the
Temple Church.' In the end, however, 'the House received the
Communion at St. Margaret's, and Dr. Usher preached the sermon.'
The Abbey was willing to yield on a point of ritual, but not on a
point of privilege; and so began the connection of the House of
Commons with St. Margaret's Church, where the Dean was ready to
allow them a freer hand.

The pews in the choir, which was completely screened off from
the transepts, were the subject of much controversy at a later time.
We may therefore note with interest an entry in Dean Neile's
accounts : 'Item, sett up in the Church about the preaching place for
the better sort to sit in in service and sermon time, ten severall large
Pewes of stronge wainscott, which with some alterations done about
the Prebends Stalls cost' over 36*l.* Simultaneously we read a
Chapter Order (1606) : 'That the prebends stalls in the Queere shalbe
made newe, to have all the prebends sitt together, halfe on one side
and halfe on the other side, every one sittinge accordinge to his
dignity and degree, the Sub-Deane still keepinge his ancient place.'

I have already mentioned Mrs. Neile's 'little pew behind the
pulpit.' The pulpit was then (and indeed till 1779) on the south
side. Near it was a great pew, with King Richard II.'s portrait
hanging up on the screen behind it. Twenty years later this pew
was a point of fierce dispute. The Prebendaries had come to prefer
it to their stalls; but Dean Williams claimed it as his preserve, and
would only allow the nobility to sit with him there.

In 1631 there was added to the Chapter a clever little man
named Peter Heylyn, who was Laud's chaplain. Besides Laud's
quarrel, he had a quarrel of his own with the Dean, who ten
years before had been ordered to call in a book of his, and
only a week before had refused him institution to a living, on the
ground that it was not in the King's gift but in his own. Peter
made a cave in the Chapter, and with three other juniors drew up a
charge of thirty-six articles against the Dean, whose enemies had
already made desperate attempts to unseat him. They preferred a
petition to the not unwilling King. But the Dean of Westminster is
notoriously difficult to get at, and a Royal Commission had to be appointed consisting of the two Archbishops, three lay Lords, and two
Secretaries of State. The articles are particularly interesting now,
as the triviality of their details afford us many glimpses into the
domestic life of the place. Lest they should scandalise the public
(perhaps rather lest their absurdity should be too obvious) Peter was

ordered to translate them into Latin. For a year and nine months nothing happened, and the Prebendaries had to present a new petition with fresh allegations, the chief of which was the matter of the pew. It is plain that they saw that they could not get rid of the Dean, and they concentrated their efforts on the recovery of the pew. The only satisfaction that they finally got was a decision that they might sit in the pew, and 'that none should sit there with them but Lords of Parliament, and Earls' eldest sons, according to the ancient custom.'

Once more the pew. The Dean has been four years in the Tower: he has just got out through a turn in the political wheel, and he returns in honour to the Abbey. He is sitting in the pew, and Peter Heylyn is in the pulpit above him, preaching moderation and charity with irritating innuendoes. At last the Dean's temper is up, and knocking loudly with his staff upon the pulpit he cries, 'No more of that point, no more of that point, Peter!' 'I have a little more to say, my Lord, and then I have done.'

And yet once again we have a sight of the pew, through the eyes of a Westminster Boy, who lived long afterwards as a Prebendary in Laud's house, the famous Robert South. William Strong is in the pulpit now (a Puritan divine whose first Abbey sermon was entitled 'Gospel Order a Church's Beauty'), and 'the leading grandees of the faction in the pew under it.' But by this time many other pews had been set up. The altar was gone from its place: the tapestries surrounding the sanctuary had been carried off to adorn the House of Commons. The sanctuary itself was occupied by a gallery of pews. The carpenter's estimates are preserved, and also a plan of the work, showing the allotment of the pews. 'Lord Bradshaw,' who then occupied the Deanery, sat on the south side, and opposite sat Dr. Busby, who had made his monitor pray for King Charles, as South bore witness, 'a few hours before his sacred head was cut off.'

'Gospel Order a Church's Beauty' may find a further illustration in the following glowing account of the services in the Abbey under the new régime:—

And about the 26 of this instant *March*, my intelligence put me in minde heere to make mention of God's admirable and most wise ordering and disposing of things to the glory of his Name, joy of his children, and vexation of his base *Brats of Rome*, and malignant Enemies of *Reformation*; in the most rare and strange alteration of the face of things in the *Cathedrall Church at Westminster*. Namely, that whereas there was wont to be heard, nothing almost but *Roaring-Boyes*, tooting and squeaking *Organ-Pipes*, and the *Cathedrall Catches of Morley*, and I know not what trash; now the *Popish Altar* is quite taken away, the *bellowing Organs* are demolisht, and pull'd downe, the *treble*, or rather, *trouble* and base Singers, Chanters, or inchanters, driven out; and instead thereof, there is now set up a most blessed Orthodox Preaching Ministry, even every morning throughout the weeke, and every weeke through the whole yeare a Sermon Preached by most learned, grave, and godly Ministers, of purpose appointed thereunto, and for the gaudy guilded Crucifixes, and rotton rabble of dumbe Idols, *Popish* Saints, and Pictures, set up, and placed, and painted thereabout, where that sinfull Singing was used; now a most sweet assembly, and thicke throng of Gods pious people, and well-affected, living teachable *Saints* is there constantly, and most comfortably, every morning to be seen at the Sermons   O our God! what a rich and rare alteration ! what a strange change is this indeed ! *

We are now at the close of our period, when, by order of the Long Parliament—'the pretended Parliament,' as the Royalists called it—a serious outrage against both Church and State was perpetrated in the Abbey cloisters.   The following documents tell the story :—

I.   *Journal of the House of Commons.*

June 2, 1643.   *Resolved,* That the Dean, Subdean and Prebends, be enjoined and required to deliver to Sir Hen. Mildmay, and Mr. Marten, the Keys of the Treasury where the Regalia are kept; that they may search that Place, and report to the House what they find there.

The Question being put, whether, upon the Refusal to deliver the Keys, the Door of that Place where the Regalia are kept shall be opened;

The House was divided :

The Yeas went forth.

Sir H. Ludlow	{ Tellers for the Yea : }	37
Mr. Strode	{ With the Yea, }	
Mr. Pierrepont	{ Tellers for the Noe : }	58
Mr. Selden	{ With the Noe }	

June 3, 1643.   The Question being put, whether the Locks of the Doors where the Regalia are kept, in Westminster Abbey, shall be opened, notwithstanding any former Order made, and Search made there; and an Inventory taken, of what things are there, and presented to the House; and new Locks set upon the Door; and nothing removed till the House take further Order.

The House was divided.

The Yeas went forth.

Sir Peter Wentworth	{ Tellers for the Yea : }	42
Sir Christ Yelverton	{ With the Yea, }	
Mr. Holles	{ Tellers for the Noe : }	41
Sir Jo. Holland	{ With the Noe }	

Resolved, etc.,   That the Locks of the Doors where the Regalia are kept, in Westminster Abbey, shall be opened, notwithstanding any former Order made, and Search made there; and an Inventory taken of what Things are . . . ; and presented to the House; and new Locks set upon the Doors, and nothing removed till the House take further Order; and that Sir Rob. Pye be there present, with that Inventory of the Regalia, that is kept in the Chamberlain's Office of the Exchequer, whether all Things be there, mentioned in that Inventory.

Sir Jo. Holland, Mr. Gurdon, Sir H. Mildmay, Mr. Marten are to take the Inventory, and to execute this Order accordingly.

From these records it appears that the keys were demanded from the Dean and Chapter by the resolution of June 2, but the proposition that in case of refusal the locks should be forced was lost by 58 votes to 37.   When the first resolution proved nugatory, it was on the next day agreed by a majority of one to break open the doors.

II.   Heylin's ' Aerius Redivivus ' (ed. 1670, p. 461; ed. 1672, p. 452).

And for a further evidence of their good intentions, a view is to be taken of the old *Regalia,* and none so fit as *Martin* to perform that Service.   Who having

---

* ' God's Ark over-topping the World's Waves' (John Vicars, 1646, p. 184).

commanded the Subdean of *Westminster* to bring him to the place in which they were kept, made himself Master of the Spoil. And having forced open a great Iron Chest, took out the Crowns, the Robes, the Swords and Sceptre, belonging anciently to K. EDWARD the Confessor, and used by all our Kings at their Inaugurations With a scorn greater than his lusts, and the rest of his Vices, he openly declares, *That there would be no further use of those Toys and Trifles.* And in the jollity of that humour invests *George Withers* (an old *Puritan Satyrist*) in the Royal Habiliments. Who being thus Crown'd and Royally array'd (as right well became him) first marcht about the Room with a stately Garb, and afterwards with a thousand Apish and Ridiculous actions exposed those Sacred Ornaments to contempt and laughter. Had the *Abuse* been *stript* and *whipt*, as it should have been, the foolish Fellow might possibly have passed for a *Prophet*, though he could not be reckoned for a *Poet*.

The earliest history of this Royal Treasury is obscure, and a recent controversy of the antiquaries has cast uncertainty on part of what we thought we knew. For it has been strongly argued that the Treasury which was robbed in the reign of Edward I. was not this chapel in the cloisters, but the vault beneath the chapter-house. I cannot now discuss the matter; but it remains certain that since the days of Edward III. certain royal treasures were kept in this little chapel. Portions of the Regalia were also there, though others were in the monastic Treasury in St. Faith's Chapel. Since the Restoration the Regalia have been in the Tower, and have only been deposited with the Dean and Chapter on the eve of a coronation. The old chapel was still used as a Treasury of treaties and records; Exchequer tallies were kept there, and also the Pyx, or 'box' containing the standards of coinage. But gradually everything has been removed. The treaties, the tallies, the Pyx itself—all have found homes elsewhere. The chapel is empty; the ancient altar stands, though somewhat damaged, and by its side is a piscina on a thirteenth-century pillar.

Let us look a little closer, and observe the construction of this tiny chapel. It is at present but 30 feet wide and 30 feet long, with a heavy round column in the centre. I say 'at present,' for originally it formed part of a long vaulted chamber beneath the ancient dormitory of the monks. This chamber was 100 feet in length, that is, as long as Henry VII.'s Chapel; and six massive columns down the middle supported the vaulting. It is the oldest remaining part of the Abbey, reaching back to the period of Edward the Confessor's building. It is now divided by partition walls, of stone or of brick, into four small compartments. Two of these are dark storehouses, a third is rented by Westminster School as an approach to their gymnasium, while the fourth is the chapel of which we are speaking. This, the most northerly portion, was walled off, perhaps two hundred years after it was built, by a rough stone wall: the other partitions are of much more recent date.

I sometimes have a vision of a new period of public usefulness for this hidden and almost forgotten site. I seem to see the ancient vaults reunited as of old in one long chamber; the old altar repaired, that we may worship once more in the one sacred portion of the

Abbey that goes back to the Confessor's period; while beneath the floor may be laid in this most venerable spot the remains of our greatest countrymen in the century that is before us, and the walls may hold memorials on a modest scale, such as in recent years have been filling the few spaces left in the main part of the church.

I dare not speak of this as a scheme, but only as a dream. Yet I see in it a possible solution of the problem that has baffled us heretofore—how to maintain the splendid tradition of Abbey burials, which otherwise will soon become merely a glory of the past. Elaborate projects of new building have been devised, only to be set aside as outrageous or impracticable. This is a possibility, worthy at least of a thought, before we resign ourselves to despair. It would involve a *minimum* of disturbance, and would reopen to public view our most ancient chamber; while in time to come an extension of the scheme might include what is called the Chapel of St. Dunstan and the ground now occupied by the gymnasium, if the interests of the School were duly provided for elsewhere.

I have put forward this suggestion quite tentatively and without any intention of pressing it. I only say that it is worth consideration and that the difficulties connected with it would not be insurmountable, if the idea should commend itself to the public mind and should obtain the sanction of the highest authorities.

Quite apart from any such suggestion for the future, I would venture to express a hope that the time has come when, with the general approbation of Englishmen, this little chapel may be restored to the custody of the authorities of the Abbey, and used as in ancient days for sacred purposes. A special interest would thus attach to it, as being the only portion of St. Edward's building which is still capable of being used for Divine service. And arrangements could be made by which reasonable opportunities for viewing it could be given at other times.

There is no reason for supposing that the King, in making use of this portion of the long vaulted chamber or chapel for the keeping of his treasures, intended to alienate the fabric from the Abbot and Convent to whom his predecessors had granted it. Had he chosen to keep his treasures elsewhere the chapel would have returned naturally to its former use.

After the second expulsion of the monks the whole property of the Abbey vested in Queen Elizabeth; and she of her Royal bounty granted the whole of it to the Dean and Chapter of her new foundation. This is shown by the words of her grant, which not only gives the whole site, but in express words 'all the chapels' (*omnes capellas*).

It would seem to be reasonable that if the State no longer requires this chapel for the purposes of a Treasury, it should revert to its ancient use.

ANNUAL MEETING,

Monday, May 2, 1904.

SIR JAMES CRICHTON-BROWNE, M.D. LL.D. F.R.S.,
Treasurer and Vice-President, in the Chair.

The Annual Report of the Committee of Visitors for the year 1903, testifying to the continued prosperity and efficient management of the Institution, was read and adopted, and the Report on the Davy Faraday Research Laboratory of the Royal Institution, which accompanied it, was also read.

Seventy new Members were elected in 1903.

Sixty-two Lectures and Twenty Evening Discourses were delivered in 1903.

The Books and Pamphlets presented in 1903 amounted to about 229 volumes, making, with 694 volumes (including Periodicals bound) purchased by the Managers, a total of 923 volumes added to the Library in the year.

Thanks were voted to the President, Treasurer, and the Honorary Secretary, to the Committees of Managers and Visitors, and to the Professors, for their valuable services to the Institution during the past year.

The following Gentlemen were unanimously elected as Officers for the ensuing year:

PRESIDENT—The Duke of Northumberland, K.G. D.C.L. F.R.S.
TREASURER—Sir James Crichton-Browne, M.D. LL.D. F.R.S.
SECRETARY—Sir William Crookes, F.R.S.

MANAGERS.

Sir William de W. Abney, K.C.B. D.C.L. D.Sc. F.R.S.
Henry E. Armstrong, Esq., Ph.D. LL.D. F.R.S.
Shelford Bidwell, Esq., M.A. Sc.D. LL.B. F.R.S.
Sir Alexander Binnie, M.Inst.C.E.
John H. Balfour Browne, Esq., K.C. D.L. J.P.
The Hon. Sir Henry Burton Buckley, M.A.
Sir Thomas Andros de la Rue, Bart., M.A.
John Ambrose Fleming, Esq., M.A. D.Sc. F.R.S.
Sir Victor Horsley, M.B. B.S. F.R.S. F.R.C.S.
The Right Hon. Lord Kelvin, O.M. G.C.V.O. D.C.L. LL.D. D.Sc. F.R.S.
Ludwig Mond, Esq., Ph.D. F.R.S.
Sir Owen Roberts, M.A. D C.L. F.S.A.
Sir Thomas Henry Sanderson, G.C.B. K.C.M.G.
Sir Felix Semon, C.V.O. M.D. F.R.C.P.
William Hugh Spottiswoode, Esq, F.C.S.

VISITORS.

John B. Broün-Morison, Esq., J.P. D.L. F.S.A. (Scot.)
John Mitchell Bruce, Esq., M.A. M.D. LL.D.
Frank Clowes, Esq., D.Sc. F.C.S.
James Mackenzie Davidson, Esq., M.B. C.M.
Francis Fox, Esq., M.Inst.C.E.
William Bolger Gibbs, Esq., F.R.A.S.
Maures Horner, Esq., J.P. F.R.A.S.
Carl E. Melchers, Esq.
Robert Mond, Esq., M.A. F.R.S.E. F.C.S.
John Callander Ross, Esq.
The Hon. Lionel Walter Rothschild, M.P.
George Johnstone Stoney, Esq., M.A. D.Sc. F.R.S.
Alan Campbell Swinton, Esq., M.Inst.C.E.
John Jewell Vezey, Esq., F.R.M.S.
George Philip Willoughby, Esq., J.P.

WEEKLY EVENING MEETING,

Friday, May 6, 1904.

SIR JAMES CRICHTON-BROWNE, M.D. LL.D. F.R.S., Treasurer
and Vice-President, in the Chair.

P. CHALMERS MITCHELL, Esq., D.Sc. Sec. Z.S.

Anthropoid Apes.

HAD it been my lot to be in control of Zoological Gardens long
before the chalk cliffs of Dover were formed, there would have been
no lack of strange creatures to excite the interest of the visitors on a
primeval Bank Holiday. But there would have been no lions and
tigers, no elephants, hippotamuses or rhinoceroses, no zebras or
giraffes, no apes or monkeys. Grotesque and gigantic flying, running,
hopping and crawling creatures would have been there, creatures
adapted for browsing, or gnawing, or flesh-eating ; but the great
group of mammalian animals, which now plays so large a part in the
economy of the earth, would have been represented by a few small
and inconspicuous creatures, probably placed in an obscure corner of
the reptile house of the Jurassic Zoo. These primitive mammals
had small skulls with very small brains. long jaws with many teeth
in a single even row along the margin, and their flat hands and feet
had five fingers or toes. From such simple beginnings the great
groups of existing mammals became specialised in different ways.
The group Primates, for instance, which contains the lemurs, monkeys,
apes and man, retained primitive characters, such as a relatively even
row of teeth, a flat foot and hand, each provided with five digits,
and each digit protected by a shield intermediate between a hoof and
a claw. On the other hand, in this group the skull and the brain
became highly specialised, and the upright posture, with its attendant
modifications of structure, was gradually assumed.

[The lecturer then described the special characters of anthropoid
apes, and gave an account of the natural history of the gibbons,
orang-utans, gorillas and chimpanzees, exhibiting a series of lantern
slides made from photographs of specimens that had lived in the
London Zoological Gardens.]

The great anatomists of last century, and in particular Huxley
and Darwin, showed clearly the essential similarity in the structure
of the great apes and man ; and from their work, and the work of later
observers, it may be taken as abundantly proved that in the primate
series there is a less gap between the great apes and man than there
is between the great apes and ordinary monkeys. I may take the

skeleton and the brain as an illustration rather than as an exposition of this anatomical fact.

[The lecturer then showed and explained a series of lantern slides of the skeletons and of the brains of anthropoid apes and man, and called special attention to a set of actual brains of anthropoid apes which had been very beautifully preserved for him by Mr. F. E. Beddard, F.R.S., the Prosector of the Zoological Society. He compared these with specimens of actual brains and models of brains of human embryos which had been lent to him by Dr. Keith, of the London Hospital Medical College.]

Since Huxley and Darwin wrote, corroborative evidence of new and unexpected kinds has been obtained respecting the relationship between man and anthropoids. This evidence may be divided into three heads :—

1. *Palæontological.*—In 1894 Dr. Eugene Dubois discovered in Java the very remarkable fossil now known as *Pithecanthropus erectus*. From the character of the bed in which the bones of this creature were found, and from the mammalian remains with which it was associated, Pithecanthropus must be referred to the Upper Pliocene series. Very animated debate has taken place amongst anatomists as to whether Pithecanthropus is to be considered human or anthropoid. That question, however, is one of classification ; the salient point is that Pithecanthropus is more ape-like than any known human type living or extinct, and more man-like than any known form of ape living or extinct.

2. *Anatomical Study of Variation.*—The most striking feature of anatomical work of the last two decades is the importance that has been attached to the study of variation. Until comparatively recently, anatomists were content to take the characters of an animal from one, or at least from a very small number of specimens. A worker who had the good fortune to dissect a specimen of a gibbon, an orang, a gorilla, a chimpanzee, and a human body, would have considered himself provided with abundant material for a discrimination of the characters of these forms. We know now, however, that the anatomical structure of a species must be estimated in percentages, in the number of times that any particular arrangement, say of blood-vessels occurs, in say 100 specimens. In the case of closely allied forms, variations of anatomical structure that occur in a high percentage in one species are found in a lower percentage in another species. Anatomically speaking, allied species are simply different centres of oscillation in a continuous series of anatomical variations. And thus it has been found that anatomical peculiarities regarded as abnormal (that is to say, occurring in a small number of individuals per cent.) in man are normal (that is to say, occurring in a large number of individuals per cent.) in the gibbon, or orang, or chimpanzee, or gorilla. Similarly the gorilla normal, or the gibbon normal, occurs in man as an abnormality.

3. *Physiological.*—A remarkable recent development of physio-

logy has been the study of the properties of blood serum. When the blood of one mammal is injected into the body of another, the blood of the latter animal shows remarkable modifications. If a few drops of the blood of a guinea-pig be added to the colourless fluid known as serum, prepared from the blood of another rodent, say a rabbit, no change takes place. If, however, the blood of the rabbit, before the serum has been prepared from it, be treated by the injection into the living animal of guinea-pig's blood, then the rabbit serum, contaminated so to speak with guinea-pig, shows new properties. It has become a delicate test for guinea-pig blood. If a few drops of such blood be placed in it, the mixture becomes opaque and reddish, as the guinea-pig blood corpuscles now dissolve in it. If clear guinea-pig's serum be added, in a few minutes a cloudy precipitate is thrown down. Such reactions are sometimes strictly specific, and form a most delicate test for the blood of an animal. It has been found, in the case of the anthropoid apes and man, that the test does not discriminate between the bloods of these forms. In the literal sense of the words, there is an extremely close blood relationship between man and the anthropoid apes. This blood relationship is shown in that a number of diseases hitherto thought peculiar to man can be shared by the anthropoid apes. At the Pasteur Institute in Paris, and at the School of Tropical Medicine in Liverpool, this physiological kinship is being the basis of a number of experiments which may lead to consequences not only of vast scientific interest, but of the greatest practical importance to man.

Finally, since Darwin and Huxley wrote, a considerable knowledge of the embryological development of anthropoid apes has been gained, and this knowledge has revealed the expected but important fact, that, until a very late period before birth, the embryos of anthropoid apes are extremely similar to those of man.

The gap between man and the lower animals is widening rapidly. In recent geological times species of anthropoid apes more man-like than the existing forms, and species or varieties of the human race more ape-like than existing man, have been blotted out. At the present day the lower races of man are disappearing, and there is more than a probability that the range of the anthropoid apes will become more restricted, and that these too are in danger of extinction. So far, the efforts to keep these apes in captivity have been uniformly unsuccessful, both in the tropics and in temperate regions. Gorillas have rarely lived more than a few weeks in captivity; orangs a few months, reaching in the extreme two or three years; there has been greater success with chimpanzees. However, none of the larger anthropoid apes have bred in captivity, and breeding must be taken as the absolute test of successful conditions.

[The lecturer then reviewed some of the difficulties that arise in connection with the keeping of apes in captivity, and stated his belief that the problem of successful confinement was capable of being solved.]

GENERAL MONTHLY MEETING,

Monday, May 9, 1904.

His Grace The Duke of Northumberland, K.G. D.C.L. F.R.S.,
President, in the Chair.

The Chairman announced that he had nominated the following
Vice-Presidents for the ensuing year:—

Sir William de W. Abney, K.C.B. D.C.L. D.Sc. F.R.S.
Shelford Bidwell, Esq., M.A. Sc.D. F.R.S.
The Rt. Hon. Lord Kelvin, O.M. G.C.V.O. D.C.L. LL.D. D.Sc.
F.R.S.
Ludwig Mond, Esq., Ph.D. F.R.S.
Sir Thomas Henry Sanderson, G.C.B. K.C.M.G.
Sir Felix Semon, C.V.O. M.D. F.R.C.P.
Sir James Crichton-Browne, M.D. LL.D. F.R.S. (Treasurer).
Sir William Crookes, F.R.S. (Honorary Secretary).

Patrick M. Brophy, Esq.
W. J. Canter, Esq., R.N.
Eyre Crowe, Esq.
Harold Thornton Ellis, Esq.
Miss Emma Morgan,
Berkeley Portman, Esq.
Burnett Tabrum, Esq., J.P.

were elected Members of the Royal Institution.

Professor Emile Hilaire Amagat (of Paris).
Professor Ludwig Boltzmann (of Vienna).
Professor Julius Wilhelm Brühl (of Heidelberg).
Louis Paul Cailletet, Esq. (of Paris).
Professor Per Theodor Cleve (of Upsal).
Professor James Mason Crafts, LL.D. (of Boston).
Professor Pierre Curie (of Paris).
Madame Sklodowska Curie (of Paris).
Professor Emil Fischer (of Berlin).
Professor Friedrich W. G. Kohlrausch (of Berlin).
Professor Hans Landolt (of Berlin).
Professor Henri Louis Le Chatelier (of Paris).
Professor Gabriel Lippmann (of Paris).

Professor Henrik Antoon Lorentz (of Leiden).
Professor Dr. George Lunge (of Zurich).
Professor Edward Williams Morley, LL.D. (of Cleveland).
Dr. H. Kamerlingh Onnes (of Leiden).
Professor Edward Charles Pickering, LL.D. D.Sc. (of Harvard).
Professor Georg Hermann Quincke (of Heidelberg).
Professor P. Zeeman (of Amsterdam).

were elected Honorary Members of the Royal Institution.

The Special Thanks of the Members were returned to Mrs. Frank Lawson for her Donation of £50, and to Sir Thomas Sanderson, G.C.B., for his Donation of £5, to the Fund for the Promotion of Experimental Research at Low Temperatures.

The PRESENTS received since the last Meeting were laid on the table, and the thanks of the Members returned for the same, viz. :—

FROM

The Secretary of State for India—Census of India, 1901: Vol. I. Part 1, Report; Vol. IA. Part 2, Tables. fol. 1903.
    Geological Survey of India: Palæontologia Indica, Ser. XV. Vol. I Part 5. fol. 1903.
Accademia dei Lincei, Reale, Roma—Classe di Scienze Fisiche, Matematiche e Naturali. Atti, Serie Quinta: Rendiconti. Vol. XIII. 1o Semestre, Fasc. 7. 8vo. 1904.
American Geographical Society—Bulletin, Vol. XXXVI. Nos. 2-3. 8vo. 1904.
American Philosophical Society—Proceedings, Vol. XLII. No. 174. 8vo. 1903
Astronomical Society, Royal—Monthly Notices, Vol. LXIV. No. 5. 8vo. 1904.
Automobile Club—Journal for April, 1904.
Bankers, Institute of—Journal, Vol. XXV. Parts 4-5. 8vo. 1904.
Boston Public Library—Monthly Bulletin for April, 1904. 8vo.
Boston Society of Natural History—Memoirs, Vol. V. Nos. 8-9. 4to 1902-3
British Architects, Royal Institute of—Journal, Third Series, Vol. XI. Nos. 12-13 4to. 1904.
British Astronomical Association—Journal, Vol. XIV. No. 6. 8vo. 1904.
Buenos Ayres—Monthly Bulletin of Municipal Statistics for Feb. 1904. 4to.
Cambridge Philosophical Society—Proceedings, Vol. XII. Part 5. 8vo. 1904.
Cambridge University Press—Mathematical and Physical Papers of Sir George Stokes, Vol. IV. 8vo. 1904.
Canada, Department of the Interior—Dictionary of Altitudes in Canada. By J. White. 8vo. 1903.
Canada, Geological Survey—Altitudes in Canada. By J. White. (With Map.) 8vo. 1901.
Chemical Industry, Society of—Journal, Vol. XXIII. Nos. 7-8. 8vo. 1904.
Chemical Society—Proceedings, Vol. XX. No. 279. 8vo. 1904.
    Journal for April, 1904. 8vo.
Civil Engineers, Institution of—Proceedings, Vol CLV. 8vo. 1904.
Clowes, Frank, Esq., D.Sc. M.R.I. (the Author)—Experimental Bacterial Treatment of London Sewage. 8vo 1904.
Cracovie, Académie des Sciences—Bulletin: Classe des Sciences Mathématiques, 1904, Nos. 1-3; Classe de Philologie, 1904, Nos. 1-3. 8vo
East India Association—Journal, Vol. XXXV. No. 34. 8vo. 1904.
Editors—Aeronautical Journal for April, 1904. 8vo
    Astrophysical Journal for April, 1904. 8vo.
    Athenæum for April, 1904. 4to.

Editors—continued.
　　Author for May, 1904. 8vo.
　　Board of Trade Journal for April, 1904. 8vo.
　　Brewers' Journal for April, 1904. 8vo.
　　Chemical News for April, 1904. 4to.
　　Chemist and Druggist for April, 1904. 8vo.
　　Electrical Engineer for April, 1904. 4to.
　　Electrical Review for April, 1904. 4to.
　　Electrical Times for April, 1904. 4to.
　　Electricity for April, 1904. 8vo.
　　Electro-Chemist and Metallurgist for April, 1904. 8vo.
　　Engineer for April, 1904. fol.
　　Engineering for April, 1904. fol.
　　Engineering Review for April, 1904. 8vo.
　　Homœopathic Review for April–May, 1904. 8vo.
　　Horological Journal for April–May, 1904. 8vo
　　Journal of the British Dental Association for April, 1904. 8vo.
　　Journal of Physical Chemistry for March, 1904. 8vo.
　　Journal of State Medicine for April, 1904. 8vo.
　　Law Journal for April, 1904. 8vo.
　　London Technical Education Gazette for April, 1904. 4to.
　　London University Gazette for April, 1904. 4to.
　　Machinery Market for April, 1904. 8vo.
　　Model Engineer for April, 1904. 8vo.
　　Motor Car Journal for April, 1904. 8vo.
　　Motor Car World for April, 1904. 4to.
　　Musical Times for April, 1904. 8vo.
　　Nature for April, 1904. 4to.
　　New Church Magazine for April–May, 1904. 8vo.
　　Page's Magazine for April–May, 1904. 8vo.
　　Photographic News for April, 1904. 8vo.
　　Physical Review for April, 1904. 8vo.
　　Public Health Engineer for April, 1904. 8vo.
　　Science Abstracts for April, 1904. 8vo.
　　Travel for April, 1904. 8vo.
　　Zoophilist for April, 1904. 4to.
Electrical Engineers, Institution of—Journal, Vol. XXXIII. Part 2. 8vo. 1904.
Florence, Biblioteca Nazionale—Bulletin for April, 1904. 8vo.
Franklin Institute—Journal, Vol. CLVII. No. 4. 8vo　1904.
Geographical Society, Royal—Journal, Vol. XXIII. No. 5. 8vo. 1904.
Geological Society—Abstracts of Proceedings, Nos. 793, 794. 1904.
Humphrys, Mr. (the Author)—Heat. 8vo. 1904.
Imperial Institute—Technical Reports and Scientific Papers. Edited by W. R.
　　Dunstan. 8vo. 1903.
Johns Hopkins University—American Journal of Philology, Vol. XXIV. No. 4.
　　8vo. 1903.
Jordan, W. Leighton, Esq., M.R.I. (the Author)—Astronomical and Historical
　　Chronology. 8vo. 1904.
Kansas University—Science Bulletin, Vol. II. Nos. 1–9. 8vo. 1903.
Linnean Society—Journal: Botany, Vol. XXXVI. No. 253; Zoology, Vol. XXIX.
　　No. 189. 8vo, 1904.
Manchester Literary and Philosophical Society—Memoirs and Proceedings, Vol.
　　XLVIII. Part 2. 8vo. 1904.
Manchester Steam Users' Association—Twenty-first Annual Report of the Board
　　of Trade on the Working of the Boiler Explosions Act 1882, Nos. 1388–1456.
　　4to. 1904.
Meteorological Office—Report of the Meteorological Council to the Royal Society,
　　1903. 8vo. 1904.
Microscopical Society, Royal—Journal, 1904, Part 2. 8vo.

Munich, Royal Academy of Sciences—Sitzungsberichte, 1903, Heft V.   8vo.   1904.
Navy League—Navy League Journal for May, 1904.   8vo.
New Zealand, Registrar-General's Office—Statistics of the Colony for 1902.   fol.
     1903.
North of England Institute of Mining Engineers—Transactions, Vol. LI. Part 7;
     Vol. LIII. Part 3; Vol. LIV. Parts 3-4.   8vo.   1904.
     Subject Matter Index of Metallurgical Literature, 1901.   8vo.   1904.
Odontological Society—Transactions, Vol. XXXVI. No. 5.   8vo.   1904.
Pharmaceutical Society of Great Britain—Journal for April, 1904.   8vo.
Photographic Society, Royal—Journal, Vol. XLIV. No. 3.   8vo.   1904.
Quekett Microscopical Club—Journal for April, 1904.   8vo.
Rome, Ministry of Public Works—Giornale del Genio Civile for November, 1903.
     8vo.
Royal Society of Edinburgh—Proceedings, Vol. XXV. No. 2.   8vo.   1904.
Royal Society of London—Philosophical Transactions, B, Nos. 227-228.   4to.   1904.
     Proceedings, Nos. 492-493.   8vo.   1904.
     Year Book, 1904.   8vo.
Selborne Society—Nature Notes for April, 1904.   8vo.
Shanks, J., Esq. (the Author)—Some neglected aspects of the Fiscal Question.
     8vo.   1904.
Smith, B. Leigh, Esq., M.R.I.—Scottish Geographical Magazine, Vol. XX. No. 5.
     8vo.   1904.
Society of Arts—Journal for April, 1904.   8vo.
Stirling, W., M.D. (the Author)—Some Apostles of Physiology.   4to.   1902.
Sweden, Royal Academy of Sciences—Handlingar, Band XXXVII. Nos. 4-6.   4to.
     1903-4.
     Arkiv för Botanik, Band I. Heft 4.   8vo.   1904.
Tacchini, Prof. P., Hon. Mem. R.I. (the Author)—Memorie della Società degli
     Spettroscopisti Italiani, Vol. XXXIII. Disp. 3.   4to.   1904.
United Service Institution, Royal—Journal for April, 1904.   8vo.
United States Department of Agriculture—Monthly Weather Review for January,
     1904.   4to.
     Annual Summary, 1903.   4to.   1904.
     Experiment Station Record for March, 1904.   8vo.
United States Patent Office—Official Gazette, Vol. CIX. Nos. 4-9.   8vo.   1904.
Upsala, Observatoire Météorologique—Bulletin, Vol. XXXV.   4to.   1903-4.
Verein zur Beförderung des Gewerbfleisses in Preussen—Verhandlungen 1904,
     Heft 4.   8vo.
Vienna, Imperial Geological Institute—Verhandlungen, 1904, Nos. 2-4.   8vo.
Yorkshire Philosophical Society—Annual Report for 1903.   8vo.   1904.

WEEKLY EVENING MEETING,

Friday, May 13, 1904.

His Grace The Duke of Northumberland, K.G. D.C.L. F.R.S.,
President, in the Chair.

M. H. Spielmann, Esq., F.S.A.

The Queen Victoria Memorial.

(Abstract.)

On the morrow of the death of Queen Victoria, it was the universal
feeling that some monument, some visible memorial of her greatness
and goodness, and of the enduring love and respect of her mourning
subjects, should be set up here, in the metropolis of the Empire. It
is, of course, true that those who deserve a statue do not need one;
but a memorial is the consolation of the survivors, and, if rightly conceived, is as noble a tribute to their own unselfish gratitude and pride
as it is to the glory of the man or woman whom they desire to
honour.

A heavy responsibility lay upon the Committee appointed to determine the exact form the memorial was to take, but there was a
general confidence that the demands and requirements would be fully
appreciated. Examples abound—examples that might be followed
and examples that must be shunned. For there exist, as were presently to be shown, a few noble monuments to great rulers departed,
and many failures, monstrous, effete, vulgar in turn, which seem to
have served less to warn the public against the false in sentiment and
the bad in art, than to accustom and reconcile them to the sight of
what is poor, pompous, and meretricious. The nation therefore
expects that the memorial to Queen Victoria shall be the finest expression of national feeling and of national art to which the race can
give utterance.

The Committee consisted of Viscount Esher, Lord Windsor, Lord
Redesdale, Sir Edward Poynter, P.R.A., Sir William Emerson, Sir
Lawrence Alma-Tadema, R.A., Mr. Sidney Colvin, and General Sir
Arthur Ellis—a committee which commanded the confidence alike of
artists and public.

Seeing the necessity of appointing a sculptor, rather than opening a competition from which probably most leading sculptors would
on principle hold aloof, the committee placed the commission outright
in the hands of Mr. Thomas Brock, R.A.; and they decided on opening a limited competition of selected architects of the greatest eminence. On the winner would fall the duty of laying out the whole
ground from end to end, and providing the architectural setting.

Three represented England—Mr. T. G. Jackson, R.A., Mr. Ernest George, and Mr. Aston Webb, R.A.; one represented Scotland—Sir Rowand Anderson; and one, Ireland—Sir Thomas Drew, P.R.H.A.

[Before the designs of the artists were examined, a series of the leading monuments of Europe were thrown upon the screen, in order that the sight of what had been done in modern times by foreign countries and peoples, might better equip the audience for forming a clear judgment in the competition which the Committee were called upon to decide; and that they might compare this English scheme with what has been done elsewhere by the greatest artists by the most glory-loving and art-loving nation of Europe.]

The following monuments were taken in order, commented upon and criticised, and were pictorially illustrated upon the screen. First those of Russia: "Peter the Great," vast in size and wonderful in balance, which occupied twelve years in execution; the "Column to Alexander I." a monolith, 84 feet high, weighing 400 tons, and costing 400,000*l.*; "Nicholas I."; and "Catherine II.," 49 feet high, the finest of the works entirely carried out by Russian artists, though cast by an Englishman. The chief Austrian work, the monument to the "Empress Maria Theresa" (62 feet high), by von Zumbusch and Hasenauer, which occupied fourteen years, and cost 72,500*l.*; the French memorials to M. Gambetta and M. Carnot; the extraordinary model for the monument now being set up to "Alfonso XII.," by Signor Querol, since simplified from the excessively baroque sketch, scarcely to be considered as serious sculpture; the attenuated Gothic canopy at Laeken, highly decorated but ill-constructed, covering a statue of "King Leopold I.," the consolidator of the Belgian monarchy —were all passed in review. Then followed others.

The noble statue memorial to the "Emperor Frederick the Great," set up in Berlin in 1851 by the sculptor Rauch—with its horse and rider, its equestrian groups, its reliefs, and its 42 feet of height—is perhaps the finest of all such modern works now in existence. The "National Monument" of "Germania," in the Niederwald, symbolising the unity and strength of the newly established German Empire, cost 55,000*l.*, and occupied six years in construction, the height being not less than 111 feet. The two latest memorials in Berlin are both by Herr Reinhold Begas, the German Emperor's favourite sculptor: the one of "Bismarck" and the other of "The Kaiser Wilhelm the Great," both colossal in size, and both, in spite of all the cleverness displayed, conspicuously lacking in certain qualities which should always distinguish the plastic art. The last-named work, 29 feet high, is backed by a long colonnade—a feature common to the "Empress Augusta" monument at Coblenz (by Professor Nives and Carlo Schmitz), to the Spanish monument already alluded to, and to Mr. Brock's memorial to Queen Victoria, and by far the most noteworthy of all modern monuments now in existence —the vast national "Memorial to King Victor Emanuel," now in course of erection at Rome. This enormous structure, a vast scenic, architectonic screen on the Capitoline Hill, by Signor Sacconi, was begun in 1884, and was estimated to cost 360,000*l.*; but, by 1898, 1,040,000*l.* had been spent, and by the time it is finished, with its colossal equestrian by Signor Chiaradia, its vast halls, terraces, etc., not less a sum than 2,000,000*l.* will have been expended—the tribute of impoverished Italy to her great King; while to our greater Queen, wealthy England has as yet subscribed but one-eighth of that amount.

With these works before our eyes, and the details of cost and periods of construction, we should be in a better position to form a fair judgment of what the Queen Victoria Memorial is to be. The conditions of the competition were verbal and necessarily vague, and wide enough to avoid hampering the architects; and as the amount that would be subscribed could not be known, no estimate or limit of cost could be suggested. A plan of the Mall was provided; a suggestion was made as to the designing of a Processional Road, opening out at the eastern end, with an architectural setting for the monument and with a necessary laying out of the ground on the west, with the further statement that Mr. Brock would design the monument, to which the architectural structures in the west would necessarily be subordinate. Mr. Brock met the architects in council, and explained his ideas by the rough model he has made.

In July 1901 the announcement of Mr. Aston Webb's success was made, and a public exhibition of the various designs was held in St. James's Palace. In view of the great skill and eminence of each competing architect, much was to be said for each individual design, but there is little doubt that the general feeling has cordially endorsed the finding of the Committee and the approval of the King.

[These designs were next considered—that of Mr. T. G. Jackson, with its noble and imposing triumphal arch, its central allée, and its colonnaded forecourt; that of Mr. Ernest George, with its beautiful five-way gate-way, and its original elongated parterre and fountains east and west, its grouped columns, and semi-circular screen; that of Sir Thomas Drew, with its re-alignment of the Mall, with its forecourt of stately buildings on the east, its triumphal arch, and particularly the reconstructed face of Buckingham Palace (which detail by itself would cost from 100,000*l.* to 120,000*l.*); that of Sir Rowand Anderson, with its great archway at the eastern end, and the elaborate design for the statue itself at the west, at the centre of a semicircular plateau—all were passed in review, in elevation and plan. In respect to the two last-mentioned architects, it was pointed out that neither the refacing of the palace, nor the designing of the statue, came within the scope of the competition.]

The design of Mr. Aston Webb has been held to comprise the greatest number of the best and most practical ideas; with the fine and eminently practical opening to the east, with good vistas, and a general symmetrical design and ensemble, relatively modest, and well calculated to be a setting to the monument and not unduly to vie with it, the appropriateness, practical and ornamental, of the design from beginning to end—especially with certain modifications now introduced—these were the merits which have secured it the support of the Committee to whom fell the difficult task of selecting among the fine designs of several of the leading architects of the day. Mr. Webb is to be credited with having the widest experience in laying out plans on a large scale, and in such work in harmonising the claims of the artist with the needs of the public and the exigencies of official requirements. He has rejoiced in the opportunity, rare in England, of bringing a fine road straight up to the great feature to be viewed. Although everyone abroad recognises this simple and obvious truth, in England it has been consistently ignored even to the present day. In other countries the road leads up to the object, or palace, or view; in England we approach them sideways. You cannot drive straight up to the Banqueting Hall at Whitehall, to the Mansion House, to the Bank, to the Houses of Parliament, to Somerset House, to the British Museum, or see them from any approaching roadway without turning your head. St. James's Palace is one of the only buildings which can fairly be driven up to. Even the recently erected Law Courts and Albert Memorial are passed by sideways on the road.

If the Processional Road were carried straight, with a view to making an outlet on the east, it would have necessitated the acquisition of Drummond's Bank at a probable cost of 150,000*l.* or 200,000*l.*, and the result would be an oblique opening into Whitehall, ending nowhere, and revealing a view of a gigantic specimen of extremely commonplace hotel architecture.

Therefore, taking the central axis of the West Strand, and having in mind the noble effect of that great thoroughfare, making it intersect with the axis of the Mall, Mr. Webb obtained a point just behind Drummond's Bank, where he masks more or less the change of axis by a large circular " place," in the centre of which he proposes to erect a statue of Queen Victoria at the time of her accession, so that the figure of the Queen at the commencement of her reign, begun in hopefulness, looks towards the end of it, accomplished in splendour. Large spaces are equally to be left opposite Waterloo Place and Marlborough Gate, until with a pleasing series of statuary at the west we come to the great central feature of Mr. Webb's design. Here we have careful and thoughtful planning, with approaches from Constitution Hill on the right and Buckingham Palace on the left, symmetrical and monumental, with a large opening into the forecourt from the Mall as well as north and south, from all three of which points an unimpeded view of Mr. Brock's monument may be obtained. There are to be fountains inside the colonnade, and the monument is backed by a screen. By the later modifications, the roadway outside the quadrant of the colonnade is brought within and nearer to the palace, and the traffic is taken inside the great forecourt, in order that the people should not be banished so far away, and that the gaiety and contiguity of passers-by may be retained. The double colonnade is therefore brought up to the water edge, and the present balustrade as it is seen to-day is but the first course of the outer part of it. Other alterations have been made, but it is hoped the charming vista of the Foreign Office and Westminster in the distance may still be secured. The high architectural screen originally designed is now replaced by a lower but still more effective balustrade nearer to the Palace, whereby Mr. Brock's memorial will not be interfered with, and the view from the palace windows will no longer be impeded.

Taking the Whitehall end, we see that the great circus, somewhat reduced from the original size, is brought slightly westward, and space is thus made for the erection of a large Government building of a decorative and monumental character, with a circular "place" on each side of it. It is intended that this new building should be devoted to Admiralty purposes, and in the centre of it will rise a great monumental archway, with which the Processional Road will end worthily, for a utilitarian Government, with a public building. The eastern and western faces of the building will be crescent-shaped, so that the slight irregularity of its site and the change in its alignment will be, it is hoped, entirely masked. From this point the Processional Road, as far as the space opposite the Duke of York's column, is narrower than the rest, owing to exigencies of the ground, but between there and the Memorial the measurements are more imposing. The total number of trees, with those newly planted, remains the same as before. The width of the chief road is 65 feet, which in relation to its length and to the traffic does not compare ill with the 90 feet of the Champs Élysées. There is an allée of trees with paths 25 feet wide on each side. These form vistas, and will be furnished at each break with groups of sculpture. The statues at the foot of the steps below the Duke of York's column will represent the Eastern Dominions—that is to say, there will be handsome gateways, and between them an important group symbolical of the "Genius of Orient." Four subsidiary groups of statuary will represent "Ceylon," "Barbadoes," the "China Ports," and "India." Opposite Marlborough Gate there will be another handsome gateway into the park, and the "place," 100 feet across, will be embellished with groups of statuary symbolical of the Western Dominions, comprising "Australia," "Canada," "Newfoundland," and "New Zealand." At the Western end of the Processional Road are the two groups of statuary representing South Africa—"the Cape" and "Natal." The ugly electric light standards now set up are temporary. An elaborate system of lamps, in groups and clusters, will be Mr. Brock's special charge, and will be in sculptured bronze, so that there will be, as regards the memorial itself, three concentric circles of light, which will mark the construction, and which, with the lamps along the stone balustrade enclosing the forecourt nearer to the Palace (the balustrade being 10 feet high), will produce a very fine effect at night. There will be an opening of the Memorial from Birdcage Walk on the south, whence a view of one of the finest groups will be afforded, and similarly, from the north, another, by a tree-skirted *tapis vert*—a grass-planted avenue of the same width as the main roads (65 feet), which will extend northward in the Green Park, and so bring the monument directly in touch with Piccadilly. It will also be approached by handsome gates and piers.

Mr. Webb has adopted a style of quiet and unpretentious Classic type, dignified, harmonious, and sufficiently monumental, and consistently in sympathy with the work of the sculptor, so that the result of this cordial collaboration will be a great work of decoration, which will be a fine ornament to London, and a worthy bequest of the twentieth century to the London of the future.

[Coming to the work of Mr. Brock, the lecturer threw upon the screen several of his most notable works in order to show the sculptor's development, during his career, in style and achievement.]

The final model for the great central monument, completed in its parts, has been elaborated as a precaution against such eventualities as fate might have in store for the sculptor before he began a single touch on the work itself. The model is 7 feet 6 inches high, and being on a one-tenth scale, represents a total height from the base to the top of the Victory's wing of 75 feet. This idea is to represent the Great Queen sitting amid the personifications of the personal qualities that made her great. At the right is a group representing "Justice," at the left "Truth," and at the back "Motherhood." "Courage" on the right and "Constancy" on the left, qualities which with the others bring about the triumph of "Victory"—"Victory" which surmounts and dominates as it were the whole structure of her virtues and surmounts her glorious reign. Round the base are freely-treated ships' prows, two bearing trophies suggestive of the Army and Navy, and two, fruits and flowers suggestive of Commerce and Prosperity. Above the slightly pyramidal core at the feet of Constancy and Courage appears an Eagle—the Eagle of Empire, a most valuable and felicitous architectural form.

The statue of the Queen, which as it sits will be 12 feet high, is draped in robes of State, the only instance in which the seated Queen wears her robes from the shoulders. On each side of the platforms is a great fountain discharging down steps into a basin 160 feet long by 28 feet across. That on the right typifies "Power," with figures representing the Army and Navy—the foundations of the Monarchy. A Sea-Nymph is below, and on the right and left are bronze reliefs elaborating the idea of the two Services. The parapet is 8 feet 6 inches high. Corresponding with this *motif* of Power, there is on the other side "Intelligence," the figures representing Science and Art. A Triton reclines below, and the figures represent Progress, Science and Art. The water is obtained from the overflow of the Serpentine, will run day and night, and fall into the ornamental waters of St. James's Park below. Around the central feature four statues accompanied by lions will be erected.   They represent " Progress" and " Peace" in front, and "Manufacture" and "Agriculture" behind.   The subjects of the two fountains to be erected in the semi-circular colonnade are " Empire " and " Progress," and gushing water will make continual play of light and sound, and add gaiety to the grass and flower-planted enclosure called " The Queen's Garden." The whole work is to be carried out in the finest and most enduring materials, both the sculptural and architectural portions.   The groups of sculpture on the Processional Road will also be under the control of Mr. Brock, who will doubtless place commissions among the most competent of our younger sculptors, and, being responsible for them, will see that they harmonise in design and in spirit with the general scheme. ,

The one unsatisfactory feature is the necessary retention of the present commonplace façade of the Palace, which can only be regarded as a miserable anti-climax.   Mr. Webb improves the sky-line by placing Mansard roofs on the centre and ends of the east front; but that is not enough.   Now that we have a magnificent Processional Road, one of the finest in Europe, to lead up to the Palace, we ought to have a fine Palace for this magnificent road to lead up to.   When that is done we shall have performed our task completely, and paid the tribute of our homage to Queen Victoria.

So far as the work has proceeded it presents a mere sketch of the imposing aspect of the completed work.   It may be finished in five to seven years, a period which is but a brief space compared with the time occupied on the Albert Memorial, and the Nelson Memorial, and the years which have been expended on the Victor Emanuel Monument in Rome.   We shall see a memorial worthy of the artists who fashioned it, and not unworthy, it is hoped, of the Nation who raised it and of the noble Sovereign whose memory it is designed to honour.

WEEKLY EVENING MEETING,

Friday, May 20, 1904.

His Grace The Duke of Northumberland, K.G. D.C.L. F.R.S., President, in the Chair.

Professor Ernest Rutherford, M.A. D.Sc. F.R.S

The Radiation and Emanation of Radium.

[No Abstract.]

WEEKLY EVENING MEETING,

Friday, May 27, 1904.

THE RIGHT HON. LORD KELVIN, G.C.V.O. D.C.L. LL.D. D.Sc. F.R.S.,
Vice-President, in the Chair.

H.S.H. ALBERT I., PRINCE OF MONACO.

The Progress of Marine Biology.

FOR some few years past the advances made by oceanography have
been very marked, thanks to the rivalry which has grown up
between different peoples. The English, the Americans, the Germans,
the Belgians, the Scandinavians, and the Russians have made great
efforts in this direction, while France, Italy, Austria and Portugal
have not remained outside of the movement. Consequently this
science in its principal features is already pretty well known.

But oceanography touches many departments of science, and
amongst them marine biology is for the moment the least advanced,
because it requires researches of a particularly difficult kind. It is
to it that I have more particularly devoted my attention, and it is of
it that I propose to speak this evening.

From the reports of many important expeditions, you are already
well aware how universally distributed life is, even in the greatest
depths of the sea; nevertheless, the means employed in this kind of
investigation have been, as a rule, too primitive to furnish very
complete results. In my own personal oceanographical work I
have, for long, employed new means and methods, which attract
different groups of marine animals, each according to its own
characteristic instincts, and I have been able in this way to add to
our knowledge of zoology.

It is not, however, enough to collect. We must also endeavour
to penetrate the mystery of the laws which regulate life in the
medium of the sea, so different in almost all respects from that of
the air. For this the oceanographer requires the collaboration of
the biologist and the physiologist.

Not unfrequently unexpected circumstances open to the observer
new horizons, to be afterwards explored by science. It is thus that
finding myself among the islands of the Azores, to which my oceanographical researches have frequently conducted me, I assisted at the
capture of a *cachalot*, or sperm whale, by the whalers of the country;
simple peasants, who launch their well appointed whale boats the
moment that the appearance of a fish is signalled by the look-out
man, who is continually stationed on a little hill in their neighbour-

hood, and I have seen how these mammals go to the intermediate depths of the ocean in search of the great cephalopods which form their exclusive nourishment. When the *cachalot* in question came to endure the convulsions of death, its stomach rejected enormous fragments of the prey which it had captured during its last sounding.

It is in this way that I have recognised the existence of a fauna remarkable for the size and the number of its components, relegated to the enormous space which separates the surface from the great depths, but whose organisation prevents its rising to the regions illuminated by the light of the sun and probably also its descending to the bottom, when this lies beyond a certain depth.

What other groups of living animals inhabit these regions? We know nothing of them yet, but we may believe that they abound, because beings as powerful as these cephalopods require much nourishment.

So soon as I understood the importance of researches capable of throwing light on the life which exists in regions inaccessible to our ordinary means, I established on board of my ship all the equipments of a whaler, namely, three whale boats each carrying a harpoon gun, several harpoons, a lance and a thousand metres of line, and I added to the complement of my ship an experienced Scottish whaler. The results of this organisation have left nothing to be desired. The cetaceans obtained already form an interesting collection, and their stomachs were abundandantly furnished with these cephalopods.

In the Mediterranean, where previously the cetaceans had never been hunted, I have taken several species of *Grampus griseus*, of *Orca gladiator*, of *Globiceps melas*, and I lost a *Balinoptera musculus*. In the Atlantic Ocean I have found several *Globiceps* and *Grampus*, as well as a very rare specimen of dolphin, *Steno rostratus*. I have also lost a cetacean of moderate size but of undetermined species.

The attack of cetaceans, especially when they are large, causes the harponeering novice an emotion which diminishes his *adresse*; and even for a good shot the use of the harpoon gun is very difficult when there is the least motion of the sea. A troop of animals has been signalled. Their presence has been revealed by their blowing, or by the regular reappearance of their backs at a greater or less distance from the ship, which is then steered towards them. If the animals are of the species already mentioned, the movement of the propeller does not trouble them ; on the contrary, they may almost always be seen to come and take up station under the stern as if retained by curiosity. But some species, and among them the *cachalot*, seem to distrust this neighbourhood, and care must be taken that they do not hear even the too marked sound of oars ; indeed in such cases it is preferable to use paddles rather than oars.

The animals have found in the depth a favourable hunting ground, and they do not leave it. They sound to this depth during a time which varies from ten to forty-five minutes, according to the species, and come to the surface again to breathe during four or five minutes.

FIG. 1.— Breaking up a Sperm Whale.

FIG 2.—Harpooning a Whale.

FIG. 3.—Towed by a Grampus

FIG. 4.—Hauling an Orca on board.

These alternations repeat themselves, sometimes for several hours consecutively, almost on the same spot, with occasional pauses, which seem to be those of repose. It is when the cetaceans appear in this way at the surface that the nearest whale-boat should make every endeavour to come up with them before they again disappear, and so soon as one of them gives a sufficiently good presentation of the part of its body near the head, the harponeer fires his shot. But this critical moment seldom arrives until after several hours of pursuit, even when the animals are full of confidence and allow the whalers to get well in amongst them. Most frequently, and under the most favourable circumstances, it happens that during the three or four seconds which the emergence of the animal at each of his eight or ten respirations lasts, the presentation is bad, or the move- of the sea has destroyed the aim ; it is then necessary to wait till after the next sound.

If the animals signalled pursue a fixed route with any speed, it is useless to attempt the attack ; it is impossible to come up with them because they are then on passage. Once I followed a large *Balinoptera* for six hours with my ship. He travelled about thirty miles in an absolutely straight line, which shows that the marine animals possess a sense of orientation more remarkable than that of the migratory birds, because these can always see the ground above which they travel.

At last, close to the boat, a powerful blow like a jet of steam comes out of the water ; the back of the animal emerges immediately afterwards ; in the movement necessary to recover the horizontal position of its head, the dorsal fin appears and finally the lumbar region, which is much curved by the action of the tail, which determines the descent. It now proceeds for several lengths, hardly submerged, whilst the steersman, who can see the lighter-coloured portions of this immense body, and sometimes a certain motion of the dorsal fin, steers the boat, driven by all the force of its crew, so as to cut the route of the cetacean. A fresh blow cuts the water, a black back presents itself at a distance of five or six metres, the shot is fired, and the eye can follow the harpoon with the attached line.

But at the first moment there is nothing to show that the animal has been touched. In a body of such size the arrival of sensation in the brain and the transmission of the will to the periphery require a sensible time. The success of the harponeer is indicated by the rapid running out of the line, which very soon produces heat and a dense smoke in the bollard round which a turn is taken in order to allow the harponeer to regulate the run of the line according to the velocity of the cetacean and the direction which it follows. This is a very delicate moment for the safety of the whaler ; nobody moves, and the turns of the line, carefully coiled in a receptacle, run out without a check. A second whaler approaches in order to take the end of this line, when it is apparent that the thousand metres in the first boat will not be sufficient, and to add it to his own line. The running out is continued from this boat, and sometimes the three whale boats are rapidly cleared of their lines. But, with the friction which such a length of line offers, and to which the resistance of the boats towed has to be added, the cetacean reduces its speed very materially, so that there is no difficulty in maintaining it. Little by little the line is got back into the boats, and after various alternations the weakened animal advances more and more slowly, and close to the surface, where it is obliged to breathe more and more frequently.

Often many hours have passed before the favourable moment arrives for despatching the unfortunate victim and terminating the drama, and this is accompanied by the most serious circumstances of the whole enterpise. The exhausted animal stretches itself on the surface, almost motionless before the boat, where the harpooneer now holds a lance which has a considerable length, because it must pass through the whole thickness of the blubber and of the muscles before it reaches the vital organs. He approaches the animal by its side, so as not to be struck by the tail, which is thrown violently into the air so soon as the cetacean receives this new wound; but it is not always possible to avoid being struck by a fin, and especially in the case of large animals, this may wreck a boat. In spite of all the skill of the crew an accident of this kind may occur, and I speak only from memory of cases mentioned by various captains in which a *cachalot*, an old and solitary individual, has seized and crushed between its jaws the boat which has attacked it. It has even been reported that two ships have been sunk by these animals in their fury, their enormous wedge-shaped head becoming in these circumstances a formidable ram.

When a cetacean of any size has been several times pierced, the red track which spreads far over the sea gives the idea of great carnage. In fact the cetaceans contain a very large amount of blood, and before the last hour when they lose it in torrents, they have already left behind them a track of ten or fifteen kilometres in length over which they have towed the boats.

I have said that apart from the interest which each species of cetacean offers of itself (and it appears that many of them are hardly known at all), it is in the first place the contents of their stomachs which occupy us. The species which I have taken differ much in the nature of their prey, and their mouths are armed correspondingly. The *right whale* is content to absorb the *Plankton* composed of extremely small animals, which in some regions form a compact mass, a real cloud; and in order to keep out objects too large to pass down its very small throat, its jaws are furnished with the well known and valuable *whalebone*, which acts as a sieve.

The Grampus, the Globiceps and the Cachalot, penetrate to a depth probably much greater in search of cephalopods, and they possess a dentition specially organised for seizing the gelatinous flesh of the cephalopods. The scars which they bear over the whole of

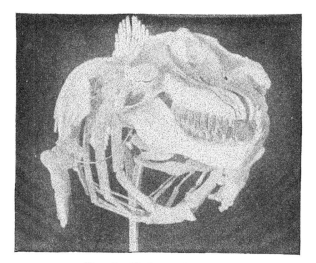

FIG. 5.—Skeleton of an Orca.

FIG. 6.—Part of the Fin of a Gigantic Cephalopod.

their bodies are evidence of the energy with which their victims defend themselves with their suckers, often armed with formidable talons.

The Orca, provided with a more compact dentition, pursues the dolphins, of which it makes scarcely more than three or four mouthfuls, showing thus a remarkable power of digestion.

The dolphins themselves are more eclectic, and I have found in their stomachs several species of fish as well as cephalopods, but in both of them the characteristics special to great depths are wanting.

The principal object which I had in view in capturing the cetaceans, the knowledge of certain beings living in the abysses, has been realised by the acquisition of a certain number of new and very rare cephalopods, some of which are gigantic, amongst which may be cited *Lepidoteuthis Grimaldii*, one of the most remarkable animals of the sea on account of its considerable size, and also because, though it is a cephalopod, it possesses scales like a fish.

The more we know of marine biology, and the more we learn from it of the links which connect the creatures spread over our planet, of the interpenetration of types, such as that shown by Lepidoteuthis, as well as of the vital force, the great power of reproduction, the number of individuals in certain species, and the high antiquity of other forms, we seem to be justified in imagining that the sea may have been the cradle of organic life when the cooling of the atmosphere determined the precipitation of the waters.

WEEKLY EVENING MEETING,

Friday, June 3, 1904.

His Grace The Duke of Northumberland, K.G. D.C.L. F.R.S.,
President, in the Chair.

Professor Svante Arrhenius, of Stockholm, *Hon. Mem. R.I.*

The Development of the Theory of Electrolytic Dissociation.

At first sight nothing seems to be more evident than that everything
has a beginning and an end, and that it is possible to divide everything.   Nevertheless, the philosophers of antiquity, especially the
Stoicists, concluded on purely speculative grounds, that these opinions
are not at all necessary.   The wonderful development of science has
reached the same conclusion as these philosophers, especially Empedocles and Democritus, who lived about 500 years B.C., and for whom
the ancients had already a vivid admiration.

Empedocles professed that nothing is made of nothing, and that
it is impossible to annihilate anything.   All that happens in the
world depends upon a change of form, and upon the mixture or the
separation of bodies.   Fire, air, water and earth are the four elements of which everything is composed.   An everlasting circulation
is characteristic of Nature.

The doctrine of Democritus still more nearly coincided with our
modern views.   In his opinion bodies are built up of indefinitely
small indivisible particles, which he called atoms.   These are distinguished by their form and magnitude, and also give different products by their different modes of aggregation.

This atomic theory was revived by Gassendi about 1650, and
then accepted by Boyle and Newton.   The theory received a
greatly increased importance by the discovery by Dalton of the
law of multiple proportions.   For instance, the different combinations of nitrogen with oxygen contain, for each unit weight of
nitrogen, $0\cdot57$, $1\cdot14$, $1\cdot72$, $2\cdot29$ or $2\cdot86$ unit weights of oxygen.[*]
Between these combinations there is no intermediate proportion.
This peculiarity is characteristic of chemistry in contradistinction to
physics, where the more simple continuous and gradual transition
from one state to another prevails.   This difference between the two
sister-sciences has often caused controversies in the domain of
physical chemistry.   The occurrence of discontinuous changes and of

---

[*] To explain this we suppose, in accordance with Dalton, that the molecules
of the different combinations of nitrogen with oxygen contain two atoms of nitrogen
and one, two, three, four or five atoms of oxygen.

multiple proportions has frequently been assumed, when a closer investigation has found nothing of the sort.

The law of multiple proportions is the one fundamental conception upon which modern chemistry is built up. Another is the law of Avogadro, which asserts that equal volumes of different gases under like conditions of temperature and pressure contain the same number of molecules. This conception, dating from the beginning of the nineteenth century, was at first strongly combated, and it was its great value in explaining the new discoveries in the rapidly growing domain of organic chemistry which led to its general acceptance in the middle of the past century, after Cannizzaro had argued strongly in its favour.

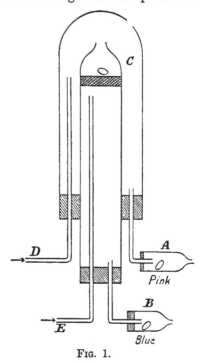

FIG. 1.

There were, however, some difficulties to be removed before Avogadro's law could be accepted. For instance, it was found that the molecular volume of sal-ammoniac, NH_4Cl, in the gaseous state was greater than might be expected from its chemical composition. This led to the supposition that the molecules of sal-ammoniac when in the gaseous state are partially decomposed into ammonia, NH_3, and hydrochloric acid, HCl. Indeed v. Pebal and v. Than succeeded in showing that this really happens. They used an apparatus that is shown in the annexed figure (Fig. 1). Two co-axial tubes are placed the one inside the other by means of a cork. The outer tube was closed at its upper end, the inner one was open, and contained at C a diaphragm of asbestos and above that a piece of sal-ammoniac. The upper end was heated by an air-bath, so that the piece of sal-ammoniac was volatilised. After this a current of hydrogen was led through both glass-tubes D and E. Now ammonia diffuses more rapidly than hydrochloric acid; if, therefore, the vapour of sal-ammoniac is partially decomposed into ammonia and hydrochloric acid, we should expect that above the asbestos diaphragm there would be an excess of hydrochloric acid and beneath it an excess of ammonia. This v. Pebal showed to be the case. The hydrogen-current from D showed an acid reaction on a piece of litmus-paper in A, and that from E showed an alkaline reaction on a similar piece of litmus-paper placed in B. It was objected that the decomposition might possibly be caused by the asbestos of the diaphragm, or by the hydrogen. v. Than, therefore, made a diaphragm of sal-ammoniac, and substituted nitrogen for hydrogen, but the effect was the same.

These experiments were performed in the years 1862 and 1864. They were based on the doctrine of dissociation, which was at that time (1857) worked out by Ste. Claire-Deville, and developed by his pupils. From the most ancient times use was made of the fact that limestone at high temperatures gives off carbonic acid, and that quicklime remains. This and similar processes were studied by Ste. Claire-Deville. He found that the same law is valid for the pressure of carbonic acid over limestone and for the pressure of water vapour over liquid water at different temperatures. On these fundamental researches the theory of dissociation was based, a theory which has subsequently played an ever-increasing rôle in chemistry, and whereby a broad bridge was laid between physical and chemical doctrines.

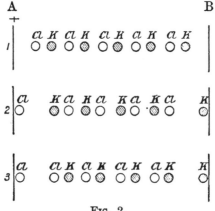

FIG. 2.

At almost exactly the same time we find in the writings of Clausius on the conductivity of salt solutions the first traces of an idea that salts or other electrolytes may be partially dissociated in aqueous solutions. Buff had found that even the most minute electric force is sufficient to drive a current through a solution of a salt. Now after the scheme of Grotthuss, at that time generally accepted, the passage of the electric current through a solution is brought about in such manner that the conducting molecules, e.g. of potassium chloride (KCl), are divided into their ions, which combine again with one another in the following manner. At first, as the current is closed, the electrode A becomes positive and the electrode B negative. All the conducting molecules KCl arrange themselves so that they turn their positive ions (K) to the negative electrode B, and their negative ions (Cl) to the positive electrode A. After this, one chlorine ion is given up at A and one potassium ion at B, and the other ions recombine, so that the K of the first molecule takes the Cl of the second molecule, and so on (Fig. 2). Then the molecules turn round under the influence of the electric force, so that we get the scheme 3 and a new decomposition can take place. This represents the Grotthuss' scheme, that supposes continuous decompositions and recombinations of the salt molecules.

As such exchanges of ions between the molecules take place even under the influence of the weakest electromotive forces, Clausius concluded that they must also take place if there is no electric force, i.e. no current at all. In favour of his hypothesis he pointed to the fact that Williamson, as far back as 1852, in his epoch-making theory of the formation of ethers, assumed an analogous exchange of the constituents of the molecules. At this exchange of ions it might sometimes, though extremely rarely, happen that an ion becomes free in the solution for a short time; at least such a conception would be

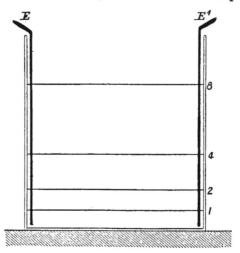

Fig. 3.

in good agreement with the mechanical theory of heat, as it was developed by Krönig, Maxwell, Clausius and others at that time.

In the meantime, Bouty, and particularly Kohlrausch worked out the methods of determining the electric conductivity of salt solutions. In 1884 I published a memoir on this subject. I had found that if one dilutes a solution—e.g. of zinc sulphate—its conductivity per molecule, or what is called its molecular conductivity, increases not infinitely, but only to a certain limit. We may figure to ourselves an experiment performed in the following manner (Fig. 3). In a trough with parallel walls there are placed close to two opposite sides two plates of amalgamated zinc, E E$_1$. On the horizontal bottom of the vessel there is placed a layer of solution of zinc-sulphate that reaches the level 1. The conductivity may be k_1. After this has been measured we pour in so much water, that after stirring the solution the level reaches 2, which lies as much above 1 as this lies above the bottom.

The conductivity is then found to be increased, and to have the value k_2. Increasing in the same manner, the volume, by addition of pure water until it is doubled, the level 4 is reached, and the conductivity is found to be greater than in the previous case—say k_4. So we may proceed further and further; the conductivity increases, but at the end more slowly than at the beginning. We approach to a final value, k_8. This is best seen in the next diagrams, which represent the newer determinations of Kohlrausch (Figs. 4, 5).

I explained this experiment in the following manner. The conductivity depends upon the velocity with which the ions (Zn and SO_4) of the molecules ($ZnSO_4$) are carried through the liquid by

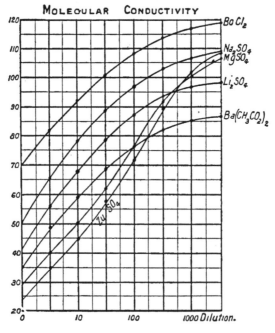

FIG. 4.

the electric force, i.e. the potential difference between E and E_1. If this potential difference remains constant, the velocity depends only on the friction that the ions in their passage through the liquid exert on the surrounding molecules. As these, at higher dilutions, are only water molecules, it might be expected that the conductivity would remain constant and independent of the dilution if it be supposed that all molecules, $ZnSO_4$, take part in the electric transport. As experiment now teaches us that the molecular conductivity increases with the dilution, even if this is very high (1000 or more molecules of water to one molecule of $ZnSO_4$), we are led to the hypothesis that not all, but only a part of, the $ZnSO_4$ molecules take part in the transport of electricity. This part increases with the dilution in the same proportion as the molecular conductivity k. The limiting value k_8 is approached at infinite dilution, and corresponds to the limit that all molecules conduct electricity. The conducting part of the molecules I called the active part. It may evidently be calculated as the quotient $k : k_8$.

If now this new conception were only applicable to the explanation of the phenomena of electric conductivity, its value had not been so very great. But an inspection of the numbers of Kohlrausch and others for the conductivity of the acids and bases, compared with the measurements of Berthelot and Thomsen on their relative strength

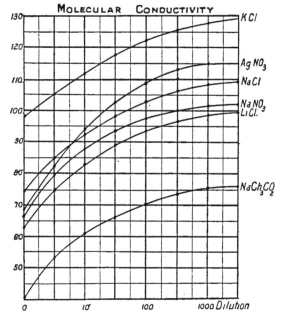

Fig 5.

with regard to their chemical effect, showed me that the best conducting acids and bases are also the strongest. I was thereby led to suppose that the electrically active molecules are also chemically active. On the other hand the electrically inactive molecules are also chemically inactive. In this connection I would mention the remarkable experiments of Gore, which were easily explained by the new manner of view. Concentrated hydrochloric acid, free from water, has no action on oxides or carbonates. Now this hydrochloric acid is almost incapable of conducting the electric current, whereas its aqueous solutions conduct very well. The pure hydrochloric acid contains, therefore, no (or extremely few) active molecules, and this agrees very well with the experiments of Gore. In the same way we explain the fact that concentrated sulphuric acid may be preserved in vessels of iron plates without destroying them, whereas this is impossible with the diluted acid.

An unexpected conclusion may be deduced from this idea. As all electrolytes in extreme dilution are completely active, then the weak acids must increase in strength with the dilution, and approach to the strength of the strongest acids. This was soon afterwards shown by Ostwald to agree with experiments.

The Norwegian natural philosophers, Guldberg and Waage, had developed a theory according to which the strength of different acids might be measured as well by their power of displacing another acid in solutions as by their faculty to increase the velocity of chemical reactions. Therefore, we may conclude that the velocity of reaction, induced by an acid, would be proportional to the quantity of active molecules in it. I had only a few experiments by Berthelot to demonstrate this proposition, but in 1884, Ostwald published a great number of observations that showed this conclusion to be true.

The most far reaching conclusion of the conception of active molecules was the explanation of the heat of neutralisation. As this is much more easily understood by means of the theory of electrolytic dissociation, I anticipate this for a moment. According to this theory strong acids and bases, as well as salts, are at great dilution (nearly) completely dissociated in their ions, e.g. HCl in $\overset{+}{\mathrm{H}} + \overset{-}{\mathrm{Cl}}$, NaOH in $\overset{+}{\mathrm{Na}} + \overset{-}{\mathrm{OH}}$ and NaCl in $\overset{+}{\mathrm{Na}} + \overset{-}{\mathrm{Cl}}$. But water is (nearly) not dissociated at all. Therefore the reaction of neutralisation at mixing a strong acid, e.g. HCl with a strong base, e.g. NaOH, both in great dilution, may be represented by the following equation:

$$(\overset{+}{\mathrm{H}} + \overset{-}{\mathrm{Cl}}) + (\overset{+}{\mathrm{Na}} + \overset{-}{\mathrm{OH}}) = (\overset{+}{\mathrm{Na}} + \overset{-}{\mathrm{Cl}}) + \mathrm{HOH};$$

or,

$$\overset{+}{\mathrm{H}} + \overset{-}{\mathrm{OH}} = \mathrm{HOH}.$$

The whole reaction is equivalent to the formation of water out of both its ions, $\overset{+}{\mathrm{H}}$ and $\overset{-}{\mathrm{OH}}$, and evidently independent of the nature of the strong acid and the strong base. The heat of any reaction of this kind must always be the same for equivalent quantities of any strong acids and bases. In reality it is found to be 13,600 cal. in all cases. This thermal equality was the most prominent feature that thermo-chemistry had discovered.

It was now asked in what respect the active state of the electrolytes differs from the inactive one. On this question I gave an answer in 1887. At that time van't Hoff had formulated his wide-reaching law that the molecules in a state of great dilution obey the laws that are valid for the gaseous state, if we only replace the gas-pressure by the osmotic pressure in liquids. As van't Hoff showed, the osmotic pressure of a dissolved body could much easier be determined by help of a measurement of the freezing point of the solution than directly. Now both the direct measurements made by De Vries, as also the freezing points of electrolytic solutions, showed a much higher osmotic pressure than might be expected from the chemical formula. As for instance the solution of 1 gram-molecule of ethylic alcohol—$C_2H_5OH = 46$ grammes—in one litre gives the freezing-point $-1 \cdot 85°C$., calculated by van't Hoff the solution of 1 gram-molecule of sodium chloride—$NaCl = 58 \cdot 5$ grammes—in one litre gives the freezing-point $-3 \cdot 26° = -1 \cdot 75 \times 1 \cdot 85°$ C. This peculiarity may be explained in the same manner as the "abnormal" density of gaseous sal-ammoniac, viz. by assuming a partial dissociation—to 75 per cent. —of the molecules of sodium chloride. For then the solution contains $0 \cdot 25$ gram-molecules of NaCl, $0 \cdot 75$ gram-molecules of Cl and $0 \cdot 75$ gram-molecules of Na, in all $1 \cdot 75$ gram-molecules. Now we have seen before how we may calculate the number of active molecules in the same solution of sodium chloride, and we find by Kohlrausch's measurements precisely the number $0 \cdot 75$. From this I was led to suppose that the active molecules of the salts are divided into their ions. These are wholly free and behave just as other molecules in the solutions. In the same manner I calculated the degree of dissociation of all the electrolytes that were determined at that time—they were about eighty—and I found in general a very good agreement between the two methods of calculation. In a few instances the agreement was not so good, I therefore made new determinations for these bodies and some others; the new determinations were all in good conformity with the theoretical prevision.

The next figure (Fig. 6) shows the freezing-points of some solution of salts, and of non-conductors. As abscissa is used the molecular concentration of the bodies, as ordinates the molecular depression of the freezing-point, divided by $1 \cdot 85$, that should be expected if no dissociation took place. As the figure shows, all the curves for the non-conductors—in this case cane-sugar, propyl-alcohol and phenol —converge towards unity with diminishing concentration. At higher concentrations there occur deviations from the simple law. As examples of binary electrolytes are chosen LiOH, NaCl and LiCl— their curves all converge towards the number 2. As ternary electrolytes are chosen K^2SO_4, Na_2SO_4, $MgCl_2$, and $SrCl_2$, they are decomposed into three ions, and their curves therefore all converge towards the number 3.

As I had taken a step that seemed most adventurous to chemists, there remained to investigate its chemical and physical consequences. The most general and wide-reaching of these is that the properties of a highly attenuated solution of an electrolyte ought to be additive, that is composed of the properties of the different ions into which the electrolyte is decomposed. This was already known to be the case in many instances, and Valson had to this end tabulated his "modules" by the addition of the one value for the negative to the other for the positive ion, we may calculate the properties of any electrolyte composed of the tabulated ions. In this way we may treat the specific weight (Valson), the molecular conductivity (law of Kohlrausch), the internal friction (Arrhenius), the capillarity (Valson), the compressibility (Röntgen and Schneider), the refractive index (Gladstone), the natural rotation of polarisation (law of Oudemans), the magnetic rotation of polarisation (Perkin and Jahn), the magnetisation (Wiedemann), and all other properties of the electrolytes hitherto sufficiently studied.

The most important of these additive properties are those of which we make use in chemical analysis. As is well known it is generally true that chlorides give a white precipitate with silver salts. It was said formerly that silver salts are reagents for chlorine. Now

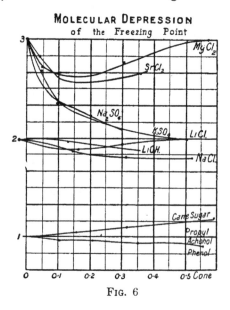

Fig. 6

we say that silver ions are reagents for chlorine ions. This expression is better than the old one, for neither all silver salts e.g. potassium silver cyanide and many other compounds of silver nor all chlorine compounds, e.g. potassium chlorate and many organic chlorides, give this characteristic reaction. The experiment succeeds only with such silver and chlorine compounds as are in a measurable degree decomposed into silver and chlorine ions. Ostwald has treated this question comprehensively, and in this way he has given a rational exposition of the general phenomena of analytical chemistry. To this fact belongs, also, the poisonous effect of some salts; this effect may be considered as a special physiologically chemical reaction of the chemical compounds. On this point there are many valuable researches by Kronig and Paul, Clarke and others.

A property that is of physical character, but is much used by the analytical chemist, is the colour of the solutions. It has been subjected to a rigorous research by Ostwald. At first we will trace how a compound, e.g. fluoresceine, $H_{12}C_{20}O_5$, behaves if one replaces its hydrogen atoms by other atoms, e.g. metals, iodine, bromine or atomic groups (NO_2). The curves in the next figure (Fig. 7) indicate absorption-bands in the spectra of the corresponding compounds. A replacement of K_2 for H_2 in the fluoresceine itself alters the absorption-spectrum in a most sensible manner. This depends upon the property that the fluoresceine is dissociated to a slight extent, which is in strong contrast to the permanganic acid which will be discussed immediately. Instead of a single absorption-band in the blue in the first case, we find two absorption-bands in the blue-green and the green part of the spectrum for the second. A similar observation

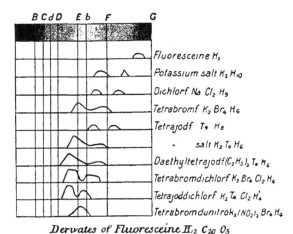

Derivates of Fluoresceine H_{12} C_{20} O_5

FIG. 7.

may be made for the tetraiodine-fluoresceine and its potassium salt. In general the figure shows that the spectrum is changed in a very conspicuous manner at the smallest chemical change of the molecule.

It might therefore be expected after the old manner of view, that the replacement of hydrogen by a metal in permanganic acid, or of one acid rest by another in the salts of para-rosaniline would wholly change the character of the spectra. This is not the case, as Ostwald has shown. The spectra are wholly unchanged, as Figs. 8 and 9 show. The spectra are all produced by the same substance, viz., the permanganate-ion in the one, the para-rosaniline-ion in the other case. Only in the case of the para-rosaniline salts we observe that the absorption is sensibly weaker in some cases than in others. The weakening depends upon the hydrolysis of the salts of the weak acids, e.g. acetic and benzoic acids. This research of

Ostwald shows in a most convincing manner the correctness of the
views of the theory of electrolytic dissociation.

It has been objected to this theory, that according to it it might
be possible by diffusion to separate both ions, e.g. chlorine and
sodium, from another in a solution of sodium chloride. In reality
chlorine diffuses about 1·4 times more rapidly than sodium. But
the ions carry their electric charges with them. Therefore if we
place a solution of sodium chloride in a vessel and we pour a layer
of pure water over it, it is true that in the first moments a little
excess of chlorine enters the water. By this the water is charged
negatively, and the solution under it positively, so that the sodium

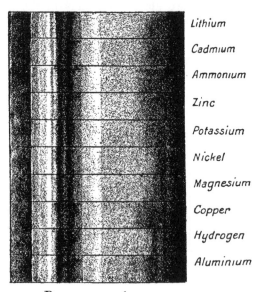

Lithium

Cadmium

Ammonium

Zinc

Potassium

Nickel

Magnesium

Copper

Hydrogen

Aluminium

Permanganates

Fig 8

ions are driven out from the solution with a greater force than the
chlorine ions. As soon as that force is 1·4 times greater than this,
the chlorine ions travel just as slowly as the sodium ions. It is not
difficult to calculate that this case happens as soon as the chlorine
ion is contained in the water in an excess of about the billionth part
of a milligramme over the equivalent quantity of sodium. This extremely minute quantity we should in vain try to detect by chemical
means. By electrical means it succeeds pretty well, as Nernst has
demonstrated experimentally for his concentration elements. Therefore, the said objection is valid against the hypothesis of a common
dissociation of the salts, but not against a dissociation into ions, that

are charged with electricity, as Faraday's law demands. Probably this objection has hindered an earlier acceptance of a dissociated state of the electrolytes, to which, for instance, Valson and Bartoli inclined.

The gaseous laws that are valid for dilute solutions, have made the calculation of the degree of dissociation possible in a great number of cases. The first application of that nature was made by Ostwald, who showed that the dissociation equilibrium between the ions and the non-dissociated part of a weak acid obeys very nearly the gaseous laws. The same was afterwards demonstrated to be true for weak bases by Bredig. The strongly dissociated electrolytes, chiefly salts, exhibit even in dilute solutions (over 0·05 normal)

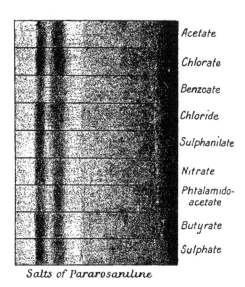

Acetate

Chlorate

Benzoate

Chloride

Sulphanilate

Nitrate

Phtalamido-
acetate

Butyrate

Sulphate

Salts of Pararosaniline

Fig 9.

anomalies, that are not yet wholly explained. Professor Jahn, of Berlin, is at work upon this most interesting question.

The equilibrium between a greater number of electrolytes has been investigated by myself, and found to be in good agreement with the theoretical previsions. This section includes the questions on the weakening of an acid by addition of its salts, and on the so-called avidity of the different acids, that is, the proportion in which two acids divide a base at partial neutralisation. Calculation gives very nearly the numbers observed experimentally by Thomsen and Ostwald. For heterogeneous equilibria between electrolytes the theory is worked out by van't Hoff and Nernst, who have in this way elucidated the common method to precipitate salts used in analytical chemistry.

By help of the gaseous laws it is also possible to determine the heat evolved at the dissociation of a weak acid or base, and in this way I was able to calculate the heat of neutralisation of acids and bases in a general manner. In an analogous way, Fanjung calculated the changes of volume at dissociation of a weak acid or base and at the neutralisation of these bodies. All these calculations gave values very nearly agreeing with the observed ones.

An important rôle is played by the water, which may be regarded as a weak acid or base. By its electrolytical dissociation, it causes the hydrolysis of salts of weak acids and bases. By observation of the hydrolysis, it was possible to calculate the electrolytic dissociation of water, and this quantity was soon after determined by electrical measurements by Kohlrausch and Heydweiller in perfect agreement with the previous calculations. For physiological chemistry this question is of the greatest importance, as is confirmed by the experimental results of Sjögvist and others. Also for the explanation of volcanic phenomena, the concurrence between water and silicic acid at different temperatures has found an application.

The catalytic phenomena in which acids and bases are the chief agents, have been investigated by many observers, and it has been found that the catalytic action depends on the quantity of free hydrogen or hydroxyl ions that are present in the solution. To this review, that makes no pretension to be complete, may also be added the wide-reaching researches of van't Hoff, Ostwald, and especially Nernst, on the electromotive forces produced by the ions. By these investigations we have now acquired an explanation of the old problem of the manner in which electromotive forces in hydro-electric combinations are excited.

I have now traced the manner in which the idea of electrolytic dissociation grew out of our old conception of atoms and molecules. Sometimes we hear the objection that this idea may not be true, but only a good working hypothesis. This objection, however, is in reality no objection at all, for we can never be certain that we have found the ultimate truth. The conception of molecules and atoms is sometimes refuted on philosophical grounds, but till he has got a better and more convenient representation of chemical phenomena, the chemist will no doubt continue to use the atomic theory without scruple. Exactly the same is the case for the electrolytic dissociation theory.

This theory has shown us that in the chemical world the most important rôle is played by atoms or complexes of atoms, that are charged with electricity. The common tendency of scientific investigation seems to give an even more preponderating position to electricity, the mightiest agent of Nature. This development is now proceeding very rapidly. Already we see not only how the theory of electrons of J. J. Thomson, in which matter is reduced to a very insignificant part, is developing, but also how efforts are made with good success to explain matter as only a manifestation of electro-dynamic forces (Kaufmann-Abraham).

To these modern developments the work of British men of science has contributed in the most effective manner.   The bold previsions of Sir William Crookes seem to be rapidly acquiring a concrete form, to the great benefit of scientific evolution.

[S. A.]

GENERAL MONTHLY MEETING,

Monday, June 6, 1904.

His Grace The Duke of Northumberland, K.G. D.C.L. F.R.S.,
President, in the Chair.

Richard Bagot, Esq.
William Carrington, Esq., M.Inst.C.E.
Ernest Gadesden Fellows, Esq.
William Robert Freeman, Esq.
Richard F. Nicholson, Esq.
Hon. C. A. Parsons, F.R.S.

were elected Members of the Royal Institution.

The Special Thanks of the Members were returned to Andrew
Carnegie, Esq., LL.D. *M.R.I.*, for his Donation of £1,200 to the
Research Fund of the Royal Institution, especially for the study of
iron and steel at low temperatures; and to Frank McClean, Esq.,
M.A. LL.D. F.R.S. *M.R.I.*, for his Donation of £100.

The Presents received since the last Meeting were laid on the
table, and the thanks of the Members returned for the same, viz. :—

FROM

The Secretary of State for India—Census of India, 1901, Vol. I. 4to. 1903.
    Report on Public Instruction in Bengal. fol. 1903.
    Records of the Geological Survey of India, Vol. XXXI. Part 1. 8vo. 1904.
    Linguistic Survey of India, Vol. V. Part 1. 4to 1904.
Accademia dei Lincei, Reale, Roma—Classe di Scienze Fisiche, Matematiche e
    Naturali. Atti, Serie Quinta : Rendiconti. 1º Semestre, Vol. XIII. Nos. 9–10.
    8vo. 1904.
American Academy of Arts and Sciences—Proceedings, Vol. XXXIX. Nos. 16–18.
    8vo. 1904.
American Geographical Society—Bulletin, Vol. XXXVI. No. 4. 8vo. 1904.
Asiatic Society of Bengal—Proceedings, 1903, Nos. 6–10. 8vo. 1904.
    Journal, Vol. LXXII. Part II. Nos. 3–4; Part III. No. 2. 8vo. 1903–4.
Asiatic Society, Royal (Bombay Branch)—Journal, Vol. XXI. No. 59. 8vo. 1904.
Astronomical Society, Royal—Monthly Notices, Vol. LXIV. No. 5. 8vo. 1904.
Automobile Club—Journal for May, 1904. 4to.
Bell, Sir Lowthian, Bart., J.P. D.L. LL.D. D.C.L. F.R.S. M.R.I. (the Author)—
    Memorandum as to the Wear of Rails. 2 vols. fol. 1896–1900.
Bennett, Frederick, Esq. (the Author)—Some Theories of the Universe. 8vo. 1903.
Berlin, Royal Academy of Sciences—Sitzungsberichte, 1904, Nos. 1–24. 8vo.
Boston Public Library—Monthly Bulletin for May, 1904. 8vo.
Botanic Society, Royal—Quarterly Record, Jan.-March, 1904. 8vo.
British Architects, Royal Institute of—Journal, Third Series, Vol. XI. No. 14. 4to.
    1904.
British Astronomical Association—Journal, Vol. XIV. No. 7. 8vo. 1904.

California University—Report of the Agricultural Experiment Station. 8vo. 1903.
    Bulletin, Nos. 149-154. 8vo. 1903.
Chemical Industry, Society of—Journal, Vol. XXIII. Nos. 9-10. 8vo. 1904.
Chemical Society—Journal for May, 1904. 8vo.
    Proceedings, Vol. XX. Nos. 280-281. 8vo. 1904.
Chicago, Field Columbian Museum—Anthropological Series, Vol. II. No. 6, Vol.
    IV; Zoological Series, Vol. III. Nos. 12-14; Geological Series, Vol. II.
    Nos. 2-4. 8vo. 1903.
Editors—American Journal of Science for May, 1904. 8vo.
    Analyst for May, 1904. 8vo.
    Astrophysical Journal for May, 1904. 8vo
    Athenæum for May, 1904. 4to.
    Author for June, 1904. 8vo.
    Board of Trade Journal for May, 1904. 8vo
    Brewers' Journal for May, 1904. 8vo.
    Cambridge Appointments Gazette for May, 1904. 8vo
    Chemical News for May, 1904. 8vo.
    Chemist and Druggist for May, 1904. 8vo.
    Dioptric Review for March–April, 1904. 8vo.
    Electrical Engineer for May, 1904. fol.
    Electrical Review for May, 1904. fol.
    Electrical Times for May, 1904. 4to.
    Electricity for May, 1904. 8vo.
    Electro-Chemist and Metallurgist for May, 1904. 8vo.
    Engineer for May, 1904. fol.
    Engineering for May, 1904. fol.
    Engineering Review for May, 1904. 8vo.
    Homœopathic Review for June, 1904. 8vo.
    Horological Journal for June, 1904. 8vo.
    Journal of the British Dental Association for May, 1904. 8vo.
    Journal of Physical Chemistry for April, 1904. 8vo.
    Journal of the Queen Victoria Indian Memorial Fund for March (No. 2), 1904.
        8vo.
    Journal of State Medicine for May, 1904. 8vo.
    Law Journal for May, 1904. 8vo.
    London Technical Education Gazette for May, 1904. fol.
    London University Gazette for May, 1904. fol.
    Machinery Market for May, 1904. 8vo.
    Model Engineer for May, 1904. 8vo.
    Mois Scientifique for April, 1904. 8vo.
    Motor Car Journal for May, 1904. 8vo.
    Musical Times for May, 1904. 8vo.
    Nature for May, 1904. 4to.
    New Church Magazine for June, 1904. 8vo.
    Nuovo Cimento for Feb. 1904. 8vo.
    Page's Magazine for May, 1904. 8vo.
    Photographic News for May, 1904. 8vo.
    Physical Review for May, 1904. 8vo.
    Public Health Engineer for May, 1904. 8vo.
    Science Abstracts for May, 1904. 8vo.
    Terrestrial Magnetism for March, 1904. 8vo.
    Zoophilist for May, 1904. 4to.
Fassig, O. L., Esq. (the Author)—Kite Flying in the Tropics. 8vo. 1904.
Florence, Biblioteca Nazionale—Bulletin, May, 1904. 8vo.
Franklin Institute—Journal, Vol. CLVII. No. 5. 8vo. 1904.
Geographical Society, Royal—Geographical Journal, Vol. XXIII. No. 6. 8vo.
    1904.
Geological Society—Journal, Vol. LX. Part 2. 8vo. 1904.¡
    Abstracts of Proceedings, Nos. 795-796. 8vo. 1904.
Gottingen, Academy of Sciences—Nachrichten, 1904, Mathematisch-Physikalische
    Klasse, Heft 1-2. 8vo.

Horticultural Society, Royal—Journal, Vol. XXVIII. Parts 3-4. 8vo. 1904.
Johns Hopkins University—Studies, Series XXI. Nos. 1-12. 8vo. 1903.
Life-Boat Institution, Royal National—Annual Report, 1904. 8vo.
Linnean Society—Journal, Vol. XXXV. Botany, No. 248. 8vo. 1904.
Literature, Royal Society of—Queen Elizabeth and the Levant Company.
    Edited by H. G. Rosedale. 4to. 1904.
Manchester Geological Society—Transactions, Vol. XXVIII. Parts 13-15. 8vo.
    1904.
Mechanical Engineers, Institution of—Proceedings, 1903, No. 4. 8vo.
Meteorological Society, Royal—Journal, Vol. XXX. No. 130. 8vo. 1904.
    Record, Vol. XXIII. No. 91. 8vo. 1904.
Meux, Lady—The Book of Paradise ot Palladius. Edited and Translated by
    E. A. Wallis Budge. (Lady Meux MS. No. 6.) 2 vols. 8vo. 1904.
Mexico, Sociedad Cientifica "Antonio Alzate"—Memorias y Revista, Tome XIX.
    No. 5; Tome XX. Nos. 1-4. 8vo. 1903.
Middlesex Hospital—Reports for the Year 1902. 8vo. 1904.
Montana University—Bulletin, Biological Series, Nos. 7-8. 8vo. 1904.
Navy League—Navy League Journal for June, 1904. 8vo.
Odontological Society—Transactions, Vol XXXVI. No. 6. 8vo. 1904.
Pennsylvania, University of—Provost's Report, 1903. 8vo.
    Catalogue, 1903-4. 8vo. 1903.
Peru, Cuerpo de Ingenieros de Minas—Boletin, Nos. 4-9. 8vo. 1903-4.
Pharmaceutical Society of Great Britain—Journal for May, 1904. 8vo.
Philadelphia, Academy of Natural Sciences—Proceedings, Vol. LV. Part 3. 8vo.
    1904.
Photographic Society, Royal—Photographic Journal, Vol. XLIV. No. 4. 8vo. 1904.
Physical Society of London—Proceedings, Vol. XIX. Part 1. 8vo. 1904.
Royal Irish Academy—Proceedings: Vol. XXIV. Sec. A, Part 4; Sec. B, Part 5;
    Sec. C, Part 5; Vol. XXV. Sec. C, Parts 1-4. 8vo. 1904.
Royal Society of London—Philosophical Transactions, A, No. 364; B, No. 229.
    4to. 1904.
    Proceedings, No. 494. 8vo. 1904.
Sanitary Institute—Journal, Vol. XXV. Part 1. 8vo. 1904.
Selborne Society—Nature Notes for May, 1904. 8vo.
Smith, B. Leigh, Esq., M.A. M.R.I.—The Scottish Geographical Magazine, Vol.
    XX. No. 6. 8vo. 1904.
Society of Arts—Journal for May, 1904. 8vo.
Stockholm, Royal Swedish Academy of Sciences—Skrifter af Anders Retzius. 8vo.
    1902.
Stonyhurst College Observatory—Results of Meteorological Observations for 1903.
    8vo. 1904.
Tacchini, Prof. P., Hon. Mem. R.I. (the Author)—Memorie della Società degli
    Spettroscopisti Italiani, Vol. XXXIII. Disp. 4. 4to. 1904.
United Service Institution, Royal—Journal for May, 1904. 8vo.
United States Coast and Geodetic Survey—Reports on Terrestrial Magnetism,
    1903-4. 4to.
United States Department of Agriculture—Monthly Weather Review for Feb. 1904.
    4to.
    Experiment Station Bulletin, 141. Losses in Cooking Meat. By H. S.
    Grindley. 8vo. 1904.
    Experiment Station Record, April, 1904. 8vo.
United States Geological Survey—Professional Papers, 9, 10, 13, 14, 15. 4to.
    1902-3.
    Water Supply Papers, 80-87. 8vo 1903.
United States Patent Office—Official Gazette, Vol. CX. Nos. 2-4. 4to. 1904.
Verein zur Beförderung des Gewerbfleisses in Preussen—Verhandlungen, 1904.
    Heft 5. 4to.
Western Society of Engineers—Journal, Vol. XIX. No. 2. 8vo. 1904.
Wisconsin Academy—Transactions, Vol. XIII. Part 2; Vol. XIV. Part 1. 8vo.
    1902-3.

GENERAL MONTHLY MEETING,

Monday, July 4, 1904.

His Grace The Duke of Northumberland, K.G. D.C.L. F.R.S.,
President, in the Chair.

Viscountess Gort,
M. H. Spielmann, Esq.
William R. W. Sullivan, Esq.

were elected Members of the Royal Institution.

The Presents received since the last Meeting were laid on the
table, and the thanks of the Members returned for the same, viz.:—

FROM

The Secretary of State for India—Madras Government Museum Bulletin, Vol. V
No. 1. 8vo. 1903.
Accademia dei Lincei, Reale, Roma—Classe di Scienze Fisiche, Matematiche e
Naturali. Atti, Serie Quinta: Rendiconti. Vol. XIII. 1º Semestre, Fasc. 11.
8vo. 1904.
American Academy of Arts and Sciences—Proceedings, Vol. XXXIX. Nos. 19-20.
8vo. 1904.
American Geographical Society—Bulletin, Vol. XXXVI. No. 5. 8vo. 1904.
American Philosophical Society—Proceedings, Vol. XLII. No. 175. 8vo. 1904.
Argentine Republic—El Crecimiento de la Población de la República Argentina
1890-1903. 8vo. 1904.
Astronomer Royal—Report to the Board of Visitors of the Royal Observatory,
1904. 4to.
Astronomical Society, Royal—Monthly Notices, Vol. LXIV. No. 7. 8vo. 1904
Automobile Club—Journal for June, 1904.
Bankers, Institute of—Journal, Vol. XXV. Part 6. 8vo. 1904.
Batavia, Royal Meteorological Observatory—Regenwaarnemingen in Nederlandsch-
Indië, 1902. 8vo. 1903.
Belgium, Royal Academy of Sciences—Bulletin, 1904, Nos. 3, 4. 8vo.
Méms. Couronnés et autres Méms. Tome LXIII. Fasc. 8; Tomes LXIV.-LXV.
Fasc. 1-2; Tome LXVI. 8vo. 1904.
Méms. Couronnés et des Savants Étrangers, Tome LXII. Fasc. 5-8. 4to. 1904.
Mémoires, Tome LIV. Fasc. 6. 4to. 1904.
Boston Public Library—Monthly Bulletin for June, 1904. 8vo.
British Architects, Royal Institute of—Journal, Third Series, Vol. XI. Nos. 15-16.
4to. 1904.
British Astronomical Association—Journal, Vol. XIV. No. 8. 8vo. 1904.
Buenos Ayres, City—Monthly Bulletin of Municipal Statistics for April, 1904. 4to
Cambridge Philosophical Society—Transactions, Vol. XIX. Part 3. 4to 1904.
Chemical Industry, Society of—Journal, Vol. XXIII. Nos. 11-12. 8vo. 1904.
Chemical Society—Proceedings, Vol. XX. Nos. 282-283. 8vo 1904.
Journal for June, 1904. 8vo.
Dax, Société de Borda—Bulletin, 1903, Part 4 8vo.
Editors—American Journal of Science for June, 1904. 8vo.
Analyst for June, 1904. 8vo.

Editors—continued.
　Astrophysical Journal for June, 1904. 8vo.
　Athenæum for June, 1904. 4to.
　Author for July, 1904. 8vo.
　Board of Trade Journal for June, 1904. 8vo.
　Brewers' Journal for June, 1904. 8vo.
　Chemical News for June, 1904. 4to.
, Chemist and Druggist for June, 1904. 8vo.
　Electrical Engineer for June, 1904. 4to.
　Electrical Review for June, 1904. 4to.
　Electrical Times for June, 1904. 4to.
　Electricity for June, 1904. 8vo.
　Electro-Chemist and Metallurgist for June, 1904. 8vo.
　Engineer for June, 1904. fol.
　Engineering for June, 1904. fol.
　Engineering Review for June–July, 1904. 8vo.
　Homœopathic Review for July, 1904. 8vo.
　Horological Journal for July, 1904. 8vo.
　Journal of the British Dental Association for June, 1904. 8vo.
　Journal of Physical Chemistry for May, 1904. 8vo.
　Journal of State Medicine for June, 1904. 8vo.
　Law Journal for June, 1904. 8vo.
　London Technical Education Gazette for June, 1904. 4to.
　London University Gazette for June, 1904. 4to.
　Machinery Market for June, 1904. 8vo.
　Model Engineer for June, 1904. 8vo.
　Mois Scientifique for May, 1904. 8vo.
　Motor Car Journal for June, 1904. 8vo.
　Motor Car World for June, 1904. 4to.
　Musical Times for June, 1904. 8vo.
　Nature for June, 1904. 4to.
　New Church Magazine for July, 1904. 8vo.
　Nuovo Cimento for March, 1904. 8vo.
　Page's Magazine for June, 1904. 8vo.
　Photographic News for June, 1904. 8vo.
　Physical Review for June, 1904. 8vo.
　Public Health Engineer for June, 1904. 8vo.
　Science Abstracts for June, 1904. 8vo.
　Zoophilist for June, 1904. 4to.
Electrical Engineers, Institution of—Journal, Vol. XXXIII. Part 3. 8vo.
　　1904.
Étard, M. A., Esq. (the Author)—Notice sur la Vie et les Travaux de Eugène
　　Demarçay. 8vo. 1904.
Florence, Biblioteca Nazionale—Bulletin for June, 1904. 8vo.
Franklin Institute—Journal, Vol. CLVII. No. 6. 8vo. 1904.
Geological Society—Abstracts of Proceedings, Nos. 797-798. 1904.
　　Geological Literature added to the Library in 1903. 8vo. 1904.
Harlem, Société Hollandaise des Sciences—Archives Néerlandaises, Série II.
　　Tome IX. Liv. 3. 8vo. 1904.
Johns Hopkins University—American Journal of Philology, Vol. XXV. No. 1.
　　8vo. 1904.
Kyoto University College of Science—Memoirs, Vol. I. No. 1. 8vo. 1903.
Leighton, John, Esq., M.R.I.—Journal of the Ex-Libris Society, Vol. XIV. Nos.
　　2-5. 8vo. 1904.
Massachusetts Institute of Technology—Technology Quarterly, Vol. XVII. No. 1.
　　8vo. 1904.
Mexico, Secretaria de Comunicaciones—Anales, Num. 10. 8vo. 1904.
Microscopical Society, Royal—Journal, 1904, Part 3. 8vo.
Milan, School of Agriculture—Ricerche, Vol. II. 8vo. 1903

Musée Teyler—Archives, Série II. Vol. VIII. Fasc 5. 8vo. 1904.
  Catalogue de la Bibliothèque, Tome III. 1889-1903. 4to. 1904.
National Physical Laboratory—Report for 1903. 8vo. 1904.
Navy League—Navy League Journal for July, 1904. 8vo.
New Jersey, Geological Survey—Annual Report for 1903. 8vo. 1904.
New York Academy of Sciences—Annals, Vol. XIV. Part 4. 8vo. 1904.
Norske, Gradmaalings-Kommission—Resultater af Vandstands-Observationer paa
  der Norske Kyst, Hefte VI. 4to. 1904.
North of England Institute of Mining and Mechanical Engineers—Tiansactions,
  Vol. LIV. Part 5. 8vo. 1904.
Odontological Society—Transactions, Vol. XXXVI. No. 7. 8vo. 1904.
Onnes, Dr. H. Kamerlingh—Communications from the Leiden Physical Labora-
  tory, Nos. 87-90. 8vo. 1904.
Pharmaceutical Society of Great Britain—Journal for June, 1904. 8vo.
Photographic Society, Royal—Journal, Vol. XLIV. No. 5. 8vo. 1904.
Rennes, University—Travaux Scientifiques, Tome II. Fasc. 3. 8vo. 1903.
Rochechouart, La Société les Amis des Sciences et Arts—Bulletin, Tome XIII. No. 3.
  8vo. 1903.
Rome Ministry of Public Works—Giornale del Genio Civile for Dec. 1903 and Jan.
  1904. 8vo.
Royal Engineers—Professional Papers, Vol. XXIX. 8vo. 1904.
Royal Society of London—Philosophical Transactions, B. No. 230. 4to. 1904.
  Proceedings, No. 495. 8vo. 1904.
Saxon Society of Sciences, Royal—
  Berichte: Mathematisch-Physische Klasse, 1903, No. 6; 1904, Nos. 1-3. 8vo.
    1903-4.
    Philologisch-Historische Klasse, 1903, Nos. 3-6. 8vo. 1903.
  Abhandlungen: Mathematisch-Physische Klasse, Band XXVIII. Nos. 6-7.
    4to. 1904.
    Philologisch-Historische Klasse, Band XX. Nos. 4 and 6. 4to.
    1904.
Scottish Microscopical Society—Proceedings, Vol. III. No. 4. 8vo. 1903.
Selborne Society—Nature Notes for June, 1904. 8vo.
Society of Arts—Journal for June, 1904. 8vo.
Sweden, Royal Academy of Sciences—Handlingar, Band XXXVII. Nos. 7-8. 4to.
  1904.
  Arkiv för Botanik, Band II. Häfte 1-3. 8vo. 1904.
  Arkiv för Kemi, Band I. Häfte 2. 8vo. 1904.
Swinburne, James, Esq., M.R.I. (the Author)—Entropy, or Thermodynamics from
  an Engineer's Standpoint. 8vo. 1904.
Tacchini, Prof. P., Hon. Mem. R.I. (the Author)—Memorie della Società degli
  Spettroscopisti Italiani, Vol. XXXIII. Disp. 5. 4to. 1904.
Transvaal Department of Agriculture—Journal, Vol. II. No. 7. 8vo. 1904.
United Service Institution, Royal—Journal for June, 1904. 8vo.
United States Department of Agriculture—Monthly Weather Review for March,
  1904. 4to.
  Experiment Station Record for May, 1904. 8vo.
United States Geological Survey—Monographs, No. XLVI. 4to. 1904.
  Bulletins, Nos. 208, 218-222. 8vo. 1903-4.
United States Patent Office—Official Gazette, Vol. CX. Nos. 5-8. 8vo. 1904.
Vienna, Imperial Geological Institute—Verhandlungen, 1904, Nos. 5-8. 8vo.
Yorkshire Archæological Society—Journal, Vol. XVIII. Part 1. 8vo. 1904.
  Index to Papers, Vol. I.-XVII. 8vo. 1904.
Zoological Society of London—Proceedings, 1904, Vol. I. Part 1. 8vo.
Zurich Naturforschenden Gesellschaft—Vierteljahrsschrift, 1903, Heft 3, 4. 8vo.
  1904.

GENERAL MONTHLY MEETING,

Monday, November 7, 1904.

Sir James Crichton-Browne, M.D. LL.D. F.R.S., Treasurer and Vice-President, in the Chair.

H.H. The Raj Bhawani Singh Bahadur of Jhalawar,
Joseph Jennens, Esq.
Edwin J. Preston, Esq.
were elected Members of the Royal Institution.

It was announced from the Chair that the Institution had received a Bequest of £100 under the Will of the late Miss Harriet Jane Moore.

The Special Thanks of the Members were returned to Dr. Ludwig Mond for his Donation of £755 for the purpose of erecting a Lift from the basement to the second floor of the Institution, and any surplus after the completion of the work to go to the Research Fund.

The Presents received since the last Meeting were laid on the table, and the thanks of the Members returned for the same, viz. :—

FROM

Lords Commissioners of the Admiralty—Greenwich Observations, 1901. 4to. 1903
  Photoheliographic Results, 1901. 4to. 1902.
  Astrographic Catalogue, 1900, Vol. I. 4to. 1903.
  Cape Observatory Annals, Vol. IX. 4to. 1903.
  Report on Cape of Good Hope Observatory for 1903. 4to. 1904.
The Secretary of State for India—Linguistic Survey of India, Vol. III. Part 2; Vol. V. Part 2. 4to. 1903.
  Archæological Survey—
      Southern India Inscriptions, Vol. III. Part 2. 4to 1903.
      Annual Report, Bengal Circle, 1904. 4to.
      Report for 1903-4. 4to. 1904.
  Geological Survey—
      Records, Vol. XXXI. Part 2. 8vo. 1904.
      Memoirs, Vol. XXXV. Part 3; Vol. XXXVI. Part 1. 8vo. 1904.
      Palæontologia Indica, Ser. XV. Vol. IV. 4to. 1903.
  Report on Administration of Central Provinces, 1902-3. 4to. 1903.
British Museum Trustees—History of the Collections in the Natural History Department, Vol. I. 8vo. 1904.
  Catalogue of Jurassic Plants, Part II. 8vo. 1904.
  Second Report on Economic Zoology. 8vo. 1904.
  Introduction to the Study of Meteorites. By L. Fletcher. 8vo. 1904.
  Catalogue of the Library (Natural History), Vol. II. E-K. 4to. 1904.
The French Government—Recueil des Chartres de l'Abbaye de Cluny, Tom. VI. 1211-1300. 4to. 1903.

Accadsmia dei Lincei, Reale, Roma—Atti, Serie Quinta. Classe di Scienze Fisiche, Vol. XIII. 2° Semestre, Fasc. 1–7.   Classe de Scienze Morali, Vol. XIII. Fasc. 1–6.   8vo.   1904.
Rendiconti, 1904, Vol. II.   4to.
American Academy of Arts and Sciences—Proceedings, Vol. XXXIX. Nos. 21–24; Vol. XL. Nos. 1–5.   8vo.   1904.
American Geographical Society—Bulletin, Vol. XXXVI. Nos. 6–9.   8vo.   1904.
American Philosophical Society—Proceedings, Vol. XLIII. No. 176.   8vo.   1904.
Aristotelian Society—Proceedings, New Series, Vol. IV.   8vo.   1904.
Asiatic Society of Bengal—Proceedings, 1903, No 11; 1904, Nos. 1–5.   8vo.   1904.
Journal, Vol. LXXII. Part I. No. 2; Vol. LXXIII. Part I. Nos. 1–2, Part II. Nos. 1–2, Part III. Nos. 1–2.   8vo.   1904.
Astronomical Society, Royal—Monthly Notices, Vol LXIV. No. 8.   8vo.   1904.
Memoirs, Vols. LIV.–LV. with Appendices.   4to.   1904.
Automobile Club—Journal for July–Oct. 1904.   4to.
Bankers, Institute of—Journal, Vol. XXV. Part 7.   8vo.   1904.
Batavia, Meteorological Observatory—Observations, Vol. XXV.   4to.   1904.
Beckenhaupt, C. Esq. (the Author)—Le Mécanisme de la Vie.   8vo.   1904.
Belgium, Royal Academy of Sciences—Bulletin, 1904. Nos. 5–8.   8vo.
Berlin, Royal Academy of Sciences—Sitzungsberichte, 1904, Nos. 25–40.   8vo.
Bohlin, Karl, Esq. (the Author)—Des Neuen Sterns Nova (3.1901) Persei.   4to. 1904.
Sur le Choc considéré comme Fondement des Théories Cinétiques de la Pression des Gaz.   8vo.   1904.
Borredon, G. (the Author)—La Grande Scoperta del Secolo XX.   8vo.   1904.
Boston Public Library—Monthly Bulletin for July–Oct. 1904.   8vo.
Botanic Society, Royal—Quarterly Record, April–June, 1904.   8vo.
British Architects, Royal Institute of—Journal, Third Series, Vol. XI. Nos. 17–20.   4to.   1904.
The Kalendar, 1904–5.   8vo.   1904.
British Association—Report of the Seventy-third (Southport) Meeting, 1903.   8vo. 1904.
British Astronomical Association—Journal, Vol. XIV. Nos. 9–10.   8vo.   1904.
Memoirs, Vol. XII. Part 3.   8vo.   1904.
Brooklyn Institute—Memoirs, Vol. I. No. 1.   8vo.   1904.
Buenos Ayres—Monthly Bulletin of Municipal Statistics for May–June, 1904.   4to.
Cambridge Observatory—Annual Reports, 1901–4.   4to.
Cambridge Philosophical Society—Proceedings, Vol. XII. Part 6.   8vo.   1904.
Canada, Department of the Interior—Report on the Great Landslide at Frank, Alta, 1903.   8vo.   1904.
Maps of Mounted Police Stations in N.W. Canada and N.W. Territories, 1904.
Topographical Map of Canada: Windsor sheet.
Railway Map of Manitoba, Alberta, Saskatchewan and Assiniboia.
Canada, Geological Survey—Reports, Vol. XIII. 1900.   8vo.   1903.
Canada, Meteorological Service—Report for 1902.   8vo.   1903.
Canadian Institute—Transactions, Vol. VII. Part 3.   8vo.   1904.
Canterbury, The Mayor and Corporation—The Ancient City of Canterbury.   By F. W. Farrar and others.   4to.   1904.
Chemical Industry, Society of—Journal, Vol. XXIII. Nos. 13–20.   8vo.   1904.
Chemical Society—Journal for July–Oct. 1904.   8vo.
List of Fellows, 1904.   8vo
Chicago, Field Columbian Museum—Anthropological Series, Vol. III. No. 4; Vol. V.–VI.; Vol. VII. No. 1. Zoological Series, Vol. III. Nos. 15–16. Report Series, Vol. III. No. 2.   8vo.   1903–4.
Civil Engineers, Institution of—Proceedings, Vol. CLVI.–CLVII   8vo   1904.
List of Members, 1904.   8vo.
Colombia, Ministerie de Relaciones Exteriores—Protesta de Columbia contra el Tratedo entre Panama y los Estados Unidos.   8vo.   1904.
Colonial Institute, Royal—Proceedings, Vol. XXXV.   8vo.   1904.

Cornwall, Royal Polytechnic Society—Seventy-first Annual Report, 1903. 8vo. 1904.

Cracovie, Académie des Sciences—Bulletin, 1904. Classe des Sciences Mathématiques. Nos. 4-7. Classe de Philologie, Nos. 4-7. 8vo. 1904.

Dax, Société de Borda—Bulletin, 1904, Fasc. 1. 8vo. 1904.

Dudgeon, J. Scott, Esq. (the Author)—Agriculture as an Industry. 8vo. 1904.

East India Association—Journal, Vol. XXXVII. Nos. 35-36. 8vo. 1904.

Edinburgh, Royal Society—Proceedings, Vol. XXV. Nos. 3-4. 8vo. 1904.

Editors—Aeronautical Journal for July and Oct. 1904. 8vo.

American Journal of Science for July-Oct. 1904. 8vo.

Analyst for July-Oct. 1904. 8vo.

Astrophysical Journal for July-Oct. 1904. 8vo.

Ateneo-Veneto, Aug. 1902 to Dec. 1903. 8vo.

Athenæum for July-Oct. 1904. 4to.

Author for Aug.-Nov. 1904. 8vo.

Board of Trade Journal for July-Oct. 1904. 8vo.

Brewers' Journal for July-Oct. 1904. 8vo.

Cambridge Appointments Gazette for Nov. (No. 17) 1904. 8vo.

Chemical News for July-Oct. 1904. 8vo.

Chemist and Druggist for July-Oct. 1904. 8vo.

Dioptric Review for Sept. 1904. 8vo.

Electrical Engineer for July-Oct. 1904. fol.

Electrical Review for July-Oct. 1904. fol.

Electrical Times for July-Oct. 1904. 4to.

Electricity for July-Oct. 1904. 8vo.

Electro-Chemist and Metallurgist for July, 1904. 8vo.

Engineer for July-Oct. 1904. fol.

Engineering for July-Oct. 1904. fol.

Engineering Review for Aug.-Nov. 1904. 8vo.

Homœopathic Review for Aug.-Nov. 1904. 8vo.

Horological Journal for Aug.-Nov. 1904. 8vo.

Journal of the British Dental Association for July-Oct. 1904. 8vo.

Journal of Physical Chemistry for June-Nov. 1904. 8vo.

Journal of State Medicine for July-Oct. 1904. 8vo.

Law Journal for July-Oct. 1904. 8vo.

London Education Gazette for July-Oct. 1904. fol.

London University Gazette for July-Oct. 1904. fol.

Machinery Market for July-Oct. 1904. 8vo.

Model Engineer for July-Oct. 1904. 8vo.

Mois Scientifique for June-Sept. 1904. 8vo.

Motor Car Journal for July-Oct. 1904. 8vo.

Motor Car World for July-Oct. 1904. 8vo.

Musical Times for July-Oct. 1904. 8vo.

Nature for July-Oct. 1904. 4to.

New Church Magazine for Aug.-Oct. 1904. 8vo.

Nuovo Cimento for April-Aug. 1904. 8vo.

Page's Weekly for July-Oct. 1904. 8vo.

Photographic News for July-Oct. 1904 8vo.

Physical Review for July-Oct. 1904. 8vo.

Public Health Engineer for July-Oct. 1904. 8vo.

Science Abstracts for July-Oct. 1904. 8vo.

Terrestrial Magnetism for June and Sept. 1904. 8vo.

Travel for July and Oct. 1904. 8vo.

Zoophilist for July-Oct. 1904. 4to

Electrical Engineers, Institution of—Journal, Vol. XXXIII. Parts 4-6. 8vo. 1904.

Florence, Biblioteca Nazionale—Bulletin, July-Oct. 1904. 8vo.

Florence, Reale Accademia dei Georgofili—Atti, 4o Serie, Vol. XXVI. Disp. 4 ; 5o Serie, Vol. I. Disp. 1-3. 8vo. 1903-4.

Franklin Institute—Journal, Vol. CLVIII. Nos. 1-4. 8vo. 1904.

Geographical Society, Royal—Geographical Journal, Vol. XXIV. Nos. 1–5. 8vo 1904.

Geological Society—Journal, Vol. LX. Part 3. 8vo. 1904.

Geological Survey of the United Kingdom—Summary of Progress, 1903. 8vo. 1904.

Göttingen, Academy of Sciences—Nachrichten, 1904, Mathematisch-Physikalische Klasse, Heft 3. 8vo.
  Geschäftliche Mittheilungen, 1904, Heft 1. 8vo.

Hellenic Studies, Society for the Promotion of—History of the Society, 1879–1904. By G. A. Macmillan. 8vo. 1904.

Henriksen, G., Esq. (the Author)—The Iron Ore Deposits on Sydvaranger, Norway. 8vo. 1904.

Historical Manuscripts Commission—Report on American Manuscripts in the Royal Institution of Great Britain. 8vo. 1904. 4 copies

Iron and Steel Institute—Journal, Vol. LXV. 8vo. 1904.

Jefferson Physical Laboratory—Contributions, Vol. I. 4to. 1903.

Johns Hopkins University—American Journal of Philology, Vol. XXV. No. 2. 8vo. 1904.

Junior Institution of Engineers—Transactions, Vol. XIII. 1902–3. 8vo. 1904.

Langley, S. P , Esq., Hon. Mem. R.I. (the Author)—On the Possible Variation of the Solar Radiation and its Probable Effect on Terrestrial Temperatures. 8vo. 1904.

Lefebure, C., Esq. (the Author)—Mes Étapes d'Alpinisme. 8vo. 1904.

Leighton, John, Esq. M.R.I.—Journal of the Ex-Libris Society for June-Aug. 1904. 8vo.

Life-Boat Institution, Royal National—Journal for Aug. and Nov. 1904. 8vo.

Linnean Society—Journal, Botany, Vol. XXXVI. No. 254. 8vo. 1904.
  Transactions, Botany, Vol. VI. Parts 7–9; Zoology, Vol. VIII. Part 13, Vol. IX. Parts 3–5. 4to. 1904.

Lisbon, Royal Astronomical Observatory—Observations d'Éclipses de Lune. By C. Rodriques.
  Corrections aux Ascensions Droites de quelques Étoiles du Berliner Jahrbuch. By C. Rodriques. 4to. 1904.

Literature, Royal Society of—Transactions, Second Series, Vol. XXV. Part 3. 8vo. 1904.

Madrid, Royal Academy of Sciences—Revista, Tom. I. Nos. 1–4. 8vo 1904.

Mallmann, J. P., Esq. (the Author)—Coke Ovens and their History. fol. 1904.

Massachusetts Institute of Technology—Technology Quarterly, Vol. XVII. No. 2. 8vo. 1904.

Mechanical Engineers, Institution of—Proceedings, 1904, Nos. 1–2. 8vo.

Medical and Chirurgical Society, Royal—Transactions, Vol. LXXXVII. 8vo. 1904.

Merck, E.—Annual Reports on the Advancement of Pharmaceutical Chemistry, Vol. XVII. 1903. 8vo. 1904.

Mersey Conservancy—Report on the Navigation of the River Mersey for 1903. 8vo. 1904.

Meteorological Office—Climatological Observations at Colonial and Foreign Stations, Part I. 4to. 1904.
  Circulation of the Atmosphere in High Latitudes. By W. N. Shaw. 8vo. 1904.

Meteorological Society, Royal—Journal, Vol. XXX. No. 131. 8vo. 1904.
  Record, Vol. XXIII. No. 92. 8vo. 1904.

Metropolitan Asylums Board—Annual Report for the Year 1903. 8vo. 1904.

Mexico, Museo Michoacano, Morelia—Relacion de Michoacan. 8vo. 1904.

Mexico, Geological Institute—Parergones, Tomo I. Nos. 1–3. 8vo. 1904.

Microscopical Society, Royal—Journal, 1904, Parts IV.–V. 8vo.

Missouri Botanical Garden—Fifteenth Annual Report, 1904. 8vo.

Navy League—Navy League Journal for Aug.-Oct. 1904. 8vo.

New Jersey, Geological Survey—Vol. V. Final Report. 8vo. 1904.

New South Wales Agent-General—Statistical Account of Australia and New Zealand. By T. A. Coghlan. 8vo. 1904.

New South Wales, Department of Prisons—Report on the Prevention and Treatment of Crime. 4to. 1904.

New York Academy of Sciences—Annals, Vol. XV. Part 2. 8vo. 1904.

New Zealand, Agent-General for—Official Year Book, 1903. 8vo.
Report on Department of Agriculture. 8vo. 1903.
Reports on Minerals, Mining Statistics, etc., of the Colony. 8vo and 4to. 1902-3.

Norfolk and Norwich Naturalists Society—Transactions, Vol. VII. Part 5. 8vo. 1904.

Odontological Society—Transactions, Vol. XXXVI. No. 8. 8vo. 1904.

Paris. Société Française de Physique—Bulletin, 1904, Fasc. 1-2. 8vo.

Pennsylvania, University of—Contributions from the Zoological Laboratory, Vol. X. 1903. 8vo. 1904.
Bulletins, Nos. 4-5. 8vo. 1904

Peru, Cuerpo de Ingenieros de Minas—Boletin, Nos. 7, 8, 12, 14. 8vo. 1904.

Pharmaceutical Society of Great Britain—Journal for July-Oct. 1904. 8vo.

Photographic Society, Royal—Journal, Vol. XLIV. Nos. 6-8. 8vo. 1904.

Physical Society of London—Proceedings, Vol. XIX. Part 2. 8vo. 1904.

Righi, A., Esq. (the Author)—Sulla Radiottivita dei Metalli Usuali. 4to. 1904.
Esperienze Dimonstrative sulla Radiottivita. 8vo. 1904.

Rome, Ministry of Public Works—Giornale del Genio Civile for Feb.-March, 1904. 8vo.

Röntgen Society—Journal, Vol. I. No. 1. 8vo. 1904.

Royal College of Surgeons—The Calendar, 1904. 8vo.

Royal Engineers' Institute—General Sir Henry Harness, R.E. By General Collinson. Edited by General Webber. 8vo. 1903.

Royal Society of Canada—Proceedings and Transactions, Vol. IX. 8vo. 1903.

Royal Society of London—Philosophical Transactions, A, Nos. 365-374; B, Nos. 231-233. 4to. 1904.
Proceedings, Nos. 496-499. 8vo. 1904.
Obituary Notices of Fellows of the Royal Society, Part III. 8vo. 1904.

Russell, W. J., Esq., F.R.S. M.R.I. (the Author)—The Action of Wood on a Photographic Plate in the Dark. 4to. 1904.

St. Bartholomew's Hospital—Statistical Tables, 1903. 8vo. 1904.

St. Petersbourg, Imperial Academy of Sciences—Mémoires, Vols. XIII. No. 6. XIV. XV. XVI. Nos. 1-3. 4to. 1904.
Comptes Rendus de la Commission Sismique Permanente, Tom. I. Liv. 3. 8vo. 1904.

Sanitary Institute—Journal, Vol. XXV. Part 2. 8vo. 1904.

Scottish Microscopical Society—Proceedings, Vol. IV. No. 1. 8vo. 1904.

Selborne Society—Nature Notes for July-Nov. 1904. 8vo.

Sennett, A. R., Esq., M.R.I. (the Author)—Across the Great Saint Bernard. 8vo. 1904.

Smith, B. Leigh, Esq., M.A. M.R.I.—The Scottish Geographical Magazine, Vol. XX. Nos. 7-11. 8vo. 1904.
Transactions of the Institute of Naval Architects, Vol. XLVI. 4to. 1904.

Smithsonian Institution—Annual Report, 1902, U.S. National Museum. 8vo. 1904.
The 1900 Solar Eclipse Expedition of the Astrophysical Observatory. 4to. 1904.
Miscellaneous Collections, Vol. XLIV. No. 1417; Vol. XLVI. No. 1441; Quarterly Issue, Vol. II. No. 1. 8vo. 1904.

Society of Arts—Journal for July-Oct. 1904. 8vo.

South Australian School of Mines—Report, 1903. 8vo. 1904.

Statistical Society, Royal—Journal, Vol. LXVII. Part 2. 8vo. 1904.

Stockholm, Royal Swedish Academy of Sciences—Arkiv för Zoologe, Band I. Häfte 3-4; Botanik, Band II. Häfte 4. 8vo. 1904.
Arsbok for 1904. 8vo. 1904.
Handligar, Band XXXVIII. Nos. 1-3. 4to. 1904.

Swedish Government—Sweden, its People and its Industry.   Edited by G. Sundbarg. 8vo. 1904.
Tacchini, Prof. P., Hon. Mem. R.I. (the Author)—Memorie della Società degli Spettroscopisti Italiani, Vol. XXXIII. Disp. 6–8.   4to.   1904.
Tasmanian Mail (the Publishers)— Centenary Souvenir of Tasmania.   fol.   1904.
Toronto University—Studies: Chemical Series, Nos. 40–43; Physical Series, No. 4; Physiological Series, No. 5.   8vo.   1904.
Turner, Prof. H. H., D.Sc. F.R.S. (the Author)—Various Papers from the Monthly Notices of the Royal Astronomical Society.   8vo.   1903.
United Service Institution, Royal—Journal for July–Oct. 1904.   8vo.
United States Department of Agriculture—Monthly Weather Review for April–July, 1904.   8vo.
      Bulletin M.   4to.   1904.
      Experiment Station Record for June–Sept. 1904.   8vo.
United States Geological Survey—Professional Papers, 11, 12, 16–20, 21–23, 28. 4to.   1903–4.
      Bulletins, 223–232.   8vo.   1904.
      Water Supply Papers, 89–95.   8vo.   1903.
      Twenty-fourth Annual Report, 1902–3.   4to.   1904.
      Mineral Resources of the United States, 1902.   8vo.   1904.
      Geologic Atlas of the U.S.A.   Folios 91–106.   fol.   1904.
United States Patent Office—Official Gazette, Vol. CXI.–CXII.   4to.   1904.
United States Surgeon-General's Office—Catalogue of Library, Second Series, Vol. IX.   4to.   1904.
Upsala, Royal Academy of Sciences—Nova Acta, Serie III. Vol. XX. Fasc. 2.   4to. 1904.
Verein zur Beförderung des Gewerbfleisses in Preussen—Verhandlungen, 1904. Heft 6–7.   4to.
      Jahrbuch, 1903, Heft 3.   8vo.   1904.
Victoria Institute—Journal, Vol. XXXVI.   8vo.   1904.
Vienna Imperial Geological Institute—Verhandlungen, Nos. 9–11.   8vo.   1904.
      Abhandlungen, Band XVII. Heft 6.   4to.   1903.
      Jahrbuch, 1903, Heft 2–4.   Svo.   1904.
Ward, H. A., Esq. (the Author)—Catalogue of the Ward-Coonley Collection of Meteorites.   8vo.   1904.
Warman, J. Watson, Esq. (the Author)—The Organ, Parts I.–IV.   8vo.   1903–4.
Washington Academy of Sciences—Proceedings, Vol. VI. pp. 1–202.   8vo.   1904.
Washington Philosophical Society—Bulletin, Vol. XIV. pp. 247–276.   8vo.   1904.
Western Society of Engineers—Journal, Vol. XIX. No. 3.   8vo.   1904.
Zoological Society—Proceedings, 1904, Vol. I. Part 2; Vol. II. Part 1.   8vo.   1904.
      Transactions, Vol. XVII. Part 3.   4to.   1904.
Zurich Naturforschenden Gesellschaft—Vierteljahrsschrift, 1904, Heft 1–2.   8vo.

GENERAL MONTHLY MEETING,

Monday, December 5, 1904.

Sir James Crichton-Browne, M.D. LL.D. F.R.S., Treasurer and
Vice-President, in the Chair.

Colonel C. W. Carr-Calthrop, I.M.S. Rtd.
Cecil Hanbury, Esq.
E. Graham Little, B.A. M.D. M.R.C.P.
Hugh Makins, Esq., B.A.
Edward Reinach, Esq.

were elected Members of the Royal Institution.

The following Resolution, passed by the Managers, was read and
adopted:—

Resolved, That the Managers of the Royal Institution of Great Britain desire
to record their sense of the loss sustained by the Institution in the decease of
Mr. Frank McClean, M.A. LL.D. F.R.S. M.Inst.C.E.

Mr. McClean became a Member of the Royal Institution of Great Britain in
1870, and has served the Institution both as Visitor and as Manager.

His important spectroscopic researches, and the interest he always took in
aiding experimental scientific inquiry, have done much to promote the objects of
the Institution.

The Managers desire to offer to Mrs. McClean and her family the expression
of the most sincere sympathy with them in their bereavement.

The Presents received since the last Meeting were laid on the
table, and the thanks of the Members returned for the same, viz.:—

FROM

The Secretary of State for India—Report on Madras Government Museum, 1903–4.
4to. 1904.
Geological Survey of India: Records, Vol. XXXI. Part 3; Memoirs, Vol.
XXXII. Part 4. 8vo. 1904.
Catalogue of the Imperial Library, Calcutta, Part I. Vols. I.–II. 4to. 1904
Accademia dei Lincei, Reale, Roma--Classe di Scienze Fisiche, Matematiche e
Naturali. Atti, Serie Quinta: Rendiconti. Vol. XIII. 2° Semestre, Fasc. 8.
8vo. 1904.
American Academy of Arts and Sciences—Proceedings, Vol XL. Nos. 6–7. 8vo.
1904.
American Geographical Society—Bulletin, Vol. XXXVI. No. 10. 8vo. 1904.
Amsterdam, Royal Academy of Sciences—Verhandelingin, 1e Sectie, Dl. VIII.
Nos. 6–7; 2° Sectie, Dl. X. Nos. 1–6. 8vo. 1903–4.
Zittingsverslagen, Vol. XII. 8vo. 1903–4.
Jaarboek, 1903. 8vo. 1904.
Astronomical Society, Royal—Monthly Notices, Vol. LXIV. No 9. 8vo. 1904.
Automobile Club—Journal for Nov. 1904.
Bankers, Institute of—Journal, Vol. XXV. Part 8. 8vo 1904.

Basel, Naturforschenden Gesellschaft—Verhandlungen, Band XV. Heft 3. 8vo 1904.
Boston Public Library—Monthly Bulletin for Nov. 1904. 8vo.
British Architects, Royal Institute of—Journal, Third Series, Vol. XII Nos. 1-2. 4to. 1904
British Astronomical Association—Journal, Vol. XV. No. 1. 8vo. 1904.
Calm, C. E., Esq. (the Author)—Sulphurous Acid and Sulphites as Food Preservatives. 8vo. 1904.
Canadian Institute—Proceedings. Vol. II. Part 6. 8vo. 1904
Chemical Industry, Society of—Journal, Vol. XXIII. Nos. 21-22. 8vo. 1904.
Chemical Society—Proceedings, Vol. XX Nos. 284-285 8vo. 1904.
　　Journal for Nov. 1904. 8vo.
Chicago, Field Columbian Museum—Publications: Geological Series, Vol. II. No. 5; Zoological Series, Vol. IV. Nos. 1-2. 8vo. 1904.
Clinical Society of London—Transactions, Vol. XXXVII. 8vo. 1904.
Editors—American Journal of Science for Nov. 1904 8vo.
　　Analyst for Nov. 1904. 8vo.
　　Astrophysical Journal for Nov. 1904. 8vo.
　　Athenæum for Nov. 1904. 4to
　　Author for Dec. 1904. 8vo.
　　Board of Trade Journal for Nov. 1904. 8vo.
　　Brewers' Journal for Nov. 1904. 8vo.
　　Chemical News for Nov. 1904. 4to.
　　Chemist and Druggist for Nov. 1904. 8vo.
　　Electrical Engineer for Nov. 1904. 4to
　　Electrical Review for Nov. 1904. 4to.
　　Electrical Times for Nov. 1904. 4to.
　　Electricity for Nov. 1904. 8vo.
　　Electro-Chemist and Metallurgist for Dec 1904. 8vo
　　Engineer for Nov. 1904. fol.
　　Engineering for Nov. 1904. fol.
　　Homœopathic Review for Dec. 1904. 8vo.
　　Horological Journal for Dec. 1904. 8vo.
　　Journal of the British Dental Association for Nov. 1904. 8vo.
　　Journal of State Medicine for Nov. 1904. 8vo.
　　Law Journal for Nov. 1904. 8vo.
　　London Technical Education Gazette for Nov. 1904. 4to
　　London University Gazette for Nov. 1904. 4to.
　　Machinery Market for Nov.-Dec. 1904. 8vo.
　　Model Engineer for Nov. 1904. 8vo.
　　Mois Scientifique for Oct. 1904. 8vo.
　　Motor Car Journal for Nov. 1904. 4to.
　　Motor Car World for Nov. 1904. 4to.
　　Musical Times for Nov. 1904. 8vo
　　Nature for Nov. 1904. 4to.
　　New Church Magazine for Nov.-Dec 1904. 8vo.
　　Nuovo Cimento for Sept. 1904. 8vo.
　　Page's Weekly for Nov. 1904. 8vo.
　　Photographic News for Nov. 1904. 8vo.
　　Physical Review for Nov. 1904. 8vo.
　　Public Health Engineer for Nov. 1904. 8vo.
　　Science Abstracts for Nov. 1904. 8vo.
　　Zoophilist for Nov. 1904. 4to.
Florence, Biblioteca Nazionale—Bulletin for Nov. 1904. 8vo.
Franklin Institute—Journal, Vol. CLVIII. No. 5. 8vo. 1904.
Geological Society—Abstracts of Proceedings, Nos. 799-800. 8vo. 1904.
　　List of Members, 1904. 8vo.
　　Quarterly Journal, Vol. LX. Part 4. 8vo. 1904.
Glasgow. Royal Philosophical Society—Proceedings, Vol. XXXV. 8vo. 1904.
Kansas University—Bulletin, Vol. IV. No. 9. 8vo. 1904.

580 *General Monthly Meeting.* [Dec. 5,

Leighton, John, Esq., M.R.I.—Journal of the Ex-Libris Society, Vol. XIV. Nos. 9-11. 8vo. 1904.
Linnean Society—Journal: Botany, Vol. XXXVII. No. 257; Zoology, Vol. XXIX. No. 190. Proceedings, October 1904. 8vo.
Massachusetts Institute of Technology—Technology Quarterly, Vol. XVII. No. 3. 8vo. 1904.
Mechanical Engineers, Institution of—General Index to Proceedings, 1885-1900. 8vo. 1904.
Meteorological Office—Wind Charts for the South Atlantic Ocean. fol. 1904.
Report of the International Meteorological Committee, 1903. 8vo. 1904.
Meteorological Society, Royal—Quarterly Journal. Vol. XXX. No. 132, Oct. 1904. 8vo.
Record, Vol. XXIV. No. 93. 8vo. 1904.
Mexico, Secretaria de Comunicaciones—Anales, Num. 11. 8vo. 1904.
Mexico, Sociedad Cientifica " Antonio Alzate"—Memorias y Revista. Tom. XIII. Nos. 7-8; Tom. XIX. Nos. 8-10; Tom. XX. Nos. 5-10. 8vo. 1904.
Natal, Commissioner of Mines—Report on the Mining Industry of Natal for 1903. 4to. 1904.
Navy League—Navy League Journal for Nov. 1904. 8vo.
New South Wales, Agent-General for—Report of the Comptroller of Prisons for 1903. 4to. 1904.
North of England Institute of Mining and Mechanical Engineers—Annual Report, 1903-4. 8vo. 1904.
Odontological Society—Transactions, Vol. XXXVII. No. 1. 8vo. 1904.
O'Halloran, G. F., Esq.—Report on Canadian Archives, 1903. 8vo. 1904.
Onnes, Dr. H. Kamerlingh, Hon. Mem. R.I. (the Author)—Het Naturkundig Laboratoium te Leiden, 1882-1904. 8vo. 1904.
Peru, Cuerpo de Ingenieros de Minas—Boletin, Nos. 11 and 13. 8vo. 1904.
Pharmaceutical Society of Great Britain—Journal for Nov. 1904. 8vo.
Physical Society of London—Proceedings, Vol XIX. Part 3. 8vo. 1904
Quekett Microscopical Club—Journal for Nov. 1904. 8vo.
Rome, Ministry of Public Works—Giornale del Genio Civile for April, 1904. 8vo.
Ross, Hugh Munro, Esq, B.A. M.R.I. (the Author)—British Railways. 8vo. 1904.
Royal Society of London—Philosophical Transactions, A. 375-376, B. 234-235. 4to. 1904.
Proceedings, Nos. 500-501. 8vo. 1904.
Obituary Notices of Fellows, Part I. 8vo 1904.
Selborne Society—Nature Notes for Dec. 1904. 8vo.
Smith, B. Leigh, Esq., M.R.I.—The Scottish Geographical Magazine, Vol. XX. No. 12. 8vo. 1904.
Smithsonian Institution—Contributions to Knowledge, Vol. XXXIII. Vol. XXXIV. No. 1438. 4to 1904.
Society of Arts—Journal for Nov. 1904. 8vo.
Sweden, Royal Academy of Sciences—Arkiv för Botanik, Band III. Hafte 1-3. 8vo. 1904
Tacchini, Prof. P., Hon. Mem. R.I. (the Author)—Memorie della Societa degli Spettroscopisti Italiani, Vol. XXXIII. Disp. 9. 4to. 1904.
Transvaal Department of Agriculture—Journal, Vol. II. No. 9. 8vo. 1904.
United Service Institution, Royal—Journal for Nov. 1904. 8vo.
United States Department of Agriculture—Monthly Weather Review for August, 1904. 4to
United States Geological Survey—Bulletins, Nos. 233-241. 8vo. 1904
Professional Papers, Nos 24-27. 4to. 1904.
Water Supply Papers, Nos. 96-98, 101, 102, 104. 8vo. 1904.
United States Patent Office—Official Gazette, Vol. CXIII. Nos. 1-4. 8vo. 1904.
Annual Report for 1903. 8vo. 1904.
Verein zur Beförderung des Gewerbfleisses in Preussen—Verhandlungen. 1904, Heft 8-9. 4to.
Vienna, Imperial Geological Institute—Jahrbuch, 1904, Heft I. 8vo.
Western Society of Engineers—Journal, Vol. IX. No. 5. 8vo. 1904.

WEEKLY EVENING MEETING,

Friday, March 25, 1904.

Sir James Crichton-Browne, M.D. LL.D. F.R.S., Treasurer
and Vice-President, in the Chair.

Professor Sir James Dewar, M.A. LL.D. D.Sc. F.R.S. *M.R.I.*

Liquid Hydrogen Calorimetry.

In the determination of quantities of heat, besides the "method of
mixtures," the various calorimeters that have been used depend
on liquefaction, evaporation, or condensation. One of the earliest,
that of Laplace and Lavoisier—a development of Black's method—
depended on the liquefaction of ice, care being taken to isolate the
ice which was the calorimetric substance from any external heat
effects, by means of a jacket of snow. In the middle of last
century Bunsen devised an exceedingly delicate instrument in which
ice was again the calorimetric substance; but instead of following
Laplace and Lavoisier's plan of measuring the weight of ice melted,
he took advantage of the reduction in volume of melted ice—namely,
about one-eleventh part—to determine with great accuracy the
quantity of heat that had been employed. As a unit of heat will
melt $\frac{1}{80}$ gramme of ice, this will produce a change of volume of
about $\frac{1}{800}$ cubic centimetre; and if the index tube of the instrument
be half a millimetre in diameter, the unit of heat will be shown by
about 6 or 7 millimetres on the scale. With such an instrument it is
therefore possible to determine very accurately one-tenth of a
gramme-calorie.

Professor Joly's calorimeter depends on condensation. In it the
quantity of steam condensed on the body, which is the subject of
experiment, is ascertained by weighing the amount of water formed.
The body is contained in a chamber of relatively large volume into
which the steam is suddenly admitted. It will be noticed, therefore,
that the uncondensed steam in the chamber acts as a jacket to the
body and prevents the passage of heat between it and external bodies;
in this manner it is isolated from disturbing causes. In Professor
Joly's hands, this instrument has been used to determine directly the
specific heat of gases at constant volume.

Other methods have been adopted for the measurement of quantities of heat. Black compared the quantities of heat in two bodies
of equal masses and temperatures, by noting the times in which they
cooled to the same temperature. Regnault and others have used the

method of mixture, in which a given quantity of the substance to be experimented on at a higher temperature, is mixed with a given quantity of a known fluid substance at a lower temperature; from the final temperature acquired by the two, the quantity of heat given up by the former can be determined. In this method the main difficulty is the isolation of the mixture from external heating or cooling influences during the experiment, and its success in the hands of Regnault was due to the skill with which he accomplished this.

Evaporation is a means of absorbing heat; no evaporation can take place without absorption of heat, and usually this absorption is relatively great. The well known experiment of the freezing of water by its own evaporation by the methods of Leslie and Wollaston are illustrative of the use of the latent heat of evaporation to produce lower temperatures. Further, by a law of Dalton, evaporation takes place most copiously into a space which is kept free from vapour of the same kind as that coming off.

It has long been known that by the passage of air through volatile liquids a considerable reduction of temperature may be effected depending on the particular substance selected, the isolation of the liquid from external heat, and the use of air at the low temperature reached in each case. The following table gives the general results of the temperatures recorded by different experimenters when ether, sulphurous acid, methyl chloride, ammonia, and ethylene, were employed.

	Temp. C. on Evaporation.	Boiling Point.	Critical Temperature.	Evaporation Temperature in terms of absolute Critical Temperature
Ether	− 34°	+ 35°	194°	·51
Sulphurous acid . .	− 50°	− 10°	155°	·52
Methyl Chloride . .	− 55°	− 24°	141°	·53
Ammonia . . .	− 87°	− 39°	130°	·46
Ethylene . . .	− 132°	− 103°	10°	·50
Liquid Nitrogen with Hydrogen passed through instead of Air as above .	− 214°	− 196°	− 146°	·47

It is interesting to notice that the limit of temperature reached by this means is in each case about half the absolute critical temperature. Thus for ethylene the absolute temperature of evaporation is 273° − 132°, or 141°, and its absolute critical temperature is 273° + 10°, or 283°, giving the ratio ·50. Now if this approximate relation holds good for a substance like liquid nitrogen, then we should anticipate that by passing a current of a gas through it like hydrogen, which at the temperature of boiling nitrogen is still a permanent gas, we should reach a temperature of $-214°$; and as this value would be just about the melting point, the nitrogen ought consequently to become solid.

The freezing of nitrogen by evaporation in a current of hydrogen at atmospheric pressure is carried out in the following way. Within a vacuum vessel A A, kept full of liquid air (Fig. 1), is inserted another vacuum vessel B B, which is held in its position by means of a cork C. About the middle of B B is fixed another cork D, which gives support to a tube E, allowing free passage between the lower part of B B and the atmosphere. F F F is a small tube coiled round the tube E as far down as the cork D, and below that continued as a coil in the lower part of B B, and ending in a nozzle at G. In the bottom of B B is placed a quantity of liquid nitrogen N. The experiment is conducted by passing pure hydrogen through the tube F, thereby cooling it to the boiling point of liquid air, and, finally, by means of the nozzle G through the liquid nitrogen. As the hydrogen passes down the spiral part of the tube F it is cooled to the temperature of the gaseous nitrogen rising from N, and bubbles through N at the temperature of the liquid nitrogen. These hydrogen bubbles are thus in the best condition, according to Dalton's law, to induce evaporation of the liquid nitrogen, without conveying unnecessary heat, and rapidly cause its temperature to fall. After a short time the hydrogen bubbles begin to move sluggishly in the cooling liquid nitrogen, and soon afterwards the nitrogen becomes solidified by its own evaporation. The appearance of the solid nitrogen as it is first formed is very extraordinary, as it deposits in long spiral tubes through which the hydrogen for a time escapes. As we now know that helium is as much more volatile a gas than hydrogen as the

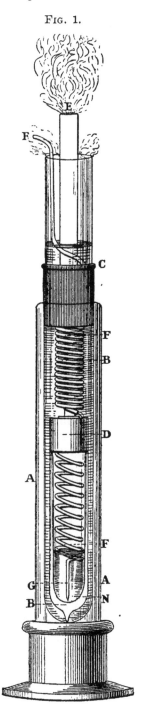

FIG. 1.

latter is than nitrogen, we may safely predict that if a current of helium were similarly directed through liquid hydrogen, the latter would be reduced in temperature until it would freeze, seeing that the melting point of hydrogen is just about half its critical temperature.

An evaporation calorimeter, where the calorimetric substance is one of the liquefied gases, would be a convenient instrument provided it could be easily constructed and was reliable in its working. The efficiency of such an instrument depends (1) on the relatively large quantity of gas given by evaporation, and (2) on the great range of temperature easily available when liquid air, oxygen, nitrogen, or hydrogen is the calorimetric substance.

The following table gives the special physical constants of the various liquid gases that are of importance in calorimetry. That calorimetric substance will be the more sensitive which gives off

Liquid gases.	Boiling point.	Liquid volume 1 gramme at boiling point in c.c.	Latent-heat in gramme-calories.	Volume of gas at 0° and 760 mm. per gramme-calorie in cubic centimetres.
Sulphurous acid	$+\ 10°\cdot0$	$0\cdot7$	$97\cdot0$	$3\cdot6$
Carbonic acid .	$-\ 78°\cdot0$	$0\cdot65$ (solid)	$142\cdot4$	$3\cdot6$
Ethylene . .	$-\ 103°\cdot0$	$1\cdot7$,,	$119\cdot0$	$7\cdot0$
Oxygen . .	$-\ 182°\cdot5$	$0\cdot9$,,	$53\cdot0$	$13\cdot2$
Nitrogen . .	$-\ 195°\cdot6$	$1\cdot3$,,	$50\cdot0$	$15\cdot9$
Hydrogen. .	$-\ 252°\cdot5$	$14\cdot3$,,	$125\cdot0$	$88\cdot9$

the larger volume of gas for a given quantity of heat. Thus oxygen gives $13\cdot2$ c.c. per calorie, while ethylene gives 7, hydrogen $88\cdot9$; hence oxygen is twice as sensitive as ethylene, and hydrogen six times as sensitive as oxygen. It is easy to detect a $\frac{1}{50}$ gramme calorie when liquid air is used, and as small a quantity as $\frac{1}{300}$ can be observed with liquid hydrogen.

In selecting the calorimetric substance of a liquid gas calorimeter, hydrogen, as giving the greatest range of temperature and sensibility, would be the best; next to it would come nitrogen, then air, and lastly oxygen. But we must remember that we are enveloped in an atmosphere of air, and have to consider its effect. Passing over hydrogen for the present, let us examine the advantages and disadvantages of the other three gases. As the boiling point of air is below that of oxygen, even if there were no layer of cool oxygen gas on the surface of the liquid oxygen, the air coming in contact with it through the neck of the calorimeter would still remain gaseous. But if we were to take liquid nitrogen as the calorimetric substance, air, being heavier than nitrogen but having a higher boiling point, would, in falling down the neck of the calorimeter, come in contact with the

cool gaseous nitrogen and be condensed. Hence nitrogen would not
be a convenient calorimetric substance. In any case, the boiling
points of nitrogen, air, and oxygen being so close together, it is
obvious that liquid air is the most convenient substance in the
neighbourhood of −180° to −200° C.

The calorimeter has been described in my paper "On the
Scientific Uses of Liquid Air,"* and later an improved form in
"Recherches sur les Substances Radio-Actives," by Madame Curie,†
further a sketch of it is given in the paper on "The Absorption
and Thermal Evolution of Gases occluded in Charcoal at Low
Temperatures." ‡ The annexed diagram shows its construction. It

Fig. 2.

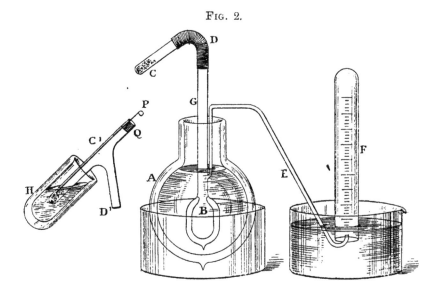

consists essentially of a large vacuum vessel A, capable of holding
two or three litres, into which is inserted the calorimeter, a smaller
vacuum vessel B, of 25 to 50 c.c. capacity, which has been sealed on
to a long narrow tube G, projecting above the mouth of A, and held
in its place by some loosely packed cotton wool. From the side of
this narrow tube, either before or after passing out of A, a branch
tube, E, is taken off to enable the volatilised gas from the calorimeter ·
to be collected in the receiver F, over water, oil, or other suitable·
liquid. To the extremity of the projecting tube G, a small test-tube·
C, to contain the portions of material experimented on, is attached by
a piece of flexible rubber-tubing D, thus forming a movable joint,

* Roy. Inst. Proc. 1894, vol. xiv., p. 398.
† 2nd Edition, p. 100.
‡ Roy. Soc. Proc. 1904, vol. lxxiv., p. 123.

which can be bent so as to tilt a few of the small pieces of substance contained in C into the calorimeter, and which afterwards assumes a position of rest somewhat like that in the diagram.

With care one can manage to tilt a single piece at a time from C into B, but an improved form of this receptacle is shown at C′D′. In it, P is a wire movable through the cork Q, fitted into the mouth of the test-tube C′, attached by a branch through the stiff rubber tube D′ to the end of G, as before. At the end of the wire P is a hook, by which one piece of the substance at a time can be pulled up and dropped into B′. When no other arrangements are made, the portions of matter experimented on are at the temperature of the room; but when lower temperatures are required initially, a vacuum vessel H containing either solid carbonic acid, liquid ethylene, air, or other gas, can be placed so as to envelop the test-tube C or C′; or if higher temperatures are required, the surrounding vessel may be filled with the vapour of water or other liquids.

Much study and handling of the instrument have brought out the following matters as essential to high efficiency. I have already pointed out that in the neighbourhood of $-180°$ C. to $-200°$ C. liquid air is the preferable substance to use, while liquid hydrogen enables observations to be made as low as $-250°$ C. The value of the vacuum of the calorimeter itself is much enhanced by making it a mercury vacuum; and further by having, previous to use, a good mercury deposit over its surface. This is attained by putting some liquid air into the calorimeter B and leaving it to stand for some time. When a quantity of liquid air has been undergoing volatilisation for a time, as the nitrogen evaporates more quickly than the oxygen, the boiling point rises slightly. Two points require attention in consequence of this; first, the maintenance of a constant temperature of the liquid air during any one series of experiments; next, the prevention of a tendency for the calorimeter B to "suck back" some of the already volatilised gas. Hence the exterior vessel A should be filled with a large quantity—some two litres—of *old* liquid air, containing a high percentage of oxygen, and the calorimeter itself should be filled with some of the *same* fluid. This will maintain very closely the constant temperature required. When any "sucking back" seems to be taking place, the calorimeter should be emptied and filled anew from the larger flask A. The tube between the calorimeter and the gas receiver should be of the size of wide quill tubing, and its lower end should be so arranged below the surface of the liquid in the collecting vessel, as to give no resultant pressure. With such precautions, results may easily be obtained correct to within 2 per cent.

The instrument having been set up and filled with liquid air, according to the above directions and precautions, an experiment is conducted by tilting up the little test-tube, previously cooled or heated, thereby dropping into the calorimeter a portion of any substance previously weighed. The substance, if left under normal conditions, in this way falls from the temperature of the room to that

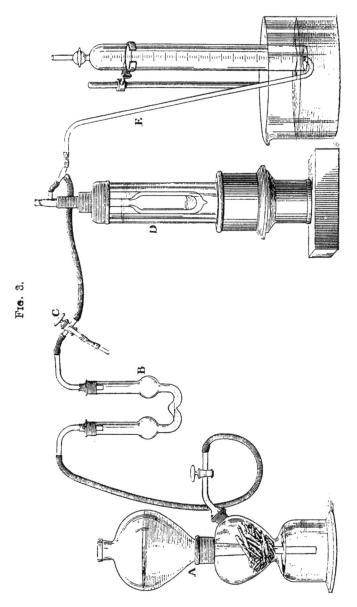

Fig. 3.

of liquid air. The heat given up by it to the liquid air volatilises some of it, which is carried off by the branch tube and measured in the graduated receiver F. Immediately preceding or following this observation, a similar experiment is made with a small portion of a selected standard substance, namely, lead. The quantity of lead is so chosen as to produce about the same volume of gas in the receiver as that supplied by the portion of substance experimented on. By this means, the circumstances of the two observations are made as similar as possible, and thereby many sources of error are eliminated.

When the hydrogen calorimeter is to be used, the temperature being so much lower than in the liquid-air calorimeter, we have to keep the ordinary atmosphere from entering the mouth of the tube G by means of a current of hydrogen. This is attained in the manner shown in Fig. 3. An ordinary Kipp apparatus A supplies hydrogen which, after being dried in the U-tube B, is allowed to pass by the stop-cock C to the calorimeter D and tube E. The hydrogen passes continually through the apparatus until the moment of beginning an experiment, when the stopcock C is turned and the hydrogen cut off.

In Fig. 4 are shown various forms of calorimeters which were experimented with in determining the best form for the calorimeter bulb. The final form adopted was that shown in D, Fig. 4. Stray globules of the liquid hydrogen might splash up on dropping the substances to be experimented on into the calorimeter, and get carried over into the gas receiver, but this cause of error was found to be negligible, provided the calorimeter was large enough.

When the body has to be transferred from solid carbonic acid or liquid air to the calorimeter, the following procedure is adopted. It is placed in the small test tube, above the indiarubber joint, which is inserted into a small vacuum vessel containing some of the substance (solid carbonic acid or liquid air), so that at the moment of making the experiment the solid, by a quick vertical movement of the vacuum vessel, is thrown into the calorimeter. A little cotton wool inserted in the mouth of the vacuum vessel prevents the carbonic acid paste, or liquid air, from being ejected.

Observations were made to determine what allowance had to be made for loss of heat while the small body was falling down the tube B ; also for similar losses by impacts on the sides of the india-rubber joint and of the glass tube. The substances used were lead, diamond, and graphite, and as the errors in any case did not exceed one-half to two-thirds per cent., they may in general be neglected.

The calorimeter may be put to various uses. Thus, on passing down the tube of the calorimeter wires of copper, iron, and german-silver, and noting the different rates at which the gas is evolved, we see that it is about six times more rapid with copper than with iron, and nine times more rapid than with german-silver, while with small rods of glass or ebonite no gas is produced in a short period of time. Again, we may measure the heat given up by condensation. Thus, if a small bulb containing some carbonic acid be inserted into the calorimeter, immediately a rapid evolution of gas takes place, until at last the carbonic acid is frozen, and after a little time the evolution ceases. The volume of gas produced from

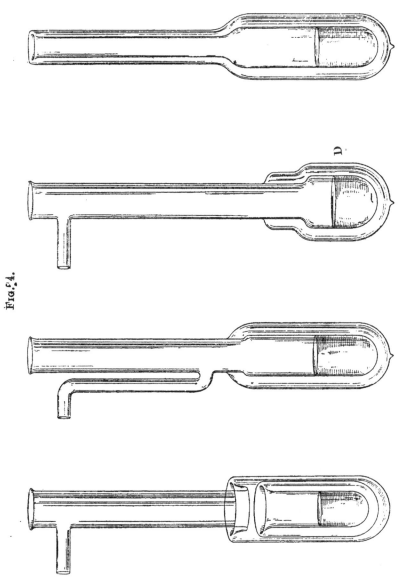

Fig. 4.

the same bulb exhausted is naturally very much less. But we may use the instrument for quite different purposes. Thus, in a small bulb is placed a mixture of hydrogen and oxygen in the propotions

2 R 2

in which they exist in water. Two insulated wires are let into the bulb, by means of which, on attaching them to an induction coil, we can explode the gases by the electric spark. The instant the explosion takes place, a corresponding vaporisation of liquid air follows, and from its amount as shown in the receiver we can determine the heat of combination of oxygen and hydrogen. In a similar manner the heat of the electric spark itself can be determined. Two fine wires, each passing through a fine glass tube, are passed down into the calorimeter until their ends are below the liquid surface, or they may be connected to a small sparking tube immersed in the liquid air. On attaching them to the induction coil and making contact, a spark passes in the liquid, or in the glass bulb, which immediately evaporates some of it, and from its gas volume in the receiver we find the heat generated by the spark.

The determination of specific heats and latents is a matter of great importance. Latent heats of evaporation are usually determined under the ordinary atmospheric pressure : alteration of pressure has, in general, but a small effect on the latent heat of solidification. Specific heats, on the other hand, vary with the temperature, so that it becomes important to measure them at various temperatures. This is usually done by observing the mean specific heats over finite ranges of temperature, as, for example, between the boiling point of water and its freezing point, between the freezing point of water and the boiling point of carbonic acid, between this latter point and the boiling point of oxygen, and between this and the boiling point of hydrogen ; or, we may go upwards, say, from the boiling point of water to the melting point of paraffin, then to the boiling point of sulphur, and so on. As so many properties of substances tend either to a maximum or a minimum, as we approach the absolute zero, there is great interest in examining the values of the specific heats of substances at low temperatures. Dulong and Petit discovered a simple, and very general law, according to which the product of the specific heat of a substance, in the solid state, and its atomic weight is constant. For example, let us take two small pieces of metal of equal weight—one, lead, with the high atomic weight of 207 ; the other, aluminium, whose atomic weight is only 27. On comparing the evolution of gas when the lead is put into the calorimeter, with that of the aluminium under the same circumstances, the very marked excess of the evaporation with aluminium over that with lead is at once apparent, and even a rough measure of the relative amounts of gas evolved shows one to be almost seven times that of the other, a ratio very nearly equal to that of their inverse atomic weights.

Two of the most interesting substances to study in this connection are carbon, in its two forms, diamond and graphite, and ice.

A large series of observations were made upon carbon in both its forms, the ranges of temperature being from about 18° C. to the boiling point of carbonic acid, thence to the boiling point of oxygen, and finally to the boiling point of hydrogen. Early determinations of the specific heat of carbon in any of its forms had shown complete departure from the law of Dulong and Petit. In 1872 Professor H. F. Weber and myself,* working independently, found that, as the temperature increased, the specific heat of carbon, whether as diamond or as graphite, continued to increase. Professor Weber found that "the specific heat of the diamond is tripled when the temperature is raised from 0° to 200°," and my experiments showed that the mean specific heat of carbon from 30° to the boiling point of zinc was 0·32, and to the temperature of the oxy-hydrogen blow-pipe (some 2000° C.) it rose to 0·42, and I added "the true specific heat at

FIG. 5.

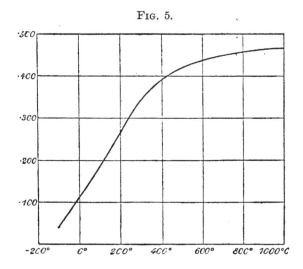

2000° must be at least 0·5; so that at this temperature carbon would agree with the law of Dulong and Petit." In 1875 † Professor Weber published the results of further experiments, proceeding by successive stages up to 1000° C., and showing "that from the point (about 600°) at which the specific heat of carbon ceases to vary with increase of temperature, and becomes comparable with that of other elements, any real difference in the specific heats of the two modifications disappears, and carbon obeys the law of Dulong and Petit."

Professor Weber plotted his results, taking specific heat as the ordinate, to temperature as the abscissa, and producing a curve like the old English ∫. He found the point of inflection for diamond

---

* H. F. Weber, "The Specific Heat of Carbon," Phil. Mag. 1872, ser. 4, vol. xliv., p. 251. J. Dewar, "The Specific Heat of Carbon at High Temperatures," l.c., p. 461.

† Phil. Mag. 1875, ser. 4, vol. xlix., p. 285.

about 60° C., and for graphite about 0° C.   But for this discovery, the trend of the curve, as shown in the diagram, Fig. 5, would have pointed to the vanishing of the specific heat of carbon about − 90° C., a result which could not be accepted as final.   The investigations carried out recently,[*] have cleared up this point, and verified the existence of the points of inflection.   The following table summarises my observations on the specific heats of diamond, graphite, and ice :—

Substance.	18° to − 78° C., or, at − 30° C.	− 78° to − 188° C., or, at − 133° C.	− 188° to − 252° C., or, at − 220° C.
Diamond . . .	0·0794	0·0190	0·0043
Graphite . . .	0·1341	0·0599	0·0133
Ice	0·463*	0·285	0·146

* This is from − 18° to − 78° in the ice experiment.

These results for diamond and graphite accord well with those of Professor Weber over common ground.   He gives 0·0806 for the diamond between $21\frac{1}{2}$° and − 80°, and 0·1301 for graphite over the same range.

The second curve in the diagram, Fig. 6, shows the sequence between Professor Weber's results and my continuation to low temperatures, and demonstrates the reversal of the curvature of the curve at low temperatures.   Similar results were got for carbon.

The results for ice follow much the same order of change.   A reference to the table above shows that the mean specific heat of ice falls from 0·463 at − 46°, to 0·285 at − 133°, and 0·146 at − 220° C.; that is, in the lowest range of temperature between the boiling points of oxygen and hydrogen, it is only one-third of its value between − 18° and the temperature of boiling carbonic acid.

It would be a matter of interest to investigate the general behaviour of various groups of substances, as regards their specific heats at low temperatures.   Without having attempted any careful systematic investigation, the following observations extracted from laboratory records are fairly representative of some classes of bodies.   In the table the specific heats of two alloys are given, which were used in the course of the investigation ; also those of sulphur, selenium, and tellurium.   Two alums, on which Kopp had made some observations, were included in the research, together with three other typical salts.   Again, naphthaline and paraffin were a pair, whose specific heats were examined ; also the chloride, bromide, and iodide of silver.   The results for the solidified gases, carbonic acid, ammonia, sulphurous acid, were of obvious interest,

* Proc. Roy. Soc., 1905.

No. of obs.	Substance.	Weight used in grammes.	Range of Temperature. Degrees Centigrade.		Vol of Gas in cubic centimetres.	Specific Heat
1	German silver	0·22	− 18 to	−188	48	0·080
1	Brass	0·627	+ 19·5	−188	166	0·099
1	,,	0·244	−188	−252·5	66 (H)	0·043
2	Tellurium	0·645	+ 18·2	−188	99·5	0·047
2	Sulphur	0·289	+ 18·2	−188	131	0·137
2	Selenium	0·353	+ 18·2	−188	80	0·068
2	Potassium alum	0·180	+ 18·8	−188	152·5	0·256
2	,, ,,	0·376	− 78	−188	130	0·223
2	Chromium alum	0·20	+ 20	−188	162	0·243
2	,, ,,	0·392	− 78	−188	135	0·222
1	Calcium chloride (hydrate)	0·184	+ 20	−188	180	0·294
1	,, ,, ,,	0·336	− 78	−188	141	0·271
3	Sodium chloride	0·105	+ 16	−188	55·2	0·187
2	,, ,,	0·253	− 78	−188	65·5	0·164
3	Ammonium chloride	0·054	+ 16	−188	45·8	0·300
2	,, ,,	0·130	− 78	−188	42·5	0·207
2	Naphthaline	0·55	+ 16	−188	31·5	0·202
2	,,	0·105	+ 16	−188	57·25	0·194
1	,,	0·090	+ 15	−188	50·5	0·204
2	,,	0·203	− 78	−188	40·6	0·126
1	Paraffin	0·08	+ 15	−188	68·5	0·312
1	,,	0·105	− 78	−188	38·5	0·176
1	Silver iodide	0·307	+ 16	−188	44·5	0·052
2	,, bromide	0·196	+ 16	−188	35·5	0·064
2	,, chloride	0·215	+ 16	−188	49·75	0·082
5	Solid carbonic acid	0·164	− 78	−188	57·1	0·215
1	,, ,, ,,	0·15	− 78	−182·5	50	0·225
1	,, ,, ,,	0·190	− 78	−182·5	62	0·223
2	Solid ammonia	0·14	−103	−188	72·25	0·519
2	,, ,,	0·156	−103	−188	77·5	0·490
3	Solid sulphurous acid	0·325	−103	−188	75·2	0·228
3	,, ,,	0·311	−103	−182·5	57·3	0·236
2	Ceylon thoria mineral	0·500	+ 15	−188	70·6	0·044

and several observations on them are given. With regard to these
bodies, it may be noted that the values found are not far removed
from those of the specific heats at constant volume in the gaseous
state, and I have no doubt that if the experiments had been
extended to temperatures between that of liquid air and hydrogen
these results would all have been below the gas constant. The other
bodies examined all show diminution of specific heat at the lower
temperatures, the most marked examples being the hydrocarbons,
paraffin and naphthaline.

While the experiments on the specific heats of diamond graphite
and ice were being carried on, the frequent determination of the
quantities of gas evaporated by lead in the same circumstances as the
diamond graphite or ice under investigation, afforded means for the
direct measurement of the latent heats of evaporation of hydrogen,
nitrogen, air, and oxygen. Lead had been selected as the metal of
comparison for the following reasons :—its low specific heat enabled

Fig. 6.

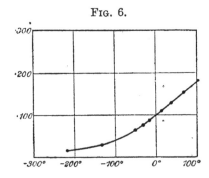

small quantities of heat to be conveyed into the calorimeter while
the mass of the metal was still considerable ; further, the variations
of the specific heat of lead with temperature are small, and its
specific heat may hence be treated as a linear function of the temperature ; and lastly, the metal is easily obtained very pure.

In this manner the latent heat of hydrogen, whether determined
by dropping small pieces of lead through a range of temperature as
great as 270°, or through as short a range as 64°, was found to be
about 121 to 122 gramme-calories. In my Bakerian Lecture,*
having assumed what at the time appeared to be the value of the
latent heat of hydrogen, namely, 200 gramme-calories, and having observed experimentally that 15 per cent. of the liquid had to be quickly
evaporated under exhaustion to reduce the temperature to the melting
point of hydrogen, I deduced the mean specific heat of the liquid
between the boiling point and the freezing point to be about 6. The
present investigation enables me to correct this statement, the value

* Proc. Roy. Soc., June 1901, vol. lxviii. p. 361.

of the mean specific heat in question being in reality about 3·4 instead of 6. This correction removes hydrogen from the list of substances which follow Dulong and Petit's law, its atomic heat being only about half the required amount. In an early communication to the Royal Society of Edinburgh,* I showed that the specific heat of hydrogen absorbed by palladium was about 3·5. It seems therefore that hydrogen in the gaseous, the occluded, and the liquid condition, has substantially the same specific heat.

It is of interest to see what support can be got for the value of the latent heat of hydrogen being about 121 or 122. From some early observations of mine with the helium thermometer on the vapour pressure of hydrogen below the boiling point, a Rankine formula was obtained,

$$\log \rho = 5\cdot5038 - \frac{53\cdot13}{t}$$

which, combined with Clapeyron's equation, where account was taken of the difference of the specific volumes of hydrogen in the liquid and the gaseous states at the boiling point, gave 120·3 as the latent heat of liquid hydrogen. Similar treatment of Travers' † smoothed results gave 119. Two Willard Gibbs' formulæ calculated from results evenly selected from these actual observations gave 123·4 and 117·5, or a mean of 120·5.

For nitrogen, the value of the latent heat was found to be about 50·4 gramme-calories at the boiling point. Other observers (Fischer and Alt, Alt, Shearer) give the values of this quantity as 48·9, 48·7 (at 718 mm. pressure), 52·07 (at 96 mm. pressure), and 49·8. The application of a Rankine formula deduced from my own observations,

$$\log \rho = 6\cdot6462 - \frac{292}{t},$$

gave 48·03 ; while the same process applied to observations of Fischer and Alt gave 49·65, and a Willard Gibbs' formula gave 51·4.

In the case of oxygen its latent heat was found to be 51·15 gramme-calories. Several experimenters have found the value of this quantity to be as high as 58·0, 60·9, 61·0. A careful direct determination of Alt's gives 52·07 (725 mm.) and 58·85 (68 mm.). By means of Rankine and Willard Gibbs' formulæ, calculated from the observations of Olzewski, Estreicher, and Travers, I have found the values 51·4, 52·53, 53·78.

These indirect methods of determining the latent heat depend on formulæ of which only the principal terms are retained. We cannot, therefore, expect more than approximate values from them, so that it is sufficient at present to show that the direct experimental values found in the above series of observations substantially agree with those obtained by indirect and approximate calculations.

---

* "The Physical Constants of Hydrogenium," Trans. Roy. Soc. Ed. 1873.
† Phil. Trans., 1902, A., vol. cc. p. 169.

In passing, it may be noticed that if the constancy of Trouton's constant is to be accepted, the latent heat of oxygen should be greater than that of nitrogen—a result in accordance with the above values.

The latent heat of air will depend on its constitution, tending towards equality with that of oxygen in the case of *old* liquid air, which is rich in oxygen. In an early experiment, I found the latent heat of air to be 49·7 gramme-calories ; in a later one, it was 53·41. I then made a series of experiments in which a succession of half-grammes of lead, 10 or 12 at a time, were dropped into old liquid air ; from these 53·63 gramme-calories were found. These values approach those of Fenner and Richtmyer's,* found by an electrical method.

During the time Professor Curie lectured at the Royal Institution, some measures of a preliminary kind were jointly made by us of the rate at which radium bromide gives out energy at low temperatures. The quantity of radium bromide was 0·42 grm. ; and it was used both in a liquid oxygen and liquid hydrogen calorimeter. The thermal evolutions are given below :

	Gas Evolved per minute.	Calories per hour.	
Liquid oxygen . . .	5·5 c.c.	22·8	Crystals.
Liquid hydrogen . . .	51·0 ,,	31·6	
Melting ice . . .	—	24·1	
Liquid oxygen . . .	2·0 ,,	8·3	After fusion.
Liquid oxygen . . .	2·5 ,,	10·3	Emanation condensed.

The apparent increase of heat evolution at the temperature of liquid hydrogen was probably due to the calorimeter being too small, so that hydrogen spray was carried away with the gas, thus making the gas volume too great, and inferentially the heat evolved.

I have to acknowledge the valuable aid I have received from my Chief Assistant, Mr. Robert Lennox, F.C.S., and that of his colleague, Mr. J. W. Heath, F.C.S.

[J. D.]

* Phys. Rev., 1905, vol. xx. p. 81.

LONDON PRINTED BY WILLIAM CLOWES AND SONS, LIMITED,
GREAT WINDMILL STREET, W., AND DUKE STREET, STAMFORD STREET, S.E.

PROCEEDINGS

OF THE

Royal Institution of Great Britain

Vol. XVII.—Part II.   No. 97

		PAGE
Jan. 16.	PROFESSOR SIR JAMES DEWAR—Low Temperature Investigations	418
Jan. 23.	TEMPEST ANDERSON, M.D.—Recent Volcanic Eruptions	231
Jan. 30.	PROFESSOR W. E. DALBY—Vibration Problems in Engineering Science	235
Feb. 2.	GENERAL MONTHLY MEETING	239
Feb. 6.	THE RIGHT HON. SIR HERBERT MAXWELL, Bart., M.P.—George Romney and his Works	243
Feb. 13.	PROFESSOR SHERIDAN DELÉPINE—Civilisation and Health Dangers in Food	247
Feb. 20.	PRINCIPAL E. H. GRIFFITHS—The Measurement of Energy	257
Feb. 27.	ADOLF LIEBMANN, Esq.—Perfumes: Natural and Artificial	258
March 2.	GENERAL MONTHLY MEETING	262
March 6.	PROFESSOR JOHN GRAY MCKENDRICK, M.D.—Studies in Experimental Phonetics	265
March 13.	PROFESSOR KARL PEARSON—Character Reading from External Signs	266
March 20.	PROFESSOR E. A. SCHÄFER—The Paths of Volition	268
March 27.	PROFESSOR W. A. HERDMAN—The Pearl Fisheries of Ceylon	279
April 3.	THE RIGHT HON. LORD RAYLEIGH, O.M.—Drops and Surface Tension	288
April 6.	GENERAL MONTHLY MEETING	290
April 24.	THE HON. R. J. STRUTT—Some Recent Investigations on Electrical Conduction	293
May 1.	ANNUAL MEETING	300
May 1.	PROFESSOR WILLIAM J. POPE—Recent Advances in Stereochemistry	301
May 4.	GENERAL MONTHLY MEETING	316
May 8.	H. RIDER HAGGARD, Esq.—Rural England	320
May 15.	D. H. SCOTT, Esq.—The Origin of Seed-bearing Plants	335
May 22.	J. A. H. MURRAY, Esq.—Dictionaries	349
May 29.	J. Y. BUCHANAN, Esq.—Problems and Methods of Oceanic Research	357
June 5.	PROFESSOR H. H. TURNER—The New Star in Gemini	375
June 8.	GENERAL MONTHLY MEETING	385
June 19 (Extra Evening).	PROFESSOR PIERRE CURIE—Radium (in French)	389
July 6.	GENERAL MONTHLY MEETING	403
Nov. 2.	GENERAL MONTHLY MEETING	406
Dec. 7.	GENERAL MONTHLY MEETING	412

LONDON

ALBEMARLE STREET, PICCADILLY

March 1905

5s.

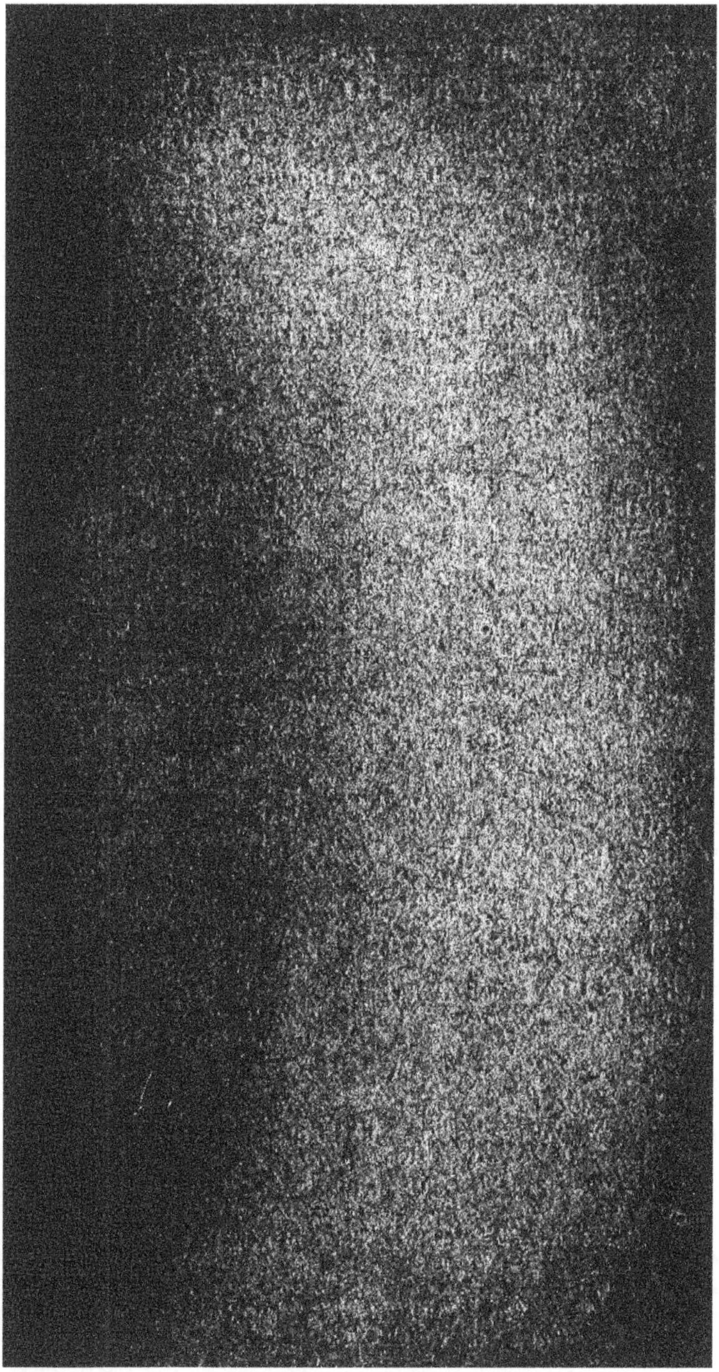

LONDON: PRINTED BY WILLIAM CLOWES AND SONS, LIMITED,
GREAT WINDMILL STREET, W., AND DUKE STREET, STAMFORD STREET, S.E.

PROCEEDINGS

OF THE

Royal Institution of Great Britain

Vol. XVII.—Part III.

LONDON

ALBEMARLE STREET, PICCADILLY

January 1906

5s.

Lightning Source UK Ltd.
Milton Keynes UK
UKHW010612110219
337000UK00006B/278/P